A HISTORY OF HUMAN ANATOMY

Second Edition

A HISTORY OF HUMAN ANATOMY

By

T.V.N. (VID) PERSAUD, M.D., Ph.D., D.Sc., F.R.C.Path. (Lond.), FAAA

Professor
Department of Anatomical Sciences
School of Medicine
St. George's University
Grenada, West Indies

Professor Emeritus and Former Head
Department of Human Anatomy and Cell Science
University of Manitoba
Faculties of Medicine and Dentistry
Winnipeg, Canada

MARIOS LOUKAS, M.D., Ph.D.

Professor and Chair
Department of Anatomical Sciences
Dean of Research
School of Medicine
St. George's University
Grenada, West Indies

R. SHANE TUBBS, M.S., P.A.C., Ph.D.

Pediatric Neurosurgery
Children's Hospital
Birmingham, Alabama, U.S.A.
Professor, Department of Anatomical Sciences, School of Medicine, St. George's University
Grenada, West Indies and Centre of Anatomy and Human Identification,
University of Dundee, Scotland

CHARLES C THOMAS · PUBLISHER, LTD.
Springfield · Illinois · U.S.A.

Published and Distributed Throughout the World by

CHARLES C THOMAS • PUBLISHER, LTD.
2600 South First Street
Springfield, Illinois 62704-9265

ISBN 978-0-398-08104-1 (Hard)
ISBN 978-0-398-08105-8 (Ebook)

First Edition, 1997
Second Edition, 2014

Library of Congress Catalog Card Number: 2014029507

With THOMAS BOOKS *careful attention is given to all details of manufacturing
and design. It is the Publisher's desire to present books that are satisfactory as to their
physical qualities and artistic possibilities and appropriate for their particular use.*
THOMAS BOOKS *will be true to those laws of quality that assure a good name
and good will.*

*Printed in the United States of America
UBC-R-3*

Library of Congress Cataloging-in-Publication Data

Persaud, T. V. N., author.
    A history of human anatomy / T.V.N. (Vid) Persaud, M.D., PH.D., D.Sc., F.R.C.Path.
(Lond.), FAAA (Professor, Department of Anatomical Sciences, School of Medicine, St.
George's University, Grenada, West Indies, Professor Emeritus and Former Head Depart-
ment of Human Anatomy and Cell Science, University of Manitoba, Faculties of Medicine
and Dentistry, Winnipeg, Canada), Marios Loukas, M.D., PH.D. (Professor and Chair,
Department of Anatomical Sciences, Dean of Research, School of Medicine, St. George's
University, Grenada, West Indies), R. Shane Tubbs, M.S. P.A.C., PH.D. (Department of
Pediatric Neurosurgery, Children's Hospital, Birmingham, Alabama, U.S.A., Professor,
Department of Anatomical Sciences, School of Medicine, St. George's University, Grenada,
West Indies). – Second edition.
        1 online resource.
    Includes bibliographical references and index.
    ISBN 978-0-398-08105-8 (epub) -- ISBN 978-0-398-08104-1 (hard)
    1. Human anatomy--History. I. Loukas, Marios, author. II. Tubbs, R. Shane, author.
III. Title.

QM11
612–dc23   2014029507

For Gisela

Vid

*To my father, Christos Loukas, who
gave me the stimulus and support
for academics and history in
particular.*

Marios Loukas

*To W. Jerry Oakes, M.D., who holds
academia high.*

R. Shane Tubbs

PREFACE

Anatomy is one of the oldest disciplines in medicine. Without a knowledge of human anatomy, the diagnosis and treatment of diseases are inconceivable. Many advances in medicine and surgery can be directly linked to improvements in understanding the structure and function of the human body. The publication of Vesalius's masterpiece *De Humani Corporis Fabrica* in 1543 ushered a new era in the history of medicine. The study of human anatomy suddenly became an objective discipline, now based on direct observations and scientific principles. The curiosity of early man regarding his own form has driven him to be inquisitive about the body's architecture. Not surprisingly, the study of human anatomy and medicine continues to be integrated and remain inseparable. The study of human anatomy has progressed to its universal acceptance and recognition as a scientific discipline, essential for the practice of modern medicine. This revised and expanded edition of *A History of Human Anatomy* presents anatomy from antiquity to the modern times.

In this book, we have presented many scholars and teachers; the time periods, places, and impact of their work; controversies in anatomy; and advances in the discipline. These topics run the gamut from early pioneers in the art to the development of techniques that have propelled the study of anatomy to its current state. Although a single source such as this work could never hope to comment on every person and place which has influenced the history of anatomy, we have attempted to present the "big picture" regarding the historic anatomists and movements that have shaped our current understanding of what we now call "medical anatomy." Perhaps we will never truly know of the early anatomists who braved passed cultural taboos to make the first examination of the human body with dissection, but we can recount those who are known to have contributed to the discipline.

To confine our topic to the history of the study of anatomy may be premature, as even now man continues to learn about the structure of his body with new and non-invasive technologies, such as MRI. With these new "eyes," we are uncovering parts of the human anatomy never before seen. As Corner (1930) has stated, "The history of a science never reaches the word finis; the work of discovery continues, each new stage growing out of the past without barrier of time or circumstance. Investigation may change its direction, but does not cease." The study of the history of anatomy, therefore, continues and alongside the study of anatomy as a scientific discipline without obvious end. In fact, Northcote (1772) has stated, "There is no condition in life or manner of the study almost, but what may be improved by the knowledge of anatomy; and to many of the liberal sciences it is assistant, to others of them absolutely necessary."

The approach we have taken in writing this book is somewhat eclectic; otherwise it would have been a daunting task to complete. The sheer volume of historical anatomy literature available, as well as many untapped archival sources, forced us to be selective in the material that we have included in this work. For this reason, we cannot claim that this book is complete or exhaustive; there are obvious gaps. Many aspects of the subject have been treated only very cursorily and others have not even been touched.

It would be impossible to compile a book such as this one without borrowing extensively from other sources. For this new edition, we are particularly indebted to Mr. Jordan Bass, medical archivist, Neil John Maclean Health Sciences Library, University of Manitoba, Canada. Of particular value are the many figures for which we are grateful. Extracts taken from the relevant scientific publications and committee reports of historical interest are also acknowledged with thanks.

Mr. Michael Payne Thomas, of our publishers, Charles C Thomas, Publisher, Ltd., gave us the encouragement that made the writing of this book a satisfying experience. Once again, our thanks must also go to the editorial and production staff for their magical skill in shaping this work.

ACKNOWLEDGMENTS

From the First Edition, *History of Anatomy: The post-Vesalian era*, 1997

I must acknowledge the kindness and generosity of the librarians and staff of the following institutions: Neil John Maclean Health Sciences Library, University of Manitoba; Cambridge University Library, England; Whipple Library, Wellcome Unit for the History of Medicine, Cambridge, England; Anatomy School Library, University of Cambridge, England; History of Medicine Division, National Library of Medicine, Bethesda; Universitdtsbibliothek, Freie Universitat Berlin; Institut fur Anatomie, Ernst-Moritz-Arndt-Universitat Greifswald; Institut fur Anatomie, Rostock Universitat; Institut fur Anatomie, Humboldt Universitat Berlin; British Library, London; Wellcome Institute for the History of Medicine, London; Bodleian Library, Oxford; and the Royal College of Physicians of London. Without access to their valuable collections of rare medical works, it would not have been possible for me to complete this book. Many of the illustrations used are photographic copies taken from these sources, and I gratefully acknowledge the help I have received.

For special assistance, I want to express my appreciation to Professor Dr. Jochen Fanghanel, Greifswald; Ms. Elizabeth Tunis, Reference Librarian, History of Medicine Division, National Library of Medicine, Bethesda, MD; Professor Michael Tennenhouse, Librarian, Ms Pam Green, Library Assistant, and Ms Carol Cooke, Resource Development Librarian, Neil John Maclean Health Sciences Library, University of Manitoba; Professor Dr. G.-H. Schumacher, Rostock; Professor Dr. H.-J. Merker, Berlin; Ms. Esther-Maria Dellmann, Librarian, Institut fur Anatomie, Humboldt Universitat Berlin; Dr. R. E. Beamish, Winnipeg; Dr. I. Carr, Winnipeg; Dr. K. L. Moore, Toronto; and Dr. D. Parkinson, Winnipeg.

Dr. Helen Brock, Cambridge, England, brought to my attention several additional papers on John and William Hunter. I am also indebted to Dr. R. Beamish, Editor of the *Journal of Cardiology*, for permission to use the material from my paper (*Can J Cardiol*, 5:12, 1989) in this book.

I should like to thank Mr. Roy Simpson and Mr. Jerry Knab for making many of the photographic prints.

I am deeply grateful to Mrs. Fran Thompson and Miss Sandra Hugill for their skilled secretarial assistance. Once again, I am indebted to my secretary Mrs. Barbara Clune for painstakingly preparing the final copy. They took a special interest in this project. My wife Gisela read the text, checked the references, and made several suggestions for improvement.

Finally, my publisher, Charles C Thomas, has demonstrated infinite patience and given me the encouragement that made the writing of this book a satisfying experience. Once again, my thanks must also go to the editorial and production staff for their magical skill in shaping this work.

ACKNOWLEDGMENTS

Second Edition

For this new edition, we are particularly indebted to Mr. Jordan Bass, medical archivist, Neil John Maclean Health Sciences Library, University of Manitoba, Canada. Gisela Persaud helped in reviewing the manuscript.

We are grateful to Mr. Ryan Jacobs and Mrs. Nadica Thomas Dominique for their skilled secretarial assistance.

We are also greatful to the medical illustrator of the Department of Anatomical Sciences, at St. George's Univeristy, Ms. Jessica Holland, for her editorial assistance with the figures.

The preparation of this text required collective efforts and collaboration of many contributors for their comments, criticism and sharing their knowledge and expertise. We are indebted and appreciative of several doctors and medical students who contributed to the book as reviewers of selected chapters, served as consultants, wrote certain sections and provided constructive criticisms and suggestions.

These individuals are:

Liann Casey

Peter Compton Craig

Elizabeth Hogan, MD

David Holmes, MD

Hima Khamar

Theofannis Kollias

Katherine Oakley

Khoan Tai

Amanda Ward

John Woytanowski

Antony Zandian, MD

CONTENTS

A HISTORY OF HUMAN ANATOMY

Chapter 1

ANCIENT RECORDS

PREHISTORIC PERIOD

From the spectacular fossil discoveries made by anthropologist Donald Johanson and his team in the Afar triangle of Ethiopia, man could retrace his origins to less than four million years (Johanson & White, 1980; Lovejoy, 1981; Johanson & Edey, 1981, 1982). The remains of this creature, affectionately named Lucy, were discovered in 1974 at a site called Hadar, and at the time were considered to comprise the oldest and most complete early hominid skeleton. Lucy belongs to a species called *Australopithecus afarensis*, hominids that walked upright, stood about 4.5 feet tall, and had brains smaller than those of chimpanzees. In 2009, paleontologists announced that a more complete hominid skeleton had been unearthed in the same region. This specimen, called Ardi, represents a species known as *Ardipithecus ramidus* and is dated 1.2 million years prior to Lucy (Fig. 1). Fossil remains from other hominids date back to at least six million years ago, extending man's origins far beyond that of what Lucy hinted. From the time of the appearance of these species, many centuries must have passed before our prehistoric ancestors began to think and act in ways that would be considered to be intelligible. Exactly when this occurred is a matter of speculation (Leakey, 1981; Lewin, 1983; Persaud, 1984; Gibbons, 2009).

The Stone Age began about 2.5 million years ago and hunting and the gathering of food had dominated man's life during this early period in history. The abundant and extraordinary prehistoric paintings, engravings, and reliefs found in different parts of the world emerged during the stone and ice ages and depict hunting scenes, animals, and in some instances, human figures (Obermaier & Kühn, 1930; Wendt, 1976; Hadingham, 1979; Leroi-Gourhan, 1982; Beltran, 1982; Sandelowsky, 1983). It is believed that prehistoric cave and rock art evolved from hunting, myths, and magic rituals.

Paleolithic cave paintings in Mas d'Azil, France depict an archer keenly aiming for the heart of his prey, suggesting that early man was conscious of specific anatomical targets, though they may not had yet fully understood how this happened (Loukas et al, 2007). Undoubtedly, the slaughtering of animals provided some crude anatomical insight, and wounds sustained by hunters might have given occasion for reflecting on the structure of the human body (Figs. 2–5).

Prehistoric paintings of 231 human hands were found in the cave of Gargas, near Aventignan, France (Janssens, 1957); 114 of the hands revealed mutilation of one or more fingers, and in only ten cases, were the hands complete. The pictures of the remaining hands have not been well preserved to determine whether they are intact or mutilated (Hooper, 1980). The age of the pictures has been estimated as possibly 30,000 years old, but the people it originated from, and the reason for the mutilation and for depicting the hands, remain a mystery. Hooper (1980) is of the opinion that these hand drawings were deliberately executed and not the results of some accidental activity. More recent analyses have offered a theory that the hands were not amputated at all, but contorted into shapes to elicit variety in the resulting images, using the hands as stencils (Guthrie, 2005).

ANTIQUITY

The great civilizations of antiquity thrived because of the complex and efficient social organization and the technological, as well as cultural, advances that were achieved. Their real accomplishments have been

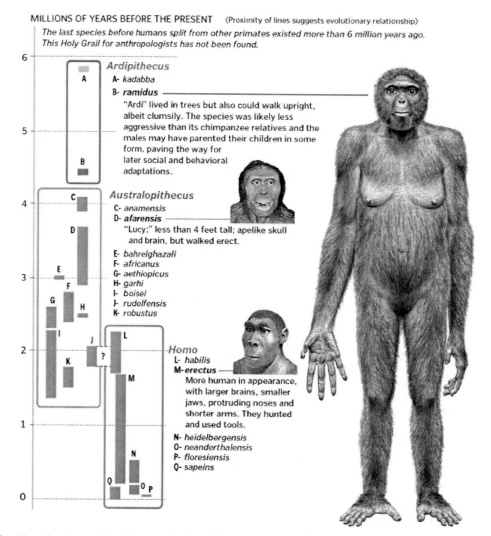

MILLIONS OF YEARS BEFORE THE PRESENT   (Proximity of lines suggests evolutionary relationship)

The last species before humans split from other primates existed more than 6 million years ago. This Holy Grail for anthropologists has not been found.

Ardipithecus
A- *kadabba*
B- *ramidus* ————————————
   "Ardi" lived in trees but also could walk upright, albeit clumsily. The species was likely less aggressive than its chimpanzee relatives and the males may have parented their children in some form, paving the way for later social and behavioral adaptations.

Australopithecus
C- *anamensis*
D- *afarensis* ————————————
   "Lucy;" less than 4 feet tall; apelike skull and brain, but walked erect.
E- *bahrelghazali*
F- *africanus*
G- *aethiopicus*
H- *garhi*
I- *boisei*
J- *rudolfensis*
K- *robustus*

Homo
L- *habilis*
M- *erectus* ————————————
   More human in appearance, with larger brains, smaller jaws, protruding noses and shorter arms. They hunted and used tools.
N- *heidelbergensis*
O- *neanderthalensis*
P- *floresiensis*
Q- *sapeins*

Figure 1. The identification of "Ardi" is a major breakthrough because her species was an early step in the process of human evolution. Scientists do not claim to know whether that Ardi's species evolved directly into modern humans, but it is an important branch on the family tree. Source: Science – AAAS, National Museum of Natural History's department of anthropology. (Illustration by J. H. Matternes; Graphic by The Washington Post – Oct. 1, 2009.)

largely neglected and still remain far from being fully appreciated because of the lack of accurate records and problems of deciphering the material that is now available (Persaud, 1984).

An example of this in more recent times is the discovery of 16,500 tablets and fragments, written some 45 centuries ago, in the ancient city of Ebla, which is located in present day Syria. The tablets and fragments were unearthed between 1974 and 1976 by a team of Italian archeologists and the ancient inscriptions have sparked heated controversy regarding the origin of man and the source of his religions (Pettinato, 1981; Matthiae, 1981).

There are many surviving artifacts, which clearly indicate that some anatomical representations were made in prescientific times. In many respects, these have paralleled the cultural evolution of man (Figs. 6–8). The oldest however, known medical text has been written in Sumerian (approximately 2200 B.C) (Fig. 9).

Mesopotamia

Magic, sorcery, and the practice of divination were distinct features of the great Babylonian civilization that flourished in the fertile valley between the rivers Tigris and Euphrates (Meissner, 1920–25; Dawson, 1930;

Figure 2 (*Left*). Venus of Willendorf. This Paleolithic limestone figurine (4 3/8" tall) is believed to be one of the earliest known representations of the human form (25,000–20,000 B.C.). The head is almost faceless; the pendulous breasts and protuberant abdomen are symbolic of a fertility goddess (Natural History Museum, Vienna).

Figure 3 (*Right*). Female figurine of baked clay (4th millennium B.C., Arpachiyah provenance). (By kind permission of the Trustees of the British Museum.)

Dhorme, 1949; Kramer, 1961, 1963; Mallowan, 1965). Here, where civilization probably began, clay tablets, several millennia old, were discovered and some described monstrous births and the internal organs of sacrificial animals with the omens they predicted (Fig. 10).

The Mesopotamian diviners even made models of these organs for instructing their disciples. The liver and lung of sheep were often used and different parts were carefully marked out with appropriate cuneiform scripts (Figs. 11 & 12), which were used for predicting the future and interpreting natural events (Jastrow Jr., 1914; Chiera,

1938; Grayson, 1980). Although used frequently for various rituals, animal dissection was not performed for scientific purposes. Mesopotamian medicine was based solely on spiritual practices and strict codes were in place to prohibit discussion or adaptation, preventing advancement for centuries (Retief et al., 2008).

Egypt

The earliest of the Egyptian papyruses (Edwin Smith Papyrus) is probably a copy of the one that was first

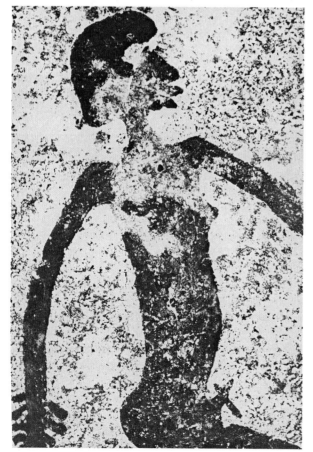

Figure 4 (*Above*). Four individuals carrying equipment are depicted in this rock painting from the Brandberg Massif in Namibia. (Courtesy of Dr. Beatrice Sandelowsky; photograph by Robert Camby.) The artist has combined beauty with fantasy to achieve a unique simplicity in depicting the frolicking group.

Figure 5 (*Left*). Prehistoric Namibian rock art depicting a male figure. (Courtesy of Dr. Beatrice Sandelowsky; photograph by Dr. Robert Camby.) Except for the anatomical misrepresentation of the right hand, other features appear realistic (Wendt, 1976; Sandelowsky, 1983).

Figure 6. Three athletes exercising (painted on a vase of Greek origin, 5th century, B.C.). Remarkable details of bodily movements and of the faces, hands and feet. (By kind permission of the Trustees of the British Museum.)

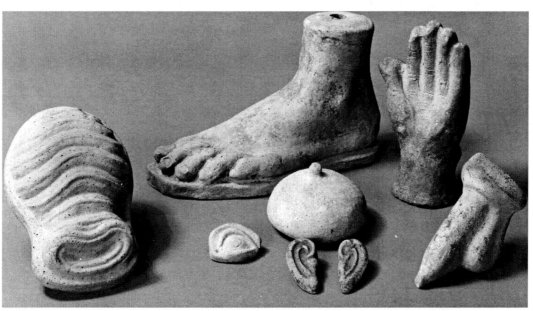

Figure 7. Terracotta models of parts of the human body (Roman, 3rd-1st century, B.C.). The uterus, left foot, hand, male external genitalia, ears, eye, and breast are represented. (Votive offerings by kind permission of the Trustees of the British Museum.)

Figure 8 (*Left*). Sculpture of Ardha Nareeswar (about 10th century A.D.) from a temple in southern India. It depicts the half-male and half-female forms of Ishwara. The basic unit of man and woman is represented by the cosmic unity of Shiva, the creator, and Shakti, the progenitor of the vital force or energy. This dual and unique concept of the human form originated from the ancient Indo-Aryan civilization several millennia before the Christian era. (Courtesy of Professor R. Padmanabhan.)

Figure 9 (*Below left*). This is the oldest known medical text, written in Sumerian. It is a collection of 15 prescriptions, most involving obscure plants and potions. Nippur 2200–2100 B.C. (Courtesy of Pensylvania museum, object #B14221. http://www.penn.museum/college-and-adults/81-press-room/press-images/521-for-press-only-archaeologists-and-travelers-in-ottoman-lands.html.)

Figure 10 (*Below right*). Clay tablet from the Royal Library of Nineveh (7th century B.C.). The cuneiform scripts are records of omens derived from examination of a sheep's liver used in divination. (By kind permission of the Trustees of the British Museum.)

written between 3000–2500 B.C. It is essentially a surgical document and was translated and published over 80 years ago (Breasted, 1930). Tumors, ulcers, abscesses, different wounds and fractures, as well as the prescribed treatments, which included suturing, cauterization, and the use of splints, are systematically listed in it. For the first time in recorded history, the word "brain" is mentioned, followed by a description of the gyri and meninges. The heart is mentioned as the center of a distributing system of vessels, which pulsates (Major, 1954). In addition to the information documented in the ancient papyri, archeological investigation of Egyptian mummies has shed light on the breadth of anatomical knowledge in ancient Egypt. The oldest prosthetics, wooden and leather toes found affixed to mummies, date back to 1069–664 B.C. (Loukas et al., 2011a).

The Edwin Smith papyrus describes 48 known pathologies, subdividing each into examination, diagnosis, and treatment. Many of the injuries addressed in the papyrus were most likely seen in the context of war. Observations made in several of the cases indicate that ancient Egyptians began amassing very specialized anatomical knowledge of the central nervous system. Trauma to specific regions of the skull and brain were associated with particular symptoms and diagnosed based on the severity of the injury. Other cases outlined in the document indicate a detailed knowledge of facial anatomy, demonstrated by treatment guidelines for nasal and mandibular fractures. In addition to shedding light on anatomy and medicine in ancient Egypt, the papyrus describes a physician ranking system that demonstrates a high organizational level from junior doctors to specialists in given fields.

The early adoption of a system of specialization is further supported by the existence of the Kahun Gynecological papyrus. Dating back to 1825 B.C., the manuscript was discovered in 1889 at the Fayum site of Lahun by Flinders Petrie. This guide to female health dealt with topics ranging from conception, contraception, infertility, birthing complications, and assessment of newborns (Loukas et al., 2011a). Treatment for other female concerns is outlined in the Edwin Smith papyrus, including instructions for surgical removal of breast masses (Loukas et al., 2011b).

The Ebers papyrus, purchased by George Ebers at Thebes in Egypt, was written about 16 centuries before the Christian era, but was compiled much earlier (Fig. 13). The author is unknown, but the name of the famous priest and physician, Imhotep, has been suggested as a source (Leake, 1952). It was published as a facsimile edition in 1875 with an introduction,

Figure 11. Babylonian clay model of sheep's liver (19–18th century B.C.) used in divination and for instruction in the temple. The cuneiform characters, inscribed in 55 sections, represent omens and magical formulae. (By kind permission of the Trustees of the British Museum.)

commentary, and notes. Although the major part of the Ebers papyrus deals with incantations, medications, and prescriptions for the treatment of diseases, the document contains numerous anatomical terms and made references to parts of the human body, so much so that it was considered to be the oldest known

Figure 12. Inscribed clay model of the lung. It was used by baru priests for teaching divination to students in the temple, by comparison with the lung of a sacrificed animal. (By kind permission of the Trustees of the British Museum.)

Figure 13. Extracts (columns 1–3) from the Ebers papyrus (about 1600 B.C.). Although the document deals with incantations, medications, and prescriptions, it contains many anatomical references (http://papyri-leipzig. dl.uni-leipzig.de/receive/UBLPapyri_schrift_00035080).

anatomical document (Macalister, 1898; Ebbell, 1937). It should, however, be pointed out that almost all of the eight discovered papyri, especially that of Ebers and Edwin Smith, mention the human body (Ranke, 1933; Grapow, 1935) and give suggestions for the treatment of various medical and surgical conditions.

Interspersed with the description of various diseases are many anatomical terms (De Lint, 1932). It is here in the Ebers papyrus that the first link between the heart and pulse is made. The details of the cardiovascular system are expounded on; however, physicians of this time believed that organ-specific elements ran through the vessels rather than blood (Loukas et al., 2011a). Some of these annotations, however inaccurate, are of historical interest and several excerpts from the Ebers papyrus are presented from the translation that was made by Macalister (1898).

There are vessels from it (the heart) to all the members. Each physician, master of healing, priest-exorcist, feels all these when he places his finger upon the head, upon the scalp (neck or occiput), upon the hands, upon the epigastrium, upon the arms or upon the legs. He traces all from the heart, because its vessels go into all his members, so he describes it (the heart) as the beginning of the vessels to all members.

There are four vessels to the nostrils, of which two carry fluid (mucus) and two carry blood.... There are four vessels to the sides of the temples, which if they carry blood to the eyes, all manner of diseases are produced in the eyes by their means, by their being open to the eyes. If water flows from them, it is the pupils of the eyes that give it. It is otherwise said that sleep causes it (the water) to come from the eyes....

If the air enters into the nostrils, it is driven into the heart and (goes) through the intestine (by these vessels), which distribute it to the whole body....

If excitement seized upon the heart, there is a rushing (of blood) to parts of the intestine and to the liver....

There are six vessels to the two arms-three to the right and three to the left, extending to the fingers....

There are two vessels to the testes, which carry semen.

There are two vessels to the kidneys, one to each kidney.

There are four vessels to the liver, to which they bring fluid and air. These give rise to all kinds of diseases when they (i.e. the fluid and air) are poured into the blood.

There are four vessels that extend to the anus, which supply it with fluid and air....

There are over the anus, opening into it, two of these vessels, one on the right and one on the left, come to it from the leg. They cause dryness of the feces.

During the period of the New Kingdom in ancient Egypt (late dynasty XVII through dynasty XX), a more accurate impression of the structure of the human body was probably first obtained because of the practice of embalming and mummification, which had reached its highest level at that time (Fig. 14). This ancient civilization believed in the immortality of the soul and resurrection of the body.

After death, only one of three spiritual elements departed, leaving two others, the *Ka* (physical features and characteristics) and the *Ba* (soul). In death, the *Ba* travelled with the sun through the underworld during the night but returned to its resting place, the body, in the morning. It therefore became necessary to preserve the body with life-like appearances and to ensure that it is provided with everything conceivable that it might need for the next world. Thus, the funerary arrangements and furnishings evolved into a complex and elaborate ritual (Harris & Weeks, 1973; Andrews & Hamilton-Patison, 1978; Cockburn & Cockburn, 1980; Leca, 1981).

Most accounts of the practice of mummification are based on that recorded by the Greek historian, Herodotus, who visited Egypt in the fifth century B.C. (Engelbach & Derry, 1942; Harris & Weeks 1973; Persaud, 1984).

It is now known that the process of mummification began before the early dynastic period, about four millennia B.C., among the inhabitants of the Nile Valley who believed in a life after death. The technique of mummification as described by Herodotus, and also 400 years later by another Greek historian, Diodorus Siculus, must have evolved to the level of perfection as practiced in the New Kingdom. More specific details of the embalming procedure, based on the painstaking research of Professor Albert Zaki Iskander, have been described (Harris & Weeks 1973).

Following death, the body was taken to the house of mummification or to the house of purification. All clothing was removed and the corpse was placed on a large wooden board. The brain was removed through

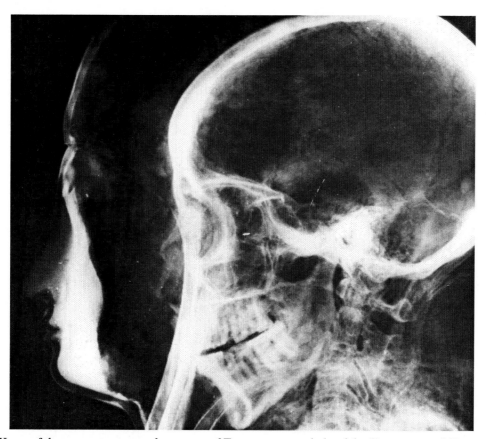

Figure 14. X-ray of the never-unwrapped mummy of Ta-pero, a court lady of the Twenty-second Dynasty, showing the perfect integration between the face and her sarcophagus (From James E. Harris and Kent R. Weeks, X-Raying the Pharoahs. Copyright ®1973, Charles Scribner's Sons. Reproduced with the permission of Charles Scribner's Sons.)

the nostrils, using a hooked probe. In rare cases, the brain was removed through a hole made in the outer skull.

The abdominal viscera, with the exception of the kidneys, were removed following an incision made in the left flank of the body. Following removal of the internal organs, the thoracic and abdominal cavities were washed with palm wine and spices. The internal organs were separately treated and placed in a container of natron for a period of 40 days.

Sprinkled with perfume and further treated with hot resin, the organs were wrapped in packages and placed in four canopic jars (Figs. 15–16). The head of a deity was carved on the lid of each jar. After the Eighteenth Dynasty these were the Four Sons of.

In the 21st Dynasty, the embalmed and wrapped organs were returned to the body cavities, each accompanied by a wax figure of one of the Sons (Fig. 17). The wrapped organs were simply placed between the legs of the corpse in the Twenty-sixth Dynasty.

The body cavities were filled with temporary stuffing in order to accelerate the dehydration process and to preserve the external features. The process of dehydration probably took half of the entire embalming period of 70 days as described by Herodotus. The

packing material was removed from the body cavities, after which they were washed with water and palm oil, and then dried. Resin or linen soaked with resin was then inserted into the cranial cavity and sawdust, myrrh, and occasionally onions were packed into the abdominal cavity. After suturing the abdominal opening, the body was treated with a mixture of cedar oil, cumin, wax, natron, gum, and possibly milk and wine, and then sprinkled with spices.

In order to retain a life-like appearance, the cheeks were stuffed with linen and the nose was plugged. The orbits were also filled with linen and the eyelids were closed. A thick coat of molten resin was applied to the entire body in order to make the skin taut and to prevent the loss of moisture. Eyebrows were painted on and often the body was decorated with jewelry.

The body was bandaged in a prescribed manner and everything that had been in contact with the mummy was then placed into six to seven large pots and buried near the tomb of the deceased.

Because of the crucial importance attached by the ancient Egyptians to the preservation of the human body after death in preparation for eternity, it comes as a startling surprise that there is no known depiction

Figure 15. Canopic jars from Egypt. Twenty-first Dynasty (about 1000 B.C.). These small wooden containers represented the deities, the four sons of Horus, who were responsible for guarding the organs and viscera which were removed from the body during mummification and stored in them. From left: Duamutef (dog-headed) protected the stomach, Imsety (human-headed) protected the liver, Qebhsenuef (falcon-headed) was responsible for the intestines, and Mapy (ape-headed) protected the lungs. The heart as seat of the soul was left undisturbed in the body. (By kind permission of the Trustees of the British Museum.)

Figure 16. Embalmers in ancient Egypt preparing a corpse for mummification. Observe the four canopic jars, which were used for the storage of the organs that were removed from the body (Persaud, 1984).

Figure 17. Frontal chest X-ray of Queen Nodjme (Twenty-first Dynasty) showing a large heart scarab and small wax statuettes representing the four sons of Horus. (From James E. Harris and Kent R. Weeks, X-Raying the Pharoahs. Copyright ®1973, Charles Scribner's Sons. Reproduced with the permission of Charles Scribner's Sons.)

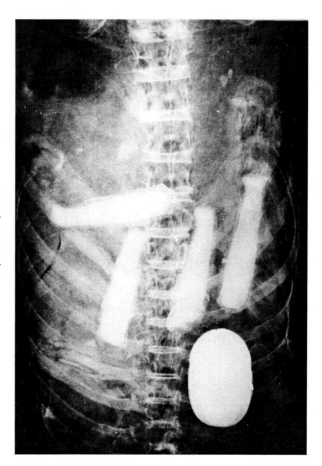

of the actual embalming process. Furthermore, there is no description of any of the internal organs or any remarks of anatomical significance among the extensive archaeological remains from this exalted civilization (Persaud, 1984).

Indeed, the Greek historian, Diodorus Siculus, remarked in his account of embalming:

> then he who is called the cutter takes an Ethiopian stone, and cuts the flesh as the law prescribes, and forthwith escapes running, those who are present pursuing and throwing stones, cursing the defilement on to his head. For whosoever inflicts violence upon, or wounds, or in any way injures a body of his own kind, they hold worthy of hatred. The embalmers, on the other hand, they esteem worthy of every honor and respect, associating with the priests and being admitted to temples without hindrance as holy men. (Harris & Weeks, 1973)

It is important to distinguish the role of the embalmer from that of a physician; the embalmer had no interest in the art of healing, but focused his work solely on the preparation of the body for the afterlife (Fig. 18). In the spiritual and material context of life in ancient Egypt, preservation of the corpse was a practical ritual that simply ensured survival in the afterlife. It would seem that beyond this there was no other implication or importance as far as the embalmers or temple priests were concerned. Despite the metaphysical nature of his intentions, the embalmer was at the forefront of knowledge and discovery of human anatomy. Ancient Egyptian embalmers exemplified the earliest use of trans-sphenoidal access to the cranial vault without disfiguring the face, a technique used in the twentieth century to remove pituitary tumors (Loukas et al., 2011a).

Mummification as practiced by the Egyptians was based on the religious conviction that life continues after death, an ancient belief that might have originated more than 50,000 years ago in the Neanderthal period. The well-preserved corpse of a hanged man, discovered in the peat bogs of Tollund in Jutland and now in the Silkeburg Museum in Denmark is considered to be that of a prehistoric sacrificial murder.

Figure 18. The first known Egyptian physician Hesy-Ra, about 2600 B.C. The three signs read *wr*, *ibkh*, *swn* – the most concise possible way to write " chief," "tooth," "physician." He is standing with the two arms along his body and the hands are free. The scribal equipments are shown on his right shoulder. He wears a short curly wig (Mariette et al., 1872).

China

Ancient Chinese medicine evolved along a unique course following the fundamental concepts of a complex balance between the two forces, *Yin* and *Yang*, which determine the *Tao* (the way). Through Buddhism, medical knowledge flowed into China from India and later contacts were established with scholars from Mediterranean countries, particularly after the fall of the Roman Empire and the expulsion of the Nestorian scholars from Constantinople.

Because of the doctrine of Confucianism, dissection was not practiced so as not to defile the human body. Physicians were taught anatomy from models and diagrams until the eighteenth century, when Wang Qinren began performing cadaveric dissections (Shoja et al., 2010). However, the medical scholars of ancient China revealed through their writings, a keen sense of awareness of the human body for the treatment of diseases. The healing art in ancient China probably began during the fourth millennium before the Christian era by three legendary emperors (Hübotter, 1929; Huard & Wong, 1968; Unschuld, 1985).

The Yi Jing or "Canon of Changes," the greatest Chinese medical classic and probably the most ancient of Chinese books, is attributed to Emperor Fu Xi. He is thought to be responsible for dividing the universe and nature into the two governing cosmic principles, *Yang* and *Yin.* Balance between these forces was considered to be the basis for harmony and good health. Whereas the hollow organs and viscera were *Yang* (light, moist, and active), the solid ones were considered to be *Yin* (dark, dry, and passive). Diseases resulting from external forces are *Yang* and those from internal forces are *Yin.*

The legendary emperor Shen Nong is regarded as the father of agriculture and herbal medicine. Not only is he credited with the discovery of the plow, but also with the compilation of the first materia medica, the Ben Cao or the great herbal. In three volumes, Emperor Shen Nong listed 365 drugs and their therapeutic benefits.

Huang di (2600 B.C.) is another legendary Chinese emperor and is considered the father of Chinese medicine. He compiled the Nei Jing or the *Canon of Medicine*, which dealt with the functions and diseases of the human body. In his book, Huang di made the remarkable assertion that "all the blood of the body is under the control of the heart. The heart is in accord with the pulse. The pulse regulates all the blood and the blood current flows in a continuous circle and never stops." He clearly recognized a relationship between blood, pulse, and the heart, which William Harvey (1578-1657) later confirmed. Other anatomical-physiological annotations in the Nei Jing were the following:

The heart is a king, who rules over all organs of the body; the lungs are his executive, who carries out his orders; the liver is his commandant, who keeps up the discipline; the gall bladder, his attorney general, who coordinates and the spleen, his steward, who supervises the five tastes. There are three burning spaces, the thorax, the abdomen and the pelvis, which are together responsible for the sewage system of the body.

The earliest Chinese medical writings are to be found in the Zuo Zhuan, which was probably written about 540 B.C., though this ancient work of Chinese prose is more of a narrative history than a medical text (web citation a). The physician Wenzi formulated an interesting theory of intrauterine development about 300 B.C. He was particularly known for his interest in diseases of women (Said, 1965).

In the first month, the liquid becomes jelly-like. In the second month the veins and blood in them are formed. In the third month, the shape of human being is formed. In the fourth month, it takes the shape of an embryo. In the fifth month, the muscles become stiff. In the sixth month, the bones become hard. In the seventh month, the figure is complete and the soul enters. In the eighth month, some movement begins with some agitation in the ninth month and early in the tenth month, the baby is born.

The great surgeon Hua Tuo (A.D. 108–208) pioneered the use of anesthetics and is credited with the preparation of anatomical charts showing the organs of the body (Fig. 19). Hua Tuo is most notably praised for his groundbreaking techniques in medicine, which were designed to enhance the performance and outcomes of the surgical procedures he conducted on his patients. Tuo astonished his colleagues with his astute anatomical understanding, which he combined with great medicinal knowledge to diagnose, advise, and predict the outcomes of his patients based on patient compliance. Chinese physicians still revere Hua Tuo for his extraordinary ability to evaluate the pulse. One anecdote credits him with diagnosing a woman who recently miscarried – with just the feel of her pulse-as having a twin pregnancy, indicating that one more fetus remained in her womb. Hua Tuo proceeded to evoke the stillbirth with acupuncture, yet another field he pioneered (Tubbs et al., 2011a).

Figure 19. Japanese woodblock of Guan Yu by Utagawa Kuniyoshi (1798–1861). In this scene he is attended to by the physician Hua Tuo. A wound from a poisoned arrow required the bone to be scraped. Without the use of anesthestic Guan Yu plays go to distract himself from the pain. (From Library of Congress http://www.loc.gov/exhibits/ukiyo-e/images.html.)

Daoism and the practice of acupuncture had an influence on anatomical thinking. The Daoists divided the human body into three regions: the upper part for the spirits, the connecting or middle part and, joined to this, the region of genital activity, represented by the paired kidneys. The drawings of the acupuncturist merely indicated the locations on the body where this procedure should be carried out.

During the Ming Dynasty, Yu Tuan (A.D. 1438–1517) edited his *Orthodox Medical Record* in A.D. 1515. This work was based on the information contained in over 30 other medical manuscripts, and listed 51 topics in need of clarification. Included were the principles of anatomy and physiology. It should be reiterated that even up to this period in the history of Chinese medicine, the human body was considered sacred and was therefore, not dissected for anatomical

studies. Even minor surgical procedures were not permitted, on account of reverence for the body, which in many respects, retarded the progress of surgery.

The physicians of ancient China recognized that their body consisted of skin, flesh, muscles, tendons, and bones. Nine orifices were identified: the eyes, ears, nose, mouth, anus, urethra, and vagina. With respect to the internal structures, they suggested, "*man was composed of the five zang or storing organs and the six fu or eliminating organs. The five zang were more important and they were the liver, heart, spleen, lungs, and kidneys. The six fu were the stomach, large intestine, small intestine, urinary bladder, gall bladder, and the three burning spaces.*"

Of fascinating interest is a rare drawing of an ancient Chinese anatomy of unknown origin. It is in the form of a wood etching and measures approximately 74 x 24 cm and depicts the structure and functions of different parts of the body. It was probably created some time during the sixteenth century or even earlier. Said (1965) stated in his book that the English surgeon, Dr. Lockhart, remarked that "*the Chinese pictures of anatomy look as if someone saw the incomplete dissection of the internal body and then has drawn the organs from memory, while he filled out the darker remaining parts from imagination and drew more what according to his own mind existed and not what actually existed.*"

The ban on human dissection was lifted in Medieval Europe, yet China maintained Confucian principles that made dissection illegal (Shoja et al., 2010). Undoubtedly, the lack of any practical knowledge of human anatomy, which during this period was only slowly evolving in Western Europe, hindered an accurate understanding of the structure of the body.

India

Similar to the Chinese, Indian civilization is one of the oldest known. From the archeological excavations and finding of Mohenjo-Daro and Harappa, it is now known that there were earlier inhabitants of the Indus Valley between the Himalaya and the Vindhya ranges, long before the Aryan conquest from the northwest in 1500 B.C. Abundant evidence suggests an advanced social organization and sanitation practice.

With cultural development in this ancient Indo-Aryan civilization, healing became entwined with religious practices (Sondern, 1936). Health and diseases were attributed to the gods with Dhanvantari as the patron. Three *Doshas* or humors (wind, bile, and phlegm) were thought to permeate the entire organism and when in harmony, led to good health. Notwithstanding their devotion to spiritual pursuits, the

Aryans surprisingly developed a rational and secular approach to the practice of healing based on keen observations, the judicious use of herbs, and surgery (Persaud, 1984).

Several phases of development within this ancient culture have been identified: the Vedic (about 1500–500 B.C.), Brahmanic (600 B.C.-A.D. 1000) and the Mughal (from A.D. 1000 until eighteenth century). The Vedas or books of knowledge were compiled during the Vedic period. It has been suggested that these revealed works of the universal spirit or creator, which embodied religion and philosophy, were formulated some 4000 years prior to the Christian era. Only four of these books have survived: The *Rig Veda*, *Sama Veda*, *Yajur Veda*, and the *Atharva Veda*. Many diseases and treatments, as well as surgical procedures, are recorded in two of these works.

The *Rig Veda* was essentially a medical treatise whereas the *Atharva Veda* (Fig. 20) was a surgical work. Thus, evolved traditional healing methods (Ayurvedic) together with practical skills, which were remarkably advanced for this period in the history of man. Much of this knowledge spread into Asia and later reached Europe during the Middle Ages as a result of translations that were made by Persian and Arab scholars in the eleventh century. For a deeper appreciation of the many remarkable accomplishments of ancient India in the field of medicine, reference should be made to the work of Bhagvat Sinh Jee (1978) and to the authoritative and monumental 12 volume series edited by Singhal and Guru (1973).

Apart from the use of medicinal plants for the treatment of a wide spectrum of diseases, the practice of surgery evolved to become one of the outstanding achievements of Indian medicine. The Laws of Manu, which probably were formulated about 3000 B.C. and compiled between 200 B.C. and A.D. 200, formed the basis of the social fabric of daily life in ancient India. For example, the nose was cut off as a punishment for adultery. Therefore, it is not surprising that rhinoplastic procedures were well advanced. Other surgical operations included the repair of torn ear lobes and cleft lip, suturing of the intestine by applying large ants followed by decapitation of the ants after they had bitten into the edges of the wound, removal of stones from the bladder and cataract extraction. As many as 101 surgical instruments, including a variety of forceps, scalpels, needles, and suturing materials have been described. Aspiration of fluid for the treatment of both ascites and hydrocele and the use of a magnet for the extraction of foreign bodies were mentioned for the first time.

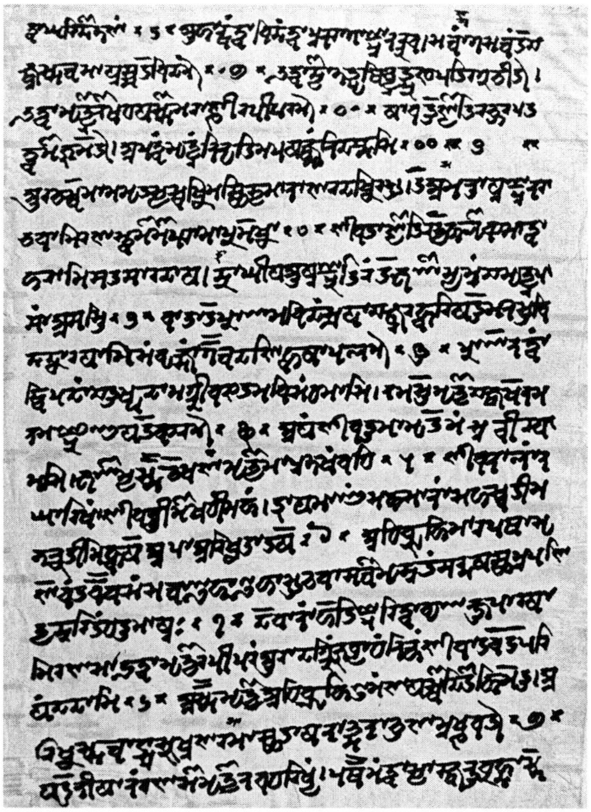

Figure 20. Page from the Paippalada Atharva Veda (on birch-bark). One of the four "books of knowledge" from the ancient Indian civilization, it deals with philosophical, religious, magical, and medical matters (Universitatsbibliothek Tubingen).

The famous physicians of Hindu medicine were Susruta, Charaka, and Vagbhata. There is still some controversy as to exactly when these scholars lived. Susruta was a surgeon who most likely lived during the sixth century B.C. and taught at the University of Kasi or Banaras. He was a younger contemporary of Atreya who taught at Taksasila or Taxila, a famous seat of learning in the West.

The medical wisdom of Atreya was compiled in the form of a compendium or the Samhita. This work was essentially a classification of diseases with some remarks on the skeleton. Susruta also produced a similar Samhita but with more emphasis on surgical matters, including surgical instruments and surgical operations. It is in this work that one finds significant anatomical considerations of the ancient Hindu. Because Susruta referred to Atreya's system of describing the bones, it is generally agreed that both these men lived and compiled their work during the sixth century B.C. In regard to Charaka, it would appear that he flourished during the reign of King Kanishka, about the middle of the second century. Vagbhata referred to both Susruta and Charaka by name and quoted their works. Vagbhata might have lived during the early part of the senenth century about A.D. 625. The work produced was a summary or Samgraha of the eight branches of medicine (Persaud, 1984).

There is compelling evidence to believe that the knowledge of human anatomy revealed at the time of Susruta was acquired not only by inspecting the surface of the human body but also through dissection (Hoernle, 1907; Keswani 1973). Prior to this, limited human anatomical knowledge was derived from physical examination of patients, animal sacrifice and the rare chance observation of improperly buried bodies (Loukas et al., 2010a). Susruta recommended to those aspiring to a career in surgery that they should acquire a good knowledge of the structure of the human body and described the method of how the body should be prepared for this purpose. Regarding the importance of dissection, he stated "*therefore the surgeon, who wishes to possess the exact knowledge of the science of surgery, should thoroughly examine all parts of the dead body after its proper preparation*" and about the method for dissecting the following is recommended (Singhal & Guru, 1973):

> Therefore for dissecting purposes, a cadaver should be selected which has all parts of the body present, of a person who had not died due to poisoning, but not suffered from chronic disease (before death), had not attained a 100 years of age, and from which the fecal contents of the intestines have been removed. Such a cadaver, whose parts are wrapped by any one of "munja" (bush or grass), bark, "kusa" and flax, etc. and kept inside a cage, should be put in a slowly flowing river (Fig. 21) and allowed to decompose in an unlighted area. After proper decomposition for seven nights, the cadaver should be removed (from the cage) and then dissected slowly by rubbing it with the brushes made out of any one of usira (fragrant root of a plant), hair, bamboo, or "balvaja" (coarse grass). In this way, as previously described, skin, etc. and all the internal and external parts with their subdivisions should be visually examined.

Susruta was able to achieve his remarkable knowledge of human anatomy in spite of religious laws that prohibited contact with the deceased other than for the purpose of cremation. Hindu law commands that persons older than two-years-old must be cremated in their natural condition at time of death, leaving Susruta to cunningly employ corpses of those less than two years old without breaking sacred law. Using a brush-type broom, he was able to scrape off skin and

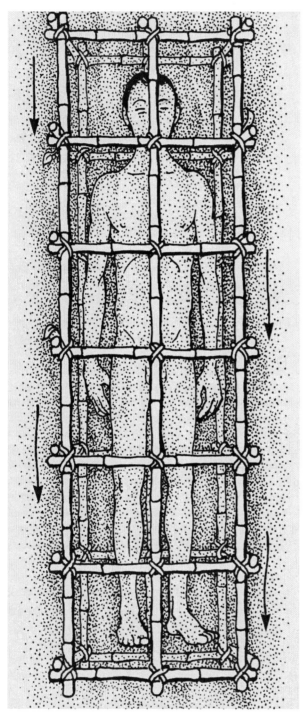

Figure 21. Preparation of a corpse for dissection as described by Susruta (Singhal and Guru, 1973). Drawing by Glen Reid, Medical Illustrator, University of Manitoba.

flesh from the macerated remains in a systematic manner without actually touching the body. Susruta's Samhita contains a fair amount of speculations and philosophical concepts organized in a system of classifications. He stated that from surgical experience, he knew of 300 bones, although 360 are recorded in the Vedas. He ascribed 120 bones to the extremities; 117 to the pelvis, flanks, back, and the chest; and 63 to the region above the neck. Susruta further described the types of bones, the importance of the skeleton, the number of joints, and the types of joints, ligaments, and muscles in different parts of the body. He assigned 20 additional muscles to the female on account of the breast and genital tract (Persaud, 1984).

Moreover, Susruta outlined surgical techniques in his Samhita that suggested that he had some knowledge of regional anatomy, particularly of the face and its blood supply. He also described the eye in great detail as a consequence of the high incidence of cataracts in his region, and delineated five Madalas, or anatomical divisions (Loukas et al., 2010a).

It has been suggested that Susruta might have arrived at the relatively large number of bones in the human skeleton because of the many dissections he carried out on children less than two-years-old, accounting for individual parts of bones that had not yet fused. Despite his erroneous account of the skeleton and other speculations, e.g., 700 veins originating from the umbilicus and distributed to all parts of the body, Susruta's knowledge of human anatomy as revealed in his account of the muscles, joints, ligaments, and even blood vessels and nerves was remarkable for the period in which he lived (Persaud, 1984; Loukas et al., 2010a).

There has been speculation as to the extent of interactions between early Indian and Greek civilizations with respect to the evolution of medical knowledge (Jee, 1978; Hoernle, 1907, Major, 1954; Stierlin, 1978). Whether the Indian concept of the human body, in particular the bones, is based on a familiarity of Greek medicine, as is the case in the Talmud, is not easily resolved. Contact between Persian and Greek scholars occurred as early as the sixth and fourth centuries B.C., respectively. According to Keswani (1973), Susruta's work was translated into Persian and Arabic in the eighth century and it was used as a surgical textbook (*Kitab-i-Susrud*) by students in medical schools under the Caliphate.

There is no evidence that Susruta acquired his knowledge of medicine from the Greeks, but we do know that about 400 B.C., Skylax and Ktesias, both in the services of Persian kings, visited India. Major (1954) believed that there must have been some exchange of knowledge between the two cultures contrary to the opinions of many scholars who have postulated an independent development with separate courses. Major (1954) commented as follows:

> After the conquest of Alexander the Great, in the fourth century B.C., commerce with India was established, and Indian medical science became part of the Greek heritage. Greek physicians became well acquainted with Hindu culture and medical science. Alexander, himself, had Indian physicians. In the later Greco-Roman period, it is obvious that Pliny and Galen borrowed heavily from Indian sources. During the Islamic period, Indian medicine was a powerful stimulus in the development of Arabic medical science. Harun al Raschid, the great caliph of Baghdad, in A.D. 773 called two Hindu physicians, Manaka and Saleah, to teach in the new medical school of Baghdad.

The answer to this controversial question of dependence and the sharing of medical knowledge among scholars of the ancient civilizations may never be known.

Chapter 2

PREHIPPOCRATIC ERA

In ancient Greece, human dissection was illegal and also considered to be a taboo, posing an obstacle for those who were interested in furthering knowledge of human anatomy. Most theories and anatomical writings of the period were derived from animal dissections, observation of humans performing certain tasks, and examination of the dead on battlefields (Marketos & Skiadas, 1999).

ASCLEPIOS

Asclepios (*Greek: Ἀσκληπιός, Latin: Aesculapius*) was the son of Apollo and Coronis (Fig. 22). According to Greek mythology, his mother died during labor and Apollo gave the baby to the centaur Chiron who raised Asclepius and taught him the art of medicine. Five of his daughters were also goddesses with powers related to health. Hygieia (*Hygiene*) the goddess of personification of health, cleanliness, and sanitation; Iaso the goddess of recuperation from illness; Aceso the goddess of the healing process; Aglaea the goddess of beauty, splendor, glory, magnificence, and adornment and Panacea the goddess of universal remedy. For the ancient Greeks, Asclepios was known as the Greek god of the healing arts, very often pictured with a knife or surgical tool. Asclepios and his daughters represented the healing aspect of the medical arts. Interestingly, the original Hippocratic oath starts with *"I swear by Apollo the Physician and by Asclepios and by Hygieia and Panacea and by all the gods...."*

Ancient writings claim that Asclepios was a surgeon and great healer. It is likely that he possessed knowledge of the cardiovascular system, for he closed and healed several wounds. It was noted by Homer that Greek doctors of his time were taught to check blood flow after bandaging a wound (Loukas et al., 2007a). Wounded or ill patients were typically treated by physicians (*Greek: ιατρός*). When physicians were not able to treat patients, they would send them to the temple of Asclepios.

From about 300 B.C. onwards, healing temples of Asclepios (*Asclepieia*) were developed in many places in Ancient Greece with the most famous located at Epidauros in northeastern Peloponnese, and the other one on the island of Kos, where Hippocrates began his career. The patient would lie down in the sacred hall called the *abaton* (*Greek: place of no walking*) and waited for the god to appear and give advice in a dream or perform his healing. The god often appeared in the form of a snake, the sacred animal of Asclepios (*Aesculapian snakes or Greek "Drakon." These snakes were nonvenomous tree climbers and constrictors of the genus Elaphe longissima longissima*). In Epidauros, a large marble column (stele) was found preserving the case histories of 70 patients. One of these cases described a man whose toe was healed by the snake. The man had been brought on a stretcher by his servants to the temple and laid out in the open. There, he fell asleep and a snake healed his wound with its tongue. Upon waking, the man remembered from his dream a young man placing an ointment on his wound (Majno, 1975).

ALCMAEON OF CROTON

Alcmaeon (*Greek: Ἀλκμαίων*) of Croton (*Greek: Κρότων is a city in Calabria, Southern Italy*) lived about 500 B.C. (Fig. 23) and was a contemporary of Pythagoras. Most likely, Alcmaeon studied at the famous medical school of (*Magna Graecia*) Croton (Debernardi et al., 2010). He is regarded as the earliest known person to have pursued actual anatomical studies (Persaud, 1984). Alcmaeon dissected animals with the sole intent of understanding their anatomy (Codellas, 1932; Erhard, 1941a; Sigerist, 1952; Arcieri, 1970; Lloyd, 1975; Debernardi et al., 2010); however, the methods he used are still unknown. Very little of his writings have survived, but it appears that he formulated theories in medicine and natural philosophy founded upon mathematics, which he learned from Pythagoras. He described health as a balance, and disease as something that offsets this balance (Codellas, 1932; Erhard, 1941a; Sigerist, 1952; Arcieri, 1970; Lloyd, 1975, Moose, 1998).

Alcmaeon was the first to mention the pharyngotympanic tube (Dolby & Alker, 1997), which was later rediscovered by Eustachius in the sixteenth century. From a study of the developing chick embryo, he concluded that the head of the fetus was the first part to be formed, and was the first to assert that the brain was the organ responsible for intelligence (Debernardi et al., 2010).

Alcmaeon carried out physiological studies aimed at unraveling the nature of sense perception. He considered man to be the only creature that has a sense of understanding along with perception, whereas other animals lack the former. He believed the senses to be connected with the functioning of the brain by *poroi* (*Greek, πόροι*), or channels (Retief et al., 2008). He thought that if the brain moved or changed its position, the senses would become incapacitated because the passages through which the sensations arrived were blocked (Magner, 1979; Debernardi et al., 2010).

Alcmaeon was also the first to mention the eyeball and described the *poroi* connecting the eyes to the brain (the optic nerve) (Retief et al., 2008). He postulated that external light and the liquid in the eyeball (vitreous humor) were essential for vision. Alcmaeon likely recognized the importance of light entering the eye for vision (Codellas, 1932; Sigerist, 1952; Arcieri, 1970; Debernardi et al., 2010). He described the ears and claimed that there is an empty space within; air

Figure 22. Statue of Asclepius, exhibited in the Museum of Epidaurus Theatre, in Greece.

enters this cavity and resonates, producing sound. Alcmaeon described that the brain is what recognizes scents as air is drawn up through the nostrils. He also believed that the tongue recognizes texture and temperature, and thus, is what gave humans the sense of taste (Debernardi et al., 2010).

Alcmaeon described the difference in blood vessels, distinguishing between arteries and veins. He attributed sleep to a transient suppression of cerebral flow and also noted that this suppression led to death when it became permanent (Loukas et al., 2007a).

EMPEDOCLES

Empedocles (*Greek:Ἐμπεδοκλῆς*) of Agrigento (a city on the southern coast of, also known as Acragas (*Greek: Ἀκράγας*), a contemporary of Alcmaeon of Croton, lived between 493 and 433 B.C. He was a gifted philosopher and orator who speculated on the functions of the body. There are many legends attributed to Empedocles, including one that he possessed supernatural powers. He claimed to have the power to heal the sick, resurrect the dead, and influence rivers and the sun (Persaud, 1984; Van der Ben, 1975; Guthrie, 2012).

Figure 23. Alcmaeon of Croton (c. 500 B.C.).

presented a theory of vision describing a light that comes out of one's eye to illuminate objects. While this theory was flawed, it did begin to pave the way for later philosophers, such as Aristotle, to develop their own theories (Johansen, 1997).

Empedocles described four elements (earth, air, fire, and water) as the root of all things (Longrigs, 1976; Loukas et al., 2007a) and also formulated the theory of the four humors, forming the basis of humoral pathology. Humoral pathology profoundly influenced medical and scientific thinking for 2000 years.

Empedocles believed that the life force was attributed to the "*innate heat*" of the body, which was distributed by the heart. Such a hypothesis undoubtedly established the central role of the heart as part of the vascular system. It was the organ that distributed the "*pneuma,*" (*Greek: πνεῦμα, breath*) the intangible life force that was more than both soul and life, throughout the body; air was the cosmic equivalent of the life-soul of man which he summarized as follows: "*As our soul, being air, sustains us, so pneuma and air pervade the whole world*" (West, 2011).

Empedocles postulated several theories of embryological (Wilford, 1968) and physiological interests (O'Brien, 1970). Although bizarre, he described the creation of many malformations, such as "*faces without necks and arms without shoulders*" as well as creatures "*with faces and breasts on both sides, man-faced ox-progeny*" (Persaud, 1984). He recognized the contribution of both the mother and father to the formation of the embryo but suggested that its gender was determined by the degree of warmth within the uterus. He also thought that all living things inhale and exhale air through bloodless channels in the body and pores in the skin (Erhard, 1941b; Cappelletti, 1975).

Using poetry, Empedocles described his theories and experiences of healing patients (Chitwood, 2004). More than 150 fragments of his two poems "*on nature*" and "*purifications,*" have survived; both are highly philosophical works dealing with the nature of the universe and transmigration of souls (Van der Ben, 1975; Guthrie, 2012).

Empedocles carried out physiological studies pertaining to the sense organs, respiration, nutrition, and growth (Siegel, 1959; Booth, 1960; O'Brien, 1970). He

DIOGENES OF APOLLONIA

Diogenes of Apollonia (around 440–430 B.C.) was the first physician to describe the vascular system. Influenced by Empedocles, he furthered the idea of *pneuma*, claiming that it moved within the vessels alongside blood. While he made no distinction between arteries and veins, referring to all vessels as *phlebes* (*Greek: φλέβες for veins*), Diogenes claimed that the body possessed two large vessels that coursed

through both sides of the abdomen, legs, and eventually the head, while branches of these vessels traveled in any space in between. He believed that there were parts of the body that contained only air: empty vessels called arteries and an empty left ventricle of the heart. Diogenes also believed that these empty, air-filled regions were part of the brain (Loukas et al., 2007a; Harris, 1973; Crivellato et al., 2006).

Chapter 3

HIPPOCRATIC CONCEPTS

Hippocrates (*Greek:* Ἱπποκράτης) (about 460–377 B.C.) was born on the Island of Cos in the Aegean (Fig. 24). He is considered to be the greatest of all physicians and the Father of Medicine, having established the healing art as a science far removed from superstition and magic. Hippocrates advised that "*diseases caused by overeating are cured by fasting... diseases caused by indolence are cured by exertion... and tenseness by relaxation*" (Olsen, 2009). Much of the so-called "*Hippocratic Corpus*," a large collection of philosophical, medical and scientific works (Fig. 25), are wrongly attributed to him and were likely written by physicians of the medical school of Cos. Between 300–200 B.C., these writings were further compiled and edited by the scholars of the famous library at Alexandria (Adams, 1939; Jones, 1945; Wake, 1952; Stroppiana, 1963; Lloyd, 1975; Finger, 2004; Persaud, 1984).

The Hippocratic Corpus built upon Alcmaeon's theory of the four elements, incorporating and describing four fluids or "*humors*" of the human body: black bile, yellow bile, phlegm, and blood, each of which were associated with certain qualities (Bujalkova et al., 2001). Hippocrates believed that these humors were internally balanced with each other; interruptions of this balance caused disease. He also believed that by noting the characteristics of the symptoms, one would be able to identify which humor was out of balance by comparing the qualities of the symptoms to the ascribed qualities of each humor. Once the cause of the disease was diagnosed, one could perform certain behaviors or tasks to increase or decrease their other humors in order to regain the natural balance. If the tasks did not cure the patient, Hippocrates explained that the balance could be restored by withdrawing measured amounts of the patient's blood, or by the administration of certain drugs (which were usually poisons) in order to flush out the imbalanced humor by induction of vomiting or diarrhea (Barlow & Durand, 2011).

Of the collection of treatises, only five are considered to be genuine. Except for the description of certain bones, what is mentioned of the structure of the human body is scant, superficial, and largely inaccurate. Even though Hippocrates postulated that anatomy should essentially be the foundation of medicine, he believed that one could learn sufficient anatomy from the observation of wounds without dissecting corpses. Indeed, there is no evidence that Hippocrates dissected a human body and his descriptions were probably made from inspection of the surface of the human body and chance observations of human wounds (Finger, 2004).

Hippocrates was apparently familiar with the bones of the skull and knew of their articulations and sutures; however, his knowledge of the internal organs and of muscles was confused and speculative. He knew of blood vessels, and from his treatise, *On the Nature of Man*, a highly speculative description of the blood vessels in the human is presented, and sites for carrying out venesection were recommended (Chadwick and Mann, 1950; Brain, 2009).

... The blood-vessels of largest caliber, of which there are four pairs in the body, are arranged in the following way: one pair runs from the back of the head, through the neck and, weaving its way externally along the spine, passes into the legs, traverses the calves and the outer aspect of the ankle, and reaches the feet. Venesection for pains in the back and loins should therefore be practiced in the popliteal fossae or externally at the ankle.

The second pair of blood vessels runs from the head near the ears through the neck, where they are known as the jugular veins. Thence they continue deeply close to the spine on either side. They

Figure 24. Hippocrates (about 460–377 B.C.). (Courtesy of Wellcome Institute Library, London.)

pass close to the muscles of the loins, entering the testicles and the thighs. They traverse the popliteal fossa on the medial side and passing through the calves lie on the inner aspect of the ankles and the feet. Venesection for pain in the loin and in the testicles should therefore be performed in the popliteal area or at the inner side of the ankle.

The third pair of blood vessels runs from the temples, through the neck and under the shoulder blades. They then come together in the lungs; the right hand one crossing to the left, the left hand one crossing to the right. The right hand one proceeds from the lungs, passes under the breast and enters the spleen and the kidneys. The left hand one proceeds to the right on leaving the lungs, passes under the breast and enters the liver and the kidneys. Both vessels terminate in the anus. The fourth pair runs from the front of the head and the eyes, down the neck and under the clavicles. They then course on the upper surface of the arms as far as the elbows, through the forearms into the wrists and so into the fingers. They then return from the fingers running through the ball of the thumbs and

the forearms to the elbows where they course along the inferior surface of the arms to the axillae. They pass superficially down the sides, one reaching the spleen and its fellow the liver. They course over the belly and terminate in the pudendal area....

The brain was considered to be a gland secreting mucous that cooled the body. The heart was described as a muscular organ of pyramidal shape and he placed within it the two auricles as reservoirs of air and the two ventricles as the fountains of life, separated by a partition (Diller, 1938; Kapferer 1951; Macdonald, 2003). Hippocrates recognized the lungs, kidneys, urinary bladder, and intestines. From the medical works attributed to him, we learn of a highly fanciful interrelationship among these organs in certain disease conditions:

If a patient over the age of thirty-five expectorates much without showing fever, passes urine exhibiting a large quantity of sediment painlessly, or suffers continuously from bloody stools, his complaint will arise from the following single cause. He must, when a young man, have been hard-working, fond of physical exertion and work and then, on dropping the exercises, have run to soft flesh very different from that which he had before. There must be a sharp distinction between his previous and his present bodily physique. If a person so constituted contracts some disease, he escapes for the time being but, after the illness, the body wastes. Fluid matter then flows through the blood vessels wherever the widest way offers. If it makes its way to the lower bowel it is passed in the stools in much the same form as it was in the body; as its course is downward it does not stay long in the intestines. If it flows into the chest, suppuration results because, owing to the upward tread of its path, it spends a long time in the chest and there rots and forms pus. Should the fluid matter, however, be expelled into the bladder, it becomes warm and white owing to the warmth of that region, it becomes separated in the urine; the lighter elements float and form a scum on the surface while the heavier constituents fall to the bottom forming pus. Children suffer from stones owing to the warmth of the whole body and of the vesical region in particular. Adult men do not suffer from stones because the body is cool; it should be thoroughly appreciated that a person is warmest the day he is born and coldest the day he dies. (Chadwick and Mann, 1950)

For the first time, Hippocrates discussed the anatomy of the spine, spinal cord, and certain diseases associated with them. He described the curvature of the spine, and knew that the spinal cord was important (even believing that it was the site of sperm production), and so he insisted that every physician should study it meticulously (Marketos & Skiadas, 1999). Hippocrates postulated that the spine consisted of vertebrae connected by anterior and posterior nerves, with what he described as mucous connections. Although these were actually ligaments, he called them nerves because there was no way to distinguish between the two. Each vertebra varied in size and had a process extending posteriorly. Hippocrates did not consider the sacrum or coccyx as part of the spine, and divided what he called the spine into three regions. The first region consisted of seven vertebrae (cervical vertebrae), which he described as being above the clavicles. The second region, he claimed, was the 12 vertebrae that articulated with the ribs (i.e., thoracic vertebrae). The final region consisted of five vertebrae (the lumbar vertebrae). He described a muscle that is attached to the spine (psoas), and also said that there existed a dense venous plexus around the spinal cord, which branched off from the veins that coursed through the lungs (Marketos & Skiadas, 1999).

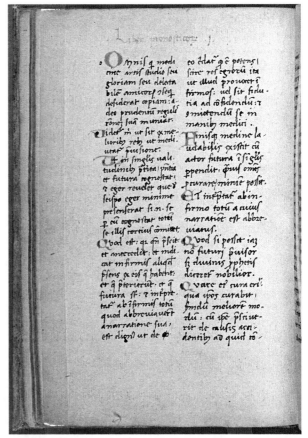

Figure 25. Corpus Hippocratisum, Prognostic. The *Prognostic* reflected Hippocrates' clinical activity which is characterized by attention to patients, the symptoms of their diseases, and their clinical description. It is shown here in a Medieval Latin translation. (Courtesy of Bethesda, MD : U.S. National Library of Medicine, National Institutes of Health, Health & Human Services http://www.nlm.nih.gov/exhibition/odyssey ofknowledge/greekmedicine.html.)

POLYBUS

Of Hippocrates's many followers, his son-in-law, Polybus of Cos (fourth century B.C.), carried out studies on the human body, dissected animals (Gray, 1999), and probably authored several of the "*Hippocratic works*." He is described as a recluse who separated himself from the world and the pleasures it afforded in order to devote himself to his two treatises, *Nature of the Child* and *On Man* (Grensemann, 1968; Jouanna, 1969; Shoja et al., 2008).

Polybus's anatomical representations were crude and erroneous. In describing the major blood vessels of the body, he too mentioned four pairs that ran from the head to the hips, lower extremities, and ankle; the jugular vessels to the loins, thighs, leg, and inner ankle; from the temples to the scapulae and lungs, from there after mutual intercrossing to the spleen and left kidney, the liver, and right kidney, and then to the rectum; and from the front of the neck to the upper extremities, the upper part of the trunk, and the reproductive organs. This fanciful description of blood vessels was the result of pure philosophical speculation and emanated from Hippocratic teachings (Shoja et al., 2008; Persaud, 1984).

HIPPOCRATIC LEGACY

In the *Timaeus* of Plato (429–347 B.C.), one of Plato's dialogues, delivered mostly in the form of a long monologue, anatomical concepts as postulated by Hippocrates and Polybus were reformulated to incorporate the doctrine of the macrocosm of the universe compared to the microcosm of the human body (Fig. 26). His

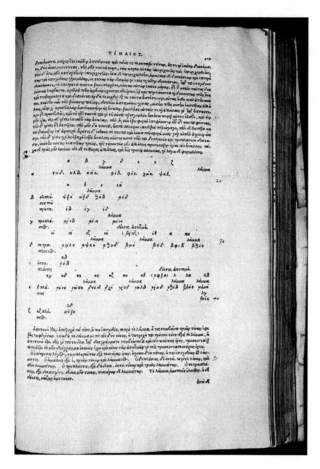

Figure 26. This volume contains the second Greek edition of the Platonic corpus, edited by Johannes Oporinus and Simon Grynaeus; the text is based upon the Aldine edition of 1513 (*Platonis Omnia Opera Cum Commentariis Procli in Timaeum & Politica, thesauro veteris Philosophiae maximo*). The book was printed at Basel by Joannes Valderum in March 1534. In addition to the text of Platonic works, the edition contains commentaries by Proclus on the *Timaeus* and the *Republic*. Notes in Latin, Greek, and French have been written throughout the volume. These notes often comment upon different readings of the text, citing sources including Cicero, Calcidius, and Proclus; printing errors are also pointed out and corrected.* The pages displayed are from the section of the *Timaeus* in which the composition of the world's body and soul is described. (Courtesy of Library of http://www.library.illinois.edu/rbx/exhibitions/Plato.)

philosophical concept of the structure of the human body and causes of diseases profoundly shaped and directed the thinking of philosophers and natural scientists (Wright, 1925; King, 1954; Miller, 1962; Cornford, 1971; Broadie, 2011). There is evidence that this dialogue may have also laid the foundation for the Triune Brain, a current model of evolution of the vertebrate forebrain (Smith, 2010).

From these Greek scholars emerged Aristotle whose systematic method of enquiry and careful observations of embryos and dissected animals laid the foundation of comparative anatomy. His many outstanding contributions to the biological sciences, politics, and philosophy provided a huge repository of knowledge (Persaud, 1984).

Chapter 4

ARISTOTLE

PHILOSOPHER AND SCIENTIST

Although Aristotle (*Greek: Ἀριστοτέλης*) (Fig. 27) was the greatest natural philosopher of his era and, even though he was not a physician, his contributions to medicine have been equaled only by Hippocrates (Jaeger, 1948; Ross, 1952; Gross, 1999; Persaud, 1984). Aristotle was born in 384 B.C., son of the court physician to King Philip of Macedonia, in the city of Stagira. He came under the influence of Plato and, following the death of his teacher, travelled extensively. After returning home, the King asked him to tutor his son, Alexander (Cooper, 2006).

Returning to Athens in 355 B.C., Aristotle established his Lyceum, which became a celebrated center for philosophical enquiries and the study of natural phenomena. He carried out extensive and fairly accurate studies, including systematic dissections on a wide variety of animals. He was also noted to have examined a 40-day-old human fetus (Crivelatto & Ribatti, 2006a; 2006b). In his work, Aristotle described fundamental concepts of organic evolution. Charles Darwin considered Aristotle as one of the world's greatest natural scientists (Lloyd, 1968; Grene, 1972); he remarked, "*Linnaeus and Cuvier were his gods, but compared to Aristotle, they were mere schoolboys*" (Tipton, 2006).

ANATOMICAL WRITINGS

Like Hippocrates, Aristotle's knowledge of the human body was based on external observations and speculative ideas obtained from dissecting animals (Aristotle, 1749;1831; Cole, 1975). Although much of it has been lost, Aristotle's diagram of the male urogenital system is probably one of the earliest known anatomical illustrations.

The four books on the *Parts of Animals (De Partibus Animalium)* and the first three books on the *History of Animals (Historia Animalium)* constitute a formidable volume of anatomical inquiries (Fig. 28). Aristotle correctly described two main blood vessels located in front of the vertebral column and mentioned the aorta for the first time in history. He compared the thickness and consistency of the aorta and the vein and stated that they, like all blood vessels, arose from the heart and not from the head and brain as previously stated by Polybus and others (Persaud, 1984).

Aristotle had a relatively extensive and somewhat accurate understanding of the cardiovascular system. He understood that there was both expansion and contraction with each heartbeat (Smith et al., 2012). In describing the heart, he placed it in the center of the chest, located more on the left side above the lungs and near the bifurcation of the trachea. The heart was considered to have a pointed apex, which is more solid that the rest of the organ and directed forward. "*In no animal does the heart contain a bone, certainly in none of those that we have ourselves inspected, ... but it is abundantly supplied with sinews (chordae tendineae), as might be expected. For the motions of the body commence from the heart, and are brought about by traction and relaxation*" (Aristotle, 2004). Three cavities were described in the heart of animals: right, intermediate, and left, with perforations leading into the lungs (Van Praagh & Van Praagh, 1983). Aristotle believed that the heart, not the brain, was the seat of intelligence (Shoja et al., 2008; Toledo-Pereyra, 2011).

Figure 27. Aristotle (384–322 B.C.). (Wellcome Institute Library, London.)

Of these cavities, it is the right that has the most abundant and the hottest blood, and this explains why the limbs on the right side of the body are warmer than those of the left. The left cavity has the least blood of all, and is the coldest; while in the middle cavity, the blood, in regards to quantity and heat, is intermediate to the other two, being however of purer quality than either. (Ogle, 1882)

He described the great cavae (vein) (i.e., the superior and inferior vena cavae) as arising from the largest compartment on the right and the aorta from the intermediate, by a thinner connection. Part of the great vein headed toward the lung (i.e., pulmonary trunk) and split into two portions, one of which headed posteriorly towards the backbone. Several veins arose from this branch (i.e., the azygos vein), and coursed along each rib and vertebra. The other portion continued towards the lung, which branched again (i.e., pulmonary arteries). These two branches each coursed towards a lung, and gave off several

branches of varying size, each of which coursed along different parts of each lung. The superior portion of the great vessel became two branches (i.e., brachiocephalic trunk), which extended laterally (i.e., subclavian veins) and superiorly towards the clavicles (i.e., internal jugular veins). These continued to course through the arms and legs. The aorta decreased in volume as it went away from the heart and gave branches that were much smaller and, for the most part, coursed alongside branches from the great vein. Coursing between these vessels, he described "*fibers*" containing "*white blood*" (lymphatics) (Shoja et al., 2008; Loukas et al., 2011c).

He described *pneuma* (similar to Empedocles) and believed that it came from the heart. Aristotle claimed that the heart was extraordinarily warm inside and could boil any food consumed. This boiling created a liquid, which would be converted to blood. He called this process "*Pneumatization*," and believed that heartbeats occurred due to a pressure created by this process that expanded the walls of the heart, which constricted again once the fluid was overflown into the vessels (Smith et al., 2012). Aristotle also claimed that palpitations occurred in disease, spasms, or during events that cause fear, when the hot substance was forced into the heart and mixed with cold waste (Shoja et al., 2008).

The trachea (windpipe) was described as located in front of the esophagus and mention was made of it being made out of a cartilaginous substance. Vessels (i.e., jugular veins) ran parallel with the trachea, one on each side. They continued until about the level of the ear, where they branched into four vessels. Some of these vessels give off several small branches throughout the head and both coursed through the surrounding brain. The other vessels branching off of the [jugular vein] turned around and coursed back towards the arm (i.e., external jugular vein) (Shoja et al., 2008). The epiglottis was described as something that prevents food from falling into the lungs during ingestion, by rising during each breath, and falling down during the ingestion of food (Crivellato & Ribatti, 2006a; 2006b). In comparison to the animals Aristotle dissected, he described the human lungs as being anomalous because they were neither smooth nor divided into lobes. He described the lung as large, rich in blood and spongy, like foam. He thought that the "*heat of the body*" was cooled in the lung by external air during breathing, but not by the "innate spirit" as in "*bloodless kinds*." The bifurcation of the trachea is mentioned, but it is considered to be united with the great vein and with the aorta. Aristotle believed that "*those viscera which lie below the diaphragm exist one and all on account of the blood vessels; serving as a band, by which these vessels, while floating freely, are held in connection with the body. For the vessels give off branches, which run to the body throughout-stretched structures, like anchor lines thrown out of a ship*" (Ogle, 1882).

As far as the gastrointestinal tract is concerned, Aristotle gave a surprisingly accurate description of the esophagus as extending from the mouth, passing through the diaphragm and terminating in the quite distensible stomach. He described the human stomach as being like that of a dog and the lower part of the abdomen as being like that of a hog because it is wide. A fatty, broad, and membranous mesentery extending over the bowels (omentum), as well as its blood supply, is mentioned. He identified the jejunum, the cecum, the sigmoid flexure, and the rectum, but not the duodenum. Aristotle noted that the contents of the gastrointestinal tract were susceptible to injury (Margolis et al., 1976).

He described the spleen, liver, kidneys, and urinary bladder quite accurately. The human spleen was described as being narrow and long, and the liver round – like that of an ox. The presence of stones, growths, and abscesses in the kidneys, lung, and liver are mentioned. He identified the porta hepatis and described veins in relation to it, but did not see any connection with the aorta. Aristotle compared the kidneys to those of oxen and placed them close to the vertebral column, with the right kidney at a higher level than the left. He identified two strong passages (the ureters) leading from the cavity of the kidneys to the urinary bladder, and also two others leading to the aorta. He believed that unlike the liver and spleen, which assist in digestion of food, the kidneys take part in the separation of the excretion, which flows into the bladder. In the center of the kidney, Aristotle identified "*a cavity of variable size*" and described the kidney as a solid organ, surrounded by more fat than other organs, and made up of numerous small kidneys. He did, however, believe that the kidneys were unnecessary for survival (Marandola et al., 1994).

Aristotle described the diaphragm with a central thin membranous (tendinous) and peripheral fleshy (muscular) part. He viewed the diaphragm as "*a kind of partition-wall or fence*," which separated the nobler (thoracic cavity) from the less noble (abdominal cavity) parts, where desire and grosser passions are located (Ogle, 1882; Aristotle, 2004).

Aristotle's description of the brain was quite interesting. He saw the brain as composed of two portions enclosed by two membranes, the outer being the

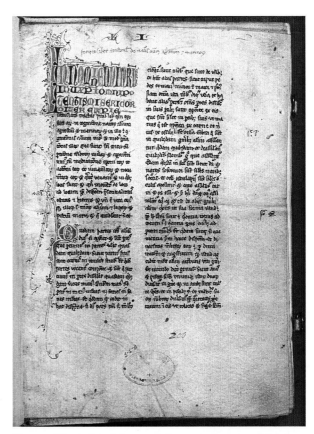

Figure 28. Aristotle's most important natural history treatise was the *Historia animalium* (*Research on animals*), shown here in Latin translation. (Courtesy of Bethesda, MD: U.S. National Library of Medicine, National Institutes of Health, Health & Human Services http://www.nlm.nih.gov/exhibition/odysseyofknowledge/diffusion.html.)

strongest. The human brain was considered to be the largest among the animals. Aristotle asserted that the brain "*is larger in men than in women*"…because the "*region of the heart and of the lung is hotter and richer in blood in man than in any other animal; and in men than in women.*" He even reported finding more sutures in the skull of men than women because "*the explanation is again to be found in the greater size of the brain, which demands free ventilation, proportionate to its bulk.*" He described a small cavity in the center of the brain linked by a membrane filled with veins but the brain itself was considered to be without blood because of which it was cold to the touch ("*for of all the parts of the body, there is none so cold as the brain*"). He claimed that the brain was essentially the pacemaker of the heart as it controlled each heartbeat. Aristotle placed the cerebellum at the caudal end of the brain and described three passages leading from the eye to the brain, but based on simple inspection, maintained that *"it has no continuity with the organs of sense"* (Clarke, 1963; Clarke & Stannard, 1963; Koelbing, 1968; Sorabji, 1970; Gross, 1999).

COMPARATIVE ANATOMY

Aristotle's vast contribution to anatomical knowledge was based on speculation from the observations he made dissecting lower animals. In his work, *De Parti-bus Animalium*, he remarked that *"the internal parts are not so well known, and those of the human body are the least known so that in order to explain them we must compare them with the same parts of those animals, which are most nearly allied"* (Cresswell, 1862).

Most remarkable were his embryological observations and his work on the *Generation of Man (De Generatione Animalium)*, which he himself designated as his masterpiece (Piatt, 1912; Preuss, 1970; Aristotle, 1982; 2004). Aristotle opened hens' eggs at different stages of development noting the characteristics at each stage. He believed that the egg supplied the ingredients necessary, while the sperm actually *"built"* the embryo. He explained that the yolk provided nourishment for the developing embryo, and that the sperm carried a *"vital heat,"* an essential part of the *pneuma* that he described (Morsink, 1982; Wolpert, 2004; Smith et al., 2012).

Aristotle laid the foundation for comparative anatomy as a result of the systematic studies of animals and on the many dissections he carried out. Drawing from his observations, he speculated about the structure of the human body (Lonie, 1964; Grene, 1972). Significant too was his influence on Alexander the Great, the son of the King of Macedonia. The most spectacular achievements in anatomy during this period (Edelstein, 1935; 1967) were to be made in the city that was founded by Alexander on the banks of the river Nile (Persaud, 1984).

Chapter 5

ALEXANDRIA

CULTURAL AND INTELLECTUAL CITY

Alexandria, the monumental city on the banks of the Nile was strategically located at the commercial crossroads of Asia, Europe, and Africa. It evolved into a great cultural and intellectual center of the ancient world (Burn, 1982; Vrettos, 2001). Encouraged by the ruling Ptolemies, eminent scholars gathered to pursue with great vigor their interest in literature and the sciences. With the fall of Greece to Rome, the great minds of the period found refuge in Alexandria. The library had an impressive collection of 700,000 volumes. It contained all the then-current knowledge of the known world. Together with the House of Muses or Museum, the library attracted Jewish, Egyptian, and Greek scholars who congregated in the pursuit of knowledge (Parsons, 1952; Canfora, 1989). In this seaside cosmopolitan community of about 600,000 people, the arts, philosophy, and medicine flourished. According to Parsons (1952), it was here that empiricism began to replace Aristotelian dogmatism as the dominant school of thought because it was based on scientific investigations, actual observations, clinical histories, and analogies.

ANATOMICAL STUDIES

The human body was dissected in order to understand more about its structure. The first anatomists credited with this distinction were Herophilus of Alexandria (*Greek: Ἡρόφιλος*) (Dobson, 1925; Potter, 1976) and Erasistratus of Cos (*Greek: Ἐρασίστρατος*) (Dobson, 1927; Lloyd, 1975) although this assertion might not be entirely accurate in view of evidence of earlier human dissection in India and the Near East (Singhal & Guru, 1973; Hoernle, 1907; Qatagya, 1982; Uddin, 1982). In a room used solely for anatomical studies, these two physicians made many discoveries through the dissection of a large number of cadavers donated by their benefactors, Ptolemy Soter and Ptolemy Philadelphus.

It is said that Herophilus was the first anatomist to have gained first-hand knowledge of the actual structure of the human body and that he had dissected more human bodies than any of his predecessors. His extensive anatomical knowledge was attributed not only to the dissection of as many as 600 corpses, but also of condemned criminals. According to Celsus, these criminals were obtained "*for dissection alive, and contemplated, even while they breathed, those parts which nature had before concealed*" (Scarborough, 1976). Both Herophilus and Erasistratus were accused of human vivisection. According to the historian Tertullian, as many as 600 criminals were vivisected and some fetuses were removed alive from the womb by Herophilus. Herophilus was described by Tertullian as "*that doctor or butcher who cut up innumerable corpses in order to investigate nature and who hated mankind for the sake of knowledge.*" To this, Celsus added "*but to lay open the bodies of men whilst still alive is as cruel as it is needless*" (Ferngren, 1982). The reputation of these eminent physicians and teachers of anatomy attracted numerous pupils to the Alexandrian medical school. The achievement of Herophilus and Erasistratus were mentioned in the works of Celsus, Galen, Oribasius, and others (Persaud, 1984).

HEROPHILUS

Herophilus was born about 300 B.C. and is often called the "*Father of Anatomy.*" Not much is known of his life and all of his writings, including his book "*On Anatomy,*" have been destroyed. He dissected extensively and made many anatomical discoveries as recorded in the works of other writers (Dobson, 1925; Potter, 1976).

Because the Ptolemies allowed dissection of the human body, Herophilus carried out many dissections, both privately and publicly. Together with Erasistratus, he was able to articulate two human skeletons, which became widely known and attracted followers from afar. Whether he actually carried out human vivisection, as alleged by Celsus and St. Augustine, is not certain and appears unfounded. Herophilus recognized the brain as the seat of intelligence and not the heart as postulated by Aristotle. He described the delicate arachnoid membranes of the brain, which he considered to be the seat of the soul. Even to this day, the confluence of the dural venous sinuses near the internal occipital protuberance is often called the *torcular Herophili.* Less known is the furrow in the inferior floor of the fourth ventricle, which he named the *Calamus Scriptorius.*

Herophilus observed the lacteals but did not know their functions. He considered the nerves originating from the brain to be the organs of sensation and differentiated them from those associated with voluntary movement, establishing that paralysis of muscles would follow damage of these nerves. In addition, Herophilus described the coverings of the eye, named the first part of the small intestine *duodenum,* and attributed the beating of the heart to pulsation of arteries. He knew of the nature of the pulmonary artery and of the mesenteric vessels and described the uterus in some detail. The discovery of the epididymis is also attributed to Herophilus. From the work of Galen, Herophilus's description of the human liver was recorded. However, he used the word *neuron* to describe tendons and ligaments. Interestingly, it has been anecdotally reported that Herophilus accepted some women as pupils, including Agnodice, who had to disguise herself as a man in order to practice medicine (Dobson, 1925; Souques, 1934; Potter, 1976; Imai, 2011).

ERASISTRATUS

Born about 250 B.C., Erasistratus (Fig. 29) was a younger contemporary and rival of Herophilus. While Herophilus emphasized treatment of the entire patient rather than individual symptoms in accordance with the Hippocratic tradition, Erasistratus argued that therapy should be directed to the local anatomical causes of diseases. He rejected the concept that knowledge of the entire body and its function in health was necessary for medical practice. In contrast to Herophilus's belief in Dogmatism, Erasistratus advocated Methodism. In spite of his outstanding contributions to anatomical studies, Erasistratus considered himself a physiologist and, as such, formulated experiments relating to bodily functions and diseases (Keele, 1961; von Staden, 1992). His pneumatic theory was based on the flow of blood and two kinds of *pneumae* through minute channels of veins, arteries, and nerves (Wilson, 1959). The veins contained blood and the arteries distributed the vital spirit, which was thought to be formed from air that would pass from the lungs into the heart. When this vital spirit reached the ventricles of the brain, it was transformed into the animal spirit, which was then distributed throughout the body by the branches of nerves, conceived as being minute channels.

Erasistratus attributed all diseases to *plethora,* an accumulation of blood from food that remained undigested and obstructed the circulation of the vital spirit. His theory of the local accumulation of blood as a cause of diseases led him to direct greater attention to the heart, veins, and arteries. He rightly saw the heart as a pump and suggested the existence of a very fine communication system between the arteries and veins. He described the auricles of the heart, and a role for the semilunar and tricuspid valves. Regrettably, his pneumatic theory of the flow of vital spirit prevented him from understanding the true nature of the circulation of blood. He also described a large number of blood vessels, including the aorta, pulmonary artery and veins, hepatic arteries and veins, renal arteries, superior and inferior vena cava, and azygos vein. Erasistratus recognized the function of the trachea (Dobson, 1927; Wilson, 1959; Lloyd, 1975, Loukas, 2011a). In regards to neurology, Erasistratus differentiated the cerebrum from the cerebellum and noted the convolutions in different species of animals, with the greatest in man, which he associated with the level of intelligence. He saw the nerves as conveying animal spirit from the brain through tiny channels

Figure 29. The Greek physician Erasistratus is shown here reclining on a couch and conversing with an assistant. From an Arabic translation of Dioscorides's De Materia Medica. Iraqi Painting: A.D. 1224, Baghdad School, written by Abdallah ibn al Fadl. (Courtesy of the Freer-Gallery of Art, Smithsonian Institution, Washington, D.C.)

and perceived the contraction of muscles as the result of distention by animal spirit originating from the nerves (Dobson, 1927; Wilson, 1959; Lloyd, 1975; Rose, 2009).

DECLINE OF ALEXANDRIA

The decline of Alexandria is traditionally attributed to the invasion and conquest of the city, first by the Romans and later by the Arabs. However, it would appear that the deterioration of Alexandria began even earlier due to internal politics, bickering, and rivalry among scholars (Dobson, 1927; Wilson, 1959; Lloyd, 1975). Even during this distant period in history, the value of basic anatomical research was questioned as to its relevance in the treatment of patients. The followers of Herophilus clashed with those of Erasistratus. The Greek scholars were at that time being persecuted by the latter Ptolemies, particularly the ninth Ptolemy (146–117 B.C.) who was known as the Second Benefactor. The work of the medical scholars came to a standstill and began to deteriorate. By the second century before the Christian era, probably, actual human dissections were not carried out but some anatomical studies were pursued using animals (Persaud, 1984).

Chapter 6

DAWN OF THE ROMAN EMPIRE

DECLINE OF ANATOMY

Following the persecution of the Alexandrian scholars and the conquest of Egypt by Caesar, Alexandria became part of the Roman Empire. The museum was destroyed and the library plundered and burned. Persecution led to the dispersion of Greek scholars within the Empire. According to Cilliers (2006), a new era of medicine and anatomy began to evolve – still under the influence of Greek scholars but becoming more and more Roman culturally. Egypt became incorporated into the Roman Empire during the reign of Augustus. Even though the government was Roman, the culture was a hybrid and complex combination of both Greek and Roman. Religion, civilization, and political power were fused. The Romans were practical in their outlook and excelled in such areas as government, agriculture, medicine, and warfare. The scientific pursuits and philosophical enquiry of the Greek scholars were thought to be less important to the more immediate and practical goal of expanding the empire through conquest. Medicine still remained the domain of the Greek physicians. It was not until 60 B.C., that a medical school called the "Asclepiadic sect in physic" was founded in Rome by Asclepiades of Bithynus (c. 120–30 B.C.). Anatomy suffered a decline because dissection became forbidden and this situation did not change until the late Middle Ages (Kevorkian, 1959; Cilliers, 2006). Asclepiades did not regard anatomy as essential for the physician (Green, 1955; Cilliers, 2006) and even with the expansion of the medical school in Rome, first by Vespasian (A.D. 9–79) and others such as Hadrian (A.D. 117–138) and Severus (A.D. 208–235), dissection of the human body was not carried out although there is evidence that this might have been sporadically performed up to the beginning of the second century of the Christian era (Fig. 30). If this was not a time for scientific inquiry, then it was one for consolidating and documentation. The great recorders of the period were Lucretius, Caius Pliny Secundus, Cornelius Celsus, and Marcus Tullius Cicero.

CICERO

Cicero (106–43 B.C.) considered superstition and dependence on religion as obstacles for the establishment of proper institutions within society even though he was fully aware of the political importance of religion for citizens and their leaders (Fig. 31). In his book, *On Divination*, he eloquently objected to the dependence on religion for all matters from the pursuit of war to the administration of government. In his work, *De Natura Deorum*, he considered the nature and origin of the universe and presented a teleological account of the structure and function of the human body (Orth, 1925; Agnfoglio, 1961). Similar to Cicero, Lucretius (96–55 B.C.) objected to religion and superstition on the basis of the atomist philosophical concepts as taught by Democritus. His great poetic work *De Rerum Natura* was concerned with all living things and the universe for which he developed a natural explanation. His theory of the origin of man, animals, and plants parallels Darwin's *Natural Selection* in many respects. On the other hand, Lucretius's philosophy was purely speculative as it was based on the teachings of Epicurus (341–270 B.C.) and of Democritus (460–370 B.C.).

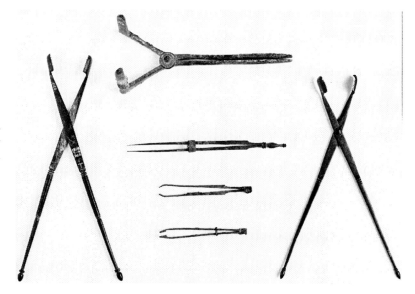

Figure 30. Surgical instruments from the early Roman era. (By kind permission of the Trustees of the British Museum.)

PLINY

Much has been written concerning Pliny (A.D. 23–79) – the most famous Roman naturalist and great encyclopedist. (Corney, 1914; Gudger, 1923; Janick, 2007). His expansive 37-volume work, *Historia Naturalis*, attempted to deal with all of human knowledge in the arts and sciences (Figs. 32–34). Pliny stated in the preface that more than 20,000 facts had been selected from 200 books written by more than 100 selected authors. Not only did he transcribe from innumerable works of Greek and Roman scholars, but he also recorded his own highly speculative philosophy and even bizarre anecdotal statements made by others. Despite the lack of any scientific merit in Pliny's astonishing interpretation of the natural world, his work remained influential until the seventeenth century (Persaud, 1984). According to Gudger (1923), "*no other work contributed so much to keep natural history alive... and following the appearance of the first printed edition in 1469, it was still the great authority, read, studied, and quoted by all students of natural history.*" Several books were devoted to the creation of the universe, the origin and description of human and animals, as well as strange mythological creatures such as the unicorn. Similar to most of what he has recorded, including the fragments of anatomical statements in the work of Pliny, were highly speculative and often erroneous statements. Most fascinating was his description of strange races of "*wonder people*" and "*monstrous births*." He mentioned Androgyni, a combination of man and woman; Abermon, individuals with feet directed backwards; and Arimaspi, people with only one eye. As a scholar of natural history, an overly enthusiastic recorder of historical and natural events,

Figure 31. Terracotta (baked clay) votives (see also Fig. 7) of abdominal cavity with internal organs and tumor masses (Roman, 3rd-1st century B.C.). In ancient Greece and Rome it was customary to dedicate models of parts of the body at the shrines of the gods responsible for healing. (By kind permission of the Trustees of the British Museum.)

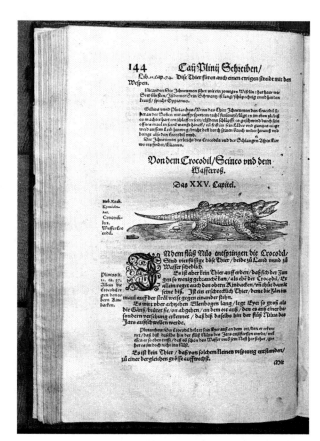

Figures 32–33. Beginning with the earliest printing of Pliny in 1469, there were at least 42 printings in numerous languages, including English, before 1536. The Kenneth Spencer Research Library houses 21 separate editions produced between 1476 and 1685. This German version contains books 7-10 and part of book 11. The translation was made by J. Heyden Eifflender von Dhaun and is illustrated with woodcuts by Jost Amman and Virgil Solis. Pliny, the Elder. Caji Plinii Secundi…Bücher und Schrifften von der Natur, Art und Eigenschafft der Creaturen oder Geschöpfe Gottes…[Naturalis historia. German.] Franckfurt am Mayn: Feyrabend und Hüter, 1565. Call Number: Summerfield E1059. Sally Haines Rare Books Cataloger Adapted from her Spencer Research Library exhibit and catalog, Slithy Toves: Illustrated Classic Herpetological Books at the University of Kansas in Pictures and Conservations (http://blogs.lib.ku.edu/spencer/crispy-critters/).

Figure 34. The Natural History of Pliny in a mid-12th-century manuscript from the Abbaye of Saint-Vincent, Le Mans, France (public domain).

and a man of influence during the most glorious days of the Roman Empire, Pliny's works demonstrated his lack of enthusiasm for the scientific study of the human body as practiced by the Greek scholars. A more influential contemporary of Pliny, whose expansive treatise in medicine contained many anatomical statements, was the Roman nobleman Cornelius Celsus (Persaud, 1984).

CELSUS

Although not a physician, Celsus (c. 30 B.C.–A.D. 45) authored *De Re Medicina*, an encyclopedic and systematic survey of current medical knowledge emphasizing surgical practices (Fig. 35). He was probably a man of great learning because he wrote in impeccable Latin on a variety of subjects such as law, philosophy, and medicine. Most of his writings have been lost and his work on medicine was not discovered until the fourteenth century (Wellmann, 1913; 1924; Temkin, 1935; Köckerling et al., 2013). *De Medicina* consisted of eight books, the last two of which dealt largely with surgery and included many anatomical descriptions. The work itself lacked originality and could be traced back to the writings of Greek physicians. Nevertheless, it proved to be an important and influential book, prompting Pope Nicholas V to order its publication in 1478 (Orth, 1925; Köckerling et al., 2013). Despite the lack of originality in his work, Celsus was rightly considered one of the most outstanding medical writers of antiquity (Castiglioni, 1940; Lipsett, 1961; Köckerling et al., 2013). In Books VII and VIII, Celsus differentiated the trachea from the esophagus and knew that the esophagus ended in the stomach. He gave an excellent description of the diaphragm, the liver, the spleen, and the kidneys. Despite earlier descriptions of the duodenum by Herophilus, Celsus was seemingly unaware of the organ because he thought that the stomach was directly connected to the jejunum by the pylorus. Celsus knew enough about the urinary bladder, the uterus, and vagina to describe surgical procedures relating to

Figure 35. Aulus Cornelius Celsus (Wellcome Institute Library, London).

these structures. Of greater merit, however, was his description of the skeleton. He described the features of the bones of the skull, including the sutures and many of its openings. His description was commendable for the details of the vertebrae, ribs, scapula, humerus, radius, ulna, tibia, fibula, metacarpals, and metatarsals. It would appear, too, that he was familiar with the semicircular canals of the inner ear and the perforated features of the ethmoid (Orth, 1925; Agrifoglio, 1961; Köckerling et al., 2013).

Celsus recognized the importance of anatomy and even advocated human dissection. His description of surgical operations for the treatment of wounds, goiter, hernias, and cataracts revealed an appreciation for anatomy in carrying out these procedures. If human dissection was carried out during this period, there was no record of it in the work of Celsus. His knowledge of anatomy was probably obtained from Greek medical scholars (Persaud, 1984). Unlike Herophilus and Erasistratus, Celsus strongly opposed vivisection on condemned criminals (Scarborough, 1976). Indeed, he wrote: "*I regard it useless and cruel to open the living body, but it is necessary for those who study to see corpses, in order to learn the position and arrangement of the single parts, a thing that is seen much better in the cadaver than in the living body*" (Castiglioni, 1941).

Chapter 7

GALEN

MEDICAL SCHOLAR AND CELEBRATED ANATOMIST

The most celebrated anatomist of antiquity was the great physician Claudius Galen (Fig. 36). He was born in Pergamon, along the coast of Asia Minor, the son of a prominent Greek mathematician, architect, and astronomer. His birth in A.D. 131 coincided with the period of greatest glory of the Roman Empire, which was marked by the great territorial expansion of Trajan as well as the implementation of his social welfare policies and architectural expansion.

Galen studied philosophy under the influence of the Stoics, the Academicians, the Peripatetics, and the Epicureans; however, later in his life, he studied medicine after his father described to his son a dream that he would become a doctor. From the age of 20, Galen travelled extensively in order to study with the greatest teachers of philosophy and medicine. He visited Greece, Palestine, Phoenicia, and Crete and finally, settled in Alexandria where he remained until he was 30 years old. He was exposed to anatomy by Satyrus, a student of Quintus, Pelops of Smyrna, and later to Heraclianus in the Alexandrian school (Green, 1951).

Galen's interests and writings went beyond medical matters and included philosophy, religion, mathematics, and grammar. He was an extremely prolific writer (Ilberg, 1889; 1892; 1896; 1897; 1902; 1930; Walsh, 1934a, b; Peterson, 1977) who even produced a guide, entitled *On His Own Books,* to his voluminous treatises. In the field of medicine alone, Galen is credited with more than 130 treatises, which became the unquestionable repository of medical knowledge for more than one thousand years after his death in A.D. 201. Only 80 of these have survived. Nonetheless, Kuhn's edition of Galen's work took 12 years to complete and spans 20 volumes (Kühn, 1821–1833).

When he was 28-years-old, Galen returned to the city of Pergamon where he served as surgeon to the gladiators for four years. During this period, Galen made anatomical observations based upon what he learned from attending to the wounds sustained by the gladiators. Human dissection was forbidden in Imperial Rome. Galen derived anatomical knowledge from the works of his teacher and his predecessors, as well as from the extensive dissections he carried out on many species of animals. Galen dissected apes, monkeys, dogs, pigs, and even bears. He even considered the internal organs of man as not very different from those of the pig. The voluminous writings of Galen contained many detailed anatomical descriptions (Withington, 1922) and there has been much debate over the years as to what constituted his personal observations versus those of his predecessors.

Galen recognized the importance of anatomy for the work of the physician. He encouraged his pupils to visit Alexandria, so they might examine the human skeleton that was there, which he himself had studied. This is evident from his work, which contains the most detailed and systematic description of the human bones (Singer, 1952). Galen's writings have provided insight to the accomplishments of the early anatomists, particularly those who had lived in Alexandria.

Despite many errors and shortcomings, Galen's anatomical writings remained unchallenged up to the time of Vesalius in the 16th century (Persaud, 1984). Rather than accepting the errors in the work of Galen, the French anatomist Jacobus Sylvius (1478–1555) remarked in response to Vesalius that, *"then, man must have changed his structure in the course of time, for the teaching of Galen cannot err."* The Royal College of Physicians of London in 1559 even made one of its members, Dr. John Geynes, retract his statement that there were 22 inaccurate passages in the works of Galen. According to Clark (1964), *"in the eyes of the College, his*

Figure 36. Claudius Galen (Wellcome Institute Library, London).

offence was double: indoors it was heterodoxy but out of doors, it brought the chosen intellectual foundation of the medical art into question if not into contempt." Dr. Geynes was subsequently examined and admitted to the College in November 1560.

The influence of Galen continued for many decades and appeared in examination notices of the College. In 1595, Dr. Edward Jordan, a medical graduate of the University of Padua, was required to read five of Galen's works before being admitted to fellowship. A Dr. Hood was forbidden to practice, because he did not read Galen's work. He confessed that the books were too expensive to purchase. In 1596, Dr. Thomas Rawlins was failed "and was admonished to work harder, particularly at Galen" (Clark, 1964).

Galen derived his knowledge of anatomy largely from the dissection of animals, but in his own work, *On Bones*, he mentioned that he was able to study human bones from tombs that were destroyed. He also observed the bones of a body that had been washed from its grave by a flood and deposited on the riverbank. He was able to observe the intact and articulated bones amidst the petrified flesh, and the skeleton *"lay ready for inspection, just as though prepared by a doctor for his pupils' lesson."* In another instance, he studied the skeleton of a robber whose flesh was eaten away by birds and after two days, it was ready *"for anyone who cared to enjoy an anatomical demonstration"* (Duckworth, 1962). The reason as to why opportunities to dissect humans during Galen's time was, most likely, that the structures of man and animals, in particular, apes and monkeys, were assumed to be fundamentally the same (May, 1968). The lack of enthusiasm for human dissection can be ascribed further to the prevailing hostility against such a pagan practice.

Galen himself marveled at the complexity and delicate arrangement of the parts of the human body. According to the Aristotelian philosophy, he saw nature as providing a form and structure in the parts of the human body, with nothing being superfluous, in carrying out their functions. Such perfection and exquisite design can only be attributed to divine providence. Christians, Jews, and Muslims accepted Galen's account of the structure of the human body and his works were translated into many languages and remained popular for more than 1,400 years. Although Galen's contributions to our knowledge of human anatomy were not significant in number, he still stands as one of the greatest medical scholars of antiquity, due to the profound influence that his voluminous writings had on medical scholars through the Middle Ages (Temkin, 1973). Despite his intellectual progress and contributions to anatomy and medical science during this period, both the fields of medicine and anatomy suffered a decline.

In A.D. 161, Galen made his first visit to Rome. It was the beginning of the reign of the Roman Emperor Marcus Aurelius (A.D. 121–180), and following the suggestion of his patron, the Consul Flavius Boethus, he began to compose his anatomical works, *On Anatomical Procedure* (Duckworth, 1962; Galen, 1956) and *On the Usefulness of the Parts of the Body* (May, 1968; Pagel, 1970; Galen, 2003). He returned to Pergamon in A.D. 163, but later joined the military expedition of Marcus Aurelius in Aquileia as a personal physician. Because of the outbreak of the plague, the army returned to Rome, and Galen was entrusted to care for the son of Marcus Aurelius, Commodus. For over 20 years (A.D. 169–192), Galen remained in Rome in pursuit of his medical work and compiled his treatises.

He returned to his home city where he died eight years later at the age of 70.

ANATOMICAL TREATISES

Galen's life, anatomical accomplishments, and other works, were critically discussed by Prendergast (1928;1930), Singer (1956), Sarton (1954), and May (1968). Sarton's monograph provides an excellent introduction to the works of Galen, including the treatises translated from Arabic and the editions that are available in English, including his work, *De Anatomicis Administrationibus*, a description of his anatomical studies in 16 books. The first nine volumes were translated by Singer in 1956 from the surviving Greek text into English, while the remaining seven books, which had existed in Arabic were translated by Max Simon (1906) and Duckworth (1962). A complete translation from Greek into English of Galen's most influential work, *De usu partium* (On the Usefulness of the Parts of the Body), was performed by Margaret May and published in two volumes. It consists of 17 books, devoted to both anatomy and physiology, which at the same time extolled the complexity and beauty of the human creation (May, 1968; Galen, 2003).

Sarton (1954) stated that Galen's treatises, *De Anatomicis Adminis trationibus*, or *The Anatomical Procedures*, was unknown in the Middle Ages; however, the translation of the first nine books from Latin to Greek text was carried out by Johann Guenther of Andernach, and it was first published in 1531 in Paris, several years before the publication of Vesalius's work. The remaining seven books became available only following their translation from the Arabic edition by Max Simon in 1906. Owsei Temkin translated portions of the German text into English, of which Duckworth created a more complete translation.

Some of Galen's other anatomical works included the following: one on the dissection of the veins and arteries (*De Venarum Arteriarumque Dissectione*); the dissection of nerves (*De Nervorum Dissectione*); a *Secundum Naturam in Arteriis Sanguis Continator* (a discussion on whether arteries are naturally filled with blood); and myology (*De Musculorum Dissectione ad Tirones*). Some of his physiological treatises included a fair amount of anatomical information, often evolving from his dissections. *De Musculorum Motu* deals with both voluntary and involuntary movements. His treatise, *De Usu Respirationis*, describes experiments on the pleura and the use of bellows for maintaining artificial respiration.

ANATOMICAL STUDIES

Galen described the bones and sutures of the skull (Singer, 1952; May, 1968). He distinguished the squamous, styloid, mastoid, and the petrous parts of the temporal bones. He also recognized the quadrilateral shape of the parietal bones. He described the essential features of the vertebral column, including the coccyx and the sacrum, which he considered to be the most important. Twenty-four vertebrae were recognized and these were divided into cervical, thoracic, and lumbar. Galen's descriptions of the ribs, the sternum, clavicle, and bones of the extremities, as well as their articulations, were fairly accurate.

Although Galen's descriptions of muscles were based on the dissections he carried out on animals, they were fairly accurate and more advanced than those made by any anatomist previously (May, 1968). Galen defined a muscle as a bundle of fibers terminating in an independent tendon. These muscles included the mylohyoid, the thyrohyoid, the six extrinsic muscles of the eye, two muscles of the eyelids, four muscles of the lips, a muscle to each *ala nasi*, the frontalis muscle and four pairs of muscles that move the lower jaw, the temporalis, masseter, digastric, and lateral pterygoid muscles. In a similar manner, he systematically described the muscles of the tongue, neck, upper and lower limbs, and trunk.

In describing the brain, which he considered to be the seat of the soul he identified seven pairs of cerebral nerves. Due to their origin from the brain, the cranial nerves were believed to be nerves of sensation, in contrast to the 30 pairs of spinal nerves, which he recognized as the nerves of motion. Galen did not recognize the olfactory and trochlear nerves. The cranial nerves were designated as follows: the optic nerve the first cranial nerve, the oculomotor and abducens nerves as the second cranial nerve, the trigeminal nerve as the third cranial nerve, the facial and vestibulocochlear nerves were combined as the fourth and fifth nerves. Galen named the glossopharyngeal, vagus, and spinal accessory nerves as the sixth pair of nerves, and the hypoglossal was considered to be the seventh. He knew of the sympathetic nervous system and of the recurrent laryngeal branches of the vagus nerve on both sides of the neck. He described experiments involving the nerves that control movement of the tongue and production of speech (Persaud, 1984).

Galen provided a detailed description of the dissection of the brains of animals, and the names of many structures that he identified are still used today, including the corpus callosum, corpora quadrigemina, fornix, pineal body, and septum pellucidum. He knew of the cerebral ventricles, their communications, and of the choroid plexus. Galen's description of the cerebral hemispheres, and the third and fourth ventricles reflected keen observations. He also described various disorders of the brain and formulated anatomical explanations for pain. For example, he believed pulsating headaches to be a result of inflammation of the brain, stating, "*If the vessels are congested, they disturb the muscles causing their detachments, which brings nerve pain and pulsation as a consequence*" (May, 1968). In Galen's descriptions of the nerves, arteries, and veins, which he considered as "*instruments common to the whole body*," he indicated that all nerves originate from the brain, the arteries from the heart and the veins from the liver (Persaud, 1984).

Galen described the body's circulatory system as divided into two separate compartments with little communication; the first was made up of the right side of the heart, and the second part was made up of the left side of the heart. He believed that the right side of the heart was a derivative of the liver, from which blood exits to pass through the vena cava either by extending to either the head and arms or to the lower extremities. The pulmonary artery, recognized by Galen as the "*arterial vein*," delivered blood to the lungs, which passed into the left side of the heart to supply blood filled with "*vital spirits*" to the body. The little communication present between the two compartments was achieved by "*minute pores*" in the septum between the ventricles of the heart. Galen considered nature "*not only just, but also skillful and wise*" (May, 1968), because the blood vessel structures passed safely to their destination. Galen failed to differentiate nerves from tendons, and his inadequate and distorted account of the blood vessels prevented him from discovering the pulmonary circulation.

Galen's description of the reproductive organs and his views on fetal development are notable as well. Unlike Aristotle, Galen observed that the ovaries were homologous to the testes, and thus both organs produced sperm. He believed that the chorion developed after the female testicles expelled semen, which passed into the uterus to provide nutrition to the male semen. Based upon his observations of the uteri of animals, Galen incorrectly described the human uterus as bircornuate, or consisting of two horns; however, he did correctly describe the cervix, vagina, and ovaries, and their accompanying arteries and veins. The suspensory ligaments of the ovaries and the round ligaments of the uterus were described as being homologous to the cremaster muscle, which he first described. His descriptions of the peritoneal relationships, particularly with respect to the urinary bladder, were accurate and remarkable. Despite the many apparent errors in Galen's anatomical treatises, one can still appreciate his meticulous anatomical descriptions of dissected animals (May, 1968).

COMPARATIVE ANATOMY

Galen meticulously dissected several species of animals, including the Barbary ape (*Macaca inuus*), the only European nonhuman primate. Galen completed his anatomical studies in a time when dissection was perceived to be a pagan practice and thus not necessary for the treatment of patients. This generation believed that human dissection was not necessary because the human anatomy could be determined from animal dissection. Galen's influential and invaluable works in anatomy contained detailed anatomical descriptions of dissected animals, which he assumed were identical to the human anatomy (May, 1968).

In describing the anatomy of the cranial nerves,

Galen stated,

"for many surgeons do not know that in his work on the roots of the nerves Marinus has enumerated only those same roots which Herophilus specifies, but Marinus has concluded that there are seven pairs, whereas Herophilus says there are more than seven, regardless of the others. Whoever does not know this is, as the proverbial expression goes, like a seaman who navigates out of a book. Thus, he reads the books on anatomy, but he omits inspecting with his own eyes in the animal body the several things about which he is reading."

When describing the brain, Galen recommended *"that the dissection is best made in apes, and among the apes in such a one as has a face rounded to the greatest extent possible amongst apes. For the apes with rounded faces are most like human beings"* (Duckworth, 1962).

For studying the male reproductive organs, Galen gave specific dissecting instructions:

> We say that in order for you to secure that the animal which you are dissecting resembles a man, you must take for that dissection an ape. But in order to achieve the effect of clarity in the appearance of such of those organs as are small and hard to see, then you must take a he-goat, a ram, bull or horse or male donkey for your dissection because that animal must necessarily possess a scrotum. (Duckworth, 1962)

For studying the larynx, Galen recommended the use of the pig.

> For in all animals which have a larynx, the activity of the nerves and the muscles is one and the same, but the loathsomeness of the expression in vivisection is not the same for all animals. Because of that for my own part, as you know already, I illustrate such vivisections on the bodies of swine or of goats, without employing apes. But it is necessary that you should extend your studies and examine the larynx. This is constructed in the same way in the bodies of apes and men, a construction which is shared by the other animals which have a voice.

You must, then, dissect a dead man and an ape, and other animals furnished with a voice, which have, besides the voice, the vocal apparatus, the larynx. For the animals which possess no voice have no larynx either. He who is not versed in anatomy thinks that in regard to the plan of the larynx great contrast sexist among the six Animal Classes, to which this our discussion refers. That is because neither the absolute dimensions nor the shapes of the parts of the larynx are precisely the same amongst all of them, a point that applies also to the number of the muscles that they have there. But in regards to the activity of each one of the parts of the larynx, and the service, which they perform, these are one and the same in all animals provided with a voice. That is because in the bodies of these animals the intention of the Creator was uniform with regard to the plan of the vocal apparatus, just as his intention was uniform also with regard to the plan of the respiratory organs in those animals provided with respiration. For the contrasts between these organs and the bodies of these animals consists solely in their absolute dimensions and their shapes. (Duckworth, 1962)

Galen's accomplishment of performing dissections in animals as well as his comprehensive descriptions of those structures he observed, continuing in the tradition of Aristotle, has made him one of the earliest comparative anatomists (Persaud, 1984).

PHYSIOLOGICAL SYSTEM OF PNEUMAS

From simple experiments Galen demonstrated that the arteries contained blood and not air as was previously thought. However, his dogmatic adherence to the pneumatic theory (Fig. 37) prevented him from understanding more about the flow of blood (Singer, 1957; Siegel, 1968). Essentially, three types of special pneuma or spirit were thought to exist which dominate the liver (*natural spirit*), the heart (*vital spirit*) and the brain (*animal spirit*). Galen thought that pneuma, which was the life force, was taken into the body during breathing. Pneuma entered the lungs via the trachea, and from the lungs it passed into the left ventricle through the *"vein-like artery"* (pulmonary vein). He demonstrated the presence of blood in the left ventricle, but believed that it originated in the liver from chyle, derived from the intestinal tract, where it became imbued with *"natural spirit."* Galen believed that this activated blood ebbed and flowed through the venous system, including through the artery-like vein (pulmonary artery), to the lungs where impurities were removed. He held the view that the venous blood in the right ventricle flowed back into the general venous system, but some of it was filtered through tiny pores in the interventricular septum to the left ventricle. Here it came into contact with the pneuma and this gave rise to the *"vital spirit,"* which was then distributed by the arteries. When the vital spirit reached the brain through its arteries, it was changed into the animal or psychical spirit. Galen believed that this vital spirit was distributed through canals in the nerves.

The natural spirit in the liver was thought to control the functions of nutrition, growth and reproduction whereas the vital spirit of the heart regulated vital functions by transporting heat and life throughout the

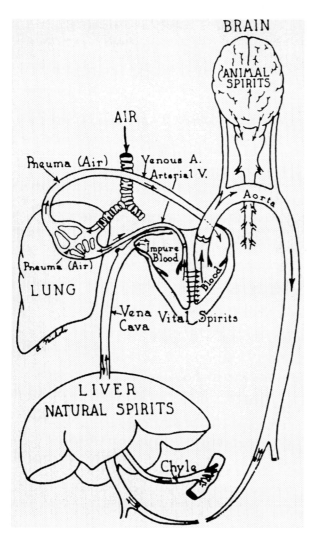

Figure 37. Galen's conception of the vascular and respiratory systems – the flow of pneuma (With permission from Major 1954).

arterial system; the vital spirit of the brain regulated the brain itself. This pneumatic theory, which originated during the Hippocratic period, was based on three dominant and regulating forces or spirits in the body on the basis of which all functions can be explained. Philosophical in its formulation and completely lacking any scientific basis, Galen's physiological system of pneumas survived up to the seventeenth century, until Harvey announced his discovery of the circulatory system (Persaud, 1984).

EXPERIMENTAL STUDIES

Notwithstanding Galen's highly fanciful concept of bodily functions, based on the theory of the three pneumas (Singer, 1957; Siegel, 1968; May, 1968), he carried out a large number of anatomical and physiological experiments, which were of a practical nature and had clinical implications. He performed experiments to elicit the function of vital organs. He examined the lungs in order to understand more about respiration, studied digestion by feeding pigs with different diets and then opened their stomachs, and he demonstrated that arteries contained blood, rather than air, by placing two ligatures on an exposed segment of artery, which he then slit open. Galen accurately described the consequences of spinal cord damage at different levels. He observed loss of sensation and paralysis of all muscles supplied by nerves originating from the spinal cord following complete resection below that level. In contrast, he found that hemisection of the spinal cord resulted in paralysis of the muscles only on the side with the lesion. Furthermore, a longitudinal medial incision of the cord did not lead to paralysis.

Galen identified that the diaphragm was not solely responsible for enabling respiration as other muscles were involved as well. Galen's detailed description of the origin and course of the phrenic nerve, as well as the accidental discovery of the recurrent laryngeal nerve, which subsequently led to his understanding of voice production in the larynx, is most impressive.

SCIENTIFIC ACHIEVEMENT

In retrospect, Galen deserves greater recognition for his work than he has received in the past (Persaud, 1984). As mentioned in the introduction to Duckworth's translation of the later books on Anatomical Procedures, the avowed aim was to *"focus the attention of scholars and scientists on one of Galen's greatest works, and to attempt to remedy the accidental, but disastrous dichotomy from which it has suffered for so many centuries."* The hope was also expressed that modern anatomists, with their interest in comparative as well as in human anatomy, and with their experimental approach towards the elucidation of anatomical problems, are likely to be more sympathetic towards Galen than have been their predecessors during the last 400 years. The eclipse of Galen's reputation at the Renaissance was perhaps inevitable. Although many recent writers have remembered Galen primarily for his errors, many now put his observations into perspective noting the era in which they were made (Duckworth, 1962).

A fitting tribute to Galen has been provided by the commentary of Hunain Ibn Ishaq of Baghdad (A.D. 809–873), a medical scholar, linguist, and Galen translator. Upon completing his translation of the fifteenth book, Ishaq stated that

> this excellent, outstanding work which is one of the compositions of a man who performed marvelously, and revealed extraordinary things, the master of the earlier surgeons, and the lord of the more recent savants, whose efforts in the practice of medicine have been unequalled by any of the prominent since the days of the learned and great Hippocrates – I mean Galen. May God Almighty be merciful to him! (Duckworth, 1962)

Galen is often described as the Prince of Physicians, second only to Hippocrates, the Father of Medicine. He was the last of the great medical scholars of antiquity, and his death, 200 years after the Christian era began, coincided with the progressive decline of Greek science. Galen's works have provided for future generations a legacy of medical accomplishments of an era that would have been all but forgotten.

Chapter 8

EARLY MIDDLE AGES

ROMAN AND BYZANTINE EMPIRES

With the decline of Greek science and culture and the fall of the Roman Empire, a new period dawned that was not fertile ground for innovation. What constitutes the Middle Ages is still a matter of debate among historians. This period probably began long before the Roman Empire ceased to exist in A.D. 476 and came to a close during the fifteenth century with the fall of the Byzantine Empire in 1453.

Because of its conformity with the philosophy of the governing stoics and the teachings of the Christian Church, Galen's work prevailed for more than 1300 years. It was believed that everything one needed to know about the structure of the human body could be found in Galen's anatomical treatises, which became established as the ultimate authoritative source. Galen's death heralded a long era with a predictable outcome. Medicine and the study of human anatomy in particular, languished in passive decline only to re-awaken in 1543 with the publication of Vesalius's *De Corporis Humani Fabrica*, a year which also coincided with the publication of Copernicus' monumental work *De Revolutionibus Orbium Coelestium.* These two works revolutionized the thinking of man with respect to the structure of the human body and the universe, respectively. Indeed, modern medicine is considered to have begun from the time of Vesalius. It is not surprising therefore for medical writers to consider the Middle Ages as the period from the death of Galen in A.D. 200 to the publication of Vesalius's work in 1543.

By A.D. 400, Christianity had spread throughout the Roman Empire and the power and influence of the church grew to the extent of becoming tyrannical, so much so that there was no interest in learning and scholarly pursuits. The work of the church took precedence with its emphasis on the saving of souls and the reinforcement of theological dogma.

Despite the efforts of the great Roman Emperors, Diocletian, who ruled from A.D. 284 to 305 and Constantine I who ruled from A.D. 312 to 337, the downfall of the Roman Empire could not have been prevented because of political problems relating to the choosing of successors, high taxes and inflation, and innumerable attacks by the Germanic tribes from Northern and Central Europe, which began in A.D. 378.

Following the fall of the Roman Empire in A.D. 476, the Germanic conquerors began to establish kingdoms, which eventually proved to be unmanageable. In contrast, the Byzantine Empire, with its capital in Constantinople, flourished. The emperor, Constantine, had moved the capital of the Roman Empire from Rome to Constantinople in A.D. 330. The city itself was the most magnificent, largest, and affluent in the whole of Europe. Its citizens were cultured, educated, and creative. Within this advanced civilization, Greek culture, as well as Roman moral values and political systems, was nurtured for almost a thousand years (Persaud, 1984).

RISE OF ISLAM AND ARABIAN MEDICINE

Western Europe was plunged into a period of darkness until about A.D. 1000 because political and military powers rested in the hands of the nobility, there was lack of a central government and democracy, and there was an overall decline in trade (Campbell, 1926). During this period, Western civilization owes an enormous debt to the Arabs for preserving the great achievements of the Greek scholars, particularly in medicine and the sciences (Moir, 1831; Brown, 1921; Campbell, 1926; Ashoor, 1984a;1984b; Hayek, 1984).

Mohammed, who was born in A.D. 570 in the city of Mecca, founded the religion of Islam based on the revelations he received while meditating in a cave when he was 40-years-old (Leiser, 1983). The teachings of the religion are found in the Qu'ran, which were divulged to him by the angel Gabriel over a period of 23 years (A.D. 610–632) (Loukas et al., 2010b). Mohammed also spread his own sayings, rulings, advice, actions, and habits orally, and these are known as the Hadiths or Hadeeths (Leiser, 1983; Loukas et al., 2010b). Both the Qur'an and Hadeeths contain information on medicine that probably influenced folk medicine, but were of lesser importance in the formal training of physicians (Leiser, 1983; Shanks 1984; Loukas et al., 2010b). A prime example of this is the three main ways of healing according to Mohammed: eating honey, cupping, and cauterizing (Leiser, 1983; Shanks, 1984). As far as human anatomy is concerned, some of the Hadiths and citations in the Qur'an with respect to intrauterine development are remarkable (Persaud, 1984). The following sequences have been described (Moore 1982; Albar 1983; Moore, 1982; Saadat, 2009):

1. Nutfa (or drop stage) – "*We created Man from mixtures of germinal drop*" and "*From a Nutfa. He hath created him, immediately programmed him.*"
2. Alaca (referring to early implanting embryo) – "*Then we made the sperm into a leech-like clot.*"
3. Mudghda (somite stage) – "*Then of that leech-like clot made a chewed-like substance*" which is mentioned as being "*partly differentiated and partly undifferentiated.*"
4. "*Then we made out of that chewed like substance bones*" and "*clothed the bones with flesh: then We developed out of it another creature.*" These are references to the formation of the skeletal and muscular systems, as well as the transition of the embryo to a new stage with more definitive and recognizable features .

Finally, with reference to differentiation between male and female fetuses (external sexual characteristics) is the Hadith "*When forty two nights have passed over the sperm-drop, Allah sends an angel to it, who shapes it and makes it ears, eyes, skin, flesh, and bones. Then he says, O Lord; is it a male or female? And your Lord decides what he wishes and the angel records it.*" The Qur'an and

Hadiths also mention the cardiovascular system: the importance of the heart, blood and its circulation, and how they are vital to the maintenance of life (Loukas et al., 2010b).

1. "*We created man – We know what his soul whispers to him: We are closer to him than his jugular vein.*"
2. "*We would certainly have seized his right hand and cut off his Al-Watin,*"– Al-Watin has been translated to words similar to aorta.
3. "*Beware! There is a piece of flesh in the body and if it remains healthy, the whole body becomes healthy, and if it is diseased, the whole body becomes diseased. Beware, it is the heart.*"

United by Islam and guided by the Qu'ran, the Arabs were eager to spread their new religion (Shanks, 1984). The Arabs dominated an empire larger than that of the Romans with the conquest of Byzantine Asia, Persia, Egypt, North Africa, and Spain. Having encountered here the dispersed Greek physicians and philosophers, as well as their great libraries, the Arab scholars began to study the works of the conquered people, particularly in medicine and the physical sciences, and encouraged their translation into Arabic. These physicians-translators were chiefly Syrians, descendants of Nestorian Christians, but also included Persians. The two most important translators during this time were Sergius Resh-Ayna (A.D. 536), who translated the works of Galen, and Hunayn-Ibn-Is'Haq (A.D. 808–873), who allowed Arabics to express difficult ideas by developing scientific terminology (Shanks, 1984). These physician translators pursued their task with great passion and vigor. As a result, many important and major works of Greek scholarship (Hippocrates, Archigenes, Dioscorides, Rufus of Ephesus, Galen, Oribasius, Philagrios, and Paul of Aegina) were lodged in the flourishing libraries of the Arab empire, particularly at the "*House of Science*" in Baghdad, its capital (Gordon, 1959; Von Grunebaum, 1963; Klein-Franke, 1982; Leiser, 1983; Ashoor, 1983; 1984a; 1984b; Shanks, 1984; Savage-Smith, 1995). The original Greek words in anatomy that were translated to Arabic, would have been lost to western civilization were it not for their rediscovery during the thirteeth century when they were translated back into Latin (Moosavi, 2009).

The intellectual efforts of Arab scholars were not limited to the translation and assimilation of the works of the Greeks but independently, they have made significant contributions in mathematics, astronomy, chemistry, physics, botany, and pharmacy. These scholars knew of the use of many herbs and chemical preparations for the treatment of diseases. Indeed, the first schools of pharmacy and dispensaries can be traced to this period (Moir, 1831).

The golden age of Arabian medicine lasted until about the middle of the twelfth century. Between A.D. 800 and 1100, Arabic was the most widely accepted language for communication among medical scholars. All major works were already translated and widely disseminated in the flourishing libraries of the Islamic empire. The library in Cordoba alone housed more than 600,000 volumes and a catalog of 44 volumes. Many of the translated works were largely commentaries based on the translations of Hippocrates and Galen's work, along with some influences from the Syrians, Persians, and Indians. This was also the period when many famous Arab physicians lived resulting in original thinking and major contributions to medical sciences, including human anatomy (Leiser, 1983; Persaud, 1984).

In order to prevent quackery and charlatanism, examinations were proposed for testing medical knowledge. One of the required subjects was anatomy (Leiser, 1983). Arab physicians pursued their work with zeal and made many meticulous observations of diseases (Browne, 1921; Leiser, 1983). Perhaps, given

Figure 38. The three founders of early medieval anatomy from the title page of the Opera Ysaac, 1515. Haly Abbas is shown on the left; Constantinus Africanus is on the right. (Corner 1927, with kind permission from the Carnegie Institution of Washington.)

the opportunity, they would have made more advances in human anatomy were it not for their religion, which like so many others, forbade human dissection (Leiser, 1983).

Arabian medicine produced many great physicians whose works profoundly influenced medical thinking when they were later translated from the Arabic into Latin (Campbell, 1926). The most famous of these were Rhazes, Haly Abbas, Avicenna, Hunain ibn Ishaq, Tabari, and Ibn Nafis (Fig. 38).

Rhazes (A.D. 860–932) was a Persian physician, alchemist, mathematician, philosopher, and astronomer who authored more than 200 books. Out of these 200 books, two books that survived and contributed greatly to medicine were: *Kitab Al Hawi* or *Liber Continens*, an encyclopedic medical compilation of 24 volumes, and *Kitab Al Mansuri* (Tubbs et al., 2007). These books were translated into Latin by Gerard of Cremona, and were widely read in Europe. Rhazes influenced many areas in medicine, especially neuroanatomy. He was the first to draw a connection between lesions in the nervous system and clinical signs, and was the first to describe the recurrent laryngeal nerve as a sensory and a motor nerve (Tubbs et al., 2007). Rhazes was also the first to describe spina bifida in *The Liber Continens* (Modanlou, 2008). Rhazes died as a poor blind man, because when he needed eye surgery, he refused any doctor who had not mastered the anatomy of the eye (Tubbs et al., 2007).

Haly Abbas (949–982) was one of the first authors to question the knowledge of Aristotle and Galen in his book *Kitab al-Maliki* (The Royal Book) (Aciduman et al., 2010). Although the Royal Book was replaced by Avicenna's Canon, it was more functional as an anatomy book. In the Royal Book, Abbas describes the two ventricles along with their openings, the valves and their functions, and pointed out that the two ventricles contract at the same time, but the left side was more forceful sending blood throughout the body (Aciduman et al., 2010).

In the pre-Islamic era, various parts of the human body were known to be important but as to their shapes and functions, the description was more poetic than factual and scientific (Hyrtl, 1879; Browne, 1921; Campbell, 1926). The heart was said to contain courage and passion, anger came from the liver, and fear and laughter arose from the lungs and spleen, respectively (Campbell, 1926).

Avicenna wrote his famous *Canon of Medicine* in A.D. 1020, which contained a fair amount of anatomy that was borrowed from the writings of Aristotle, Hippocrates, and Galen in particular. The Canon presented a classification of organs and their functions using a systems-based approach, which is now used as the basis for modern clinical-based anatomy (Gruner, 1930; Shoja et al., 2011). This was the most used book to study medicine during that time, and was just as influential and widely used in Europe after its translation into Latin by Gerard of Cremona (Erolin, 2012). In the *Canon*, which is divided into five parts, Avicenna describes: anatomy of simple organs, that the aorta has three valves, which open when blood leaves heart during contractions and close when it relaxes to prevent blood from entering, muscle movements may be possible due to nerves, the six muscles which are responsible for movements of the eyeball the optic nerves, and described the anatomy of the cerebellum and caudate nucleus. The descriptions of various organs in the *Canon* were novel for his era, but whether Avicenna performed human dissections or not is controversial (Shoja et al., 2007).

The most original contribution to the descriptive and functional anatomy of the eye was made later by an Arab mathematician and physicist, Al-Hazen or Ibn-al-Haytham (A.D. 965–1040) in his work *Kitab ai-Menazir*. He deviated from humoral and teleological traditions in order to formulate a theory of vision based on his understanding of the eye as an optical system (Russel, 1982).

Hunain (A.D. 803–873) translated as many as 129 of Galen's books and a variety of Greek medical and scientific works into Syriac and Arabic with the help of his son Ishaq and nephew Hubaysh al-As'am. Hunain also compiled several medical treatises including a survey of medical knowledge and a book dealing specifically with eye diseases (Bergstrasser, 1925; Meyerhof, 1926; 1928).

Ali ibn Rabben at-Tabari, a physician of the ninth century, came from Merv of Tabaristan (old Persia) to Samarrah (present day Iraq) after the death of the Merv's king. Tabari's main work was *Firdous al-Hikmat* (Paradise of Wisdom), which dealt with embryology, as well as the anatomy of the brain, nerves, heart, and liver (Shoja et al., 2011).

Ibn Nafis or Nafis (A.D. 1210–1288) grew up in a village near Damascus in Syria. After completing his studies in Damascus, Ibn Nafis made his way to Cairo, Egypt where he became a specialist in ophthalmology and wrote the *al-Muhaddab* (The Perfected Book). The *al-Muhaddab* has two sections, the first section focused on anatomy, physiology, and pathophysiology of ophthalmology and the second section focused on various diseases of the eye. Ibn Nafis's many interests led him to discover the pulmonary or lesser circulation in his book *Kitab al-Mujiz*. Contradictory to Galen's model, Ibn Nafis asserted that the interventricular septum of the heart had no openings and that blood flowed from the heart to the lungs and from the lungs to the left side of the heart by the pulmonary veins (Klein-Franke, 1982; Ghalioungui, 1982; Persaud, 1989; Loukas et al., 2008b). Thus, for the first time, the pulmonary circulation was accurately described. Ibn Nafis also alluded to the idea of pulmonary capillaries, which were not discovered until 400 years later by Marcello Malpighi (1628–1694) (West, 2011).

Whether dissections were carried out or not in medieval times in the Arab world continues to be debated. Neither the Qur'an, nor the Hadith, specifically prohibited dissections, but there was a general sentiment within the community that dissections were looked down upon, especially dissections of Muslims (Savage-Smith, 1995). Additionally, the hot and dry climate of the countries in the Arabian Peninsula did not favor the preservation of the body for the purpose of dissection. Although there is no concrete evidence, recent data point to the fact that early Arab physicians, including Ibn Nafis, practiced dissection (Nasr, 1968; 1976; Qatagya, 1982; Ahmed, 1982; Uddin, 1982; Savage-Smith, 1995).

In this regard, attention should be directed to the anatomical illustrations preserved in several medieval treatises, all perhaps copies derived from Islamic sources (O'Neill, 1969; 1977; 1982). It is now believed that the original *Fünfbilderserie*, consisting of five anatomical drawings found in two Bavarian monastic manuscripts and reported in 1907 by Sudhoff, can be traced to an original manuscript with nine sets of drawings. The complete drawings, as well as the sequence according to the anatomy of Galen, have been preserved as part of a medical codex in the library of Gonville and Caius College in Cambridge, England (http://www.iranicaonline.org/articles/ebn-elyas).

As pointed out by O'Neill (1982), the famous Ashmolean anatomical drawings (Figs. 39–42) of the Bodleian Library in Oxford, England are essentially the same as those present in Cambridge, except for a drawing of the heart emerging from the sheath-like lungs. In addition, the normal sequence of the drawings has been disrupted. The Cambridge drawings reveal Arabic influences, which are not evident in the manuscript that was studied by Sudhoff (1908; 1916; 1930). According to O'Neill (1982), the Gonville and Caius manuscript presents "*the most complete and graphically intact Latin copy of the treatise known.*" O'Neill (1982) suggested that "*an illustrated Arabic anatomical manual found its way from Spain into an English Benedictine*

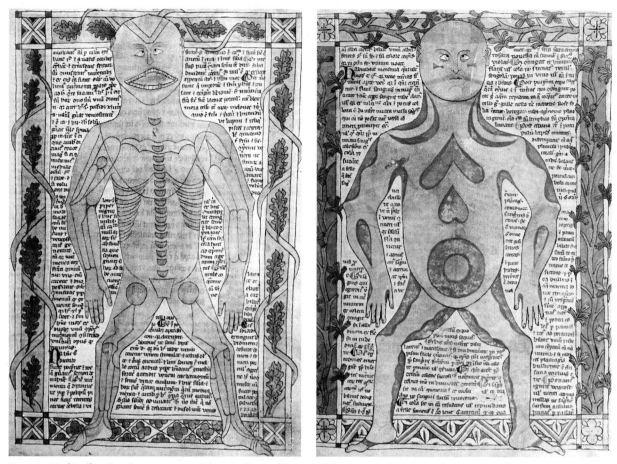

Figure 39 (*Left*). Crude drawing of the skeleton from a 13th century English manuscript. (Ms. Ashmole 399, folio 20; Bodleian Library, Oxford.)

Figure 40 (*Right*). The muscular system; an imaginary representation from a 13th century English manuscript. (Ms. Ashmole 399, folio 22; Bodleian Library, Oxford.) Peculiar arrangement of muscles, probably an attempt to relate various bodily movements to specific muscle masses.

monastery" and thereby reached European scholars. Fanciful and interesting as they are, these anatomical illustrations are among the earliest known. Galenic in concept, these crude illustrations reveal no practical knowledge of the inside of the human body (Figs. 43 & 44) (Persaud, 1984).

EUROPEAN UNIVERSITIES

Europe was experiencing a gradual transformation because of a change in societal values, improvement in trade, expansion of the cities, and the establishment of strong governments. With the fall of the Roman Empire, the authority of the Pope diminished and religion became more unified. The "discovery" of America in 1492 by Christopher Columbus and of the art of printing with moveable type by Gutenberg in 1454 were all momentous achievements. There was renewed interest in every aspect of human endeavor and the time for exploration of new ideas and discoveries was eminent. It was also the period that gave birth to the great universities and medical schools (Gordon, 1959), which fostered a renewed interest in the anatomy of the human body. Prior to this period anatomy was considered to be pagan science and looked upon with contempt. Such an attitude was a reflection of the psychology of the times (Persaud, 1984). There was a great preoccupation with religion, the occult, mysticism, and astrology. The human body

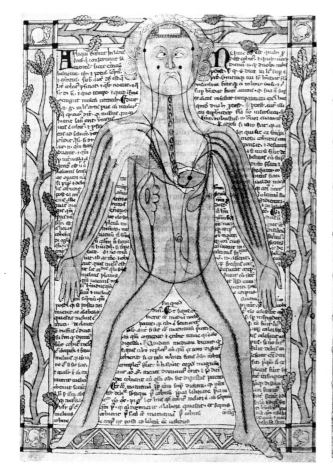 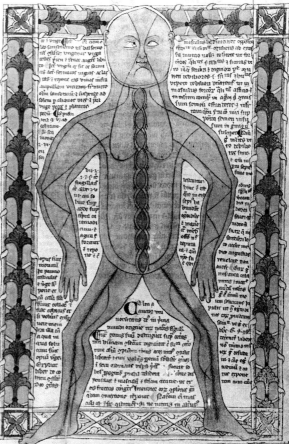

Figure 41 (*Left*). Drawing of the arterial system from a 13th century English manuscript. (Ms. Ashmole 399, folio 19; Bodleian Library, Oxford.) The black spot in the heart was conceived as the site of the origin of the arteries. The anastomosis of the two arteries in the head probably represented the rete mirabile. Observe the characteristic squatting or "*frog-like*" posture of these medieval anatomical figures.

Figure 42 (*Right*). Speculative concept of the nervous system from a 13th century English manuscript. (Ms. Ashmole 399, folio 21; Bodleian Library, Oxford.)

and its functions, as well as man's fate, were thought to be under the magical influence of the planets. Because every part of the body was governed by one of the heavenly bodies, then it followed that diseases could be accounted for by the movements of the planets. The zodiacal man became popular and accounts for the many anatomical drawings, decorated with astrological signs (Figs. 45 & 46), that have emerged from this period (Major, 1054; Lassek, 1958)

The renewed interest in human anatomy was intimately linked with the emergence of the universities, but it still had to be legally approved by both church and government. In Italy alone, 15 universities were established between A.D. 1200 and 1350. The first center of learning to be established was the University of Salerno, but as far as human dissection was concerned, it would appear that this occurred at the University of Bologna. The reason was not for the teaching of anatomy but for ascertaining the cause of death in a suspected case of poisoning (Lassek, 1958).

The University of Bologna began for the teaching of law and the medical faculty was established in 1156. It is stated that by 1320 there were at least 15,000 students there. One of the first anatomists to have taught at Bologna was Taddeo Alderotti, who carried out human dissections in the teaching of anatomy, according to the records of his disciples Bartolomeo da Varignana, Henri de Mondeville and Mondino de Liuzzi. It is also that postmortem autopsies were carried out on the body of a nobleman, Azzolino, who died in 1302, under the direction of Bartolomeo da Varignana (Simili, 1951; Siraisi, 1981),

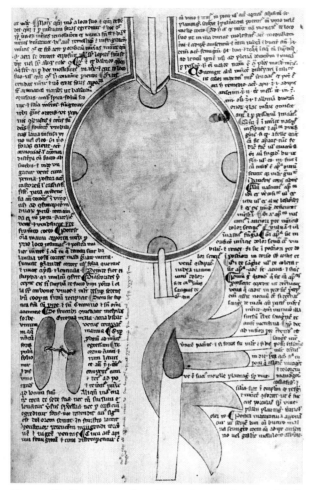

Figure 43. Drawings of a spherical stomach, paired kidneys, and a five-lobed liver from the Ashmole 399 (folio 23) 13th century manuscript (Bodleian Library, Oxford). These illustrations were copied from other sources and bear little or no relationship to the actual organs.

a student of Taddeo and also by the surgeon William of Saliceto, a contemporary of Taddeo. The authorization to carry out human dissections is attributed to Frederick II (1194–1250), Emperor of the Holy Roman Empire and King of Jerusalem.

As many as 80 universities were established in Europe before the end of the Middle Ages, but nowhere was the study of human anatomy pursued with such vigor as in the universities of Italy. The most celebrated of them was the University of Bologna. The University of Salerno is the earliest known university in Europe (Harrington, 1920). It is believed that the monks from the Benedictine Monastery of Monte Casino influenced the scholastic organization of the Medical School of Salerno. St. Benedict founded the Benedictine Monastery in the sixth century. At the

end of the sixth century, the Monks fled to Rome because the Lombards destroyed their monastery. They returned in A.D. 720 to rebuild, but the monastery was once again destroyed in 884, but this time, it was destroyed by the Saracens. They rebuilt the monastery once again 70 years later, and it continued to exist until its dissolution in 1886. When the monks were at the monastery, it was also regarded as a hospital and a medical school (Packard, 1920).

This was the monastery that Constantinus Africanus fled to after being accused of sorcery in Carthage. Africanus was born in Africa and then travelled to Egypt

Figure 44. Position of the infant in the womb as depicted in this 13th century English manuscript. (Ashmole 399, folio 15; Bodleian Library, Oxford.) Above are twins, male and female; below, the child lies transversely with attached umbilical cord.

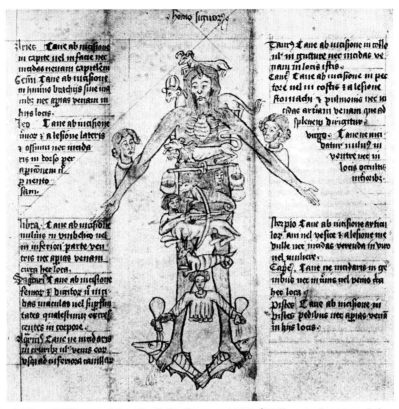

Figure 45. Astrological Man. Wellcome Ms. 40. Calendarium 1463. (Wellcome Institute Library, London.) In the Middle Ages and during the Renaissance, astrology was taught in several European universities and many physicians were also astrologers. It was believed that the planets ruled different parts of the human body. Thus, Jupiter – lungs, liver and feet; Mars – left ear and genitalia; Sun –right side of heart; Venus – neck and abdomen; Mercury-arms, hands, shoulders and hips; Moon-left half of body and stomach.

and China for 39 years searching for knowledge. It was when he came back from this journey that he was accused of sorcery. Before going to the monastery, he was appointed secretary to Robert Guiscard, but he resigned soon after and went to the monastery. Here, he translated and made commentaries on the works of authors such as Hippocrates, Galen, Haly Abbas, and Avicenna (Packard, 1920).

The work of Haly Abbas was entitled *Al Maleki* (The Royal Book). He attempted to summarize the entire system of classical medicine. In Constantine's translation of the work, which he entitled *the Pantegni* or *The Whole Art*, a systematic account of the human body based on the teaching of Galen is presented. In 1539, a collected edition of his books was published in Basle. These works of Africanus brought Arabian medicine to Europe (Packard, 1920).

In A.D. 1075, Robert Guiscard captured the city of Salerno and it prospered under Henry VI until 1194 after which the city was ruined. This was when the monks of the Benedictine Monastery of Monte Casino

came into the city to set up monasteries, as they realized the importance of the city as a place where people stopped by to get cured (Packard, 1920).

Literary activity began at the Medical School of Salerno in the eleventh century. Some of the earliest authors were Gariopontus, Petrocellus, and Petronius. The most famous author from the Medical School of Salerno was Trotula, a woman, who wrote books on obstetrics, hygiene, and other medical topics. Her most famous books were *De Mulierum Passionibus* (*Trotula Major*) and *Trotula Minor*. Another well-known author was Cophon the Younger who wrote *De Anatomia Porci* in the twelfth century, which served as a standard work of reference for the learning of anatomy, and *Ars Medendi*, which was a book on the practice of medicine (Packard, 1920).

In 1224, Frederick II sent out a decree that all medical students had to pass an examination to gain the title of Magister, which was the title given to doctors. To be eligible for the examination, one had to be a 21-years-old and have studied medicine for seven

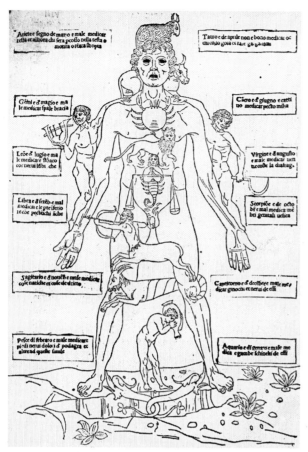

Figure 46. "*Zodiac man*" from Ketham's Fasciculus Medicinae (1522). The astrological symbols are shown in relationship to the different parts of the body (Instituti Ortopedici Rizzoli, Bologna). The human body as the microcosm or parvus mundus was viewed and interpreted in relationship to the universe or macrocosm. The 12 signs (constellations) of the Zodiac were represented as follows: Aries, the ram; Taurus, the bull; Gemini, twins; Cancer, the crab; Leo, the lion; Virgo, the virgin; Libra, the balance; Scorpio, the scorpion; Sagittarius, the archer; Capricorn, the goat; Aquarius, the water-bearer; and Pisces, the fish.

years. If they were eligible, they would be tested on the works of Hippocrates, Galen, Avicenna, and Aristotle. Frederick II also passed another decree for surgeons, which stated that all surgeons had to study

anatomy for a year at either the School of Salerno or the University of Naples, and pass an examination. The influence of the University of Salerno, as well as other factors, contributed to the development of similar centers of higher learning in other parts of Italy, France, Germany, and England (Packard, 1920).

The University of Montpellier apparently had *magistri physici*, medical teachers, as early as A.D. 1000, which would make it the oldest university in Europe. It was founded by Jewish physicians from Spain and like Salerno, which it rivaled, was the center for the dissemination of Greek medicine throughout Europe. The Bishop appointed teachers and in A.D. 1230, the passing of an examination conducted by two of the masters was a requirement for practicing medicine. Here, the learned Catalan Arnold of Villa Nova (1235–1311) translated Avicenna's book on the heart and wrote other medical treatises. He was an independent thinker, highly regarded, and knew many languages. As physician to several popes and the Kings of Arragon and Sicily, he was a man of considerable influence (Packard, 1920).

After 1314, dissections were carried out in Montpellier and these were opened to the public after the payment of a fee. In 1376, the Duke of Anjou ordered that the body of an executed criminal should be made available for dissection annually. The famous surgeon, Guy de Chauliac (1300–68) studied in Montpellier and Henri de Mondeville (1260–1320), another surgeon who encouraged the study of anatomy (MacKinney, 1962), was a professor there.

In medieval France, medicine was taught only at Montpellier and later at the University of Paris, which was also renowned for theology and philosophy (Cooper, 1930). The University was founded during the latter part of the twelfth century and attracted students from all over Europe. The teaching of medicine probably began in 1210 when Gilles de Corbeil of Salernum was appointed physician to King Philip.

The University of Oxford began in 1167 and the University of Cambridge in 1217, both patterned themselves after the University of Paris where the courses in medicine were the same and at the beginning, did not include instruction in surgery.

Chapter 9

MONDINO DE LUZZI

RESTORER OF ANATOMY

Human dissection was an intricate part of medical study, but declined after A.D. 200 when the Catholic Church prohibited dissection, arguing for the sanctity of the human body. As such, outside the work of Galen, the Early and High Middle Ages showed limited advancement in anatomy. It was not until the Late Middle Ages, when the Catholic Church authorized human dissection of executed criminals, that Mondino de Luzzi (Mondino, 1493) was able to perform one of the first public human dissections in over 1,100 years (Fig. 47). Mondino is revered as the "*Restorer of Anatomy*," not because he conducted one of the first human dissections in centuries, but rather because his unique style of professorship and his detailed documentation of dissections, which culminated into one of the first modern anatomy textbooks, lead to the incorporation of the first systemic study of anatomy in a medical curriculum (Castiglioni, 1941; Crivellato and Ribatti, 2006c).

MONDINO'S PROFESSORSHIP

Mondino's introduction of anatomy into the medical curriculum was so widely accepted because his approach to dissection was novel for his day. Guy de Chauliac, a distinguished French surgeon, noted that Mondino performed the dissections himself rather than dictating dissection procedures to assistants who would ultimately perform them, as was common practice in the past (Beasley, 1982). Mondino compiled his first-hand dissection procedures to form one of the earliest modern anatomy textbooks, *De Omnibus Humani Corporis Interioribus Membris Anathomia* (*Anathomia*) (Figs. 48 & 49). Although a pioneer in dissection methods, Mondino admitted that his dissection findings were to confirm rather than to disprove the previous findings of Galen (Beasley, 1982).

Even though Mondino received much respect for his first-hand accounts of human dissections in *Anathomia*, he would go on to use dissection assistants later in his professorship. In Ketham's work *Fasciculus Medicinae* (Fig. 50), there is an illustration of Mondino sitting removed and elevated from the dissection, providing instructions to his assistants (Ketham, 1941) (Fig. 51).

Mondino's use of dissection assistants is noteworthy because one of his assistants, Alessandra Giliani of Persiceto, possessed such remarkable skill in dissection that she was able to create detailed and elaborate wax models, further increasing Mondino's reputation and enhancing his teaching methods. According to the chronicle:

> Alessandra Giliani of Persiceto became the most valuable to Mondino because she would clean most skillfully the smallest vein, the arteries, all ramifications of the vessels, without lacerating or dividing them, and prepare them for demonstration she would fill them with various colored liquids, which, after having been driven into the vessels, would harden without destroying the vessels. Again, she would paint these same vessels to their minute branches so perfectly and color them so naturally, that, added to the wonderful explanation and teachings of the master, they brought him great fame and credit. (Walsh, 1911)

The title page of the 1493 edition of the *Anathomia*, published in Leipzig, depicts Mondino in full robes

55

Figure 47. Mundino, the Italian anatomist, making his first dissection in the anatomy theatre at Bologna, 1318. Oil painting by Ernest Board (1977–1934). (Wellcome Institute Library, London.)

and cap sitting remotely from the dissection and reading from the opened book in his left hand to direct the assistant, who is suspected to be Alessandra Giliani of Persiceto.

Mondino's first-hand dissection accounts culminated into one of the earliest modern anatomy textbooks, and his use of novel resources, such as wax models, enhanced the teaching of the subject.

There is speculation as to why Mondino was the only anatomist successful in this endeavor. Mondino's standing as a member of an influential Florentine family was an important factor in his success, by providing him the opportunity and giving him the political clout to institute such a change. Mondino's father, Nerino, and grandfather, Albizzio, were both pharmacists, and his uncle, Luzio, was a professor of medicine in Bologna (Castiglioni, 1941). Mondino's upbringing provided him the privilege to study in the College of Medicine at the University of Bologna. Following graduation, Mondino established himself as an expert in his field while employed as a public lecturer in practical medicine and surgery at the university. Beyond his expertise, Mondino wielded a good deal of political influence. He served as a highly regarded diplomat for the city government and served as an Ambassador of Bologna to the son of King Robert of Naples. Mondino's affiliation with such an influential family in the community, his expertise in medicine, and his role as Ambassador jointly influenced his success at introducing anatomy into the medical curriculum (Castiglioni, 1941).

MONDINO'S *ANATHOMIA*

Mondino's efforts to introduce anatomy into the medical curriculum were made tangible by the compilation of his first-hand dissection documentations into *De Omnibus Humani Corporis Interioribus Membris Anathomia* (*Anathomia*). The *Anathomia* was written in 1316, yet it was not published until 1478 in Pauda, Italy (Castiglioni, 1941). It was the most widely used anatomical text for over 250 years and evolved over time with more than 40 editions (Castiglioni, 1941; Crivellato & Ribatti, 2006c; Wischhusen & Schumacher, 1970). Although much of the medical information provided in *Anathomia* was derived from commentaries of Hippocrates, Aristotle, and Galen, Mondino's dissection approach was to support rather than to refute Galen; *Anathomia* was significant because Mondino's dissection procedures guided and instructed fellow professors and students on the dissection process (Wilson, 1989).

The first anatomical dissection and demonstration in Rostock took place in 1513; for some other European cities the following years have been cited: Bologna (1302), Montpellier (1315), Padua (1341), Prague (1348), Venice (1368), Florence (1388), Vienna (1404), Paris (1478), Cologne (1479), Tubingen (1482), Leipzig (c. 1500), Edinburgh (1505), Strassburg (1517), Basel (1531), Marburg (1535), Oxford (1549), Salamanca (c. 1550), Lausanne (c. 1550), Zurich (c.1550), Amsterdam (1555), Cambridge (1557), London (1564), Bern (1571) (Wischhusen & Schumacher, 1970).

The *Anathomia* begins with the justification that human beings are superior to all other creatures because of their intellect, reasoning-ability, tool-making abilities, and upright stature (Siraisi, 1990). Therefore, human beings are worth being studied. Mondino segmented the body into three distinct parts by physiological activity: (1) the superior ventricle (skull), which

Figure 48 (*Left*). Title page of the 1493 edition of the *Anathomia* published in Leipzig. The man in full robes and cap sitting on the chair is undoubtedly the professor. He is reading from the opened book in his left hand and directing the young bareheaded assistant. The intestines are shown.

Figure 49 (*Right*). An anatomical demonstration in Rostock at the beginning of the 16th century. This woodcut illustration appeared in the Rostock edition (1514) of the Anatomia Mundini by Nicolaus Thurius Marschalk (Universitatsbibliothek Rostock; from Wischhusen and Schumacher, 1970). The first anatomical dissection and demonstration in Rostock took place in 1513; for some other European cities, the following years have been cited (see Wischhusen & Schumacher, 1970): Bologna (1302), Montpellier (1315), Padua (1341), Prague (1348), Venice (1368), Florence (1388), Vienna (1404), Paris (1478), Cologne (1479), Tubingen (1482), Leipzig(c. 1500), Edinburgh (1505), Strassburg (1517), Basel (1531), Marburg/L (1535), Oxford (1549), Salamanca (c. 1550), Lausanne (c. 1550), Zurich (c.1550), Amsterdam (1555), Cambridge (1557), London (1564), Bern (1571).

contained the "*animal members*," (2) the middle ventricle (thorax), which contained the "*spiritual members*," and (3) the inferior ventricle (abdomen), which contained the "*natural members*" (Singer, 1957). Mondino based his dissection procedures on his assertion that the compartments of the human body arranged in order of inferiority to superiority are: "*natural members*," "*spiritual members*," and "*animal members.*"

Mondino's dissection procedures began with the "*natural members*" or organs of the abdominal cavity. A

vertical incision from the abdomen to the pectoral muscles and horizontal incision above the navel were made to access the abdominal cavity. According to the text, the stomach was the first organ dissected and was described as spherical with the internal lining depicted as the "*seat of sensation*" and external coat explained as being involved in digestion. The dissection procedure continued with the removal of the "*false ribs*" in order to access the spleen. The spleen was described as secreting black bile into the stomach

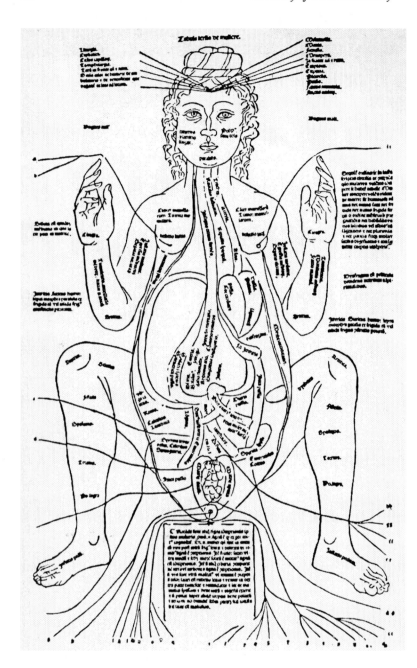

Figure 50. Situs figure from Ketham's *Fasciculus Medicinae*, Venice, Johann et Gregorio 1491 (From Major, 1954).

through imaginary canals (Singer, 1957). Mondino described the liver as having five lobes, the gallbladder as being the seat of yellow bile, and also the cecum with no mention of the appendix (Singer, 1957). Though Mondino provides little explanation of the pancreas, the pancreatic duct is described in great detail. Lastly, Mondino presents his observations of the urinary bladder and enlargement of the uterus during menstruation and pregnancy (Castiglioni, 1941).

Mondino next explains the dissection of the "*spiritual members*" or the thorax. The description of the heart is detailed, yet inaccurate. The heart is described as having three chambers: a right ventricle, middle ventricle, and left ventricle. The right ventricle is noted to have a significant opening through which blood originating from the liver enters. Despite the anatomical inaccuracies, Mondino provides a detailed and accurate description of the superior vena cava. The left tricuspid valve is noted to have three leaflets. After dissection of the heart, Mondino continues with the dissection of the lungs. Mondino thoroughly examined the course of the pulmonary artery and vein and notes the pleura and importance of pulmonary pathologies including pneumonia (Castiglioni, 1941). To conclude the thorax dissection, Mondino provides a very elementary explanation of the larynx (Singer, 1957).

Finally, Mondino provides discussion of his dissection of the most superior ventricle of the body containing the "*animal members*" or the skull. Mondino divided the brain into three anatomical vesicles: (1) anterior vesicle responsible for the senses, (2) middle vesicle housing the imagination, and (3) posterior vesicle containing the memory. Mondino concluded that the choroid plexus controlled the mental processes by opening and closing the passages between the ventricles. Mondino also provided commentary on the cranial nerves, yet his explanation was derived from Galen's *Uses of the Parts of the Body of Man* (Singer, 1957).

One of the noted challenges for Mondino's dissections was the timeliness of dissection, since there was no way of preserving a cadaver for a long period. Dissections typically took place over the course of four days and even continued overnight, when necessary. Since there was rarely time for the dissection of the limbs, Mondino suggested a sun-dried body as a means for studying the muscles, arteries, veins, and nerves of the limbs (Kornell, 1989).

The *Anathomia* was primarily an anatomical text with little surgical information. However, one of the most noteworthy descriptions was the closure of an intestinal wound by having large ants bite the edges of the incision. The ant's then had their heads removed with the result of an intestinal anastomosis, foreshadowing modern-day surgical stapling (Beasley, 1982). Additionally, Mondino provided thorough descriptions on surgical treatment for hernias and cataract surgery (Singer, 1957).

Although Mondino's dissection findings shared the same inaccuracies as those of Galen from centuries

Figure 51. Anatomical dissection scene from Ketham's edition of Mondino's *Anathomia* (Venice, 1500). Mondino is shown seated on the elevated chair reading from a book and presiding over the dissection. The assistant is about to make an incision. (Instituti Ortopedici Rizzoli, Bologna.)

before, the anatomical inaccuracies were overshadowed by his detailed instructions on how to approach human dissection.

MONDINO'S LEGACY

Mondino became Professor of Anatomy at the University of Bologna in 1306 where he remained for the next twenty years until his death. In 1315, the Catholic Church granted him permission to perform one of the first public human dissections in over 1100 years. Mondino has been called the "*Restorer of Anatomy*" because he introduced a rational approach for the systematic study of the human body through novel dissection techniques. Moreover, his *Anathomia* can rightly be considered the first modern manual that dealt exclusively with anatomy. It remained popular among medical students for several centuries (Persaud, 1984).

Chapter 10

LEONARDO DA VINCI

RENAISSANCE ART AND HUMAN ANATOMY

Many famous artists of the early Renaissance period pursued the study of the human body, including performing dissections in order to depict the beauty of the human form in an accurate and realistic manner (von Toply, 1903; Richer, 1903; Janson, 1977; Parker, 1983). This union of art and science brought the study of human anatomy to a new promising course.

Not only was the human body dissected and its skeleton studied, but also, more emphasis was placed on the laws of geometry and mechanics as applied to the human form. Indeed, the greatest of these anatomical artists, Leonardo da Vinci, went on to perform experiments that sought to explain the human form as it related to function. His outstanding accomplishments as an artist, scientist, and inventor have been attributed to his originality, creative ingenuity, fertility of invention, and his passionate search for new knowledge. He also had a great interest in botany, mathematics, geology, astronomy, and philosophy and in all of these fields, he excelled (Persaud, 1984).

Contemporary artists of Leonardo, such as Albrecht Dürer (1471–1528), Antonio Pollaiuolo (1432–98), Verocchio (1435–88), Luca Signorelli (1442–1524), Michelangelo (1475–1564), Titian (1477–1576), and Raphael (1483–1520) followed the same principle but from the standpoint of an artist in order to depict the human form both naturally and accurately by combining artistic ability with keen observation.

LEONARDO: THE RENAISSANCE MAN

It is written that Leonardo remarked, *"let no man read me who is not a mathematician. No human investigation can lay claim to being true science unless it can stand the test of mathematical demonstration. The man who undervalues mathematics nourishes himself upon confusion"* (Chamberlain, 1914). Several of Leonardo's petitioned artworks have been closely analyzed and interpreted to contain figures and elements based on mathematical proportions. Arguably his most famous painting, the Mona Lisa, is set up to display what was referred to as the golden section/ratio in which the composition displays multiple geometric consistencies to draw in the viewer (Stakhov, 1989).

Leonardo's background and versatile interests provided him with the ability to draw connections among what was originally viewed as separate disciplines. Frank Netter describe da Vinci as one who fervently searched for unity in all knowledge (Netter, 1956; Calkins et al., 1999). This search for unity allowed da Vinci to progress his understanding and expertise of multiple subjects simultaneously. Examples of da Vinci relating the human body to the physical laws of nature's rivers and streams are uniquely described by Kemp (2006). Da Vinci's method of comparative understanding was most brilliantly described by Roger Bacon as the vision of a *"universal science"* where each branch of knowledge contributed to the whole (Kemp, 2006).

Anatomical Studies

Perhaps Leonardo's most popular and widely known reputation comes from his work as an artist. His observations and eventual dissection of the human body were preliminary necessities to him that he could depict it more vividly. Eventually,

Leonardo's enthusiasm for human dissection led him to the study of anatomy (Belt, 1955; Keele, 1978; 1983).

Nowhere does Leonardo's observational comprehension stand out so clearly as in his magnificent, enthralling, and accurate anatomical drawings (Clark, 1935a; 1935b; 1968; Herrlinger, 1953; Saunders & O'Malley, 1982). These anatomical drawings were in the tradition that began with Giotto, which displaced conventionalism and aimed at a more natural and realistic representation. Not satisfied with descriptive anatomy, he forged ahead to explain the function of various structures. Two years before his death, Leonardo was visited by Cardinal Luis of Aragon and his secretary who recorded the following:

> This gentleman has written of anatomy with such detail showing by illustrations the limbs, muscles, nerves, veins, ligaments, intestines, and whatsoever else there is to discuss in the bodies of men and women, in a way that has never yet been done by any one else. All this we have seen with our own eyes; and he said that he had dissected more than 30 bodies both of men and women of all ages. (McCurdy, 1932)

Cardinal Luis's record of Leonardo's sketches shed light on the innovation that Leonardo was bringing to the field of anatomy. It is speculated that the disarticulated technique known today as *Beauchene* may have been influenced by the anatomical drawings of Leonardo where he shows a disarticulated skull's parts and their relation to one another in Weimar Sheet (Spinner, 2011) Throughout his career, da Vinci's works reveal a departure from the widely accepted theories of the human body to a direct imitation of nature as he saw it in dissection. For example, it is unlikely that for an artist such as Leonardo who could draw the axial section of the brain, that he could miss the cross-sectional structures of the ventricles. At this point in Leonardo's development, it is not clear whether he was drawing the brain he observed, the brain he remembered, or the brain he read about in anatomical works of his time. Leonardo's sketches from a few years later show that he had again returned to the subject of the ventricles, representing their true anatomy with great success (Tascioglu & Tascioglu, 2005). It is shown by his sketches that Leonardo was very interested in cross-sectional dissection and analysis of various structures. (O'Rahilly, 1993; Saunders & O'Malley, 1982). The depiction and appreciation of cross-sections would not become popular again until the nineteenth century.

Leonardo was familiar with the authoritative anatomical works of his day. It is written that Leonardo may have used Mondino's *Anothomia* as a dissecting manual in his work and he is even referred to with reference to the extensors of the toes.

In 1497, Alessandro Benedetti (1450–1512), an Italian scholar and Professor of Medicine at the University of Bologna and Padua, published an anatomical guide consisting of five books (Anatomice). It dealt with the structure and dissection of the human body but lacked originality. Also popular during this period was the work of the surgeon Guy de Chauliac of Montpellier. His book, *Cyrurgia Magna*, was written in 1363 and a French translation of it appeared in Paris in 1478. Translations of Galen and Avicenna from the Arabic were also available. From his notes, it is also known that Leonardo was familiar with Albertus Magnus's (1193/1206–1280) work in anatomy and zoology, *De Animalibus* (Persaud, 1984). It is important to note that Da Vinci was not fully educated in that he had not mastered Latin, and considered himself "*a man of no letters.*" Most of the scientific literature at the time was written in Latin so Da Vinci may or may not have understood the importance of some of his discoveries (Lydiatt & Bucher, 2011).

Leonardo's approach to the study of the human body was that of an astute anatomist (McMurrich, 1930; Keele, 1978; 1983). He always attempted to combine anatomical structure with function. Leonardo also picked up on abnormal structures such as anomalous pectoral musculature (pectus excavatum) and vestigial structures such as the levator claviculae muscle, which he illustrated in his notes (Ashrafian, 2013; Mosconi & Kamath, 2003; Capo & Spinner, 2007). His dissections were carried out in the hospital of Santa Maria Nuova in Florence and later in Santo Spiritu Hospital in Rome. He dissected at night, perhaps in secret, and made use of a sharp knife, chisel and a bone saw. His own account, according to Keele, of this process is inspiring:

> And you who say that it is better to look at an anatomical demonstration than to see these drawings, you would be right, if it were possible to observe all the details shown in these drawings in a single figure, in, which with all your ability you will not see, nor acquire a knowledge of more than a few veins, while in order to obtain an exact and complete knowledge of these I have dissected more than ten human bodies, destroying all the various members, and removing even the very smallest particles of the flesh which surround these veins,

without causing any effusion of blood other than the imperceptible bleeding of the capillary veins. And as one single body did not suffice for so long a time, it was necessary to proceed by stages with so many bodies as would render my knowledge complete; and this I repeated twice over in order to discover the differences. But though possessed of an interest in the subject, you may perhaps be deterred by natural repugnance, or if this does not restrain you then perhaps by the fear of passing the night hours in the company of these corpses quartered and flayed, and horrible to behold, and if this does not deter you then perhaps you may lack the skill in drawing essential for such representation; and even if you possess the skill it may not be combined with a knowledge of perspective, while if it is so combined you may not be versed in the methods of geometrical demonstration, or the methods of estimating the forces and strength of the muscles, or you may perhaps be found wanting in patience so that you will not be diligent. Concerning which things, whether or not they have all been found in me, the one hundred and twenty books, which I have composed, will give their verdict yes or no. In these I have not been hindered either by avarice or negligence, but only by want of time. Farewell. (Keele, 1951; 1952)

This journal entry gives a very detailed insight into the working conditions experienced from the late fifteenth through early sixteenth centuries: "*There were no chemicals to preserve the cadavers, which began to decompose as he was drawing them*" (Perloff, 2013). The ability to attain knowledge relied on the thoroughness of the investigator, and many times Leonardo continued dissecting multiple bodies in one sitting to "*render the complete knowledge.*" It is seen that the "*art*" of anatomical dissecting and recording required a strong stomach, a keen eye, and much endurance. Da Vinci's great respect for the process of dissecting and the education gained there allowed him to thrive at such tasks. (Perloff, 2013) It is written that Leonardo said, "*if you find from your own experience that something is a fact and it contradicts what some authority has written down, then you must abandon the authority and base your reasoning on your own findings*" (Tarshis, 1969; Calkins et al., 1999).

From the annotations in Leonardo's notebooks, it is learned that he had planned on writing a treatise on anatomy that was probably never completed. His large number of anatomical drawings and extensive notes would have probably found their place in the

work he had conceived. He had already mapped out, at the beginning of his anatomical studies, the scope and content of the book. In his notebook, he had written under the heading "*Of the Order of the Book*" the following:

This work should begin with the conception of man and describe the nature of the womb, how the child lives in it, and up to what stage it dwells there, and the manner of its quickening and feeding, and its growth, what interval exists between one stage of growth and another, and what drives it forth from the belly of its mother before the proper time. Then describe which are the members, which grew more than the others after the child is born, and give the measurements of a child of one year. Next describe a grown male and female and their measurements, and the nature of their complexions, color, and physiognomy. Afterwards describe how he is composed of vessels, nerves, muscles and bones. (Saunders & O'Malley, 1982)

Such an expansive plan for a book would have encroached upon Leonardo's time and prevented him from pursuing his diverse interests. In 1508, he was already concerned about his voluminous notes, manuscripts, and drawings. In his own handwriting, Leonardo recorded

begun at Florence in the house of Piero Di Braccio Martelli, on the 22nd of March 1508. This makes the collection, without order, taken from many sheets, which I have here copied hoping to arrange them later, each in its place according to the matters of which they treat. I believe that before I make an end of this I shall have to repeat the same things many times, for which, O reader, do not blame me, for the subjects are many and the memory cannot retain them and say....This I do not need to write, since I have written it before. (Saunders and O'Malley, 1982)

Two years later, the work was still in progress because he himself noted again that he was hoping to finish it in spring. For obvious reasons, Leonardo, the artist, must have directed his initial efforts at understanding the musculoskeletal system. His knowledge of mathematics and mechanics is most evident in the numerous drawings showing bones in their correct proportions and limbs in motion. Muscles are revealed in layers and there is no doubt left as to their correct shape and attachments to the bones. Many of the drawings indicate a deep insight as to the opposing functional grouping of muscles and the movement of bones acting as levers, such as in the movement of the forearm in pronation and supination and the movements of the bones of the leg and foot during walking (Figs. 52 & 53) (Persaud, 1984).

Experiments and Discoveries

Many of Leonardo's discoveries were due to the technical procedures he developed and experiments he carried out both in cadavers and in animals. It is noted that Leonardo had, on several accounts, transitioned anatomical concepts taken from animal dissections and extrapolated them to human anatomy (Clarke & Dewhurst, 1972). In one instance, he demonstrated the shape and extent of the ventricles of the brain in the ox by filling them with melted wax. He was the first person to use such a technique for demonstrating

the size and shape of any body cavity (McMurrich, 1930; Keele, 1979). This technique was used in the field of sculpting and serves as another example of how da Vinci carried over his work to multiple disciplines (Gross, 1997).

The abolition of spinal reflexes in the frog by pitting its spinal column was demonstrated by Leonardo. He carried out experiments on the mechanisms of voice production and described the role of the intercostal muscles during respiration and the events associated with swallowing. His drawings revealed the correct inclination of the pelvis, the papillary muscles, chordae tendineae, and the moderator band in the right ventricle.

Leonardo's drawings and writings on the heart, blood vessels, and movement of the blood have been extensive and it is quite a surprise that his work did

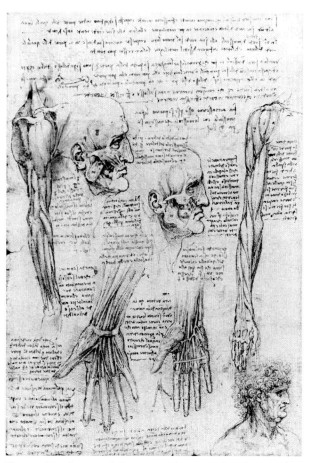
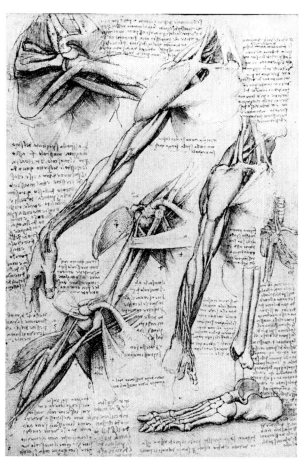

Figure 52. Series of drawings showing the clavicle (and even the subclavius muscle), superficial and deep facial muscles, arm and forearm muscles, the ulnar and median nerves, and the ulnar artery (superficial palmar arterial arch). (Reproduced by Gracious Permission of Her Majesty Queen Elizabeth II.)

Figure 53. Leonardo's drawings of the muscles of the shoulder region, arm, and forearm. The bones of the ankle region and foot are shown below on the right. (Reproduced by Gracious Permission of Her Majesty Queen Elizabeth II.)

not lead him to discover the circulation of the blood. With around 800 anatomical illustrations made by da Vinci, a large number were devoted to the cardiovascular system. In particular, Leonardo directed his attention to the valves of the heart and their movements during systole and diastole (Persaud, 1984). His notions of the cardiac cycle and the heart's asynchronous contraction and dilation were original. He also showed that the heart was indeed a muscle and was the first to deduce the importance of currents in the blood as a mechanism of closing the heart valves (Keele, 1951; 1952). His drawings showed that he knew of the origin and distribution of the coronary arteries, as well as of the atria (Fig. 54).

Leonardo knew of the condition of arteriosclerosis (Keele, 1973) as a result of dissecting an old man who had just died. One of his beautiful drawings compared the blood vessels with that of a younger person. These studies helped to provide an insight into arterial disease in the elderly (Belt, 1955). The artery and the vein, which in the old extend between the spleen (splenic) and the liver (porta hepatis), acquire so great a thickness of skin that they contract the passage of the blood that comes from the mesenteric veins, through which this blood passes over to the liver and the heart, and to the two greater vessels, and in consequence to the whole body. And apart from the thickening of the skin (or coat), these veins grow in length and become tortuous like a snake, and the liver loses the humor of the blood, which was carried there by this vein, and consequently, the liver becomes dried and becomes like a congealed brain (*crusca congelata*) both in color and substance (Keele, 1952). Leonardo's conception of eddy currents along with the process of arteriosclerosis may have been shaped by his interest in fluid dynamics (Keele, 1973)

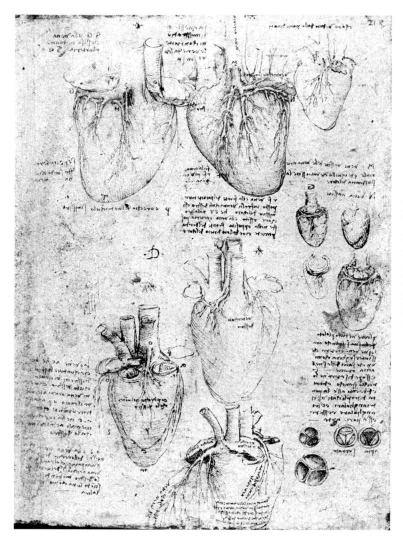

Figure 54. Sketches showing the distribution of the coronary vessels, viewed from different aspects. On the lower right-hand corner are the pulmonary and tricuspid valves seen from above. Leonardo's observations were based on dissection of the ox heart. (Reproduced by Gracious Permission of Her Majesty Queen Elizabeth II.)

Leonardo was the first to depict the appendix in his drawings of the gastrointestinal tract (Fig. 55). However, he conceived the diaphragm and muscles of the abdominal wall as the forces controlling movement of the gut. His lack of appreciation of peristalsis in the wall of the gastrointestinal tract was evident too in his description of the ureters and of the flow of urine from the kidney to the bladder. He viewed the ureter as a simple tube through which fluids flowed as a result of gravity and even demonstrated in a series of diagrams the effect of various bodily positions on the flow of urine from the kidneys to the bladder.

For an understanding of early development, he studied incubated chick eggs and the embryos of lower mammalian animals. The placenta and fetal membranes were drawn and described. Unlike his predecessors, Leonardo drew the uterus as containing a single cavity, complete with the ovaries and the uterine tubes, as well as their related blood vessels. One of his magnificent drawings shows a fully developed fetus within the bisected uterus with the umbilical cord attached to the placenta (Fig. 56) (Persaud, 1984).

Leonardo was also able to display the precise anatomical structure of the skull along with its separate components. In his depiction of the skull, entitled "*A Skull Sectioned*," Leonardo provided illustrations of sagittal and partial coronal planes of the skull. The ability to cut a skull in such a manner with the archaic tools then available to Leonardo contributes to the presentation. Leonardo's sketch provides an unprecedented amount of detail on the osteology of the skull together with the correct dental formula (Gerrits & Veening, 2013).

The first representation of the thyroid gland as an anatomical organ was made by Leonardo from his

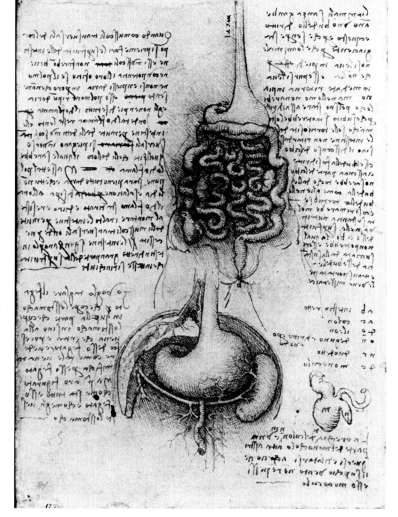

Figure 55. Drawing of the gastrointestinal tract, showing for the first time the vermiform appendix. Above, the esophagus, stomach and intestine are depicted; below is a shrunken liver, enlarged spleen and a prominent splenic vein in relation to the stomach. The vermiform appendix is represented in the diagram above and again more prominently on the lower right. (Reproduced by Gracious Permission of Her Majesty Queen Elizabeth II.)

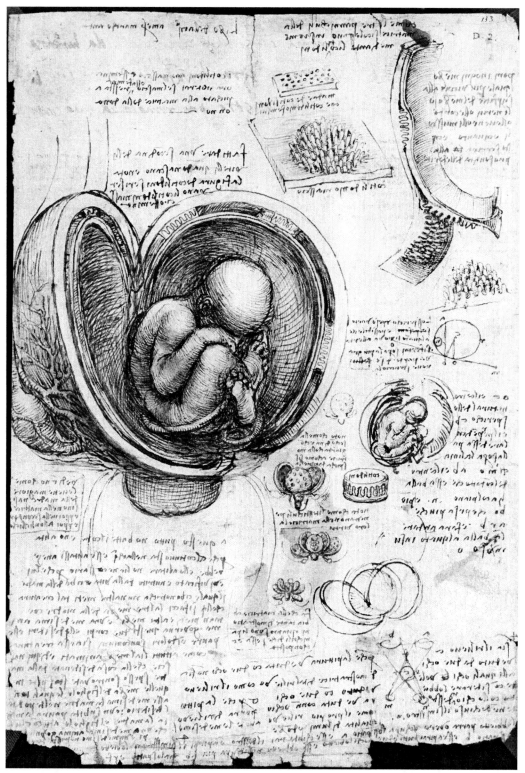

Figure 56. The pregnant uterus bisected to show the fetus in its natural position. The placenta, however, is that of a cow. The sketches on the right and below reveal the different layers of fetal membranes. In one of the small diagrams, the fetus can be seen through the transparent amnion following removal of the chorion and uterine wall. (Reproduced by Gracious Permission of Her Majesty Queen Elizabeth II.)

dissections. Authors have written that Leonardo commented about the thyroid, "*these glands are made to fill the space where the muscles are absent and hold the trachea away from the hyoid bone*" (Lydiatt & Bucher, 2011).

Anatomical Masterpieces

In more than 750 anatomical drawings of the muscular, skeletal, vascular, nervous, and urogenital systems, Leonardo produced a memorable body of work of unchallenged artistic beauty and scientific accuracy. He depicted parts of the human body from different perspectives. For Leonardo, it was not enough to illustrate but also to explain his drawings, which were often accompanied by questions and remarks relating to functions (Figs. 57 & 58) (Persaud1984). It is because of his insight to form and function that he has been described by Borowitz (1986) as being more than an artist or anatomist but as a "*philosopher-anatomist*" (Borowitz, 1986; Smith, 2006). So much of what Leonardo had discovered and recorded remained unknown for centuries to emerge again only in relatively recent years. Much of what he had accomplished was not to be surpassed for centuries.

Leonardo died on the 2nd of May, 1519. He bequeathed all his manuscripts and drawings to his beloved disciple and friend, Francesco Di Melzi, who kept them secured for almost half a century in admiration of his great master. Following the death of Melzi in 1570, the manuscripts were passed on to his nephew Orazio. The manuscripts eventually were disassembled and dispersed, passing through many hands, including Pompeo Leoni and Thomas Howard, Earl of Arundel. Several facsimile editions of Leonardo's anatomical drawings have been published but most of the originals are now in the Royal Library at

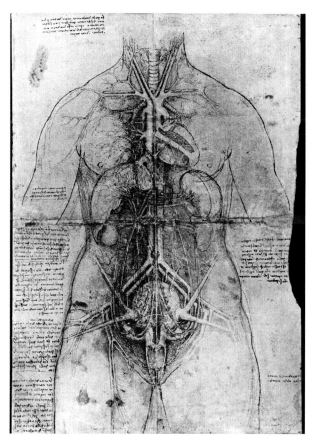

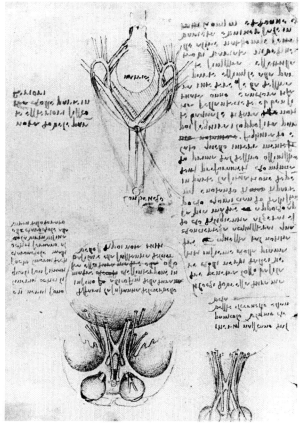

Figure 57. This elaborate and impressive drawing is undoubtedly an earlier work of Leonardo. There are many inaccuracies and the anatomical representations suggest that the drawing was based on the dissection of animals. An attempt has been made to consolidate the structure of the female anatomy with Galenic concepts of function. (Reproduced by Gracious Permission of Her Majesty Queen Elizabeth II.)

Figure 58. The sketches above show the optic chiasm and several cranial nerves. Below is a drawing of the uterus with paired vessels supplying it from either side. (Reproduced by Gracious Permission of Her Majesty Queen Elizabeth II.)

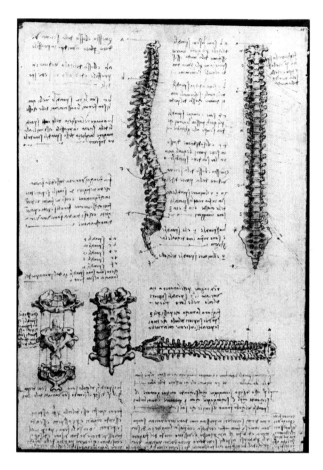

Figure 59. Leonardo's drawings of the articulated vertebral column (viewed from the lateral, anterior, and posterior aspects). The spinal curvatures are evident. (Reproduced by Gracious Permission of Her Majesty Queen Elizabeth II.)

Windsor Castle (Clarke, 1968; Keele, 1979; Saunders & O'Malley, 1982).

Despite some errors and misconceptions, Leonardo did make significant discoveries in anatomy. Not only did he describe certain anatomical structures for the first time, but also, he recognized the true curvature of the spinal column (Fig. 59) and the true position of the fetus in utero. It remains, however, undisputed that Leonardo's work is of epic importance. Unlike his contemporaries, it was Leonardo alone who pursued the study of the human body with such thoroughness that he quickly transcended the needs of the artist and drifted into the scientific pursuit of anatomy. His disciplined mind and scientific rectitude brought human anatomy into its first modernization phase. His knowledge of dissection and anatomy achieved through the use of simple tools and methods would only come to light hundreds of years later with technological advances and modern imaging procedures, such as MRI.

Chapter 11

EUROPEAN RENAISSANCE AND HUMAN ANATOMY

The period of European history known as the Renaissance inspired a new way of critical thinking, free of scholastic precepts and theories, which increased man's awareness of himself. It placed him at the epicenter of the universe. During the Renaissance, classical scholarly manuscripts that were scattered, especially works of Greek antiquity, were rediscovered, translated, and published. This revival of classical learning led to changing ideas of man and nature (Nutton, 1988; 1993a). It was believed that nature and experience were the real sources of truth. Man began to make direct observations from nature. Observational evidence had to be tested, and so began the experimental approach. Any personal experiences and facts derived from nature fostered intellectual development (Debus, 1978; Cameron, 1991; Hale, 1993).

When the Renaissance began and at what particular time it ended are not well defined even among classical scholars and science historians. It was undoubtedly a gradual process, starting in the fifteenth century and progressing to the eighteenth century. For many, the publication by Sir Francis Bacon (1561–1626) of his book called *The Proficience and Advancement of Learning* in 1605 marked the end of the Renaissance and the beginning of the modern scientific era. Bacon himself felt that his work heralded in a new scientific epoch (Hodges, 1985). The relationship of the Renaissance to the scientific revolution is of importance when one considers the scientific achievements, new discoveries and advances (Butterfield, 1965). Debus (1978) suggested that the time span from about the middle of the fifteenth through the middle of the seventeenth centuries because it covered the period during which major scientific and technological changes occurred.

The earlier date indicates the beginning of the new humanistic interests in the scientific and medical works of classical antiquity, and the later date marked the period that led to the general acceptance of the mechanistic principles of Descartes (1596–1650), Galileo (1564–1642), Borelli (1608–1679), Boyle (1627–1691), and Newton (1642–1727).

As remarked by Field and James (1993), "*the terms 'Renaissance' and 'Scientific Revolution' should be used with some degree of circumspection because these fluid terms have the advantage of not imposing a spurious unity on the products of a particular time or a particular place.*" For their purposes, they suggested that the period of the Scientific Revolution ended in the mid-eighteenth century even though the changes, which occurred in this period, were not all yet completed (Cohen, 1994).

The spiritual and technical emancipation of Europe was not achieved only through the revival of classical learning but also through the invention of printing with moveable type and the voyages of discovery to distant lands. Johannes Gutenberg's invention of printing with moveable type made it possible for books to be produced more easily and less expensively. The mechanical production of books made it possible for precious knowledge and new ideas to spread throughout Europe, as well as to other parts of the world. The use of the compass by navigators facilitated long journeys to foreign countries, which led to the flow of riches, and new ideas back to Europe. Even warfare between nations had evolved to include the use of gunpowder with devastating consequences. The tide of human cultural development during the Renaissance quickly gathered momentum and the changes that occurred affected every aspect of man's existence (Persaud, 1997).

The most prominent individuals who deeply influenced Renaissance science and medicine were Nicholas Copernicus (1473–1543), Andreas Vesalius, and Phillippus Aureolus Theophrastus Bombastus von Hohenheim, also known as Paracelsus, as well as Archimedes, Galen, and Ptolemy, through their writings, which were being critically re-examined (Debus, 1978). With respect to the study of the human body (French, 1993), the first great anatomist of the Renaissance was Jacobus Sylvius (1478–1555), who was one of the most influential teachers of anatomy (Fig. 60).

JACOBUS SYLVIUS

Jacobus Sylvius (1478–1555) devoted his early years to mathematics and classical philology and later became interested in anatomy after studying the Greek translations of Hippocrates and Galen. At a mature stage in his life, Sylvius decided to study medicine. He graduated from the University of Montpellier in 1529 with his doctorate degree when he was already 51 years old. Having obtained the M.D. degree, Sylvius returned to Paris, but the Faculty prevented him from teaching because he did not have a baccalaureate degree, which he subsequently obtained.

Sylvius collaborated with the noted surgeon Jean Tagault, who was also the Dean of the Faculty of Medicine. His knowledge of practical anatomy increased as a result of the dissections he carried out. Sylvius was industrious, eloquent, and influential. He was an ardent follower of Galen and interpreted his writings, who as the first professor in France to have taught anatomy through human dissection, refused to accept any deviation from the writings of Galen even when his personal observations were at variance. In his opinion, it was more likely that the human body had changed from Galen's time (Persaud, 1984).

We are indebted to Sylvius for introducing anatomical nomenclature, particularly with respect to the muscles. He described the technique of color injections, parts of the sphenoid bone, including the sphenoidal sinus, and the ventricles of the human brain.

Sylvius's *In Hippocrates et Galeni Physiologiae Partem Anatomicam Isagoge* was published in 1555, but apparently it was completed in 1542. The book was essentially a systematic account of anatomy based on the teachings of Galen and included some personal observations as a result of human dissections. Sylvius was greatly influenced by the work *Liber introductorius anatomiae* of Niccolo Massa (Fig. 61), which although dated 1559 on the title page showed 1536 on the colophon (Persaud, 1984).

Because of Sylvius's greed for money and his vindictive and arrogant personality, his contemporaries disliked him. The following epitaph was placed on his gravestone: *"Here lies Sylvius, who never did a thing without a fee. Even in death, he grieves that you read this inscription free."* It should be noted that the name Jacobus Sylvius is often confused with that of another far greater anatomist, Franciscus Sylvius (1614–1672). No less than 1,300 anatomical parts have been named after the latter (Haeger, 1989), and it was Franciscus Sylvius who was the first to discover the aqueduct or canal between the third and fourth ventricles of the brain (Anon, 1962).

Andreas Vesalius, the greatest anatomist of the era, and *"the most commanding figure in Europe in medicine between Galen and Harvey"* (Garrison, 1929), was a

IACOBVS SYLVIVS
Regius Lutetia Profesor

Figure 60. Jacobus Sylvius (1478–1555). (Wellcome Institute Library, London.) Sylvius was the first professor in France to have taught anatomy through human dissection. From him, Vesalius learned dissection techniques, as well as the teachings of Galen.

student of Sylvius. From Sylvius, Vesalius learned the technique of dissection, as well as the traditional exposure to Galen's teachings. In the preface of his book, Vesalius was effluent in his praise for Sylvius; although, after the publication of Vesalius's book, the two men fell apart, and Sylvius, filled with jealousy and rage, became vindictive. In his works, Vesalius referred to Sylvius as "*my erudite teacher Jacobus Sylvius*" and as "*the never to be sufficiently praised Jacobus Sylvius*" (Persaud, 1984). Later in his Letter on the China Root, published in 1546, Vesalius clearly minimized the influence of Sylvius on his work by stating that:

> I don't know if Sylvius paid any attention to the remark in my books...that is that I have worked without help of a teacher, since perhaps he believes that I learned anatomy from him, he who still maintains that Galen is always right. Sylvius, whom I shall respect as long as I live, always started the course by reading the books On the Use of Parts to us.... He brought nothing to the school except occasionally bits of dogs.... It happened one day that we showed him the valves of the orifice of the pulmonary artery and of the aorta, although he had informed us the day before that he could not find them. Since Sylvius omitted a chapter dealing with the vertebrae, as well as many others, during his so-called course which he was then giving for the 35th time and read nothing else anatomical except the books *On the Movement of Muscles* in which he everywhere agreed with Galen, it is not astonishing that I write that I have studied without the aid of a teacher. (O'Malley, 1964)

According to Nutton (1993b), during the sixteenth century, one can speak as far as the study of human anatomy is concerned of "*The triumph of observation over book-learning, and of the penetration into Northern Europe of techniques, ideas and discoveries first formulated in Bologna and, above all in Padua.*" Even earlier, the statutes of 1405 of the University of Medicine and Arts at Bologna stipulated regulations for the study of anatomy. Permission had to be obtained from the rector of the university before any "*anatomy*" could be carried out, after which other qualified persons were invited to attend. For the dissection of a male cadaver, no more than twenty persons were permitted; in the case of a female cadaver, thirty persons were allowed because these occurred infrequently. Third-year students were allowed to attend only one anatomical demonstration in any year. Furthermore, a student was permitted to observe no more than one female and two male anatomical demonstrations (Bullough, 1958).

Figure 61. Niccolo Massa (1485–1569). (Wellcome Institute Library, London.) Massa recognized the need for human dissection. He was the author of an important anatomical work, first published in 1536.

Students from many "*nations*" studied in Padua during the Renaissance. These included Germans, Hungarians, Poles, Bohemians, Netherlanders, and English. In the second half of the sixteenth century, 6,060 Germans graduated from the University of Padua. Two prominent men of English medicine, John Caius (1510–1573) and William Harvey (1578–1657), were educated at Padua during this period. Subsequently, both made important contributions to medicine, and to human anatomy in particular, as a result of their personal encounters there.

After leaving Cambridge, John Caius went to Padua where he lodged in the same house as Andreas Vesalius. With letters of introduction he travelled later through Italy, Germany, and France before returning to England. When he became president of the Royal College of Physicians in 1565, he obtained, after considerable lobbying of Parliament, permission for the bodies of executed felons to be dissected for anatomical studies, as in Padua (Hale, 1993). Later, William Harvey established the scientific method through the formulation of a hypothesis, experiments, and sound reasoning. His revolutionary theories that the blood circulates continuously throughout the body and that the heart functions as a pump evolved as a fundamental concept in medicine.

ANDREAS VESALIUS – ARCHITECT OF
THE NEW ANATOMY

Andreas Vesalius was born on December 31, 1514 in Brussels. Between 1528 and 1533, he studied at Louvain at the Pedagogium Castri and later at the Collegium Trilingue. Vesalius learned Greek, Latin, and Hebrew in the best humanistic tradition because at that time knowledge of the ancient languages was an essential prerequisite for all scholarly pursuits.

We learn from Vesalius himself that during his early years he became interested in the medieval works of

Figure 62. Title page of Vesalius's *De Humani Corporis Fabrica* 1543. (Courtesy of Dr. F.D. Bertalanffy.) A public dissection is in progress, conducted by Vesalius himself. This engaging woodcut also adorned his Epitome. The significance of the frontpiece and the identity of individuals depicted have been the subject of much scholarly debates (Speransky et al., 1983). In the second edition, published just two years later, a re-engraved title page, with many minor modifications, was used. Vesalius is now more prominently represented, the nude figure holding the left column is clothed, a goat has been added to the dog in the right foreground, etc.

scholarship and in anatomy. It was during this period also that he began to study anatomy in a very practical way by dissecting mice, moles, rats, dogs, and cats. He referred to one of his fellow students, Gisbertus Carbo, to whom he had given the first skeleton he had articulated from bones he obtained by robbing the gallows, in his work.

In 1533, Vesalius entered the distinguished but extremely conservative medical school in Paris. He was to remain here for the next three years, which in retrospect, proved to be a crucial period in the development of his personality and critical approach to the study of human anatomy. What is known about Vesalius during this period has been derived from his own treatise and from the writings of others (Ball, 1910).

By the time Vesalius had reached the University of Paris, the works of both Galen and Hippocrates were available, having been translated directly from the Greek into Latin. Only a few years earlier, the learning of medicine was largely from the translations and commentaries of medieval and Muslim scholars. As remarked by Saunders and O'Malley (1982),

> Physicians, seeing for the first time the works of Galen and Hippocrates stripped of their dross, believed that now they had captured the essence and spirit of the great classical authors and were at last about to enter a new Golden Age. As yet medicine had not developed a philosophy of progress but tended to look upon the present as inferior in knowledge and achievement to the past with the resultant enslavement to the literal word, and in particular to that of Galen.

Around this period, there were many eminent scholars who taught medicine at the University of Paris and attracted students from all parts of Europe. However, as far as anatomy was concerned, teaching continued to be based on the traditional enclave of Galenism and relatively few dissections were carried out (von Staden, 1992), notwithstanding an appeal made by the medical faculty in 1526 to Parliament for more material to dissect.

The publication of Vesalius's *De Humani Corporis Fabrica* in 1543 ushered a new era in the history of medicine and marked the beginning of modern anatomy (Fig. 62). The work emanated from an analytical mind who knew that in order to describe the true structure of the human body, he had to first dissect it. Vesalius's observations were not always in agreement with the established teachings of Galen, which were slavishly accepted for more than 12 centuries. Nevertheless, he had the courage to describe what he saw, for in science lies only truth.

In order to pursue his work, Vesalius was faced not only with scientific matters, but he also had to surmount sensitive moral and philosophical problems. During the Middle Ages, the art of healing was highly regarded, but anatomy, which was concerned with the dead body, occupied a very low position. The dissection of the dead and decaying human body stood in stark contrast to the classical humanism of the Renaissance. It was an era, too, when spiritual values were held supremely over material values.

Vesalius's work was profoundly original. He was not only "*engaged in a continuous struggle against philosophical authorities*" but he also "*denied and defied the scholastic tradition.*" Science and philosophy were to experience profound changes with human anatomy evolving into a pure empirical branch of science, having survived its intellectual crises (Cassirer, 1943).

Vesalius's masterpiece appeared in folio size (16" × 11") and consisted of 659 numbered pages with a magnificent woodcut title page and a portrait of the author by Jan Calcar (Fig. 63). There are 277 plates, including 22 full-page woodcuts, and smaller sizes. The book is divided into seven sections in accordance with Galen's approach to dissection. The preface of six pages is directed to Emperor Charles V and is dated August 1, 1542. This is followed by Vesalius's letter of August 24 from Venice to the publisher Oporinus. The work is set in Roman type and there are many instances of typographical errors, as well as inconsistencies in the numbering of pages (Cushing, 1943; Rollins, 1943; Vesalius, 1543).

The book is copiously illustrated (Figs. 64–66). Of some interest is the continuous panorama of a background landscape that is revealed when plates 1, 2, 6, 5, 4, and 3 are viewed adjacent to each other. This area has been identified as the Euganean Hills, just a few miles out of Padua.

The first book of 40 chapters deals with the skeletal system. In the second book, muscles and ligaments are covered in 62 chapters. Fifteen chapters are devoted to the circulatory system in the third book, and in the fourth book, 17 chapters are devoted to cerebral and peripheral nerves. In the fifth and sixth books, the abdominal and thoracic organs are described, respectively. The seventh book of 19 chapters is devoted to the brain and organs of special senses. The final chapter deals with vivisection and animal experiments. It includes an illustration of a pig strapped to a

Figure 63. The only known genuine portrait of Andreas Vesalius (from his *De Humani Corporis Fabrica*), a work of Jan Calcar done when Vesalius was 28 years old. (Neil John MacLean Health Sciences Library, University of Manitoba.)

board in preparation for experiments.

Repeating Galen's dramatic experiments on the recurrent laryngeal nerve, Vesalius outlined how he strapped the animal to the board,

so that the head may be immobilized but at the same time the animal may breathe and cry fully. But before the animal is bound in this way, I usually rehearse to the spectators...those things which are to be revealed by the present cut, so that long drawn out explanation may not be needed during the dissection, to disturb it. And soon I make the long incision in the neck, which divides the skin and the muscles lying below it to the arteria aspera (trachea) being careful that the cut should not slip to one side and injure veins worth seeing. Then I seize the arteria aspera in my hands and separate it from the surrounding muscles with my forefingers

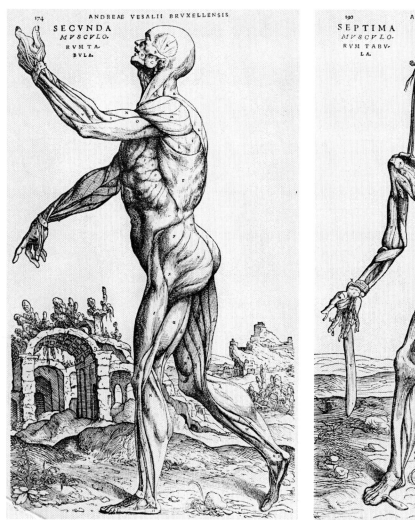
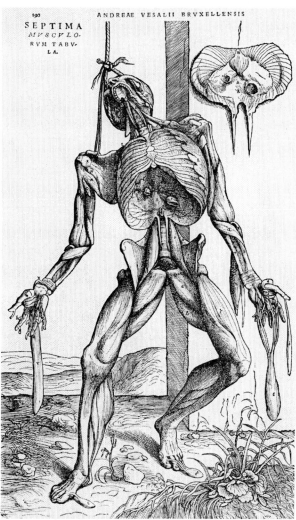

Figure 64. Superficial muscles, viewed from the side. (Courtesy of the Neil John Maclean Health Sciences Library, University of Manitoba.)

Figure 65. Another view of the muscles together with the diaphragm. For dissection, the cadaver was suspended in an upright position by a rope. Many muscles are readily recognizable and the openings for the inferior vena cava, esophagus and aorta in the diaphragm are precisely located. (Courtesy of the Neil John Maclean Health Sciences Library, University of Manitoba.)

only I look for the soporales arteries (the carotid arteries) by its side and close to them the sixth pair of cranial nerves. Then I see the recurrent nerves close besides the trachea, and these I sometimes tie and sometimes cut, at first on one side and then on the other, so that the voice fails and disappears wholly when the nerves are affected on both sides. However, if I release the ligatures the sound returns....

Vesalius's book, "*De Humani corporis fabrica*" was the most modern and advanced for that time because of the new discoveries he made about the body. Within

a few years of its publication, the book or plates were plagiarized by well-known anatomists, including Thomas Geminus (1510–1562), Juan di Hamusco (ca. 1525–ca. 1588), Volcher Coiter (1534–1576), and others. Vesalius had presented a fairly accurate concept of the human body. He boldly corrected Galenic teachings and also rejected its tradition and authoritarianism. However, the physiological considerations of Vesalius were no more advanced than those of Galen (Roth, 1892; 1895; Lambert, 1936a;b; Cushing, 1943; Persaud, 1984).

With respect to the magnificent illustrations in

Vesalius's book, it will perhaps never be resolved whether they were all drawn by Jan Calcar (Feyfer, 1933). The work has been attributed to one or more of several artists, including Jan Calcar, Titian and others of his school, Vesalius, and even Leonardo da Vinci. Strong evidence in favor of Jan Calcar for many of the illustrations, if not all, was presented by Cushing (1943). He referred to the bareheaded young man in the front row with the open sketchbook having the initials S.C. on the cover.

For reasons discussed by the Vesalian scholars Saunders and O'Malley (1982), the conclusion was reached that the drawings for both the *Fabrica* and its accompanying *Epitome* were done by students of Titian. Not only Jan Calcar but also several other artists worked under the supervision of Vesalius. It is undeniable that Vesalius himself must have drawn some of the sketches.

Just 29 years old and at the height of his fame, and only a few months after the publication of his great book, Vesalius succumbed to the harsh criticisms and abuse of his contemporaries by relinquishing his chair at the University of Padua and burning all of his valuable unpublished papers. Of this act during December 1543, he was to later write in his *China Root* epistle that "as to my notes, which had grown into a huge volume, they were all destroyed by me; and on the same day there similarly perished the whole of my paraphrase of the ten books of Rhazes to King Almansor,

> ...I was on the point of leaving Italy and going to Court; those physicians of whom you know had given the Emperor and the nobles a most unfavorable report of my books and of all that is published nowadays for the promotion of study; I therefore burnt all those works mentioned, thinking at the same time it would be an easy matter to abstain from writing for the future. I have since repented more than once for my impatience, and regretted that I did not take the advice of the friends who were then with me. (Cushing, 1943)

Vesalius became court physician to both Philip II and Charles V (*medicos familiaribus ordinaribus*) as a result of which he also participated in military expeditions.

For reasons unknown, Vesalius undertook a pilgrimage in 1564 to Jerusalem and on his return journey was shipwrecked on the small Greek island of Zakynthos where he died on October 15, 1564 (Wharton, 1902; Cullen, 1918a;b Lambert, 1936a;b). There

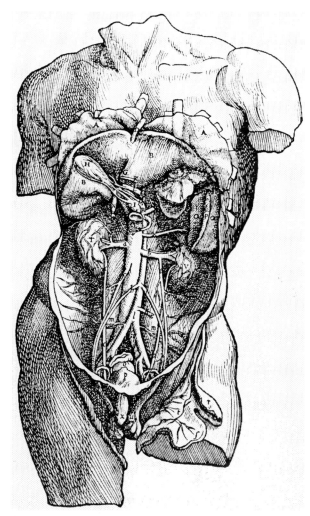

Figure 66. Dissection of the abdominal cavity. The intestine has been removed to reveal structures on the posterior abdominal wall. Observe the position of the kidneys, distribution of blood vessels and nerves, and the opened left scrotum. (Courtesy of the Neil John Maclean Health Sciences Library, University of Manitoba.)

are many varied accounts of his death and none might be entirely accurate. Certain, however, is the scientific legacy he nurtured with obsessive devotion, which made him "*The first man of modern science.*" His pristinely preserved masterpiece, the *De Humani Corporis Fabrica*, is universally recognized as "*one of the greatest treasures of Western civilization and culture,*" which "*established with startling suddenness the beginning of modern observational science and research*" (Saunders & O'Malley, 1982).

Chapter 12

THE IMMEDIATE SUCCESSORS OF VESALIUS AT PADUA

The Immediate successors of Vesalius at Padua were Matteo Realdo Colombo and Gabriele Falloppio (Fallopius). Between 1540 and 1544 Vesalius was frequently away from Padua during which Colombo apparently gave demonstrations in anatomy, and in 1544 when Vesalius left Padua, Colombo was appointed as his successor. Colombo was a student of Vesalius, and Fallopius was so greatly impressed by the work and teaching of Vesalius that he also considered himself to be one of Vesalius's pupils. The relationship of these two men to Vesalius took different courses.

REALDO COLOMBO

Colombo was born in 1516 in a small village not far from Cremona. At 14 years of age, he began an apprenticeship in an apothecary and already during this period he was dissecting animals. From Cremona, Colombo moved to Milan and later in 1535 to Venice where he learned surgery with Joannes Antonio Plato who encouraged him to attend the medical lectures at the University of Padua.

Colombo attended all of Vesalius's lectures and anatomical demonstrations. It would seem that he also assisted Vesalius because Vesalius mentioned Colombo in his *De Fabrica* with respect to the hyoid bone and that Colombo was once his student. In the famous front piece of Vesalius's book, Colombo is depicted on the right-hand corner turning to an assistant who was holding a goat, probably to be dissected for comparative studies. It is not certain whether he obtained a medical degree, but when Vesalius was away from Padua, Colombo carried on with the anatomical demonstrations.

For whatever reasons, Colombo left Padua for Pisa in 1544 and in 1549 he was invited to become professor in the *Sapienza* in Rome. Here, he was friendly with Michelangelo Buonarrtoti (1475–1564), the great sculptor and painter. Colombo also performed an autopsy on the body of Ignatius Loyola, founder of the Jesuit order of missionaries, who died in 1556. Colombo referred to the autopsy in his book *De re Anatomica*, published in 1559.

Even though it has been suggested that Colombo might have been envious of Vesalius (Koch, 1972), it cannot be denied that he was in his own right a brilliant anatomist who made many discoveries and wrote an influential anatomical treatise that was widely used (Persaud, 1997). Colombo's book, *De re Anatomica* (Figs. 67 & 68) was an entirely different work from that of Vesalius's *De Fabrica*, in that the text was concise and well written with clear and accurate descriptions in most instances.

Colombo's treatise was not illustrated like that of Vesalius's *De Fabrica*. Nonetheless, it was the preferred work among anatomists of that era. For example, Helkiah Crooke (1576–1648) and Thomas Winston (1575–1655) both quoted from the works of Colombo. Crooke had compiled a well-illustrated and comprehensive textbook of anatomy from the work of others, which he acknowledged. It was extremely popular with six editions printed. According to Russell (1987), Crooke produced *"the first comprehensive textbook in the Vesalian manner to be compiled by a British author in the new century."* Winston was also the author of an anatomical work published in 1659 with a second edition in 1664.

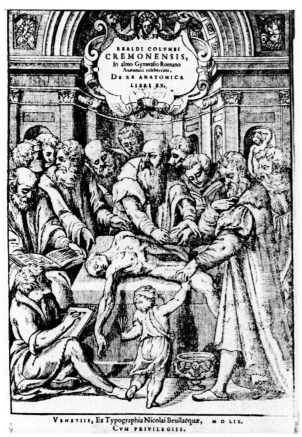

Figure 67. Dissection scene from the frontispiece of Realdo Colombo's book *De re Anatomica,* published in Venice, 1559. Colombo is standing behind the table, holding a scalpel in his right hand and looking in the direction of the person reading from an anatomical treatise. The abdominal wall of the cadaver on the table is partly dissected, revealing some of the viscera. The artist sitting on the floor is observing the dissection and making sketches from it. (Courtesy of Bethesda, MD: U.S. National Library of Medicine, National Institutes of Health, Health & Human Services.)

Figure 68. Title page of Realdo Colombo's treatise *De re Anatomica.* The work contained no drawings, but many new findings were described, including the pulmonary circulation.

A painting of the English physician John Banister (1533–1610), presenting an anatomical demonstration to the Barber-Surgeons in 1581 (Fig. 69), shows an opened copy of Colombo's treatise. Also, Colombo's work on the pulmonary circulation was described in Banister's book *The Historie of Man...* (Fig. 70) (1578). Finally, on pages 12 and 41, Harvey, in his famous book, *Exercitatio Anatomica De Motu Cordis Et Sanguinis In Animalibus*, published in 1628, referred to Colombo's theories describing the flow of blood from the right ventricle, through the lungs, to the left side of the heart. In Colombo's book, *De re Anatomica*, published posthumously in 1559, one finds a clear description of the pulmonary circulation, which he based on clinical observations, animal experiments, and dissections.

Colombo's book contained detailed and accurate descriptions of the human skeleton, especially of the cranial bones. For the first time, one finds an excellent account of the inner ear, the tympanic membrane, the ear ossicles, the two fenestrae, and their relationship to the vestibule and the cochlea. Contrary to Galenic teaching, Colombo placed the lens of the eye behind the iris and not at the center of the eyeball.

Colombo corrected many misconceptions about the muscles as found in Vesalius's work. He described

Figure 69. John Banister (1533–1610) demonstrating the abdominal viscera and skeleton to his colleagues at the Barber-Surgeons' Hall, London, in 1581. (Reproduced with the permission of Glasgow University Library, Department of Special Collections.) The book is that of Realdo Colombo (Paris Edition, 1572), which Banister used for the lecture. Just above the treatise is the following text: *Anatomia scientiae dux et aditumque ad dei agnitionem praebet*, which implies that a knowledge of the structure of the human body leads one to a better understanding of God. John Banister was a well-known physician in Nottingham but was given a restricted license to practice in London on account of his poor performance before the examiners.

the atrial and ventricular cavities, and also the larynx. Even though Colombo obviously exaggerated the number of dissections he carried out and of the large number (600,000) of skulls he had examined, he will be remembered as one of the most outstanding anatomists to have succeeded Vesalius. In his footsteps followed an anatomist of no lesser stature, the great Gabriele Falloppio (Fallopius).

GABRIELE FALLOPPIO

Gabriele Falloppio (Fallopius) succeeded Colombo at Padua. According to Koch (1972), Falloppio did not participate in any of Vesalius's anatomical demonstrations, and it is doubtful whether they had ever met, yet Falloppio greatly admired and respected Vesalius from his works. Vesalius regarded Colombo as one of his students, which he mentioned in his book, but which Colombo never acknowledged. It was Falloppio who publicly proclaimed himself a student of Vesalius. Like his adopted master, Falloppio excelled through his teachings, book, and anatomical discoveries.

Falloppio was born in 1523 in Modena where he first studied to become a priest. In 1542, he was named a Canon, but three years later, he left Modena for Ferrera to study medicine (Fig. 71). It is recorded that he dissected in the surgical school in Modena on December 13 and 14, 1544, the body of a criminal who was hanged. By then Vesalius's magnificent work was already published and it must have deeply impressed Falloppio. At the age of 26 years, Falloppio received an invitation to occupy the chair in Pisa, which was vacated by Colombo who left for a similar position in Rome. In his anatomical studies, Falloppio tried to understand the function of structures he saw and it was this rational approach to anatomy that led to so many discoveries.

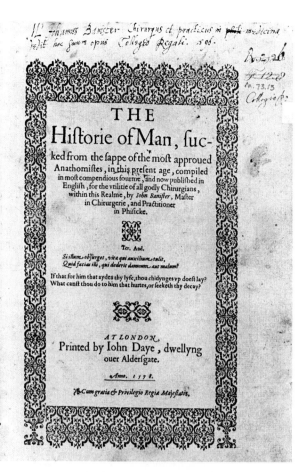

Figure 70 (*Right*). Title page of John Banister's *The Historie of Man, sucked from the sappe of the most approued Anathomistes*, published in London, 1578. Banister's book was compiled from the "*latest anatomical works by Continental authors*" and it has been described as "*the most pretentious anatomical work by a British author*" published during that period (Russell, 1987). Colombo's observations on the pulmonary circulation are described in this work: "*But very wyde they wander, sayth Collumbus. For the bloud through the arteriall Veyne is carried to the lunges, whence, being attenuated, it is caried by the veniall arterie into the left Ventricle of the hart together with ayre which no man before his tyme noted, or at least have left ertant.*" (King's College Library, Cambridge.)

Figure 71 (*Below*). Gabriel Falloppius explaining one of his discoveries to the Cardinal Duke of Ferrara. (Oil painting by Francis James Barraud. Wellcome Library, London.)

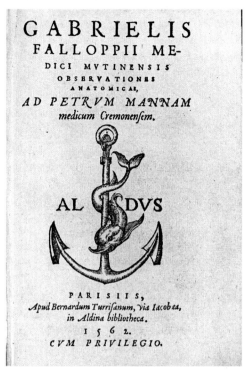

Figure 72. Title page of Gabriele Fallopius's *Observationes Anatomicae*, 1562. This edition was issued in Paris a year later, after it was first published in Venice.

From Pisa, Falloppio was invited to the chair of anatomy, surgery, and botany in Padua where he carried out most of his scientific studies. In 1561, Falloppio's work, entitled *Observationes Anatomicae*, was published in Venice (Fig. 72). The book contains no illustrations. In it, Falloppio described the tubes attached to the uterus which are associated with his name (the uterine or Fallopian tubes), the ovaries, hymen, clitoris, round ligament of the uterus, the canal which carries the facial nerve and bears his name (canal of Fallopius), the cochlea and labyrinth of the inner ear, the paranasal sinuses, and the chorda tympani nerve. The inner ear was never before so clearly described. Falloppio described the stylomastoid foramen and the ethmoid bone and air cells, as well as cranial nerves IV, V, VII, and IX. Falloppio even corrected some misconceptions Vesalius had relating to the muscles, and he introduced the names vagina and placenta for these structures. The discovery of the stapes, one of the ear ossicles, was also claimed by Falloppio, as well as by the anatomists Bartolomeo Eustachius and Giovanni Filippo Ingrassia. The uterine tubes, which are generally named after Falloppio, were already described by Herophilus (c. 300 B.C.), Galen (c. 130–201) and Rufus of Ephesus (c. A.D. 100). Falloppio, however, was the first to recognize their functions. He described them as the "*trumpets of the uterus.*" Falloppio died on October 9, 1562, at the early age of 39 years. He left us an impressive body of new knowledge and a worthy successor.

HIERONYMUS FABRICIUS

Falloppio was followed by Hieronymus Fabricius (1537–1619), who was one of William Harvey's teachers at Padua. He made important discoveries relating to the valves of the veins, which he described in his book *De Venarum Ostiolis*, published in 1603. It is believed that this work was the final clue that led Harvey to the experimental demonstration of the circulation of blood (Persaud, 1997). Hieronymus Fabricius (Fig. 73) was born in Aquapendente near Orvieto in 1537. He first studied philosophy but then changed to medicine. Falloppio was one of his teachers at Padua and he eventually succeeded him as professor of anatomy in 1562. He held this position for close to 50 years from 1562 to 1613.

Similar to his predecessors, Fabricius dissected tirelessly in order to reveal accurately the structures of the human body. Moreover, he carried out diverse studies on fetuses and in animals. His scientific observations were in the areas of comparative anatomy, embryology, and functional anatomy. Fabricius's treatise *Totius Animalis Fabrkiae Theatrum*, dealing with comparative anatomy, was never published. His *Opera Omnia* was such an impressive work that it was republished in 1738 by Bernard Albinus of Leiden.

Fabricius's book *On the Development of Eggs of Birds* was a monumental work and an important landmark in the history of embryology. His remarkable observations, which were very well illustrated, are a clear testimony of the many meticulous studies Fabricius carried out in an era when the microscope did not exist. His second embryological treatise *On the Formed Fetus*, published in 1604, contained the best drawings of the human fetus, membranes, and of the placenta up to that time in history (Adelmann, 1942).

William Harvey stated in the preface of his book *De Generatione* (1662) that it was Fabricius's work that inspired him to carry out embryological studies.

Fabricius also wrote a book on the eye and vision, with also very fine illustrations of the internal structure of the eyeball. He was the first to realize the true structure of the lens, but he erroneously believed, like other Galenists, that the image was within the lens itself (Persaud, 1997).

It was the discovery of the membranous folds in the veins, which Fabricius named valves, which brought him the greatest fame. Several anatomists, including Jean Fernel, Jacobus Sylvius, Andreas Vesalius, Canano (1515–1579) and Amatus, Lusitenus (1511–1568), had described these structures. Fabricius was the first to trace them throughout the body and he eventually demonstrated their presence in all the veins of the extremities.

Fabricius's book *De Venarum Ostiolis* (On the Valves of the Veins) was a small tract of 23 folio pages and 8 plates with figures of the heart and venous systems. The valves are shown in drawings for the first time (Figs. 74 & 75). Fabricius mentioned in the book that he had already observed these "*Ostiolee*" in 1574 and that they were arranged at intervals in the veins in such a way that their openings were directed towards the heart. This important observation was the crucial element missing for the demonstration of the circulation of blood. Fabricius did not recognize the true functions of the venous valves. He believed that nature had created these structures to prevent too much blood from flowing outwards from the heart to the tissues where it will only accumulate, causing over distension of the veins, swelling and further damage. According to Fabricius, the valves slowed the outward flow of blood in the vessels, or else the extremities would become so filled that there would not be enough blood to supply the vital organs (Fig. 76).

Fabricius died in 1619 at the age of 86 years. In his long professional career he made many important discoveries, which added to our store of anatomical knowledge and set the scene for new ideas (Fig. 77). He was highly respected in his lifetime and he received many honors. Fabricius was Galileo's personal

Figure 73. Fabricius ab Aquapendente (1537–1619). (From *Opera omnia anatomica et physiologica*, Patavii, 1625.) (From Wegner, R.N.: *Das Anatomenbildnis. Seine Entwicklung im Zusammenhang mit der anatomischen Abbildung.* Basel, Benno Schwalbe and Co., Verlag, 1939. Courtesy of Institut fur Anatomie, Ernst-Moritz-Arndt-Universitat, Greifswald, Germany.)

physician and it is believed that William Harvey lived in his house at one time.

GIULIO CASSERIUS

Giulio Casserius succeeded Fabricius in Padua. Casserius was born in 1561 in Piacenza in Italy and studied medicine at Padua where he came under the influence of Fabricius. Casserius served as an assistant to Fabricius, who became his mentor, and during this period he developed expertise in the anatomical techniques that were prevalent then. Both Fabricius and Casserius were teachers of William Harvey. Casserius private anatomy classes were so popular that Fabricius became jealous and forbid him from lecturing. He was able to devote more time to dissection and anatomical studies.

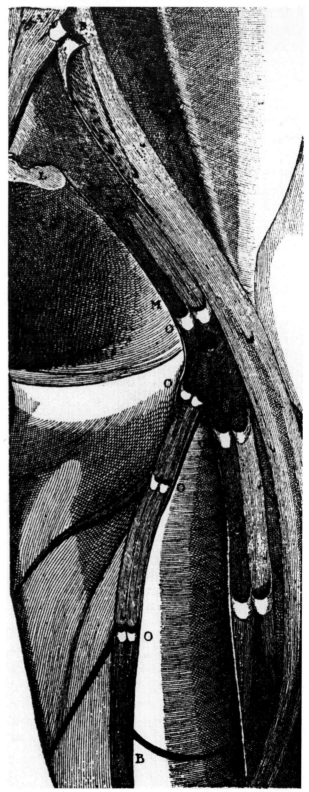

Figure 74 (*Above left*). Cover page of the "De *Venarum Ostiolis*." Patavii: Ex Typographia L. Pasquati, 1603 (https://library.missouri.edu/exhibits/anatomy/pre17th. htm).

Figure 75 (*Right*). Illustration of the venous valves in the iliac, femoral, and saphenous veins. From Girolamo Fabrici d'Aquapendente's *De Venarum Ostiolis*, Padua, 1603. (Courtesy of Institut fur Anatomie, Ernst-Moritz-Arndt-Universitat, Greifswald, Germany.)

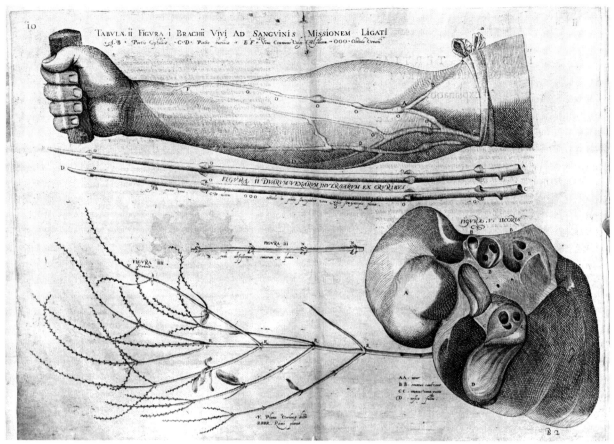

Figure 76. Plate II of *De Venarum Ostiolis* illustrates the presence and structure of the veins in the human forearm using a ligature. The veins of the arm are also dissected out and opened lengthwise to show the anatomical structure of the valves. Fabricius, ab Aquapendente. *De Venarum Ostiolis*. Patavii : Ex Typographia L. Pasquati, 1603 (https://library.missouri.edu/exhibits/anatomy/pre17th.htm).

Casserius was a skilled anatomist with attention to accuracy and details. He was the first to describe the glands found in the eyelid, some 60 years before Heinrich Meibom, professor of medicine, history, and poetry, at Helmstadt in Germany, rediscovered them.

Casserius is best remembered for his work on comparative anatomy of the ear and the larynx. His treatise *De Vocis Auditusque Organis* was published in Venice in 1601. It contained many magnificent illustrations (Figs. 78 & 79) made from copper engravings. These included both the ear and larynx. The English surgeon, John Browne (1642–1702), completely copied the illustrations, as well as the title, from Casserius's *Tabulae Anatomicae* (Figs. 80 & 81) for his own book *Compleat treatise of the muscles* (1681). Moreover, the text of John Browne's book was copied almost verbatim from William Mollins's (1617–1691) book "*MYEKOTOMIA: or, the anatomical administration of all the muscles of the humane body, as they arise in dissection.*

As also an analitical table reducing each muscle to his use and part (1648)." Several editions of this plagiarized work by Browne were printed in English, and it was also translated into Latin and German (Russell, 1959a; 1959b).

Casserius died in 1616, and he was succeeded by Adrian van der Spiegel (Spieghelius) (1578–1625) at the University of Padua. Van der Spiegel included in his book, "*De Formato Foetu,*" published in 1627, the most complete set of Casserius tables with oversized stylistic figures showing the pregnant uterus, placenta, and fetus. With recent advances, the fetus can be visualized at all stages before birth. Real images of the fetuses are now readily available and stand in contrast to Casserius unrealistic engravings, which were done in a different era. Yet, Casserius's figures have prevailed over the centuries and continue to influence current perception of the fetus-in-utero (Heilemann, 2011).

Figure 77 (*Left*). Title page of the surgical treatise *Opera Chirurgica*...(1723), by Hieronymus Fabricius of Aquapendente (1533–1619). (Courtesy of the Neil John Maclean Health Sciences Library, University of Manitoba.) Fabricius was Professor of Anatomy and Surgery at Padua, and he was responsible for the construction in 1594 of the famous anatomical theater.

Figure 78 (*Right*). Engraved plate showing the penis and dissection of the perineal region, taken from the work of Giulio Casserius (1561–1616). (From a later edition of the *Tabulae anatomicae*, Frankfurt, 1632; Courtesy of the Neil John Maclean Health Sciences Library, University of Manitoba.)

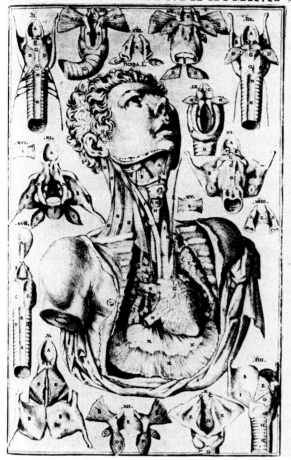

Figure 79 (*Left*). Dissection of the neck and thorax showing the heart, lungs, cervical muscles, and other related structures. This plate is taken from the anatomical treatise (*De vocis auditusque…*) of Giulio Casserius (1561–1616), which was published in 1601. (Courtesy of Bethesda, MD: U.S. National Library of Medicine, National Institutes of Health, Health & Human Services.)

Figure 80 (*Above right*). Frontispiece of Tabulae anatomicae by Giulio Casserius. This work was first published posthumously by Daniel Bucretius in 1627, Venice. The above edition was published in Frankfurt in 1632. (Courtesy of the Neil John Maclean Health Sciences Library, University of Manitoba.)

Figure 81. Tabvlae anatomicae LXXIIXX: omnes nouae nec ante hac visae. Frankfurt : Impensis & Coelo Matthaei Meriani, 1632. (http://doc.med.yale.edu/historical/about/founders/donors.html)

Chapter 13

FROM PADUA TO LEIDEN

About the same time as the flame of anatomical revival was flickering in Padua, another spirited center was emerging in the Netherlands. The University of Leiden (Leyden) with its famous medical school became the most influential force in the training of physicians, as well as in anatomical studies. Like the role of Padua in the formal education of physicians from many countries, so was it at Leiden.

FOREIGN MEDICAL QUALIFICATIONS

Robb-Smith (1974) noted that during the first part of the seventeenth century, physicians who had their foreign medical degrees incorporated in Cambridge came *"about equally from Padua and Leiden, while the number of Cambridge men going to other continental universities was negligible."* The situation at Oxford was similar, where according to Sinclair (1974) *"many graduates in Arts went abroad for their medical studies because these were so poor in Oxford...."* Escaping the still prevailing influence of Galenism, prospective medical students sought *"the newer learning that was so firmly established in Holland, Italy, and France."* By the second half of the seventeenth century, the majority went to Leiden, which *"became the leading continental university, for Sylvius and Lucas Schact were excellent teachers...."* (Robb-Smith, 1974).

Whereas some individuals went to Padua, Leiden, and other continental universities on account of the quality of teaching, others saw it as *"an economical and easy way to obtain an M.D. for incorporation in one's English alma mater, thus opening the doors of the Royal College*

of Physicians" (Robb-Smith, 1974). Many physicians and surgeons in Britain, who later became famous, also followed this circuitous path before they could have established themselves professionally. At the same time, the records indicate that, in most cases, the students did not spend much time in the foreign medical schools so as to receive a proper and formal training. It was often a short period, determined by finances and as a matter of convenience.

The purpose of the Medical Act of 1511 was to stop *"ignorant persons"* and others who *"partly use sorcery and witchcraft"* from practicing medicine by granting this privilege only to graduates of the Universities of Oxford and Cambridge or those who were licensed by the church. Essentially, no one was allowed to practice medicine unless he had graduated from one of these two universities. The College of Physicians, following its establishment in 1518, was also granted the right to examine and approve candidates suitable for medical practice.

UNIVERSITY OF LEIDEN

When William I (1533–1584) of Nassau, Prince of Orange, founded the University of Leiden in 1575, there was no indication from its unremarkable origin that it would emerge as the most famous and influential institution in all of Europe (Fig. 82). After the crushing defeat of the Spaniards in the protracted and bloody religious war that was fought against the invaders, the Dutch monarch rewarded the brave people of Leiden for their courage and the considerable support they had given him. Offered the alternative options of either a university in the city or the freedom from paying any taxes for many years, the

Figure 82. The University of Leiden in the seventeenth century. (From Meurs, J.: Athenae Batavae, Leyden, Elsivir, 1625; with permission, from Major 1954.)

surviving citizens opted for the university (Lindeboom, 1974). It was the intellectual sanctuary of many great and creative minds, including the humanist Desiderius Erasmus (c. 1466–1536), the jurist and theologian Hugo Grotius (1583–1645), the renowned physician Hermann Boerhaave (1668–1738), the physicist Christiaan Huygens (1629–1693), Antoni van Leeuwenhoek (1632–1723), and Alexander Monro, Primus (1697–1767).

In 1578, John James (Jacobus Jaimes), an Englishman, was the first student to be matriculated in the Faculty of Medicine. He graduated in 1581, the second person to do so. Students attending the university came from many countries; half were foreigners, most British. Between 1709 and 1738, 1,919 students were enrolled in the Faculty of Medicine: 353 from England, 244 from Scotland, 122 from Ireland, and several from the British Colonies in America (Beukers, 1992).

HERMAN BOERHAAVE

The reputation of the medical school at the University of Leiden was largely due to Hermann Boerhaave (1668–1738), the great reformer in medical education, as well as "*the most successful clinician and medical teacher of the century*" (Ackerknecht, 1982). In 1701 Boerhaave (Fig. 83) was appointed to the faculty where he had a small clinic with 12 beds for the care of his patients. He introduced the clinical bedside teaching still practiced worldwide today. Boerhaave emphasized the importance of following the progress of the patient *right up* to recovery or, in the cases of deaths, at autopsy where he would demonstrate the relationships between clinical symptoms and pathological lesions.

Boerhaave was not only one of the most gifted in clinical medicine, but he was an accomplished chemist, botanist, and anatomist (Burton, 1743; Lindeboom,

1968). Boerhaave republished in Latin (*Biblia Naturae*), at his own expense, the book of his countryman, Jan Swammerdam, *Bybel der Natuure* (Bible of Nature), which first appeared in 1669. This scientifically important work described instruments for testing muscle function, which are still in current use. Boerhaave's own books (Lindeboom, 1959), especially the *Institutiones medicae* and his *Aphorisms* (Fig. 84), were the standard works in Europe at that time. He kept an ornately bound volume, which he declared contained the secrets of medicine. When he died, the book was anxiously opened by his colleagues and pupils to reveal only blank pages but with one on which was written, "*Keep the head cool, the feet warm, and the bowels open.*"

Although a fervent clinician, "*Boerhaave was thoroughly convinced of the importance of anatomy*" (Lindeboom,

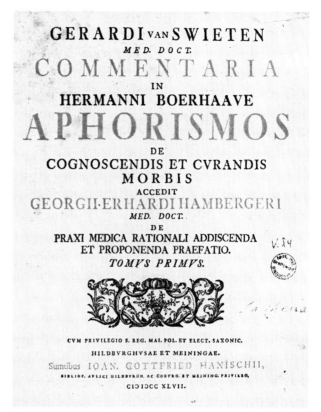

Figure 83. Hermann Boerhaave (1668–1788). The greatest physician of the eighteenth century. He was professor of medicine and botany at Leiden University, but he also greatly influenced the teaching of anatomy in Leiden. (From *An Account of the Life and Writings of Herman Boerhaave*, 1743. Courtesy of the Neil John Maclean Health Sciences Library, University of Manitoba.)

Figure 84. The Aphorisms of Herman Boerhaave (*Aphorismi de cognoscendis et curandis morbis.*) This is the title page of Boerhaave's work with the commentaries of Gerard van Swieten (1700–1772). This edition appeared in Germany in 1747. (Courtesy of Institut fur Anatomie, Ernst-Moritz-Arndt-Universitat, Greifswald, Germany.) Boerhaave published the *Aphorismi*...in 1709, a year after his *Institutiones*. Following the style of Hippocrates, the Aphorisms were a "*collection of short pronouncements concerning the diagnosis and therapy of various diseases.*" It was an extremely popular work and "*numerous editions spread the fame of the author*" (Lindeboom, 1959).

1959). He even found time to assist the anatomists Anton Nuck (1650–1692), W. Rau (1668–1719), and Albinus with anatomical demonstrations and he personally carried out autopsies on patients who died. Boerhaave wrote an anatomical treatise on the glands (*Epistola ad Ruyschium, pro Sententia Malpighiana de Glandalis*, 1722) in support of Malpighi who "*taught that the glands are separate and anatomically bounded formations in the body, with the purpose of secretion. Ruysch...cherished the idea that the glands were not anatomically distinct formations but only vascular differentiations as he did not see them in his praeparates....Ruysch answered in an epistle, afterwards printed together with that of Boerhaave*" (Lindeboom, 1959). At that time, Ruysch

was 84-years-old but still active with his anatomical studies.

Boerhaave studied diseases as a natural outcome and phenomena of life; this pragmatic approach made him the most famous physician of his era in Europe. Moreover, the immense contribution of Boerhaave to modern medicine was possible because "*with all his accomplishments Boerhaave was better able than any man of his time to achieve something like a medical synthesis, to bring all the sciences to the service of the patient....To him the debt of British medicine, and through it of British wellbeing, is quite incalculable. Through his pupils he is the real founder of the Edinburgh Medical School, and through it of the best medical teaching in the English-speaking countries of the world*" (Singer and Underwood, 1962; Bower, 1817–1831). Boerhaave's influence on the rise of medicine in other European countries was no less. For example, through many of his pupils, including von Haller, the foundations were established for the reform of medical education in Germany.

Anton Nuck

The Dutch anatomist Anton Nuck (1650–1692) first described a small peritoneal pouch (the processus vaginalis), which extends into the labum majus (labia majora). Remnants of this canal, if not obliterated, can enlarge and become cystic, develop into an indirect inguinal hernia, or lead to a hydrocele. In clinical practice, eponyms that are commonly associated with Anton Nuck's name include *"Nuck's canal,"* *"Nuck's cyst,"* and *"Nuck's hydrocele."* Anton Nuck did his medical studies in Harderwyjk and Leiden, where he obtained the Doctor of Medicine degree in 1677 for studies on diabetes. In 1683, Nuck was appointed to teach anatomy at the Collegium Anatomicum Chirurgicum in Den Haag, and in 1687 he returned to Leiden to become Professor of Medicine and Anatomy.

Nuck was highly regarded as a teacher of practical anatomy and he attracted large number of students. Moreover, he reported important discoveries relating to the secretory glands, lymphatics, the eye, and teeth. Nuck developed the technique of injecting marker substances into the salivary gland (sialography), investigated the function of the aqueous humor of the eye, and designed special dental forceps suited for extracting teeth according to their shape (Figs. 85 & 86).

Pieter Pauw

In 1597, long before Boerhaave was born, an anatomical theater had been built in Leiden by Pieter (Petrus) Pauw (1564–1617). It was the first anatomical theater to be established in the Netherlands following the practice that was then prevalent (Richter, 1937) in Italy and Germany (Fig. 87). Pauw (Paaw) was born in Amsterdam. He studied at first in Leiden and Paris; later in Rostock from where he obtained his doctorate degree. Pauw also spent some time in Padua during which he was influenced by both Giulio Casserius (1561–1616) and Fabricius ab Aquapendente (1537–1619). Pauw wrote an anatomical

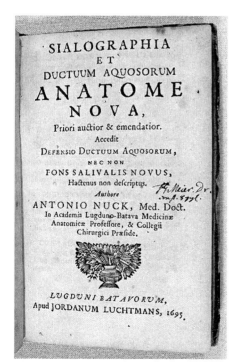
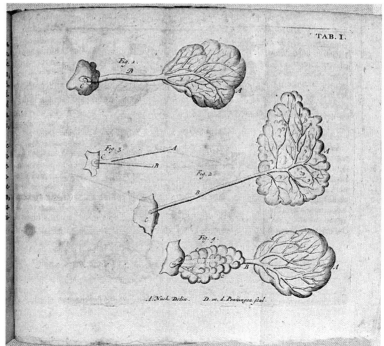

Figures 85–86. This text by Anton Nuck, recently acquired by The Lilly Library. Nuck used a technique similar to that used by Frederik Ruysch, injecting colored markers into salivary glands via not only their ducts, but also through their blood vessels. He called this method for studying the glands "sialographia" (sialography). Anton Nuck. Sialographia et Ductuum Aquosorum Anatome Nova…Leiden: Apud Jordanum Luchtmans, 1695. http://www.indiana.edu/~liblilly/anatomia/viscera/nuck.html.

Figure 87. The Anatomical Theater in Leiden. It was designed after the anatomical theater in Padua and built in 1597 under the direction of Pieter Pauw. This engraving (1610) by Swanenburg shows the interior of the theater. A corpse lies on the centrally placed table, and one of the two men inspecting it lifts the covering to show the dissection that is in progress. The abdominal viscera are revealed. An assortment of surgical instruments can be seen in the background. There are several well-dressed men and women in the room, conversing or looking at the odd collection of skeletons and mounted specimens. (Courtesy of Leiden University.) According to Kennedy and Coakley (1992), the engraving is "heavily didactic and moralistic.... The classic *memento mori* or 'reminder of death' iconography, a putto seated on the cupboard leans on a skull and holds an hour-glass...the human skeletons, each with a flagged motto reminding us of the transience of life on earth.... A woman is shown the flayed skin of a man...and another holds a mirror, the symbol of vanity."

treatise entitled *Primitiae anatomicae de humani corporis ossium*, published in Leiden in 1615 (Fig. 88), and he was responsible for a new edition of Vesalius's Epitome (1616). The Swiss anatomist Albrecht von Haller (1708–1777) held Pauw in high regard and described him as *Hominus melancholicus, valuit acri ingenio et memoria valida.* The engraving of Andreas Stog depicting Pieter Pauw dissecting in the *Theatrum anatomicum* in Leiden (c. 1617) remains an enduring tribute to a gifted anatomist (Schumacher & Wischhusen, 1970). Pieter Pauw was the first professor of anatomy in Leiden, but the great reputation of its University as a center for the teaching of anatomy was ushered with the arrival of the outstanding German anatomist Albinus (Persaud, 1997).

Bernhard Siegfried Albinus

Bernhard Siegfried Albinus (1697–1770) was attracted to Leiden from the University in Frankfurt/Oder by Boerhaave and the fame of Leiden (Fig. 89). A similar invitation made to him by Prussia for anatomy teaching was evidently less attractive. Albinus was appointed to the chair of anatomy and surgery at the age of 24 years. (Punt, 1983.)

Boerhaave and Albinus excelled in their fields and as teachers. Both their reputations, but especially that of Boerhaave in clinical teaching, lured prospective medical students to Leiden, which at that time was the leading University in continental Europe. A survey of Cambridge physicians showed that between 25% and

Figure 88. Frontispiece to Pieter Pauw's *Primitae anatomicae*, 1615. It shows a dissection scene in the Anatomical Theater in Leiden. Several men are depicted around a table on which a cadaver is in the process of being dissected. Observe the complete skeleton, prominently placed in the background, and two dogs sitting on the floor in front of the dissecting table. Fragrant herbs were scattered on the floor to alleviate the strong stench from the cadaver. Pieter Pauw (1564–1617) is shown teaching and demonstrating the parts of the body. (Courtesy of Bethesda, MD: U.S. National Library of Medicine, National Institutes of Health, Health & Human Services.)

Figure 89. Bernhard Siegfried Albinus (1697–1770). Engraving by Johann Jacob Haid from a painting by Carel Isaack de Moor. (From Wegner, R.N.: *Das Anatomenbildnis. Seine Entwicklung im Zusammenhang mit der anatomischen Abbildung.* Basel, Benno Schwalbe and Co., Verlag, 1939. Courtesy of Institut fur Anatomie, Ernst-Moritz-Arndt-Universitat, Greifswald, Germany.)

50% had spent some at Leiden during the eighteenth century (Rob-Smith, 1974; Rock, 1969).

In 1725, Boerhaave and Albinus published in two volumes the collected works of Vesalius (*Opera omnia anatomica & chirurgica cura*). The anatomical nomenclature used in this work was the most modern of the era. Albinus also edited the works of Harvey (1736), Fabricius (1737), and Eustachius (1744) (Fig. 90).

Albinus's own treatise on the skeleton and muscles of the human body "*Tabulae sceleti et musculorum corporis*" was published in 1747. This was a work of high scholarship, which introduced new standards in depicting the human body. The illustrations, comprising 28 plates, were large, more accurate than before, and beautifully presented. Albinus monitored the work of Jan Wandelaer, the artist, at every stage during the preparation of the illustrations. By combining his observations from several cadavers, he depicted the different parts of the body in a more accurate manner. Perspective was an important feature of the illustrations and total quality was added by presenting the anatomical drawings against decorative and artistic backgrounds. This fascinating book was extremely popular and it was translated from Latin into other languages, including English, in several editions (Persaud, 1997).

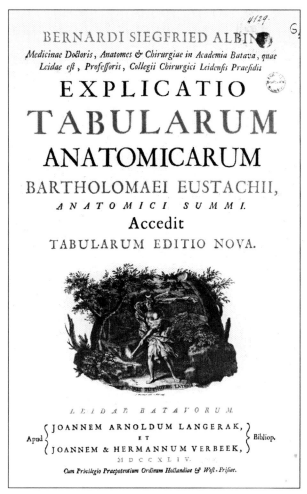

Figure 90. Albinus's edition of Eustachius's anatomical plates, published in 1744. (Courtesy of Institut fur Anatomie, Ernst-Moritz-Arndt-Universitat, Greifswald, Germany.)

Albinus described for the first time several muscles, and he was the first person to classify the muscles in a systematic manner. Many of the names he assigned to the muscles are still part of our anatomical nomenclature. Albinus also made immense contributions to the fields of microscopy and physiology (Punt, 1983).

Frederik Ruysch

At least one other eminent Dutch anatomist should be mentioned because the technique he developed would have profoundly influenced not only Albinus but anatomists everywhere else. Frederik Ruysch (1638–1731) was born in Amsterdam where he was appointed Professor of Anatomy in 1665 (Fig. 91). Ruysch developed the techniques of injecting blood and lymphatic vessels to a level of perfection that allowed one to demonstrate precisely the tiniest of vessels in almost every organ and part of the body. This was one of the greatest achievements in anatomical research, which pushed the limits of anatomy as an observational science to a level that was then unimaginable.

It would be difficult to give an account of all of Ruysch's studies, but it is safe to say that there is hardly a region of the body he did not meticulously investigate, including the blood vessels with his injection technique. Only a few will be mentioned here: the valves of the lymphatic vessels; pulmonary blood vessels, including the smallest branches; vessels of the skin and bones, as well as the growth and union of the epiphyses; the spleen; reproductive structures, including the uterus; blood vessels in the teeth; and the vascular arrangement in the brain and of the meninges (Persaud, 1997).

Figure 91. Frederik Ruysch (1638–1731). Engraving by Jan Wandelaar. (From Wegner, R.N.: *Das Anatomenbildnis. Seine Entwicklung im Zusammenhang mit der anatomischen Abbildung.* Basel, Benno Schwalbe and Co., Verlag, 1939. Courtesy of Institut fur Anatomie, Ernst-Moritz Arndt-Universitat, Greifswald, Germany.)

Because of the special embalming and injection techniques Ruysch had perfected, he was able to prepare and preserve a large number of diverse and unique anatomical and biological specimens (Figs. 92 & 93), which attracted visitors from all of Europe. His museum was described "*as a perfect necropolis, all the inhabitants of which were asleep and ready to speak as soon as they were awakened.*" Ruysch's museum was bought by Peter the Great of Russia for 30,000 florins. The czar had visited the museum "*and was so struck with the lifelike countenance of a child that had been preserved by Ruysch for many years after it had ceased to breathe, that he actually laid aside both imperial dignity and Muscovite severity, and kissed the apparently animated features*" (Lonsdale 1870). At that time, Ruysch's museum was by far the largest and most impressive museum in Europe. Prior to 1664, anatomical museums contained only dried specimens, until the Irish physicist and chemist, Robert Boyle (1627–1691), showed that animal tissues could be preserved in alcohol. Well-known museums of note during this period were those of the Danish physician and scholar, Olaus Worm, Albertus Seba and others. (For an account of the early methods that were used to prepare anatomical specimens, and of some of the early medical museums, reference should be made to Paul, 1937; Edwards and Edwards, 1959).

Figure 92. Ruysch's artistic display of anatomical specimens, from the *Thesaurus Anatomicus*, Part 1, 1701. (From Edwards, J.J. and Edwards, M.J.: Medical Museum Technology, Oxford University Press, 1959; by permission of Oxford University Press.)

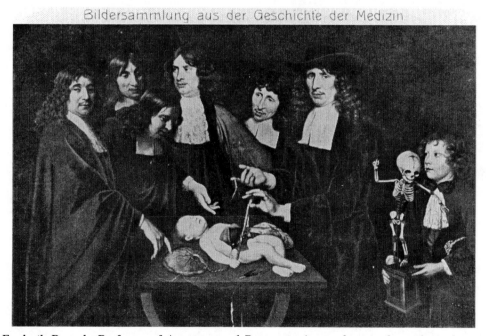

Figure 93. Frederik Ruysch, Professor of Anatomy and Botany in Amsterdam, is shown here demonstrating the placenta and umbilical cord in a stillborn infant. (National Library of Medicine, Bethesda, MD.) see; Nuland, 1992 (for details regarding the iconological and anatomical aspects of this work).

ANATOMICAL MODELS

Because of the scarcity of human cadavers for anatomical studies, anatomists were forced to explore other alternatives, such as anatomical models made in wood and in wax. At first, these were crude representations of the body parts, which, with increasing accuracy and technical skill of the anatomists, evolved into realistic scientific masterpieces, as well as works of art. In Bologna and Florence, schools were established where students could learn the art of anatomical wax modeling. By the early eighteenth century, most medical museums had a substantial number of these lifesized anatomically flayed models of men or écorché in wax on display. The most famous pioneers of this art form were the Italian anatomists and sculptors (Pucetti, 1997; Riva et al., 2010), including Gaetano Giuilo Zumbo (1656–1701), Ercole Lelli (1702–1766), Anna Morandi Manzolini (1716–1774), Felice Fontana (1730–1803), and Clemente Susini (1754–1814). Their creative legacy has endured up to the present time.

Many of these historic wax sculptures can still be seen as part of a collection at the *Instituto di Anatomia*, University of Bologna (Rosito et al., 2004; Messbarger, 2010), in the *Museo la Specola* in Florence (Lanza et al., 1979; Bani, 1986; Bonuzzi & Ruggeri, 1988; Riva et al., 2010), and at other institutions (Hilloowala & Goldstein, 1982; Hilloowala, 1984).

Gaetano Giulio Zumbo

The Sicilian artist and master wax sculptor Gaetano Giulio (Zummo) Zumbo (1656–1701) is mostly remembered for his fascinating series of wax models that bordered on the macabre. Zumbo's extraordinary work in wax included the distorted faces of plague victims (Fig. 94), floating eyeballs and intestines, decomposing syphilitic victims, the tortured agony of dying, and decaying cadavers with maggots feasting on them. Reflecting on Zumbo's morbid work, a

Figure 94. This is a wax statue made by Zumbo depicting the effects of the plague and is found in La Specola, Florence.

memento mori, one is harshly reminded of man's mortality and that death itself is inevitable.

Zumbo came from a poor aristocratic family, and his early education was for a life in the clergy. Recognizing his artistic skills, he decided to become an artist and taught himself how to paint and to make wax models. In 1691, he presented for the first time his artistic work in Bologna. This led to an employment with Grand Duke Cosimo III de' Medici during which Zumbo created many of his terrifying spectacles. Five years later, he left the Duke for other opportunities.

In 1695, Zumbo then went to Genoa where he met the French surgeon Guillaume Desnoues who was impressed with Zumbo's artistic skill. Both men began a collaboration to produce anatomical models for teaching. Desnoues carried out the dissection; and Zumbo, using bright colored wax, made exact models of the dissected specimens. As a result, they both discovered the technique for preparing polychromatic wax specimens, a major technical innovation in anatomy. The collaboration between Zumbo and Desnoues eventually soured and came to an end. In 1700, Zumbo left for Paris where he obtained royal privilege granting him the rights to manufacture anatomical models in colored wax. Zumbo died in 1701 and was buried in the cemetery of Saint Sulpice, the second largest church in Paris (Puccetti, 1997; Puccetti et al., 1995; Orlandini & Paternostro, 2010).

Anna Morandi Manzolini

The spectacular wax models of human anatomy that were made by the gifted sculptress Anna Morandi Manzolini (1716–1774) (Fig. 95) are now housed in the Department of Anatomy and in the Museo di Palazzo Pogi at the University of Bologna. These anatomical models are astonishingly realistic, mostly lifesize, and convey considerable reliable information. Because of the scarcity of cadavers for dissection, models of various parts of the body were made for anatomical studies. Anna Manzolini began her professional life as an artist. She had no formal training in anatomy but her anatomical creations have attracted dignitaries, scholars and tourists for centuries.

Anna Manzolini married Giovanni Manzolini, a professor of anatomy at the University of Bologna. At home, she assisted her husband in making wax models of anatomical specimens from actual dissections. Following the untimely death of her husband, and left with two children, she was happy when the university offered her a position to teach anatomy and

Figure 95. A portrait of Morandi Manzolini (Wellcome Library, UK).

also to continue making anatomical models for the museum. Manzolini acquired a thorough knowledge of human anatomy, which she gained from the dissections she carried out.

With extraordinary skill as a sculptress, Manzolini made in wax a large number of realistic models that revealed all parts of the body. Anatomically correct and not lacking in detail, these wax sculptures brought Anna Manzolini recognition and rewards. The University of Bologna made her a professor, and she was invited by many universities to give lectures. Anna Manzolini received commissions from Catherine the Great of Russia and the King of Poland. In 1769, Joseph II, Emperor of Austria, saw the collection and was greatly impressed. Returning to Vienna, he established an anatomical museum because of its educational value for the public. Anna Manzolini is remembered as a highly accomplished and extraordinary woman for the era during which she lived (Rosito et al., 2004; Messbarger, 2010).

THE ANATOMY LESSON OF DOCTOR NICOLAAS TULP

The first public dissection of a human cadaver in Amsterdam occurred during the middle of the sixteenth century. The surgeons established an independent guild following separation from the Guild of Clog and Skatemakers. Human dissections were regularly carried out by the Amsterdam Company of Surgeons as a result of permission granted in 1555 by King Phillip II of Spain, Count of Holland.

Rembrandt and the Amsterdam Surgeons

The Dutch artist, Rembrandt van Rijn, through his famous painting of the Dutch physician and anatomist, Doctor Tulp and his colleagues, has left us one of the best known dissecting scenes ever painted (Fig. 96). The 26-year-old Rembrandt started this painting in 1628 and it was completed in 1632. The painting shows Doctor Nicolaas Tulp (1593–1674) giving an anatomical demonstration to seven members of the Guild of Surgeons in Amsterdam (Heckscher, 1958; Schupbach, 1982; Scherer, 1990).

It was fashionable for surgeons during this period to commission such portraits, perhaps with the desire to immortalize themselves. According to Nuland (1992),

Aert Pietersz (in 1603), Michiel and Pieter van Miereveld (1617), Thomas de Keyser (in 1619), and Nicolaes Eliasz (in 1625) had already produced such works, but their paintings were essentially group portraits of expressionless surgeons positioned around a corpse, staring out at the viewer. (Nuland, 1992)

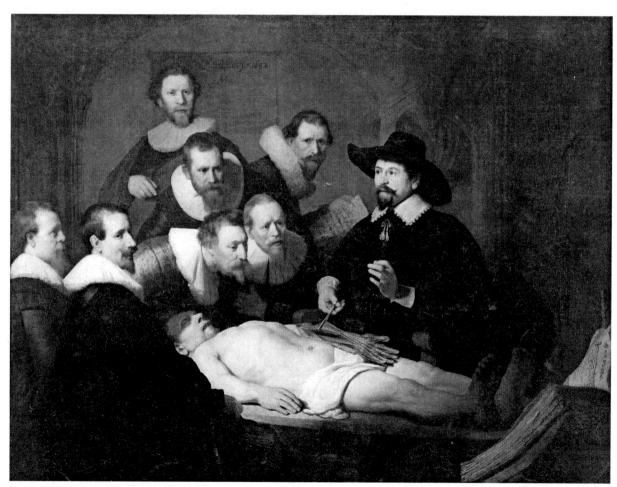

Figure 96. The Anatomy Lesson of Dr. Tulp (1632) by Rembrandt van Rijn. (Courtesy of Stichting Johan Maurits van Nassau, Mauritshuis, Netherlands.)

Nicolaas Tulp

Doctor Nicolaas Tulp (1593–1674) (Fig. 97) was a follower of Vesalius, and interestingly Rembrandt's painting of him and his colleagues depicted characteristics that are present in Stephen van Calcar's painting of Vesalius on the title page of his *Fabrica*. Moreover, Doctor Tulp is shown demonstrating the flexor muscles and tendons of the left forearm and hand to an alert group of surgeons.

Doctor Tulp was a hardworking and dedicated doctor who had a lucrative practice. He was a man of considerable influence in the city of Amsterdam and was elected the mayor for a period of four terms, In 1628, the city council and judges appointed Tulp to the position of Demonstrator in Anatomy (*Praelector anatomiae*) at the Guild of Surgeons. During winter, he carried out dissections on the available bodies of hanged criminal. Prominent citizens, council members, and surgeons were invited to observe and learn from the dissection. Those attending the dissection had to pay a fee.

The ileocecal valve, as well as its function, was first described by Nicholas Tulp. He also recognized lymph vessels (lacteals) in the human, which had already been discovered in the dog by the Italian physician and anatomist Gasparo Aselli (1582–1626) in 1622. Tulp described close to 200 illnesses in his book "*Observationeum medicarum libri tres*" (1641), largely based on autopsies he himself carried out. In addition, his "*Pharmacopoe Amstelodamensis*" (1636) was the first attempt in standardizing pharmaceutical practice in the Netherlands.

Figure 97. Nicolaas Tulp. Terracotta bust by Artus Quellinus (Wellcome Library, UK).

Chapter 14

THE ERA OF WILLIAM HARVEY

By the beginning of the seventeenth century, scientific observations were made and recorded based on actual experiments. For example, Santorio (1561–1636), also known as Sanctorius, who had studied medicine at Padua, published in 1614 his *De Medicina Statica*. This famous work reported the results of experiments that Sanctorius carried out, sometimes on himself, over a period of 30 years. Sanctorius's studies, carried out with respect to respiration and fluid loss through the skin under different conditions, were the beginnings of modern metabolic research. He also published a monograph in 1612 describing a thermometer, which was used for studying diseases. The thermometer was constructed on the principle of Galileo's thermometer. Using the length of a piece of string from a thimble pendulum, Sanctorius was able to obtain information on the rate of the pulse.

A NEW ERA IN SCIENTIFIC MEDICINE

The greatest achievement in experimental medicine during this period was the demonstration by William Harvey (1578–1657) of the circulation of the blood. It remains as one of the most outstanding achievements in science and medicine (Snellen, 1984). The publication in 1628 of his book *Exercitatio Anatomica De Motu Cordis et Sanguinis in Animalibus (De Motu Cordis)*, on the movements of the heart and the circulation of blood in animals, represents a milestone in the history of medicine (Fig. 98). Revolutionary in itself, this discovery had a wider impact on the future development of medicine.

There is good evidence that William Harvey knew of the existence of the circulation of the blood, which was contrary to the prevailing Galenic concepts (Hall, 1960), but only after many years of thoughtful deliberations, dissections, clinical observations, and experiments did he feel confident enough to publish his findings. Thus began a new era in medical research where a hypothesis is formulated and tested through experiments.

Only after critically evaluating the findings against existing knowledge is one permitted to make a judgment or a conclusion. Such was Harvey's approach to medical problems, which introduced for the first time a model that is common today for experimental work in scientific medicine. Similar to Andreas Vesalius, William Harvey was a skillful anatomist as well as a practicing physician. From the time of the Renaissance to the seventeenth century, Vesalius and Harvey stand as landmarks in an era during which medicine progressively moved from an observational to an experimental science.

WILLIAM HARVEY

William Harvey (Fig. 99) is one of the best known names in the history of medicine. So much has already been written about him throughout the world, and his scientific achievements are constantly being evaluated in the light of newer discoveries (French, 1994; Andrews, 1981). Our account on the life and achievements of William Harvey can only be an extract from the varied and voluminous literature already existing on him (Moore, 1918; Wyatt, 1924; Bayon, 1938a; b; 1939a; b; Chauvois, 1957; Keynes,

1966; Bylebyl, 1979; McKenna, 1987a; b), in addition to Harvey's publications which were available to us in original or as facsimiles (Willis, 1847; Harvey, 1628 (1953; 1957); O'Malley et al., 1961).

Early Years and Cambridge

William Harvey was born on April 1, 1578, in Folkstone, which is a small town in the south of England. From a family of seven sons and two daughters he was the eldest. William's father, Thomas Harvey, was a prominent citizen of the town of which he was at one time the mayor. His first wife died of childbirth in 1576, and after a few years, he was remarried to a Joan Halke (or Hawke). From the first marriage there was a daughter and William was the eldest from the second marriage. At ten years of age, William was sent to King's School, Canterbury, where he remained for

Figure 99. William Harvey (1578–1657). (Department of Anatomy, University of Manitoba.)

Figure 98. Title page of William Harvey's book *Exercitatio Anatomica De Motu Cordis et Sanguinis in Animalibus* (The anatomical exercises concerning the motion of the heart and blood in living creatures), Frankfurt, 1628. (Courtesy of Bethesda, MD: U.S. National Library of Medicine, National Institutes of Health, Health & Human Services.)

the next five years. In May 1593, William matriculated at Gonville and Caius College where he probably received a good education in the humanities, including Latin and Greek, and in the natural sciences. The college had a long history and a good reputation. It was founded in 1348 as Gonville Hall and, in 1558, became Gonville and Caius College in recognition of the substantial endowments and contributions made by Doctor John Caius (Latin form of Keys, Kayer, Kays).

John Caius and Thomas Linacre

Doctor Caius was born in Norwich. He attended Cambridge University before going to Italy in 1529 where he studied medicine at the University of Padua (Fig. 100). Caius published two medical works based on the teachings of Galen (Nutton, 1985) and his former teacher Montanus. Returning to England in 1544, Caius was appointed personal physician to the king. When he died in 1573 his remains were placed in a tomb in the north wall of the college's chapel (Rolleston, 1932; Langdon-Brown, 1946). It is of interest that prior to Caius, another Englishman, Thomas Linacre (1460–1524) (Fig. 101), had also pursued, much

Figure 100. John Caius. Line engraving by C. Ammon, junior, 1650. (Wellcome Institute Library, London.)

earlier in 1485, medical studies in Italy. Linacre was a personal physician to King Henry VIII and founded a chair in medicine at both Oxford and Cambridge. He also took part in the founding of the College of Physicians of which he was the president in 1518.

It is possible that the examples of John Caius and Thomas Linacre, two distinguished physicians of the period (Moore, 1918), had influenced William Harvey in going to Padua for his medical training. Through the influence of John Caius, Gonville College in Cambridge received for dissection two bodies of criminals hanged at Cambridge. Even though the king granted this special privilege, there is no record as to whether anyone carried out human dissection in the college or elsewhere in Cambridge (Macalister, 1891; Rolleston, 1932; Pratt, 1981).

William Harvey in Padua

William Harvey received his B.A. degree in 1597 and left for Italy to continue his studies. Apparently

he arrived in Padua in 1598 where he spent four years pursuing medical studies (Wyatt, 1924; Keynes, 1966). It could have been the prestige of the university or the reputation of some of its teachers that attracted William Harvey to Padua (Randall, 1940; Underwood, 1963). In any case, students from Protestant countries felt that they were more welcomed here and less persecuted. They felt at ease and safe compared to other universities in the north of Italy. The arms of the Inquisition seemed far away and a spirit of tolerance and liberalism prevailed in Padua.

The great reputation of the University of Padua (Randall, 1940; Underwood, 1963), the prevailing spirit of liberalism, the distinguished professors, and the intellectual atmosphere all combined to attract students from many countries in Europe (Premuda, 1974, 1986; Rossetti, 1983). At that time, Padua was a

Figure 101. Thomas Linacre. Ink drawing with wash. (Wellcome Institute Library, London.)

part of the republic of Venice, which was an influential cultural center and had strong commercial and trading institutions (Fig. 102).

According to the records of the University of Padua, during Harvey's time, there were large contingents of students from many countries who were attracted to Padua so much so that they were grouped into different "*nations.*" Indeed, the students made up as much as a third of the population of the city. As early as 1222, a center of scholastic learning or "*stadium*" was started in Padua by a group of professors and students following their exodus from Bologna. It was one of the earliest medieval "*studium,*" following the traditions of the medieval centers of learning, to be established in Europe. Because the professors and students lodged in an inn known as the Bo (Ox), this name has been traditionally associated with the university over the ages.

In the traditional "*magistral*" university, the professors who represented the governing body made decisions. In contrast, the students themselves in the studium generale made all major decisions. At that time, this was not an uncommon practice in many European countries. The Paduan Studium was represented by a universitas juristarum (law) and a universitas artistarum (divinity, medicine, philosophy). The former had more students and more money in contrast to its sister institution. The College of Doctors of Medicine was established in about 1250 (Rossetti, 1983). Students were admitted to both branches of the institution according to nationalities. There were representatives (conciliarii), who, together with the two rectors, formed the executive of the university. The executive of the university, the congregation of students and the executive formed the governing body (Rossetti, 1983).

Shortly after his arrival in Padua, Harvey was elected a member of the governing council in 1600 to represent the English "*nation*" of students. There are several reminders of Harvey's residence in Padua, including entries in the university records of his election as a conciliarius of the English nation of students, as well as two stone tablets ("*stemmata*") in the courtyard of the Bo. The tablet shows an overshield with indented border and two heads, one above and the other below. In the center is a left hand holding a lighted candle around which two green serpents are shown at the sides. Apparently, this stemma of "*William Harvey, the Englishman*" was discovered only in 1893 after a long search.

The Physician William Harvey

Having received his doctoral degree on April 25, 1602, Harvey left Padua for England sometime before

Figure 102. The facade of the University of Padua. This early seventeenth century lithograph shows several groups of students in front of the building. Inscribed over the portal is the following: *Gymnasium omnium disciplinarum* (University of all disciplines). (Courtesy Department of Anatomy, University of Manitoba.)

March 24, 1603. It seems that during the intervening period he had remained in Padua. In 1604, Harvey settled in London, where he was to spend a greater part of his professional life. Harvey applied to the University of Cambridge for a license to practice medicine. This was granted on the basis of which he then applied for admission to the College of Physicians. On October 5, 1604, Harvey was admitted a candidate of the College of Physicians and subsequently, on May 16, 1607, he was elected a fellow of the college. Two years later, on October 14, 1609, Harvey was appointed by the board of governors at St. Bartholomew's Hospital in London as a physician. On November 24, 1604, Harvey married Elizabeth Browne, the daughter of a London physician. There were no children.

Lumleian Lecturer and Anatomists

The College of Physicians was founded in 1518. King Henry VIII had conferred on the college the sole right to grant license for the practice of medicine in London, a privilege that was eventually extended throughout the entire country. The distinguished English physician Thomas Linacre was the first president of the college and he made available for meetings of the college two rooms in his home, which he bequeathed to the college at his death in 1524. The college established the Lumleian lectures in anatomy and surgery in 1582 from funds that were given to the college by Lord Lumley, a benefactor of the college. Associated with these lectures was the election of a Lecturer from among the fellows of the college. This prestigious and highly regarded position was held by Harvey from 1615 until he resigned in 1656 (Moore, 1918).

The lecturer of anatomy during the sixteenth and seventeenth centuries was a person of considerable importance and highly regarded. Very few bodies were available for dissection, and on account of putrefaction, dissection had to be completed within a few days following a specific sequence. The abdomen and thorax were first dissected, and these were followed by the cranial cavity and the limbs last. In England, the Act of Parliament in 1540, which united the Corporation of Barber-Surgeons and the Surgeons, authorized the masters of the new guild to receive each year the bodies of four "*felons*" condemned and put to death "*for their further and better knowledge, instruction, insight, learning and experience in the science and faculty of surgery.*" In 1565, the president of the College of Physicians received a similar privilege from Queen Elizabeth I to receive up to four bodies for dissection from persons executed in London or within a vicinity of 16 miles.

Harvey's first lecture, given in April 1616, was on the anatomy of the viscera. The handwritten notes (*Prelectiones anatomie universalid*) for this first course of lectures are now in the British Museum (Sloane 230). The manuscript consists of close to one hundred foolscap pages with notes on both sides. The notes are rough, concise, full of abbreviations, alterations, and annotations, both in Latin and common English. The manuscript was part of the library of Sir Hans Sloane, and according to the terms of his will, it passed on to the nation in 1754 (Whitteridge, 1964).

The notes Harvey prepared were for his Lumleian lectures and "*judging by their scope, by the research into the literature which is revealed in the citations, and by the personal observations briefly referred to, Harvey must have spent much time in compiling them*" (O'Malley et al., 1961). Moreover, it is now believed that the notes were written over a period of years, starting prior to April 1616 and with several noticeable amendments in the years following. In addition to the historical importance of these lecture notes, they also provide some insight on the life, professional activities, interests, and characters of Harvey.

According to D'Arcy Power, "*The first set of notes deal with the outside of the body, and the abdomen and its contents. The second portion contains an account of the chest and its contents; while the third portion is devoted to a consideration of the head with the brain and its nerves.*" Only nine of the 98 pages, which the manuscript contains are allotted to the heart. The scheme of the lectures was to give a general introduction in which the subject was arranged according to different headings, and then to consider each part under a variety of subheadings. Harvey's playfulness is shown even in the introduction. A roughly drawn hand indicates each main division, and each hand is made to point with the eight different fingers. The first hand points with its little finger, and has the other fingers bent, so the thumb is outstretched as if applied to the nose of the lecturist. The next heading is indicated by an extended ring finger, whilst the later ones are mere "*bunches of five*" or single amputated digits. In his description of the abdomen Harvey shows himself fully alive to the evils of tight lacing, for, in speaking of the causes of difficult respiration, he says, "*young girls by lacing: undercut their laces*" (Power, 1897).

The notes, which Harvey used for the anatomy lectures, revealed that he was a remarkably keen observer of nature and that he had assiduously studied the works of scholars from ancient times to his contemporaries (Moore, 1918; Whitteridge, 1964). More than 50 times he had mentioned Aristotle in his notes, more often than any other scholar; and on the very first page, he recommended, like Aristotle did in his *History of Animals*, that studying the viscera of animals would help in understanding the human anatomy. Of the medical authors, Harvey quoted from the works of both Hippocrates and Galen, as well as from Avicenna and Averrhoes.

Harvey had read the revolutionary work of Vesalius with its "*new*" anatomy, and he was most familiar with the anatomical writings of Caspar Bauhin (Fig. 103), Nicolas Massa, and Realdo Colombo to which he often referred. For example, Nicolas Massa's *Liber Introductorius Anatomiae* was quoted by Harvey with respect to the length of the ileum and of the colon, as well as to the position of the kidneys. From the work of Colombo, he knew of the length of the splenic vein and the pulsation of the abdominal aorta, and from Fallopius he learnt about the intestinal wall and gallstones. Harvey even quoted from the writings of one of his adversaries, Jean Riolan, who fervently rejected Harvey's monumental work on the circulation of blood.

Moore (1918) commented that the anatomy lecture notes "show us Harvey in every part of his daily work: examining and treating his patients, *performing autopsies with his own hands, listening to the remarks of other physicians, watching every peculiarity of mankind whether within the hospital or outside it, lecturing at the College of Physicians with a dissected body on the table before him, the parts of which he pointed out with the little rod of ebony and silver still to be seen there.*" For that period in history, *Harvey had a sufficient supply of human material, "not only for the general survey of anatomy,"* which he gave at the end of the first year, but also for the more detailed study of the trunk and the head, which was given at the close of the second and third years, respectively (O'Malley et al., 1961).

As a lecturer, Harvey's style was "*devoid of all formality or pretension*" and his sole aim was to make the subject understandable. This is most evident in his concluding remarks on the heart, which clearly indicated that Harvey had already formulated his views on blood circulation long before it was formally published in 1628.

Mention should be made of another of Harvey's manuscript (*De musculis*) he had prepared on the muscles of the human body. This handwritten manuscript is also at the British Museum (Sloane 486). Mostly with writing on both sides, it consists of 121 sheets, and it is dated 1627, the year before Harvey's great discovery was published. The notes, based on personal dissection, covered the anatomy of the muscles and

Figure 103. Caspar Bauhin (1560–1624) at 29 years of age. Woodcut from a drawing by Tobias Stimmer. (From Wegner, R.N.: *Das Anatomenbildnis. Seine Entwicklung im Zusammenhang mit der anatomischen Abbildung.* Basel, Benno Schwalbe and Co., Verlag, 1939. Courtesy of Institut fibAnatomie, Ernst-Moritz-Arndt-Universitat, Greifswald, Germany.)

their action, as well as clinical problems. "*Sixty-eight leaves are occupied by the anatomy of the muscles, and 450 muscles are enumerated. A note De passionibus musculorum shows how carefully he had attended to nervous and muscular symptoms,*" differentiating between "*complete paralysis of a limb and partial loss of power accompanied by tremor…*" (Moore, 1918).

Following the custom then, Harvey prefaced his notes with some general rules and guidance with respect to the anatomical dissection and demonstrations he carried out. His audience was made up of both physicians and surgeons. The directives were probably derived from many sources. The extract that follows is taken from the scholarly and useful translation of Whitteridge (1964):

GENERAL RULES FOR ANATOMY

1. Show as much in one viewing as can be, for instance, from the whole belly or from the

whole of some other part, those things which are interesting, and thereafter make division (according to the position and connections of the parts).

2. Point out the peculiarities of the particular body [you are dissecting], and the things that are new or but newly discovered.

3. To supply only by speech what cannot be shown, on your own credit and by authority.

4. Cut up as much as may be in the view of all, that practical skill may be learned together with theoretical knowledge.

5. Review your own and other people's observations in order to consider carefully your own opinion, or, in the strictest form, deal with other animals according to the rule of Socrates where it is fairer written. From this follow observations outside the field of anatomy but relating i. to the causes of diseases – of the greatest use to the physicians, ii. to the verity of Nature – of the greatest use to the philosophers. iii. Observations for the purpose of refuting errors and solving problems and, iv. for discovering the uses and actions of the parts, their rank and, on account of this, their classification. For the end of Anatomy is knowledge of the part, why it exists, for what purpose it is necessary and what is its use. For philosophers the chief use of Anatomy is to learn what parts are required in each action and what is of most importance. For the physicians it is to learn what is the natural constitution of the body, that is, the general rule, by which means they may single out those who are sick, and then know what to do in diseases.

6. Not to praise or dispraise, for all did well (as beholden to those who concluded erroneously for they lacked opportunity).

7. Not to dispute [or] confute other than by visible evidence, for to do otherwise would need more than three days.

8. Briefly and plainly, yet not letting pass any one thing unspoken of which is subject to the view.

9. Not to speak anything which without the carcase may be delivered or read at home.

10. Too eager an enquiry is not suitable in the dissection of each member and the time does not allow it.

11. To serve in their three courses according to the [hour-]glass: i. the lower belly, nasty yet recompensed by admirable variety, ii. the parlour, iii. the divine banquet of the brain.

Circulation of Blood: Establishing the Facts

Long before the publication of his work, *De Motu Cordis*, announcing his discovery of the movements of the heart and the circulation of the blood, Harvey had already formulated his ideas, which changed the future course of medicine.

It is plain from the structure of the heart that the blood is passed continuously through the lungs to the aorta as by the two clacks of a water bellows to raise water.

It is shown by the application of a ligature that the passage of the blood is from the arteries into the veins.

Whence it follows that the movement of the blood is constantly in a circle, and is brought about by the beat of the heart. It is a question therefore whether this is for the sake of nourishment or rather for the preservation of the blood for the limbs by the communication of heat, the blood cooled by warming the limbs being in turn warmed by the heart.

The preceding remarks, extracted from Harvey's lecture notes of 1616, show that he had already discovered the circulation of the blood and undoubtedly presented his ideas to the physicians and surgeons who attended his lectures. Whitteridge (1990) suggested that this "*famous passage*" might have been added to the manuscript "*at any time up to 10 years later*" and rejected the 1616 date. For reasons that are not clear, Harvey waited another 12 years before publishing his discovery. Perhaps he needed to carry out more experiments in order to confirm what was then a revolutionary idea and contrary to the then prevailing teachings of Galen (Hall, 1960). Furthermore, at that time, it was dangerous to entertain such heretical views, which questioned established dogma that was acceptable to both the church and scholars (Huntley, 1951; French, 1994). Perhaps it was prudent to wait for a more appropriate time. In his great work, *De Motu* Fig. 104), Harvey concluded in a precise and succinct manner as follows:

Now then in the last place we may bring our opinion, concerning the circulation of the blood, and propound it to all men. Seeing it is confirm'd by reasons and ocular experiments, that the blood does pass through the lungs and heart by the pulse of the ventricles, and is driven in and sent into the

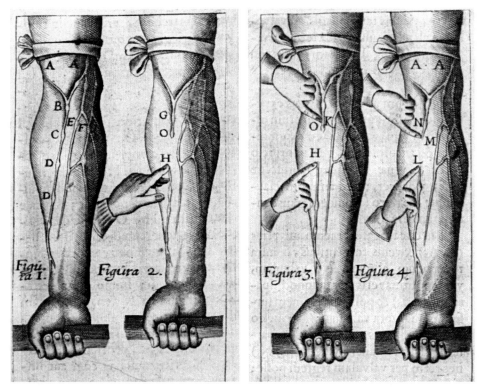

Figure 104. The anatomical plate in *De Motu Cordis* is an adaptation of the diagram used by his teacher and friend Fabricius in *De Venarum Ostiolis.* Figure 1 shows distended veins in the forearm and the position of valves. Figure 2 shows that if a vein has been "milked" centrally and the peripheral end compressed, it does not fill until the finger is released. Figure 3 shows that blood cannot be forced in the "wrong" direction.

whole body, and does creep into the veins and porosities of the flesh, and through them returns from the little veins into the greater, from the circumference to the centre, from whence it comes at last into the vena cave, and into the ear of the heart in so great abundance, with so great flux and reflux, from hence through the arteries thither, from thence through the veins hither back again, so that it cannot be furnished by those things which we do take in, and in a far greater abundance, than is competent for nourishment: It must be of necessity concluded that the blood is driven into a round by a circular motion in creatures, and that it moves perpetually; and hence does arise the action and function of the heart, which by pulsation it performs; and lastly, that the motion and pulsation of the heart is the only cause. (Keynes, 1928)

When one considers the discovery of Harvey, it is evident that it follows from the work of many others and at the same time there were aspects that still needed to be resolved (Snellen, 1984). Despite the controversies surrounding this discovery (Wyatt,

1924; Bayon, 1938a; b; 1939a; b; Hall, 1960; French, 1994), it remains firm that the credit for the recognition and demonstration of the circulation of the blood must be credited to William Harvey (Persaud, 1997).

Harvey and the College of Physicians

In addition to lecturing in anatomy and surgery for the College of Physicians, Harvey served the college on several occasions as a censor who, together with three other fellows, was responsible for monitoring those who practiced medicine in London and the surrounding areas. As well, Harvey was one of eight elects whose duty was to examine those desiring to practice medicine throughout England. For a while, he was also the treasurer of the college (1628 and 1629). Thus, we find Harvey carrying out many administrative duties on behalf of the College of Physicians, but the position he held as lecturer of anatomy was one of special importance (Keynes, 1966).

Bearing in mind that in the Middle Ages skills in the healing arts were acquired through a system of apprenticeship, a master and pupil relationship, rival groups evolved and formed guilds. On the one hand were the physicians who diagnosed and treated internal problems, whereas the barber-surgeons were responsible for wounds, external injuries, and other external bloody acts such as amputations. The apothecaries held an even lower position because they merely administered the treatments (douches, enemas, etc.) that were prescribed by the physicians. The physicians regarded themselves as better educated for the practice of medicine, and they distanced themselves from the barber-surgeons, who were more involved in *"menial and manual"* tasks. Nonetheless, both the physicians and the barber-surgeons realized that they needed to know something about the structure of the human body. During this period, instruction in anatomy was largely based on lectures, with occasional demonstrations from bodies that were obtained clandestinely. Sometimes available were the bodies of criminals that were hanged. The bodies were rapidly dissected within a few days on account of putrefaction. There was no embalming practiced at that time.

In 1540, through an Act of Parliament, the Company of Barber-Surgeons was united after representations to King Henry VIII. With the Royal Charter of the new guild, the surgeons were now authorized to receive the bodies of four *"felons condemned and put to death"* for their further and better knowledge, instruction, insight, learning, and experience in the science and faculty of surgery. A similar privilege was granted by Queen Elizabeth I in 1565 to the president of the College of Physicians. They too had the right to receive up to four bodies for dissection from persons executed in London, or within a vicinity of 16 miles.

Scientific Studies

As a physician, Harvey was well known and respected. He was appointed on February 3, 1618, a personal physician to King James I, and in 1630 he became Physician-in-Ordinary to Charles I. Harvey often accompanied the king on his hunting trips and various expeditions. They became friends and the king took an interest in Harvey's comparative anatomical and embryological studies. In addition to his work on the movements of the heart and the circulation of the blood, Harvey compiled two other works, one dealing with insects and the other on the generation of animals (Andrews, 1981; Meyer, 1936; 1939; Needham, 1959).

Harvey, a royalist, accompanied the king on his military expeditions during the years of the Civil War (1639–1649). This was a very unsettled period with the religious uprising in Scotland, civil strife, and considerable unrest everywhere. During absence from his quarters in Whitehall Palace, his rooms were ransacked by parliamentary soldiers in 1642, which resulted in considerable loss to Harvey. The deep anguish Harvey must have felt can best be revealed by his own words:

> Let the gentle minds forgive me, if recalling the irreparable injuries I have suffered, I here give vent to a sigh. This is the cause of my sorrow: Whilst in attendance on His Majesty the King during our late troubles, and more than civil wars, not only with the permission but by the command of parliament, certain rapacious hands not only stripped my house of all its furniture but, what is a subject of far greater regret, my enemies abstracted from my museum the fruits of many years of toil. Whenst it has come to pass that many observations, particularly on the generation of insects, have perished with decrement, I venture to say, to the Republic of Letters.

Regrettably, the work on the insects was lost forever, but the manuscript for the other treatise (*De Generatione Animalium*) survived and he later entrusted it to his former student and good friend Sir George Ent for publication.*

Harvey at Oxford

Following the Battle at Edgehill in 1642, the King's troop arrived in Oxford on October 29, 1642. Harvey was well known for his scientific work and as a physician in Oxford (Frank Jr., 1980), and within a few months, he received incorporation as a doctor of physics at Oxford University. In 1645, he was elected Warden of Merton College, a position he held for a year but obviously had to resign after the surrender of Oxford in 1646. An interesting anecdote recorded by Aubrey (1898) during this period is that he recalled seeing Harvey at Oxford in 1642 during visits to a Doctor George Bathurst, a fellow of Trinity College. Apparently Doctor Bathurst had a *"hen to hatch eggs in his chamber which they opened daily to see the progress and way of generation."* At Oxford, Harvey did not only continue his comparative anatomical studies and dissections, but he closely collaborated with several distinguished men of science (Charles Scarborough,

---

*For a commentary on this work, see Meyer (1936).

Highmore, Willis, and others) who eventually paved the way for the establishment of the Royal Society in 1660 (Hall, 1974).

Final Years

At the time Harvey had retired from his position of warden of Merton College, he was 68-years-old and he was a sick man, suffering from gout. The king, his friend and patron, was in prison. Harvey had already suffered tremendous personal losses through the ransacking of his quarters, the destruction of his museum, and loss of his notes that recorded the scientific observations he had made over the years.

Three of his brothers had already died, and he spent most of his last years in the house of his brother Eliab at Roehampton where he died on June 3, 1657. His body, wrapped in lead ("*lapt in lead*"), was entombed in the family chapel at Hampstead Parish Church, which lies in Essex.

Harvey was a physician and an anatomist, as well as a physiologist. His research, however, was based on anatomical principles and this was reflected in the title of his book as an anatomical treatise (*Exercitatio Anatomica De Motu Cordis...*). Harvey belonged to all of medicine and to assign him to any particular discipline would be contrary to the spirit of the era during which he lived (Persaud, 1997).

ANDREAS CESALPINO

Cesalpino and the Circulation of Blood

Unlike many of his eminent contemporaries among the great anatomists of Italy, Andrea Cesalpino (1519–1603) (Fig. 105) is often ignored and not sufficiently recognized for his scientific work, especially with respect to his theory of the circulation of the blood (Persaud, 1989). Moreover, his contributions to the field of botany were highly original, based on a revival of the work of the Greek philosopher Theophrastus (c. 372–c. 287 B.C.).

The claims of Italians that Cesalpino had discovered the circulation of blood before William Harvey are based on his discussion of blood flow through the lungs and heart, as well as on the use of the word "*circulatio*" in 1559 (Bayon, 1938a; Keynes, 1966; Clark et al., 1978). Cesalpino stated that:

> The orifices of the heart are made by nature in such a way that the blood enters the right ventricle by the vena cava, from which the exit from the heart opens into the lungs. From the orifice of the aorta. Certain membranes are placed at the openings of the vessels to prevent the blood from returning, so that the movement is constant from the vena cava through the heart and through the lungs to the aorta.

Indeed, Cesalpino described the general circulation of blood, or he was not far from the truth. At the same time, however, he also believed the Galenic theory of the "*flux and reflex*" (the to-and-fro movement according to the doctrine of Galen) of the blood in the veins, which he alluded to in his book, published after his death. Cesalpino was the first person to describe the

swellings of veins, which appear below sites of ligatures. Whether he knew of the venous valves is uncertain. In some ways, he also anticipated the exchange

Figure 105. Portrait of Andrea Cesalpino (1519–1603), Italian physician, philosopher, and botanist. (Line engraving 1765 by Giuseppe Zocchi.)

of gases across a thin membrane in the lungs. Cesalpino believed that the blood reaching the lungs from the heart is distributed in fine branches which lie close to very tiny air passages without any mixing of blood and air as taught by Galen (Persaud, 1997).

Life and Other Achievement of Cesalpino

Cesalpino was born in Arezzo, Tuscany, in 1519. He studied medicine at the University of Pisa and was probably a student of Realdo Colombo who taught there from 1545 to 1549. Cesalpino received his medical degree in 1551, and four years later he was appointed professor of medicine and botany, a position he held for more than four decades. In 1592, Cesalpino became professor of medicine in Rome, following an invitation from Pope Clement VIII to serve as his personal physician.

In addition to being an anatomist and physician, Cesalpino was also a naturalist, botanist, and philosopher and in all these areas he made notable scientific and scholarly contributions. He wrote influential books on demonology and medicine, metals, the classification of plants, as well as several treatises on religion and medicine.

Cesalpino was the director of the Botanical Gardens in Pisa, and his interest in Botany led him to formulate the first scientific botanical system, which became widely adopted and used for at least a century. The plants were classified according to natural affinities, fruits, and their products. He wrote a treatise entitled *De Plantis*, published in 1583.

As a philosopher, Cesalpino identified God with the universe or macrocosm, where man was merely a mirror image or microcosm of it. Because of his views, he had to appear before the authorities of the church and defend himself. Cesalpino died in 1603, "*a very learned man who combined a great reverence for the ancients, especially Aristotle, with an appreciation of the more modern aspects of natural history and medicine*" (Magner, 1992).

LYMPHATICS AND CAPILLARIES

The lymphatic channels and the capillaries are relevant to the circulatory system. During Harvey's time, the capillaries were not yet known and he believed that the lymphatics absorbed and transported nutrients from the gut. Already in the sixteenth century, Eustachius had observed in a horse a vessel that was filled with a white fluid that was connected with the internal jugular vein. It was located on the left side of the vertebral column. It is likely that Eustachius had discovered the thoracic duct. Regarding the question of priority in the discovery of the lymphatic vessels, reference should be made to Nielsen (1942).

Gasparo Aselli

Lymphatic channels were described by Fallopius and the Venetian anatomist Niccolo Massa (1499–1569). The functional importance then of these vessels was not known. In July 1622, Gasparo Aselli (1581–1626), professor of anatomy at Pavia (Persaud, 1984), observed many white delicate cords (*venae albae et lacteae*), which traversed the mesentery from the intestines in a dog that was recently fed and killed. He observed the numerous valves in these vessels, which erroneously traced to the pancreas and liver (Fig. 106).

Aselli's discovery and his work on the lacteals were published posthumously, a year later, as a book *De Lactibus Sive Lacteis Venis*. This was the first printed medical treatise with colored plates. The coloring of the illustrations was done by hand on every single copy. Aselli's discovery was confirmed as a result of an autopsy carried out on a convicted felon who had a very good meal prior to execution. It is of some interest that Aselli speculated in his book about the circulation of blood. He believed that the blood was carried by the pulmonary artery to the lung, and after mixing with air it returned to the left ventricle through the pulmonary vein. He reasoned that there was really no need for the pores in the septum as described by Galen and others (Persaud, 1997).

OTHER OBSERVATIONS ON THE LYMPHATICS

Simon Paulli

Other physicians have commented and added to our understanding of the lymphatics (Wyatt, 1924; Nielsen, 1942). In 1629, Simon Paulli (1603–1680) also publicly demonstrated the vessels (Fig. 107). Simon Paulli, a Danish physician, was born in Rostock, Germany. He obtained the M.D. degree from

Figure 106. Gasparo Aselli (1581–1624). From an engraving made by Cesare Bassano in 1627. (From Wegner, R.N.: *Das Anatomenbildnis. Seine Entwicklung im Zusammenhang mit der anatomischen Abbildung.* Basel, Benno Schwalbe and Co., Verlag, 1939. Courtesy of Institut fur Anatomie, Ernst-Moritz-Arndt-Universitat, Greifswald, Germany.)

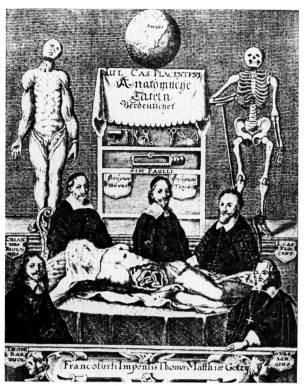

Figure 107. Copper plate engraved title page from the German edition of Casserio's anatomical plates, published in Frankfurt, 1656. The anatomists Simon Pauli (1603–1680), Jean Riolan (1577–1657), Giulio Casserio (1561–1616), Thomas Bartholin (1616–1680), and Johann Vesling (1598–1649) are depicted standing near the dissecting table. The abdomen of the cadaver is opened revealing the viscera. An articulated skeleton and a model of a muscle figure stand at a higher level in the background. Observe the globe and the assortment of surgical instruments on display. (Courtesy of Institut für Anatomie, Ernst-Moritz-Arndt-Universität, Greifswald, Germany.)

the University of Wittenberg, after pursuing medical studies in Rostock, Leiden, Paris, and Copenhagen. Paulli became professor of anatomy, surgery, and botany at the University of Copenhagen. He was also court physician to Denmark's Frederick III and later Christian V. With the support of Frederick III, Paulli established the first anatomical theater (Domus Anatomica) in Copenhagen in 1645, where he lectured and conducted public anatomical demonstrations.

Paulli made many important contributions in both medicine and botany. He was concerned about the basic skills of barber-surgeons and pharmacists, which he tried to improve. His published works included: a treatise on medicinal plants, *Quadripartitum Botanicum, de Simplicium Medicamentorum Facultatibus* (1639); a textbook on Danish plants, *Flora Danica* (1648); and

on the harmful effects of tobacco, coffee, tea, and chocolate, *Commentarius De Abusu Tabaci Americanorum Veteri, Et Herbæ Thee Asiaticorum in Europe Novo* (1661).

Jean Pecquet

The French anatomist Jean Pecquet (1622–1674) in his work *Experimenta nova anatomica*, published in 1651 in Paris, described the observations he had made in the dog a few years earlier (Figs. 108 & 109). In 1647, he was the first to recognize the thoracic duct. Pecquet described the common receptacle of the lacteals and lymphatics and of how the thoracic duct emptied its contents into the venous system where the jugular and subclavian veins meet. A more detailed account of the thoracic duct in man was published in

Figure 108. In 1651, Jean Pecquet discovered the thoracic duct and thus disproved the traditional Galenic idea that chyle was carried by the vessels of the intestine to the liver. He published his results in this text, *Experimenta Nova Anatomica*, a recent acquisition of The Lilly Library. http://www.indiana.edu/~liblilly/anatomia/viscera.html.

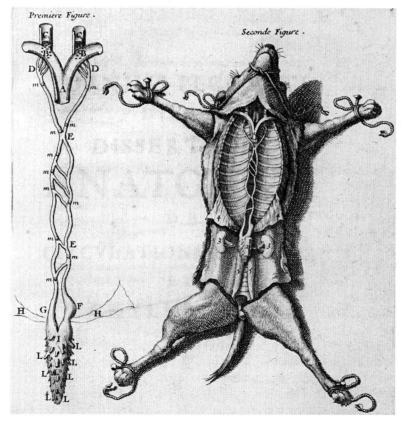

Figure 109. A picture of a dissected dog depicting the thoracic duct, from *Experimenta Nova Anatomica*. http://www.indiana.edu/~liblilly/anatomia/viscera.html.

Figure 110. Thomas Bartholin (1616–1680). Painting by Heinrich Dittmers; Original in Frederiksborg Palace. (From Wegner, R.N.: *Das Anatomenbildnis. Seine Entwick lung im Zusammenhang mit der anatomischen Abbildung.* Basel, Benno Schwalbe and Co., Verlag, 1939. (Courtesy of Institut für Anatomie, Ernst-Moritz-Arndt-Universität, Greifswald, Germany.)

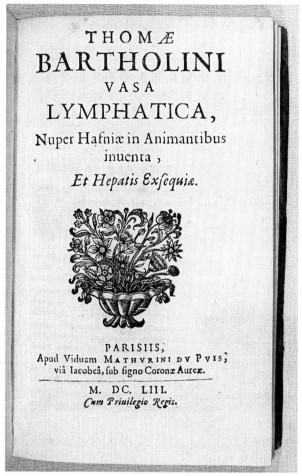

Figure 111. The cover page of Bartholin's book *Vasa Lymphatica, Nuper Hafniae in Animantibus Inventa, et Hepatis Exsequiae.* Paris: Apud Vicuam Mathurini du Puis, 1653. http://www.indiana.edu/~liblilly/anatomia/viscera.html.

1653 by the Danish physician Thomas Bartholin (1616–1680), professor of anatomy at Copenhagen (Fig. 110), in his book *Dubia anatomica de lacteis thoracis in homine brutisque nuperrime observatis, historia anatomica* (Fig. 111). As yet the lacteals were not clearly delineated from the lymphatic vessels and the termination of the lymphatics was not known.

The Swedish anatomist Olof Rudbeck (1630–1702) observed in 1651 lymphatic vessels in the liver and intestines, which he traced to the thoracic ducts. He published his findings in a work entitled *Nova exercitatio anatomica, exhibens ductus hepaticos aquosos, et vasa glandularum serosa* in 1653.* Rudbeck was from Uppsala, Sweden, and had studied in Uppsala and

Leiden. A similar claim in priority for this discovery goes to George Jolyffe or Joyliffe (1621–1658), an Englishman who received his M.D. degree in Cambridge in 1652.

Francis Glisson (1597–1677), in his book *Anatomia hepatis* (Figs. 112 & 113), published in 1654 in London (Cunningham, 1993), referred to Jolyffe's observations, which Jolyffe had mentioned to Glisson, as well as described in his thesis. Claims of priority for the discovery of the lymphatic vessels have been discussed by Nielsen (1942) and others. It was suggested that both Thomas Bartholin and Olof Rudbeck should be considered as the discoverers of the lymphatic system with equal right (Persaud, 1997).

---

*For an English translation of Rudbeck's monograph, see Nielsen (1942).

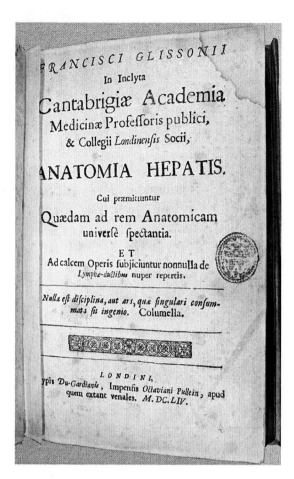

Figure 112. *Anatomia Hepatis* was the first text printed in England to present a detailed anatomical and physiological account of just one organ based on original research. http://www.indiana.edu/~liblilly/anatomia/viscera.html.

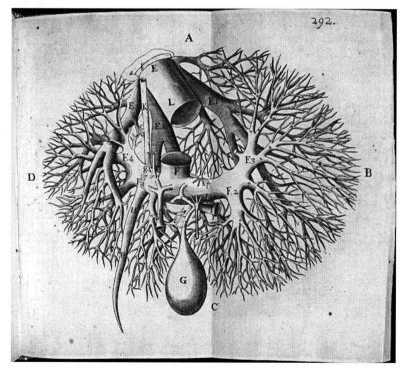

Figure 113. Glisson used advanced anatomical methods, such as casts, the injection of colored fluids, and microscopy to illustrate the liver more accurately than had been done before. He delineated the passage of blood from the portal vein to the vena cava and also showed that lymph does not flow to the liver but rather away from it, passing to the recently discovered *capsula communis*. This fibrous capsule, which Glisson was the first to correctly describe, is now referred to as "Glisson's capsule." http://www.indiana.edu/~liblilly/anatomia/viscera.html.

CAPILLARIES

How arteries and veins communicate was still to be resolved. Harvey believed that there cannot be a direct communication between arteries and veins, and one of the reasons he gave was the difference in the size and structure of the vessels. He suggested that blood flowed from one type of vessel to the other through *porositates* or "*hidden Meanders*" through which arteries and veins communicate. In 1649, he stated that "*I have been unable to trace any connection between the arteries and veins by a direct anastomosis of their orifices....I can therefore boldly affirm that there is neither anastomosis...of the hepatic arteries with the hepatic veins....*"

Leonardo da Vinci (1452–1519) used the word *chapillari* in his anatomical notes to describe the minute endings of the blood vessels. Through the injections of dyes and pigments, Leonardo da Vinci and others, including Domenico Marchetti (1626–88), Robert Boyle (1627–1691), Jan Swammerdam (1637–1680), Regnier de Graaf (1641–1673), Stephan Blankaart (1650–1702), Christian Johann Lange (1655–1701), and Friedrik Ruysch (1638–1731), have advanced our understanding of the communications between arteries and veins in various organs. However, it was Marcello Malpighi (1628–1689; Fig. 114) who first observed the capillaries linking the arteries and veins in the lungs.

Marcello Malpighi Discoveries of the Capillaries

Such a remarkable discovery was achieved by Malpighi with the simplest of approaches. The observations he made in 1660, just three years after Harvey's death, he communicated in two letters to Giovanni Borelli (1608–1679), who was professor of mathematics at the University of Pisa, as well as his friend and teacher. With the help of a magnifying glass, Malpighi described the microscopic structure of alveoli in the dog, which he communicated in the first letter. In the second letter, he described how blood flowed through the capillary network in the living frog's lung. These observations were published by Malpighi in a monograph entitled *De pulmonibus, Observationes anatomicae* in Bologna in 1661 (Figs. 115 & 116). However, Malpighi did not realize the significance of his findings.

Scientific Studies and Achievements

In a small work entitled *De Omento...*, published in 1665, Malpighi described what must have been blood corpuscles he had observed in the blood vessels of the omentum of a hedgehog. Already in 1658, blood corpuscles where observed by Jan Swammerdam (1637–1680). Further insights and a better understanding of the capillary system were obtained through the studies carried out by the Dutch microscopist Antoni van Leeuwenhoek (1632–1723), the Italian biologist Lazaro Spallanzani (1729–1799), and others. With more discoveries relating to the mechanisms of breathing, the behavior and exchanges of gases, and the physiology of respiration, a new course was charted, which saw the demise of the vital spirits, that was characteristic of Galenic dogma.

Similarly to the discovery of Harvey, the discovery of the capillary system was not readily accepted. In his *Theoria Medica Vera*, published in 1708, the German chemist and physician Georg Ernst Stahl (1660–1734) was doubtful of its existence. Even John Hunter (1728–1793), the renowned British surgeon and anatomist, in his book *A Treatise on the Blood...*(London, 1794, p. 14), remarked that "*These early observers probably imagined more than they saw.*"

Figure 114. Marcello Malpighi (1628–1694). (Department of Anatomy, University of Manitoba.)

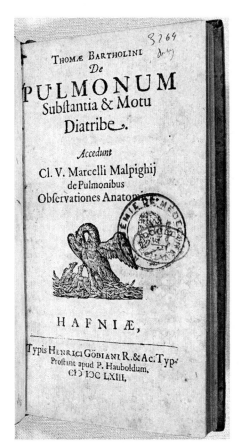

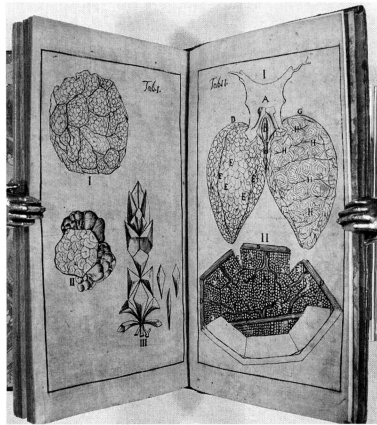

Figure 115 (*Left*). *De Pulmonibus* is Malpighi's first publication, and according to Luigi Belloni, his "fundamental" work. Originally sent to Borelli in two short letters, it was published first in 1661. The 1663 edition, owned by The Lilly Library, is the second edition and is itself a rarity. http://www.indiana.edu/~liblilly/anatomia/viscera.html.

Figure 116 (*Right*). Tables I and II are depicting the lungs as were seen through Malpighi's microscope in 1659. Malpighi also discovered that the lungs are composed of membranous alveoli surrounded by an extensive capillary network that open into their ultimate tracheobronchial ramifications. http://www.indiana.edu/~liblilly/anatomia/viscera.html.

Malpighi was the first person to make important scientific observations using then what was described as a microscope. Malpighi was born in 1628, the same year William Harvey published his monumental work *Exercitatio Anatomica De Motu Gordis Et Sanguinis In Animalibus.* He studied philosophy at Bologna where he later changed to the study of medicine. Malpighi can be rightly considered as the father of microanatomy because of the many minute structures he observed with the microscope. The glomeruli in the kidneys were discovered by Malpighi and described in his book *De Viscerum Structura Exercitatio Anatomica*, published in 1659. In the same year he described the spleen more accurately than anyone before and also the corpuscles in it, which bear his name.

It was Robert Hooke (1635–1702), the brilliant and inventive British scientist, who actually gave us the structure of a cell, a name that he also coined. In his book, *Micrographia*, published in 1665, Hooke observed tiny compartments "*all cellular or porous in the manner of a honeycomb, but not so regular*" in sections of cork he examined. Malpighi also described the cell, but far more accurately, in his book *Anatomy of Plants*, which appeared in 1670.

Most impressive were Malpighi's studies on the development of the chick embryo. This was a continuation of the work of William Harvey, and he communicated his findings, describing the earliest stages of chick embryology, to the Royal Society in London in 1672. In his work *De formatione pulli in ovo*, published in 1673, he described the aortic arches, the neural groove and cerebral vesicles. No significant information was added to this knowledge until the nineteenth century. The Royal Society recognized

Malpighi's scientific achievements by electing him as a fellow (Persaud, 1997).

Malpighi's Final Years

Malpighi was appointed professor of theoretical medicine at Pisa University in 1658, and after moving to Bologna and Messina, he returned to Bologna as professor of anatomy where he remained for the rest of his academic life. His discoveries and theories evoked strong opposition from his colleagues, particularly the Galenists. Some turned against him, including his teacher and patron, Borelli. Malpighi was physically assaulted by two masked men, and he lost his manuscripts and microscopes in 1684 when his house was burned down. Following an invitation from Pope Innocent XII, Malpighi left Bologna for Rome in 1691 where he died three years later.

THE PULMONARY CIRCULATION

Many epoch-making advances have not been accorded proper recognition at the time of their exposition because the discovery might have been made long before the world was prepared to receive it. Often, the original publication is forgotten and only unearthed long after someone else has been credited for the discovery. Such is the case of the description of the flow of blood through the lungs, which was known centuries before the publication in 1628 of William Harvey's monumental work *Exercitatio Anatomica De Motu Cordis Et Sanguinis In Animalibus* (Persaud, 1989).

From Aristotle to Galenic Belief

Aristotle (384–322 B.C.) believed that the heart was the primary organ of the body, the seat of the soul, for it was not only central but also mobile and hot with extensive connections to the rest of the body. He conceived the human heart as having three cavities, the middle of which contained the purest and thinnest blood and was connected to the air-filled aorta and arteries. Thus, the left ventricle was identified. Also, communications between all three cavities, in particular the largest (right ventricle), with the lungs were described. Aristotle believed that blood was manufactured in the heart and that the lungs served the purpose of bringing blood and air together. His limited account of the vascular system nevertheless provided an essential basis for Galen's physiological concept (French, 1979), which survived up to the seventeenth century, when Harvey announced his discovery of the circulatory system.

The Greek physician Herophilus (355–280 B.C.) found that only arteries pulsate, and he distinguished arteries from veins on the basis of the thickness of their walls. The pulmonary veins he considered to be conduits for the passage of air from the lungs to the left side of the heart. That the arteries contained blood and not air was demonstrated by Galen. He even contemplated the possibility of blood and air flowing together, but it was not clear to him how air from the lungs entered the arterial system. Based on a theory first formulated by Erasistratus (300–250 B.C.), a contemporary of Herophilus in Alexandria, Galen believed that the life-bearing principle or "*pneuma*" was taken into the body through the trachea. From the lungs it was carried to the left ventricle via the pulmonary vein where it mixed with the blood, giving rise to another and higher form of the "*pneuma*" the "*vital spirit.*" This "*vital spirit*" was then transmitted with the blood in the arteries to different organs. The "*injurious vapours*" or residual wastes formed from the combustion of blood and "*vital spirit*" passed back through the pulmonary vein into the lungs to be expelled (Fleming, 1955; Furley & Wilkie, 1984; Wilson, 1959).

Galen (A.D. 130–200) described the valvular flaps (*membranarum epiphyses*) in the heart but was not aware of their functions. To explain the mixing of the "*pneuma*" and blood in the left ventricle (seat of the soul) he claimed that there must be pores in the interventricular septum. How else could blood that was continuously being manufactured in the liver, consumed by the tissues and delivered into the right ventricle, receive its share of pneuma? Indeed, the pits he saw from the right ventricle on the surface of the interventricular septum had to perforate into the left ventricle, "*otherwise the pits would be without function; but Nature does nothing in vain, and nothing in the body is without function*" (French, 1979).

The "Christianismi Restitutio" of Michael Servetus

More than 12 centuries later the first known printed account of the lesser circulation appeared in a book by Michael Servetus (c. 1511–1553), the Spanish theologian, humanist and physician (Fig. 117). The book, entitled *Christianismi Restitutio*, was published in

Figure 117. Michael Servetus. (Reproduced with permission from an engraving; Courtesy of the School of Anatomy Library, University of Cambridge, England.)

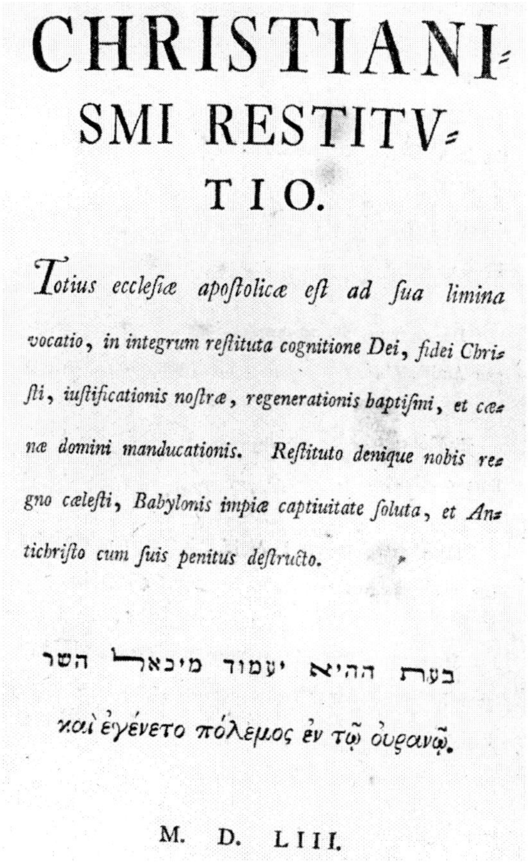

CHRISTIANI-
SMI RESTITV-
T I O.

Totius ecclefiæ apoftolicæ eft ad fua limina vocatio, in integrum reftituta cognitione Dei, fidei Chrifti, iuftificationis noftræ, regenerationis baptifmi, et cænæ domini manducationis. Reftituto denique nobis regno cælefti, Babylonis impiæ captiuitate foluta, et Antichrifto cum fuis penitus deftructo.

בעת ההיא יעמוד מיכאל השר

καὶ ἐγένετο πόλεμος ἐν τῷ οὐρανῷ.

M. D. LIII.

Figure 118. Title page of the *Christianismi restitutio.* (Courtesy of the Bibliotheque Nationale, Paris.)

1553 (Fig. 118). It was not intended as a physiological treatise but a theological work containing proposals for the reformation of Christianity. In discussing the divine spirit, Michael Servetus described the flow of blood through the lungs in relation to the path taken by the "*spiritus*" and the "*anima.*" He believed that the soul of man resided in the bloodstream, and for any understanding of the soul one would need to study the origin and movement of blood itself (Fulton, 1953).

Servetus claimed that

> vital spirit has its origin in the left ventricle of the heart, and the lungs assist greatly in its generation. It is generated in the lungs from a mixture of inspired air with elaborated, subtle blood, which the right ventricle of the heart communicates to the left. However, this communication is made not through the middle wall of the heart, as is commonly believed, but by a very ingenious arrangement the subtle blood is urged forward by a long course through the lungs; it is elaborated by the lungs, becomes reddish-yellow and is poured from the pulmonary artery into the pulmonary vein.

> Then in the pulmonary vein it is mixed with inspired air and through expiration it is cleansed of its sooty vapors. Thus finally the whole mixture, suitably prepared for the production of the vital spirit, is drawn onward from the left ventricle of the heart by diastole. (O'Malley, 1953)

Invisible Passages in the Interventricular Septum

Having denied the existence of the interventricular pores, the alternative pulmonary route soon became obvious, which perhaps led Servetus to his discovery. The view of an impervious interventricular septum was an entirely new concept that challenged established Galenic dogma.

Andreas Vesalius (1514–1565), in the first edition of his revolutionary *De Humani Corporis Fabrica*, published in 1543, wrote that the septum of the ventricles, as I said, formed the thickest substance of the heart, abounds in pits impressed into both sides of it; for this reason the surface which faces the ventricle is uneven. None of these pits (at least insofar as may be observed) penetrates from the right ventricle into the left, so that we are compelled to wonder at the industry of the Creator of all things by which the blood sweats from the right ventricle into the left through invisible passages.

In the second edition of his book, published in 1555, Vesalius seemed even more doubtful when he stated the following:

> And though these pits (foveae) are most conspicuous, none of them, as far as one can make out with one's eyes, passes from the right ventricle into the left through the septum of these ventricles. And I have not found even the most hidden passageways by which the septum of the ventricles could be penetrated, though they are mentioned by the professors of anatomy who are definitely convinced that the blood is taken over from the right into the left ventricle. As I shall demonstrate more fully later, I am therefore very much in doubt about the function of the heart in this particular part. (Furley & Wilkie, 1984)

Vesalius was most skeptical about the openings in the interventricular septum of the heart, but at the same time he did not completely deny their existence. One first encounters among the anatomists an unequivocal denial for the existence of interventricular pores in the books of Juan Valverde de Hamusco (1520–1588) and his teacher Realdo Colombo (1516–1559). The Spanish anatomist Valverde was a student of Realdo Colombo, successor to Vesalius at Padua. The influence of Colombo on his former pupil and prosector was great, and it is believed that the concept of an impervious interventricular septum might have originated from Colombo (Coppola, 1957).

In his book *De Re Anatomica* (Colombo, 1559), Colombo emphatically denied the existence of openings in the interventricular septum which will permit the flow of blood from the right ventricle to the left, but instead affirmed that "*the blood is carried through the pulmonary artery to the lung and is there attenuated; then it is carried along with air, through the pulmonary vein to the left ventricle of the heart. Hitherto no one has noticed this or left it in writing, and it especially should be observed by all.*"
In his *Anatomia Del Corpo Humano*, published in 1556 in Spanish (de Hamusco, 1556), and three years later in Italian, Valverde acknowledged his debts to Colombo

by stating "*all that is useful that will result from this book of mine, is not less to be attributed to Andreas Vesalius, than to Realdo Colombo, my preceptur in this faculty.*" Regarding the interventricular septum he too was unequivocal in his views: "*the partition, like the rest of the substance of the heart which the left [ventricle] forms is hard, thick and strong, as we have said the heart to be.... But the [surface of the] partition between the ventricles is somewhat irregular by reason of some rivulets or sulci which are made in the substance of the heart, and they are much more manifest in the left ventricle than in the right; but no blood passes from one ventricle to the other, as they say who have written of it until now*" (Coppola, 1957). From the writings of Valverde and Colombo one can only speculate that they were not aware of Michael Servetus's *Christianismi Restitutio*, the publication of which was to have sealed his fate.

Martyrdom of a Heretic

Michael Servetus was burnt as a heretic at the stake with almost all copies of his book on October 27, 1553. His final work, which questioned contemporary theological opinion and tradition, in particular the doctrine of the Trinity, greatly offended both Catholics and Calvinists. Only three copies of Servetus's book have survived and with them the legacy of a humanist and physician who became a martyr on account of his progressive views (Fulton, 1953; O'Malley, 1953; Bainton, 1953; Hall, 1960). According to a Mr. La Roche, "*After the sentence was pronounced upon this unfortunate Spaniard, he appeared sometimes quite confounded and motionless, sometimes fetching deep sighs, and then crying after the Spanish manner, Misericorde! Misericorde!*" (cited in British Library's Copy of Reprinted *Christianismi Restitutio*, with annotations).

The copy of *Christianismi Restitutio* in the British Library [G.14227 (1)] is apparently a reprint made in 1790 in Nuremberg. In a handwritten note at the beginning of the book it is stated, "All the copies, except one, were burned along with the Author, by the implacable Calvin. This copy was secreted and saved by D. Colladon, one of the Judges." Eventually it came into "the hands of Dr. Mead who endeavoured to give a Quarto Edition of it; but before it was nearly completed; it was seized by John Kent...on the 27th May 1723, at the instance of Dr. Gibson, Bishop of London, & burnt, a few copies excepted."

Claims of Priority

William Harvey was not aware of either Servetus's or Colombo's description of the pulmonary circulation even though the English surgeon John Banister (1533–1610) referred to Colombo's account of the pulmonary circulation in his anatomical compendium *The Historie of Man*...which was published in 1578. Bannister remarked as follows: "*But very wyde they wander, sayth Columbus. For the bloud through the arteriall Veyne is carried to the lunges, whence, being attentuated, it is caried by the veniall arterie into the left Ventricle of the hart together with ayre: which no man before his tyme noted, or at least have left ertant*" (Banister, 1578).

The next reference to Servetus's account of the pulmonary blood flow again appeared in the book *Reflections on Ancient and Modern Learning* by the antiquarian and chronologist William Wotton, published in 1694 (Wotton, 1694). In his commentary on the circulation of blood, William Wotton noted on pages 211 and 212 that

the first Step that was made towards it, was, the finding that the whole Mass of the Blood passes through the Lungs, by the Pulmonary Artery and Vein. The first that I could find, who had a distinct Idea of this Matter, was Michael Servetus, a Spanish

Figure 119. Ibn al-Nafis. (From Hamdard Medical Digest, the Organ of the Institute of Health and Tibbi Research 10:15, 1966.)

Physician, who was burnt at Arianism, at Geneva, near 140 years ago....In a book of his, titled, "Christianismi Restitutio," printed in the year MDLIII he asserts, that the Blood passes through the Lungs, from the Left to the Right Ventricle of the Heart; and not through the Partition which divides the two Ventricles, as was at the Time commonly believed.

That there were already contenders in claiming priority for describing the pulmonary circulation is clear from the following paragraph, also taken from William Wotton's book:

Realdus Columbus, of Cremona, was the next that said anything of it, in his 'Anatomy,' printed at Venice, 1559 in Folio; at Paris, in 1572 in Octavo; and afterward elsewhere. There he asserts the same (a) Circulation through the Lungs, that Servetus had done before; but says that no Man had ever taken notice of it before him, or had written any Things about it: which shows that he did not copy from Servetus; unless one should say that he stole the Notion, without mentioning Servetus's Name; which is injurious, since in these Matters the same Thing may be, and very often is observed by several Persons, who never acquainted each other with their discoveries.

Realdo Colombo was a brilliant anatomist and in his book, published a few months after his death, one also finds detailed and accurate descriptions of the pulmonary, cardiac and aortic valves, which undoubtedly emanated from studies carried out in living animals. In describing the circulation of blood through the lungs, both Servetus and Colombo revealed a remarkable understanding of the function of the pulmonary vessels in relation to the heart which was relatively advanced for that period and not even known to Vesalius nor his contemporaries (Persaud, 1997).

Ibn Al-Nafis and the Lesser Circulation

More recent historical evidence, however, brought to light that the lesser circulation had already been described in the thirteenth century by Ibn al-Nafis (Fig. 119). The manuscripts of Ibn Nafis were rediscovered by Muhyi-ad-Din at-Tatawi, an Egyptian scholar, who had been studying them in the Prussian State Library in Berlin for his doctoral dissertation. The work was submitted to the medical faculty of Freiburg University in 1924 (El-Tatawi, 1924; Meyerhof, 1935; Bittar, 1955; Persaud 1989).

Ibn Nafis was born in Damascus in 1210 and educated in Egypt where he practiced medicine until his death in 1288. He was the author of a large number of medical works, including a commentary on the anatomy of Avicenna, also known as Ibn Sina (A.D. 980–1037), which was probably completed sometime during the second part of the thirteenth century. In his work, one finds an almost identical description of the pulmonary circulation as presented three centuries later by Servetus (Meyerhof, 1935).

"As the production of the [vital] spirit is one of the functions of the heart, and as the spirit consists of much-refined blood with a large admixture of airy substance, it is necessary that the heart should contain both refined blood and air so that the spirit may be generated out of the substance produced by their mixture; this takes place where the spirit is generated, viz. in the left cavity of the heart. It is moreover, indispensable for the heart of man and of such animals as have lungs like him, to possess another cavity in which the blood is refined in order to become apt for the mixture with air.…This cavity is the right cavity of the heart."

The blood, after it has been refined in this cavity, must be transmitted to the left cavity where the [vital] spirit is generated. But there is no passage between these two cavities; for the substance of the heart is solid in this region and has neither a visible passage, as was thought by some persons, nor an invisible one which could have permitted the transmission of blood, as was alleged by Galen. The pores of the heart there are closed and its substance is thick. Therefore, the blood, after having been refined, must rise in the arterious vein to the lung in order to expand in its volume and to be mixed with air so that its finest part may be clarified and may reach the venous artery in which it is transmitted to the left cavity of the heart.

The idea of invisible pores in the interventricular septum of the heart was firmly refuted by Ibn Nafis. He reasoned correctly that blood from the right ventricle reached the left ventricle only after passing through the lungs because from his anatomical studies he found no evidence for the presence of pores in the interventricular septum. Yet, the physiological concept of the circulation of blood through the whole body eluded the discoverer of the lesser circulation, for Ibn Nafis too believed that the animal spirit is transmitted by the aorta to all organs in the body and the tissues received their nutrition from the venous channels in the ebb and flow manner.

Andrea Alpago (1450–1521) of Beluno had lived in the Middle East collecting, editing, and translating the works of various Arabic medical scholars. He also translated into Latin Ibn-Nafis's commentary on the fifth cannon of Avicenna, which was eventually published in Venice in 1547 (Bittar, 1955; Coppola, 1957). Even though the section dealing with the pulmonary circulation was not included, it is not unreasonable to assume that Alpago would have been familiar with it.

Whether Servetus, Colombo and Valverde knew of Ibn Nafis's description of the lesser circulation through Alpago or from other sources is a matter of speculation (Schacht, 1957). The many similarities found in the texts of Ibn Nafis and Servetus are indeed remarkable. At the same time, the possibility cannot be ruled out that Servetus and Colombo, as well as Valverde, arrived independently at the same discovery as Ibn Nafis (Izquierdo, 1937; Temkin 1940; Coppola, 1957).

COMPLETING THE GREAT CIRCLE

Richard Lower

Richard Lower (1631–1691), a pioneer of blood transfusion, showed in 1669 that blood becomes oxygenated in the lungs. With the main purpose of the circulation of the blood now established, this discovery brought Harvey's work to a rational end (Wyatt, 1924; Fishman and Richards, 1982). Richard Lower was born in Tremeer, near Bodmin, Cornwall. In 1649, he was admitted to Christ Church College, Oxford, where he received his medical degree and further training. As assistant to Thomas Willis (1621–1675),

Lower became well known for the studies he carried out on the brain and cerebral circulation (Tubbs et al., 2008). His extraordinary dissecting skill was acknowledged by Willis. In 1666, Lower left Oxford for London where he established a medical practice. He became a member of the Royal Society and a Fellow of the College of Physicians. Lower was one of the most distinguished physicians of his time and made many outstanding contributions to anatomy and physiology. He employed innovative experimental and operative techniques in animal studies in order to understand clinical problems and to test his

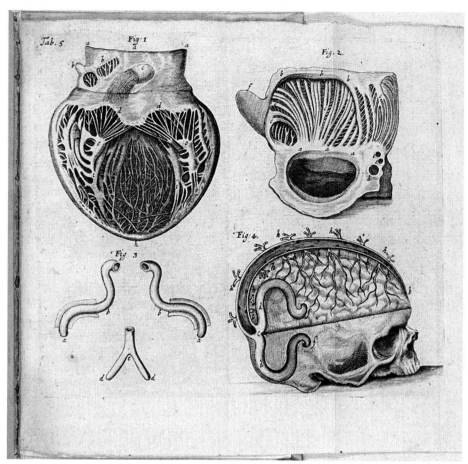

Figure 120. A picture from his book *Tractatus de Corde. Item de Motu & Colore Sanguinis et Chyli in Eum Transitu.* London: Typis J. Redmayne, Impensis Jacobi Allestry, 1669. http://www.indiana.edu/~liblilly/anatomia/viscera.html.

hypotheses. In 1665, Lower carried out the first successful transfusion in a dog at Oxford, "*from an artery of one animal into a vein of a second,*" and in 1667 he was successful in doing the same in a human.

Lower's earliest work was with Thomas Willis, which Willis acknowledged in the preface of his book *Cerebri Anatome* – "*I employed the assistance of cooperation of Richard Lower, a doctor of outstanding learning, and an anatomist of supreme skill. The sharpness of his scalpel and of his intellect, I readily acknowledge, enabled me to investigate better both the structure and the functions of bodies, who secrets were secretly concealed....*" In 1669, Lower's major work *Tractatus de Corde...* was published in London (Fig. 120). It contained an account of the heart's structure and movement, as well as his personal observations, which added to Harvey's description of the circulation of blood. Lower estimated the quantity and velocity of the circulating blood. Moreover, he described the structure and function of the pericardium, the unique arrangement of the muscle fibers in

the heart, and of the differences in color between arterial and venous blood. Lower attributed the red color of arterial blood to its mixing with air in the lungs, and the darker color of venous blood to the loss of air as the blood flows through the body. In addition, Lower also described his transfusion studies (Fig. 121), as well as the physiological observations he made in his anatomical and clinical work (Boyle, 1665–1666). Several editions of Lower's book were published, and it was widely used.

In his *Tractus de Gordis,* published in 1669 (Lower, 1932), Lower remarked that "*Harvey described the structure of the heart and the movement of blood in a way that left practically nothing to be added....But there are points not mentioned in Harvey's circulation which need consideration.*" Lower found that the contraction of the heart stopped when the vagus nerve is ligated, and that the origin of the heartbeat is located in the cerebellum. Moreover, following up on the drawings published by Nicolaus Steno, he carried out studies in dissections of

Figure 121. In 1666, Boyle gave a somewhat gruesome account of one of the first instances of blood transfusion between animals, in this case two dogs. The following year, Richard Lower also performed the first transfusion of blood from a sheep into a human. He was one of the foremost surgeons of his day and was involved in pioneering discoveries in blood circulation and breathing. Although Lower understood the usefulness of blood transfusions following injury or other blood loss, in humans it often caused severe and sometimes fatal reactions, and 10 years later it was banned by Parliament. (http://trailblazing.royalsociety.org/photos/1666SA1.jpg).

the heart musculature so as to understand the arrangement of the spiraling fibers and the manner of its contraction. He concluded that the contraction of the heart is "*like the wringing of a linen cloth to squeeze out water.*"

With the accession of King James II, Richard Lower's career and clinical practice suffered because he was highly critical of the Catholic Church. He spent increasingly more time in Cornwall where at 60 years of age he died. His benefactors were French and Irish Protestant refugees, as well as St. Bartholomew's Hospital in London.

Chapter 15

PHYSICIANS, SURGEONS, AND ANATOMISTS

BLENDING NEW BELIEFS WITH OLD DOGMA

All advances and changes take place from accepted doctrine and established practices, and the discoveries in human anatomy during the Vesalian renaissance also followed this pattern. The blend of the new with the old, and eventually the acceptance of new ideas, which later became established facts (McGirr & Stoddart, 1991), is most evident from the medical treatises, atlases, and textbooks that were published over the two centuries after Vesalius's masterpiece. No doubt this has been a gradual process without sudden discontinuity (Conrad et al., 1995; Jolin, 2013).

Even when William Harvey briefly announced his discovery of the circulation of blood in his Lumleian lectures on anatomy, given in 1618 to the College of Physicians in London, Harvey waited another ten years obtaining more evidence before he took the bold step to publish his work.

The authoritarianism of the period and the influence of Galenism did not tolerate any radical thinking. Just a few decades earlier, in 1559, John Geynes, M.D. of Oxford, had to recant his criticisms of Galen before he was made a fellow of the college. Also, a Thomas Fludd, M.D. of Cambridge, was given a license to practice even though his performance in the examination was poor, but he was required to read specific sections from the work of Galen in order to improve his knowledge.

For Harvey, it was a bold rejection of Galenic anatomy with respect to the notion of the flow of blood in the body (Brain, 1986), but as he remarked in Chapter 8 of his book, "*the die has now been cast, and my hope lies in the lover of truth and the clear-sightedness of the trained mind*" (Franklin, 1961). The same can be said for the discovery of the pulmonary circulation of blood (Persaud, 1989). Both of these discoveries can be viewed as a "*Galenic answer to a Galenic problem*" (French, 1975).

The kind of anatomy that was taught to students can easily be imagined from a perusal of the atlases and textbooks that were then in use (Jolin, 2013). Where lecture notes, either from the professor or a student from this period, are available, then one probably will gain a more accurate insight from these regarding the prevailing level of anatomical knowledge.

Of relevant interest are the lecture notes of John Moir, a student who attended Marischall College in Aberdeen and the anatomy classes in 1619 and 1620. The small, well-preserved notebook (Aberdeen University Library, Ms. 150), containing the anatomy lectures, were translated from Latin into English and with useful annotations added by French (1975). French indicated that the notes were merely a "*brief explanation of anatomy*" which served in providing a concise account of the human body.

Everything had to be in accordance with the views of the reformers, within the framework of the philosophy course. Moir's notes therefore represented a part of a general educational arts course. Moreover, the notes showed the "*considerable distance between the practical medical man, whether he was a foreign educated physician or a locally trained surgeon, and the universities the traditional home of medicine.*" The anatomy lecture notes were largely derived from the works of Caspar Bauhin and Andre du Laurens who were extremely influential anatomists during the late sixteenth and early seventeenth centuries. Also, there was much in common between John Moir's notes and the anatomy course medical students probably received (French, 1975). At the same time, new scientific ideas and concepts of the structure of the human body had been displacing old dogma.

FRANCIS GLISSON

Francis Glisson was one of the great physicians of the seventeenth century who made many other discoveries both in medicine and anatomy. In medicine, he is probably best remembered for his classical publications on rickets and on the liver. Glisson's work *Anatomia hepatis* (1654) established him as an outstanding anatomist because of the depth and clarity of anatomical details it contained, as well as its originality (see Fig. 112). The capsule around the liver is named after him, but he had also described the circular ring of muscle fibers around the openings of the pancreatic and bile ducts. This is generally called the sphincter of Oddi, after the Italian anatomist Ruggero Oddi, who described it much later in 1887.

Glisson first studied classics at Cambridge and later medicine under Harvey. He was appointed Regius Professor of Physik at Cambridge when he was only 39 years old. He succeeded William Harvey as Lumleian lecturer at the Royal College of Physicians and was one of the founders of the Royal Society of which he became president in 1667. Pursuing knowledge with a physiological basis, he advanced a new doctrine that he called "*Irritability*" based on the assumption that the human body reacted in a characteristic manner to external influences. Seeking to formulate a unifying concept of all bodily functions, Glisson believed that "*Every part that suffers from an incommodity seeks to disembarrass itself.*"

THOMAS WILLIS: DISCOVERING THE NERVOUS SYSTEM

Prior to the sixteenth century our knowledge of the structure and functions of the central nervous system was largely based on speculation and misrepresentation. Even the external features of the brain itself and the spinal cord, together with the peripheral nerves, were almost invariably crudely depicted. The earliest major changes began through the clinical work of the brilliant physician Thomas Willis (1625–1675). He was one of the first to give a reasonably accurate description of the anatomy of the brain, the spinal cord, the peripheral nervous system, and the autonomic nervous system (Feindel, 1964; Hughes, 1992).

In addition to describing the accessory nerve and the ophthalmic branch of the trigeminal nerve, Willis's description of the arteries at the base of the brain forming an arterial circle, now known as the circle of Willis (Symonds, 1955), bears testimony to only some of his achievements (Feindel, 1978). Thomas Willis was foremost an astute clinician (Isler, 1968) who carefully recorded his clinical findings, which even today stand as models of case histories. Thomas Willis was the first to discover myasthenia gravis, a debilitating neuromuscular disorder, as a distinct clinical entity. He published this work in 1672, but it remained largely unknown until 1903, because the report was in Latin. He should rightly be considered a pioneer in neuropathology because he was one of the first to correlate clinical findings in the patient with the changes found in the brain at autopsy.

Hughes (1992), in his biography of Thomas Willis, remarked that "*Willis's liking for order led him to gather up anatomical and physiological facts into systems, a most*

helpful concept in his time." He made many original discoveries relating to the nervous system and described for the first time several diseases. Moreover, he was one of the pioneers of comparative anatomy having dissected a wide variety of animals in order to compare their structures with those of the human.

Willis was not only very observant, but he had a critical and enquiring mind which provided the basis for the many clinical and scientific observations (Isler, 1968) he made from his patients. Hughes (1992) placed Willis as a medical scientist "*somewhere near the first place in the seventeenth century.*" He concluded Willis's biography with the observations that "*a large body of important work in anatomy, some in the new subject physiology, and a great deal, and possibly his major contribution, in clinical medicine, is formidable evidence of his pre-eminence as a medical scientist.*" Furthermore, Willis was a great physician and medical scientist, but "*William Harvey surpasses Willis in distinction by a large margin, but this Harvey does in comparison with any doctor in any century.*"

In 1543, Vesalius had published his monumental work *De Humani Corporis Fabrica* that was based on human dissections he himself carried out. This book ushered the beginnings of modern anatomy. Just over 80 years later, in 1628, William Harvey published his *Exercitatio Anatomica De Motu Cordis Et Sanguinis*, a work that, like Vesalius's magnificent treatise, was to revolutionize the science of medicine. By the time Willis was born and receiving his medical training at Oxford, medicine had progressed rapidly from an empiric practice to one that embodied rationality, conservatism, and all the revolutionary discoveries of the preceding years.

Thomas Willis was born on January 27, 1621 at Great Bedwyn in Wiltshire.* The family moved to North Hinksey, a small village not far from Oxford, where his mother had inherited a small farm of 50–70 acres. Willis's mother died in 1631, and his father remarried the widow of a friend who also farmed in the area. From his home in North Hinksey, Willis started school at Oxford about the time of his mother's death. Both buildings can still be recognized at their original sites. The school is now a tobacconist. At the age of 16 years, on March 3, 1638, Willis matriculated at the University of Oxford where he was at Christ Church.

Medical Studies at Oxford

Willis began his medical studies at Oxford in June, 1642 and graduated as a bachelor of medicine (B.M.) in December 1646, just after the city was surrendered by the Royalists. During these years there must have been considerable disruption in the life of the city, including formal instruction at the university. When Willis began to practice medicine in his rooms at Christ Church and in the marketplaces in the vicinity of Oxford he must have missed out on what would have been required for a formal education in medicine.

The medical course lasted 14 years, of which three were spent attending the lectures on Galen and in anatomy. For the bachelor of medicine degree, the student had to attend the lectures on Galen, participate in anatomical dissections, and take part in two public disputations. For the doctor of medicine degree, four years of further studies were required. Willis was mostly influenced by William Petty (1623–1687), who was the deputy to Sir Thomas Clayton (1612–1693), the Regius Professor of Medicine and Tomlin's Reader in Anatomy. Sir Thomas was, in contrast to his father who held the same position, a poor doctor and anatomist. It was remarked that he "*possesst with a timorous and effeminent humour, could never endure the sight of a mangled or bloody body.*" His appointment of the gifted Petty was considered to be the most worthwhile thing he did during his tenure. Willis's knowledge of medicine and anatomy was extensive, and it is most likely that he was inspired to pursue anatomical studies by the genial William Petty.

William Harvey, Thomas Clayton, and Robert Boyle were also present in Oxford at the time Willis was a student there. Harvey was well known and held prominent positions as personal physician to the King, Charles I, and as warden of Merton College. Willis might have been influenced by Harvey's experimental studies in medicine and his view that autopsies are necessary for studying diseases. Thomas Clayton (1575–1647) was Regius Professor of Medicine and also Tomlin's Reader in Anatomy, as well as master of Pembroke College. He was considered to be an excellent physician and teacher with the main duties to read the text of Hippocrates and Galen twice weekly. Robert Boyle (1627–1691) arrived in Oxford in 1664 where he set up his own private laboratory and carried out experiments over the next four years that paved the way for great discoveries. Boyle was one of the leading scientists of the seventeenth century. It is not known to what extent Willis was influenced by these gifted scholars.

Professor of Natural Philosophy

In 1657, Willis married. In addition to his medical practice, he pursued anatomical studies and carried out scientific experiments, particularly relating to the anatomy, physiology, and diseases of the nervous system. Unlike the quacks that were prevalent then, Willis, like his medically trained colleagues, used and experimented with remedies that appeared to be rational. In fact, this paved the way for the scientific study of diseases and their treatment. Willis had his own apothecary where his remedies were secretly prepared and widely valued. His practice grew enormously and he eventually became wealthy with an income that was the highest in Oxford at that time

Willis's reputation grew and, probably helped by his Royalist loyalty, he was appointed Sedleian Professor of Natural Philosophy in 1668 at Oxford. It is certain that Gilbert Sheldon through his influence helped Willis in securing the appointment, which was decisive for his career. The professor appointed was expected to read two days weekly at 8:00 A.M. from selected books of Aristotle. Students were fined for not attending the lectures, and the professor would also have been fined heavily if he did not give it. Willis lectured as required on Wednesdays and Saturdays at 8:00 A.M. during full-term, but from his own material, which moved away from Aristotelian tradition and incorporated the new Baconian Philosophy. The format and content of Willis's lectures we know from the notes that were taken by John Locke during at least two sessions. Willis's textbooks were based on

---

*For biographical details, see Feindel (1964; 1978) and Hughes (1992).

these lectures, and they have helped in no small part to spread his discoveries, as well as his reputation, in England and the rest of Europe. The lectures on the anatomy and physiology of the nervous system incorporated his own observations on dissected material, as well as his speculations.

Scientific Discoveries and Publications

Willis's clinical findings and scientific discoveries, as well as theories of functions and diseases (Rather, 1974), were published in his numerous books. Most of his discoveries on the nervous system are to be found in the *Cerebri Anatome* (1664) (Fig. 122), *Pathologiae Cerebri* (1667), and in *De Anima Brutorum* (1672), which was reprinted several times. Willis gave us not only

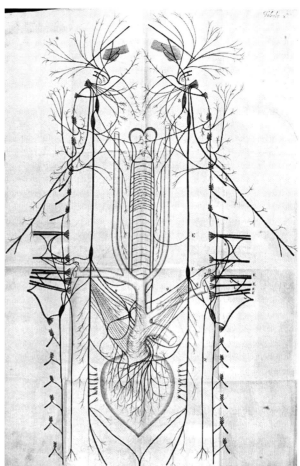

Figure 123. A depiction of cardiac innervation. Thomas Willis. *Cerebri Anatome: Cui Accessit Nervorum Descriptio et Usus.* London: Typis T. Roycroft, Impensis J. Martyn & J. Allestry, 1664. http://www.indiana.edu/~liblilly/anatomia/head.html.

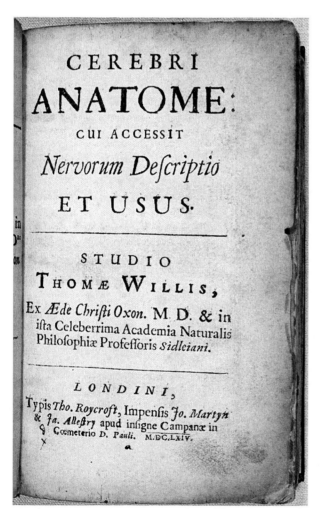

Figure 122. Front page of Thomas Willis book. *Cerebri Anatome: Cui Accessit Nervorum Descriptio et Usus.* London: Typis T. Roycroft, Impensis J. Martyn & J. Allestry, 1664. http://www.indiana.edu/~liblilly/anatomia/head.html.

his original neuroanatomical discoveries but also clear, vivid, and illustrated descriptions of the human nervous system (Fig. 123). Also, we find in his works a classification of the nervous system in its component parts, and a nomenclature of terminology, which has established itself in all areas of medicine relating to the nervous system.

The *Cerebri Anatome* was published in London in 1664 (Fig. 124). The work, which went through several editions, was dedicated to Willis's patron, Doctor Gilbert Sheldon, Archbishop of Canterbury. Christopher Wren and Richard Lower did some of the fine drawing, which vivified the text. Willis acknowledged the help he received from both Wren and Lower and Millington. The work was highly original and it established itself as a definitive textbook of the nervous

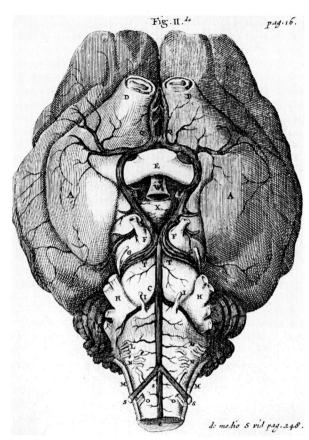

Figure 124. The base of a sheep's brain. Note the arteries and anastomoses, forming a complete circuit. This engraving was done by Christopher Wren (1632–1723). He was the architect responsible for the design of many beautiful churches and buildings in England, including the new St. Paul's Cathedral after the Great Fire in London. (From Willis, Thomas: *Cerebri Anatome.*, 1664; Courtesy of the School of Anatomy Library, University of Cambridge.)

system until the beginning of the eighteenth century. Willis's description of the brain, the spinal cord, and of the peripheral and autonomic nervous system, as well as the related blood supply, was unprecedented. Christopher Wren's drawing of the base of the human brain and Lower's depiction of the autonomic nervous system have remained as masterpieces of anatomical drawings.

In the *Pathologiae Cerebri*, many case reports of various diseases were either described for the first time or in a manner never before. This included epilepsy and asthma.

Willis considered the *De Anima Brutorum* to be his most important work. One finds here many original observations in comparative anatomy, clear descriptions of syndromes and diseases (Rather, 1974) described

for the first time. In the first part of the book, "*the nature, parts, powers, and affections of the soul*" were discussed, and this was followed by "*the unfolding of diseases which affect the soul and its primary seat, namely, the brain and nervous stock.*" This book remains as one of the monumental treatises in the field of medicine.

Most medical students today will associate Willis's name only with the arteries and their branches at the base of the brain. These blood vessels, together with their anastomoses, form a pattern that is widely known as the circle of Willis. The cerebral arteries were first described in detail by Gabrielle Fallopio (1523–1562), but he did not recognize the significance of these cerebral vessels. Guilio Casserio (1561–1616), one of William Harvey's professors at Padua, and Johannes Vesling (1598–1649) also published accounts with drawings of an incomplete circle, but lacking the important anastomoses between the two anterior cerebral arteries. In 1658, Johann Jacob Wepfer (1620–1695), in his account of a patient who had died as a result of apoplexy, gave the first accurate description of the cerebral arteries at the base of the brain. Cerebral arteries, their branches, and anastomoses to form an arterial circle, were accurately described. Willis's account of these arteries was accompanied by clear and accurate drawings done by Christopher Wren (Feindel, 1978), which was lacking in the account given by Wepfer (Symonds, 1955; Hughes, 1992).

Galen (A.D. 131–201), in describing the brain, which he considered to be the seat of the soul, identified seven pairs of cranial nerves. Because these nerves originated from the brain, they were considered to be nerves of sensation in contrast to the 30 pairs of spinal nerves, which he recognized and designated as nerves of motion. Galen did not recognize the olfactory and trochlear nerves. He designated the optic nerve as the first and considered both oculomotor and abducent to be the second. Galen saw two divisions of the trigeminal nerve as the third and fourth pair of nerves, and his fifth pair of cranial nerves was the facial and auditory combined. Galen named the glossopharyngeal, vagus, and spinal accessory nerves as the sixth pair of nerves, and the hypoglossal was considered to be the seventh. He knew of the sympathetic nervous system and the recurrent laryngeal branches of the vagus nerves on both sides of the neck. Willis recognized and described the olfactory, optic, oculomotor, and trochlear nerves. He described the ophthalmic division of the trigeminal nerve, as well as the abducens nerve. He distinguished the facial nerve from the auditory nerve, which he described as two branches. The spinal accessory nerve, which Willis was the first to describe, intrigued him because of its "*very irregular*" origin and branches.

The first modern account of the autonomic nervous system can be found in Willis's *Cerebri Anatome*. The impressive drawing that was done of the autonomic nervous system by Richard Lower contributed greatly to a better understanding of this part of the nervous system. The branches of the vagus nerves, and of the sympathetic trunk, were dissected and traced throughout their distribution to the heart, lungs, and abdominal organs. He described the nerves from the vagus and sympathetic trunk, which innervate the blood vessels, including the branch of the vagus nerve that is distributed to the arch of the aorta.

Willis observed and described many new features of the spinal cord, including the cervical and lumbosacral enlargements, as well as the anterior and posterior routes through which he believed that nerve impulses passed to and from the brain. As a result of studying dissected fetal specimens and injected material, Willis was able to provide a far better account than anyone before of the blood supply of the spinal cord. The drawings, made by Richard Lower, illustrate the two anterior spinal branches from the vertebral arteries uniting to form the anterior spinal artery. The tributaries accompanying the nerve roots, the posterior spinal arteries and the spinal veins, are all described and clearly depicted.

Final Years

Willis and his family moved from Oxford to the city of Westminster in late 1667. He was probably encouraged by Gilbert Sheldon, who was now the Archbishop of Canterbury, or because of the greater opportunities, including financial, that would be available in a larger city. On account of his reputation, Willis's practice grew and he became extremely wealthy in time. Although Willis was generous and provided free medical services for the poor, he amassed such wealth that he was able to buy a large estate in Buckinghamshire, apart from his mansion in the country and a townhouse.

About the time of his move to Westminster, Willis suffered great personal losses through the death of his sons, his daughter, and in October 1670, his wife. Two years later Willis remarried to a Mrs. Elizabeth Collier (Dame Elizabeth Calley). Three years later, at the age of 54 years, Willis became seriously ill with some form of respiratory disease.

And when this most expert person was not relieved by frequent bleeding and diligent taking of remedies, himself perceived the period of his life to approach (his friends hoping better), and after three days his household affairs being settled, and having taken the viaticum of the Holy Eucharist, and being received into the peace of the church, he commended his blessed soul to God, having his senses entire to his last breath, and finished his most exemplary life with the like death (John Fell, 1675).

Thomas Willis died on November 11, 1675 and was buried in the North Transept of Westminster Abbey. At the time of his death, Willis had a high reputation in England and the rest of Europe. He made significant discoveries in medicine and science, apart from his outstanding description of many important diseases and the discovery of new ones. Willis's legacy will also include the foundations he established for the scientific study of the structure, function and pathology of the nervous system (Frank Jr., 1976; Hughes, 1992).

JAMES DRAKE

The English physician and political writer James Drake (1667–1707) was the author of a popular medical treatise entitled *Anthropologia Nova* or *A New System of Anatomy* (Fig. 125). The two volumes of this work were published in 1707. In the preface, Drake acknowledged the assistance he received from the surgeon and anatomist William Cowper for both the text and plates. Many of the figures were taken from the work of Stephan Blanchard (1650–1720). Blanchard's book, "*A New Anatomy with Concise Directions for [the] Dissection of the Human Body with a New Method of Embalming*" was published in 1688. After Drake's death, two further editions were published in 1717 and 1727 by his wife who was a physician. By then William Cheselden's "*The Anatomy of the Humane Body*," published in 1717, and was the established text (Tubbs et al., 2012).

James Drake was born in Cambridge and entered Caius College, Cambridge in 1684. An exceptional student, he studied medicine and received the M.B. (1690) and M.D. (1694) degrees. Drake left for London where he established a medical practice and was engaged in writing political pamphlets. In 1701, Drake was elected a Fellow of the Royal Society and

Figure 125. A portrait of James Drake. Line engraving by M. van der Gucht, 1717, after T. Foster. (Wellcome Institute Library, London.)

admitted a Fellow of the College of Physicians. A tory supporter, he wrote a pamphlet in 1702 entitled "*The History of the Last Parliament,*" which was critical of the government and bordered on sedition. It was felt that his remarks vilified the future Queen Anne. For this, he was summoned to appear before the House of Lords, prosecuted, tried, and acquitted. The following year, Drake published "*Historia Anglo-Scotica*" which did not please the Presbyterians. With his supporters, Drake continued to publish controversial and defamatory pamphlets. He was convicted for libel but eventually acquitted due to a technical error. Drake became bitter and disappointed because of "*ill-usage from some of his party.*" It is recorded that this "*threw him into a fever, of which he died at Westminster, March 1707*" (Stephen, 1888).

WILLIAM CHESELDEN

The distinguished surgeon and lithotomist William Cheselden (Fig. 126) was born on October 19, 1688 in Leicestershire to a wealthy family. At 15 years of age, he was already a pupil of William Cowper, the famous anatomist, having spent some time before that as an apprentice to a barber-surgeon in Leicester. After a further apprenticeship to a surgeon at St. Thomas's Hospital, Cheselden passed the examinations of the Barber-Surgeons Company in 1710, which gave him the right to practice surgery. A year later, at the age of 23, he had established an anatomy school in his own house where he also dissected human bodies for classes. Cheselden gave a course of 35 lectures, four times each year, which became extremely popular. Later, Cheselden continued his lectures in anatomy at St. Thomas's Hospital in London.

Private Anatomy School

Cheselden can be regarded as one of the pioneers of the private anatomy schools in England, and the successes of his enterprising spirit inspired many others, including William and John Hunter. An extract from an advertisement in March 1721 of his anatomy lectures announced that

A course of anatomy of which will be shown all the known mechanisms of the human body, together with the Comparative Anatomy of Birds, Beast and Fishes with the various contrivances for the different ways of life. The whole to be illustrated by Mechanical Experiments, there being a new apparatus made for this purpose. To be performed by William Cheselden, surgeon, F.R.S., and Francis Hawksbee at his house in Crane Coat, Fleet Street, where such creations are taken in. To begin Tuesday, March 28, at 6:00 in the evening. N.B. This quote being chiefly intended for gentlemen, such things only will be omitted as are neither instructive nor entertaining, and care will be taken to have nothing offensive.

Among Cheselden's pupils, was Thomas Cadwallader (1708–1799) of Philadelphia. After completing his medical studies, he returned home in 1730 and from 1745 to 1751 he gave the first course in anatomy in America (Lassek, 1958). Cadwallader was "*so proficient in dissection...that students and physicians alike urged him to give a public course of lectures on the cadaver*" (Marks & Beatty, 1973).

Surgeon and Anatomy Teacher

Cheselden quickly established himself as a skilled surgeon and as a good anatomy teacher. At the early age of 24 years, in 1712, he was elected a fellow of the Royal Society. His initial attempts to secure for himself a position as surgeon at St. Thomas's Hospital was apparently blocked by the influential Richard Mead, who was Reader in Anatomy with the Company of Barber-Surgeons, and was a surgeon of importance at the hospital. Eventually, in 1718, Cheselden was appointed an assistant surgeon and a year later became one of the principal surgeons. It has been suggested that Cheselden's difficulties with Mead made him a strong supporter of the movement, which led to the separation of the barbers from the surgeons. He became one of the first wardens of the new surgeon company and in 1746 was elected its master.

Figure 126. William Cheselden (1688–1752). Painting by Jonathan Richardson (1665–1745). (From Wegner, R.N.: *Das Anatomenbildnis. Seine Entwicklung im Zusammenhang mit der anatomischen Abbildung.* Basel, Benno Schwalbe and Co., Verlag, 1939. Courtesy of Institut für Anatomie, Ernst-Moritz-Arndt-Universität, Greifswald, Germany.)

Lithotomist

Cheselden is best known as a surgeon who became one of the greatest experts in carrying out "*the operation for removal of the stone.*" At first, he used the suprapubic approach, but because of the technical problems and dangers he adopted a lateral approach based on that of Frere Jacques. In 1723, Cheselden published a small book, 36 pages long, entitled *Treatise on the High Operation for Stones*, which he surprisingly dedicated to Doctor Richard Mead. He himself stated that he carried out at St. Thomas's Hospital 213 of these operations with relatively few losses.

Of his surgical experience, Cheselden published relatively little. He stated that

What success I have had in my private practice I have kept no account of because I had no intention to publish it, that not being sufficiently witnessed. Publicly, in St. Thomas's Hospital, I have cut 213:

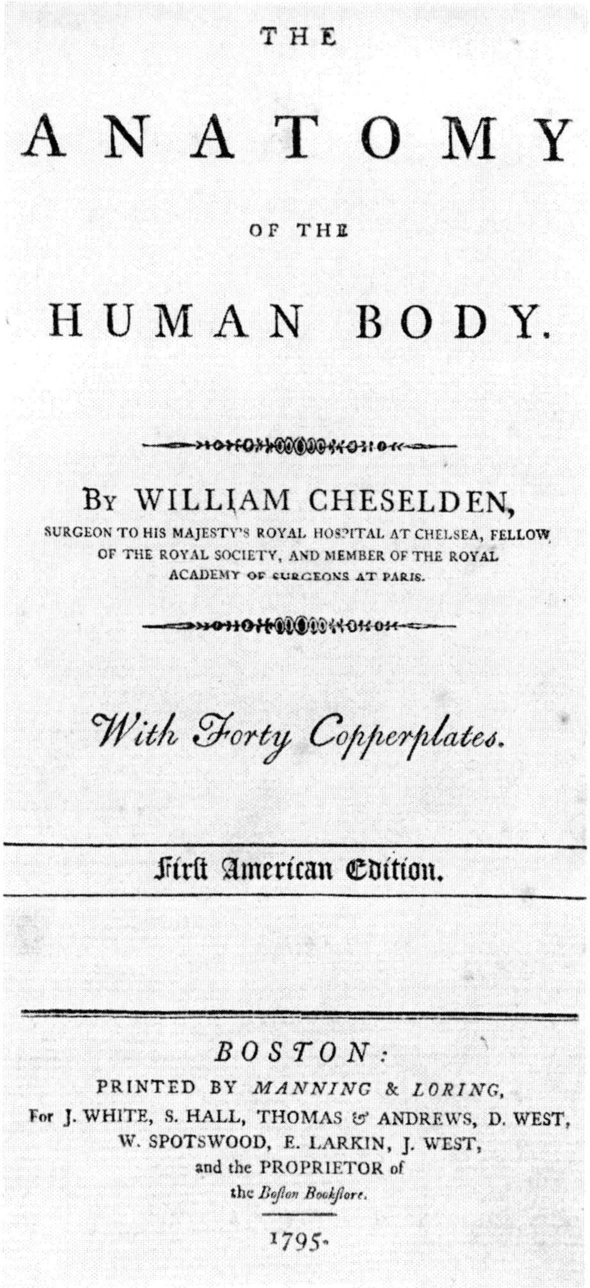

THE

ANATOMY

OF THE

HUMAN BODY.

———>=0+=0+=0@@@@=+=0=:0=<———

By WILLIAM CHESELDEN,

SURGEON TO HIS MAJESTY'S ROYAL HOSPITAL AT CHELSEA, FELLOW
OF THE ROYAL SOCIETY, AND MEMBER OF THE ROYAL
ACADEMY OF SURGEONS AT PARIS.

———>=0+=0+=0@@@@=+=0=+=0=<———

With Forty Copperplates.

——————————————————————

First American Edition.

═══════════════════════════

BOSTON:

PRINTED BY *MANNING & LORING,*

For J. WHITE, S. HALL, THOMAS & ANDREWS, D. WEST,
W. SPOTSWOOD, E. LARKIN, J. WEST,
and the PROPRIETOR of
the *Boston Bookstore.*

1795.

Figure 127. Title page of the first American Edition of William Cheselden's book on *The Anatomy of the Human Body*. Boston, 1795. (Courtesy of the Neil John Maclean Health Sciences Library, University of Manitoba.)

of the first 50 only three died; of the second 50, three; of the third 50, eight; and of the last 63, six. Several of these patients had the smallpox during their care, some of which died but I think not more in proportion than what usually die of that distemper; these are not reckoned among those who died of the operation.

Cheselden was an extremely modest man and about his reputation he remarked:

If I have any reputation in this way I've earned it dearly, for no one ever endured more anxiety and sickness before an operation, yet from the time I began to operate all uneasiness ceased and if I have had better success than some others I do not impute it to more knowledge but to the happiness of mind that was never ruffled or discontented and a hand that never trembled during any operation.

Cheselden ranks among the famous of British surgeons; he was both brilliant and skilled, known for performing the lateral lithotomy within a minute. In 1727, he was appointed surgeon to Queen Caroline and his patients included the Pope, Sir Isaac Newton, and Sir Hans Sloane. In 1729, he was the first foreign associate to be elected at the Royal Academy of Surgery in Paris. Leaving St. George's Hospital in 1737, he was appointed surgeon to Chelsea Hospital, and it was during this period, in the summers of 1749 and 1750, that the young John Hunter came under his influence.

The Anatomy of the Humane Body

In 1713, Cheselden's book *The Anatomy of the Humane Body* was published (Fig. 127). It was well illustrated with 23 copper plates (Figs. 128–131), and the work incorporated the second edition of his *Syllabus*. The book was not purely descriptive, but it emphasized the functional and practical aspects of anatomy. Not surprisingly, the book proved extremely popular, and it remained in print for close to a century. It was the anatomy textbook used by most medical students in Britain. The first American edition of the book was printed in Boston in 1795, and in 1790, a German translation was published in Gottingen.

Cheselden's other work *Osteographia* or *The Anatomy of the Bones*, which appeared in 1728, was less successful. Several editions of the *Osteographia* were published, but apparently there were too few subscribers as a result of which Cheselden suffered some financial loss. He complained about this in the 1740 edition of his original *Anatomy*, where he stated that of the 300 copies of his Osteographia, only 97 were sold to subscribers.

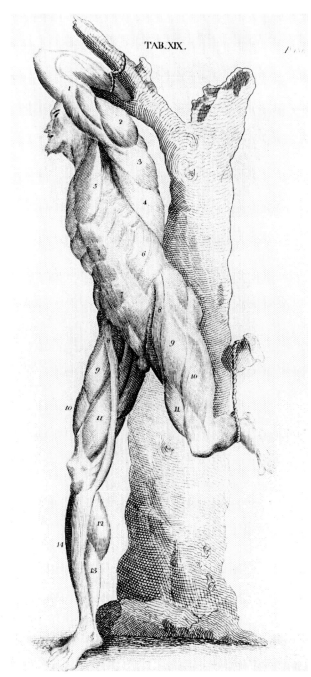

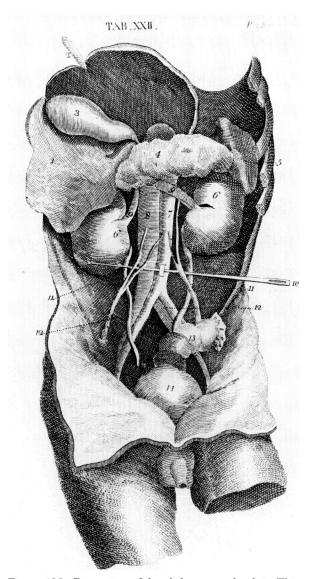

Figure 128. Copper-engraved plate from William Cheselden's work. (Courtesy of the Neil John Maclean Health Sciences Library, University of Manitoba.)

Figure 129. Dissection of the abdomen and pelvis. This elegant copper-engraved plate, taken from William Cheselden's *The Anatomy of the Human Body* (First American Edition, Boston, 1795), shows the relationships of the major organs. Observe the liver (1), gallbladder (3), pancreas (4), spleen (5), kidneys (6), abdominal aorta (7), inferior vena cava (8), ureter (11), iliac vessels (12), rectum (13), and bladder (14). (Courtesy of the Neil John Maclean Health Sciences Library, University of Manitoba.)

PERCIVALL POTT

In the early seventeenth century, the increasing interest in human dissection provided a more rational basis for the practice of surgery. The two great masters of British surgery who continued the tradition of carrying out human dissections and from their observations improved operative procedures were Percivall Pott (1714–1788; Fig. 132) and John Hunter (1728–1793). Whereas Percivall Pott was a clinical surgeon,

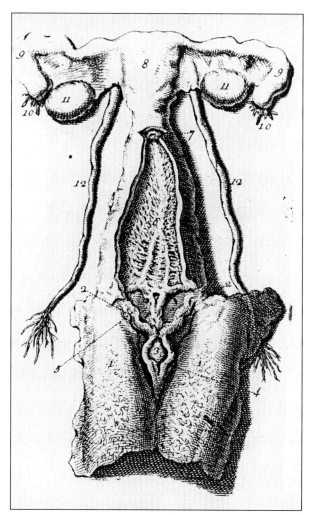

Figure 130. Specimen of the female reproductive tract. Observe the uterus (8), the dissected vagina (6), the uterine tubes (9), and ovaries (11). (From Cheselden, W.: *The Anatomy of the Human Body*. Boston, 1795. Courtesy of the Neil John Maclean Health Sciences Library, University of Manitoba.)

Figure 131. The portal venous system. The veins were demonstrated by injecting liquid wax into the vessels, which later hardened (see Edwards and Edwards, 1959, for details of this and other earlier techniques that were used in making anatomical specimens and for revealing the blood vessels). Observe the hepatic portal vein itself (2), one of the mesenteric veins (4), branches in the mesentery (1), and the intricacy of the veins in the liver (3). (From Cheselden, W.: *The Anatomy of the Human Body*, 4th Ed., W. Bowyer, London, 1730. Courtesy of the Neil John Maclean Health Sciences Library, University of Manitoba.)

John Hunter was a natural scientist and a tireless dissector, in addition to being a great surgeon. Both, however, were men of boundless energy who like so many other surgeons of the era recognized the importance of anatomical knowledge for safe and rational surgery.

Percivall Pott is one of the great figures in the history of British surgery. His writings and operations he carried out advanced the field of surgery. Percivall Pott was a surgeon-anatomist, which was at that time quite common, and much of his success in the treatment he carried out and operations he performed can be attributed to his earlier experiences in anatomy.

Surgeon-Anatomist

The well-known London surgeon Edward Nourse (1701–1761), who was one of the two surgeons at St. Bartholomew's Hospital in London, had his own private school in Aldersgate St. where he lectured in anatomy and surgery.

At 16 years of age, Percivall Pott became an apprentice to Doctor Nourse, and it was his duty to dissect the specimens that were to be used for demonstrating the lectures. In addition, he assisted Nourse in the treatment of patients on the wards and during operations. After several years of apprenticeship, which involved carrying out many meticulous dissections and observations, Percivall Pott acquired a love for anatomy and dissection that was to remain with him throughout his professional life.

In 1736, Percivall Pott passed the examination and received the Great Diploma *"testifying his skill and empowering him to practice"* from the Company of Barber-Surgeons. In 1745, Pott received an appointment as assistant surgeon at St. Bartholomew's Hospital. Four years later he became a full surgeon and remained in that position at the hospital until his retirement in 1787, two years before his death on December 22, 1789. Pott apparently continued his interest in anatomy at the Aldersgate St. school, which, according to an announcement appearing on October 17, 1734 in the *London Evening Post*, was relocated to the hospital. The announcement stated that *"Desiring to have no more lectures at mine own house, I think it proper to advertise that I shall begin a course of Anatomy, Chirurgical Operations and Bandages, on Monday, November 11, at St. Bartholomew's Hospital. Ed. Nourse, Assistant Surgeon and Lithotomist to the Hospital."* In later years, Percivall Pott's name was added to the advertisement.

Surgical Books

Pott was the author of several surgical works which contained his innovative surgical approaches based on his own observations in the dissecting room and on operations he carried out. Best remembered over the years for the fracture and dislocation of the ankle joint, which is known as Pott's fracture, he wrote a book on hernias where he emphasized the need for early surgery in order to relieve the strangulated hernia. Priority in the description of hernia led to controversial exchanges with the great William Hunter (1718–1783), brother of John Hunter, both famous surgeons and widely respected for their anatomical

Figure 132. A portrait of Percival Pott. Stipple engraving by J. Heath, 1790, after Sir J. Reynolds. 1790 By: Joshua Reynolds after: James Heath. (Wellcome Institute Library, London.)

studies and discoveries. Pott's other books included: *Fistula Lachrymalis* (1758), *Head Injuries* (1760), and *Fractures and Dislocation* (1768). From his anatomical studies, Pott knew that the muscles have to be relaxed before the bone can be properly aligned. He also recommended that the fracture should be immediately reduced. Physicians will also be familiar with the clinical condition he so accurately described that is now known as *"Pott's disease."*

LORENZ HEISTER

Lorenz Heister (1683–1758), who was Professor of Anatomy and Surgery at Altdorf and later at Helmstedt, was one of the great pioneers in surgery. During the eighteenth century, his books were popular, especially among surgeons. Of Heister's books, the most influential was his *Chirurgie*, which presented the subject in a systematic manner. The work appeared in seven German and ten English editions, as well as in several translations. A skilled and innovative surgeon, Heister believed that "*it is necessary for a surgeon to have complete, or at least very good, knowledge in anatomy as well as in medicine, so that he has enough judgment and understanding to study all the causes and circumstances, and to draw his conclusions from them*" (Haeger, 1989).

AMBROISE PARÉ

Barber-Surgeons

Medical schools had been flourishing in France from the fourteenth century, and, not unlike the situation in Britain, the university-trained medical doctors distanced themselves from the barber-surgeons (Riesman, 1935; Dobson and Walker, 1980). Describing the seventeenth century French surgeons, Haeger (1989) remarked that

the surgeons' competence really was poor, and most practitioners had a lower social status than anywhere

else. In 1634, there were 40 barber-surgeons in Paris, and by the middle of the 18th century their numbers had increased to more than 300. Apart from luminaries like Ambroise Paré (Fig. 133) and some obstetricians of merit, the art was conducted largely as a hobby of the barber-journeymen, with correspondingly poor quality.

Indeed, there were few surgeons with university training and only in 1645 did they join with the barbers' guild. Much earlier, in 1540, the English Guild of Surgeons had joined up with the Company of Barbers to form the United Barber-Surgeon Company (Dobson & Walker, 1980).

Ambroise Paré (1510–1590), the greatest surgeon of the Renaissance (Shah, 1992), published in 1561 his book on anatomy entitled, "*Anatomie Universelle du Corps Humain.*" Based on the work of Vesalius, "*it had the enormous significance of making Vesalius well-known among surgeons*" (Haeger, 1989). Trained as a barber-surgeon, Paré grasped the importance of anatomy for his craft as a military surgeon. He introduced humane methods of treating wounds. To prevent loss of blood, a profusely bleeding wound was cauterized, or the wound was sealed with hot oil. Paré instead used a soothing dressing made of egg whites, rose oil, and turpentine to cover the wound. He was one of the first who used ligatures in order to stop the bleeding following surgery, and he also designed artificial limbs for injured soldiers.

On Monsters and Marvels

Paré wrote a book "*On Monsters and Marvels*" which was essentially an illustrated encyclopedia of curiosities, monstrous human and animal births, bizarre beasts, and natural phenomena. Although he described many cases of anomalies, his explanation for them reflected the era in which he lived: the glory of God; the wrath of God; too much seed; too little seed; mingling of the seed; a narrow womb; blows to the womb; demons, devils; and the mother's imagination.

Figure 133. Ambroise Pare (c. 1510–1590), "*the father of modern surgery.*" (From *Les oeuvres d'Ambroise Pare*, Lyon, Rigaud & Obert, 1633.)

JEAN-LOUIS PETIT

Jean-Louis Petit (1674–1760) stands out as one of the most outstanding surgeons of his time (Fig. 134). Already as a child he was assisting the great surgeon and anatomist Alexis Lure (1658–1726). Later as a trained surgeon, he began to give, privately, lectures and demonstrations in anatomy and surgery, which attracted students from everywhere. Petit is remembered for the muscular gap (trigonum Petiti) he described in the lumbosacral region.

PIERRE-JOSEPH DESAULT

Petit was succeeded by another precocious and brilliant surgeon, Pierre-Joseph Desault (1744–1795; Fig. 135), who regarded anatomy as a practical subject. He advocated that anatomy should be learnt in the dissecting room and related to functions (Haeger, 1989). Desault was born to a poor family in Magny-Vernois in Franche-Comte, France. He received his early education from the Jesuits hoping to become a priest, but because of his interest in medicine he became an apprentice to a local barber-surgeon.

In 1776, Desault moved to Paris with some knowledge of human anatomy and surgery. Not yet 20 years of age, and without a university degree, he opened a school of anatomy where he lectured and introduced new techniques in anatomy and in the practice of surgery. In 1782, Desault was admitted to the Academy of Surgeons, and he received an appointment as a surgeon-major to the De la Charite Hospital in Paris.

Desault became the most prominent surgeon in Paris. He was appointed to the Hôtel-Dieu where he founded a school of clinical surgery. Because of his clinical skill and reputation as a teacher, he attracted a

Figure 134. Jean-Louis Petit. (Courtesy of Bethesda, MD: U.S. National Library of Medicine, National Institutes of Health, Health & Human Services.)

Figure 135. A portrait of Pierre-Joseph Desault. Coloured mezzotint by Gautier after F. P. J. Kymli. (Wellcome Institute Library, London.)

large number of students from France and other countries. There were as many as 600 students and fellows attending Desault's lectures and surgical clinics. A gifted teacher and a skilled surgeon, he instilled a high level of professionalism among his pupils. In 1791, Desault founded and edited a surgical journal, *Journal de chirurgerie*, which published interesting cases from the surgical clinic.

Apparently for no serious reason, Desault was arrested and imprisoned in 1793 by the revolutionary government. Following protests by his patients and students he was released after three days. Desault's death two years later was caused by pneumonia and not by poisoning as was suspected. Desault authored a three volume book on surgical techniques (*Oeuvres chirurgicales de Desault*, 1798–1799), and Xavier Bichat (1771–1802) edited his treatise on fractures and dislocations (*Traite des maladies chirurgicales*, 1799) which was translated into English (Walsh, 1908).

XAVIER BICHAT

Anatomy found another admirable representative, Xavier Bichat (1771–1802), who followed Desault (Fig. 136). He was an early founder of histopathology and it has been stated that he dissected six hundred corpses during a single winter (Haeger, 1989).

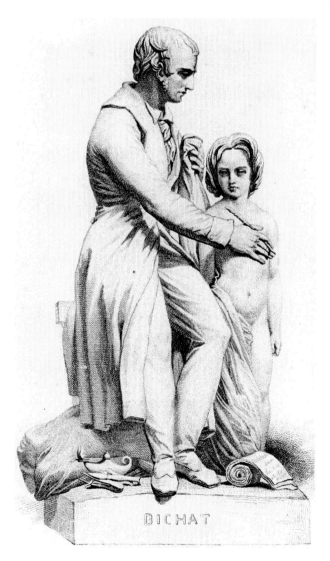

Figure 136. Marie Francois Xavier Bichat (1771–1802). Bichat is depicted here examining a child. (From Bichat's *Recherches Physiologiques sur La Vie Et La Mort.* Courtesy of the Neil John Maclean Health Sciences Library, University of Manitoba.)

Figure 137. Antonio Scarpa (1752–1832). Engraving by Faustino Anderloni from a painting by Gaetano Cattaneo. Frontispiece, Scarpa's *Saggio di osservazioni e d'esperenze sulle principali malattie degli occhi* (Pavia, 1801). (From Wegner, R.N.: *Des Anatomenbildnis. Seine Entwicklung im Zusammenhang mit der anatomischen Abbildung.* Basel, Benno Schwalbe and Co., Verlag, 1939. Courtesy of Institut für Anatomie, Ernst-Moritz-Arndt-Universität, Greifswald, Germany.)

ANTONIO SCARPA

Another brilliant surgeon-anatomist of this period was Antonio Scarpa (1752–1832) whose name is eponymously associated with several anatomical structures. He was professor of anatomy in Pavia (Fig. 137). Scarpa wrote on the nervous system, the anatomy of the eye, and on hernias. He pioneered operations for clubfoot and hernias. From contemporary sources it appears that Scarpa was a difficult and complicated person. According to Zimmerman and Veith (1967),

he (Scarpa) was endowed with a brilliant and calculating intellect, enormous energy, boundless ambition, a passion for accuracy and exactness in his research, and a broad range of cultural and artistic interests.... Lamentable aspects of his personality brought him hatred and opposition, which could only be held in check by the general fear of his power and ruthlessness.... (cited in Haeger, 1989)

JOHN HILTON

Before anesthesia, analgesics, radiology, electrocardiology, bacteriology, and antibiotics became a part of medical practice, doctors depended on their five senses for making a diagnosis. Invariably, the treatment that followed was simple, based on previous experience, except for a few crude surgical procedures.

The greatest anatomist in Britain at that time was John Hilton (1805–1878) who was born at Castle Hedingham in Essex (Fig. 138). From humble beginnings, he studied anatomy intensively, became a respected professor of anatomy, and later reached the highest echelon as a surgeon. Today, his name is immortalized

in several eponyms and a series of lectures on "*rest and pain*" which he delivered to the Royal College of Surgeons in England (Shenker and Ellis, 2007; Brand, 2009).

Anatomical John

Hilton entered Guy's Hospital in London in 1824, and four years later he was appointed Demonstrator of anatomy. Nicknamed "*Anatomical John,*" he acquired a thorough knowledge of anatomy from the meticulous and detailed dissections he carried out "*for an hour or*

Figure 138. John Hilton. Photograph by Barraud & Jerrard, 1873. (Wellcome Institute Library, London.)

two every morning" over 16 years. The moulageur and sculptor Joseph Towne created wax models from Hilton's preserved anatomical preparations. These unique specimens have survived the years and are kept in the Guy's Hospital medical school museum.

Surgeon-Anatomist

Hilton was appointed an Assistant Surgeon at Guy's Hospital in 1844 and a year later he became lecturer of anatomy. He was a good teacher, interesting, and admired by the students. In 1839, Hilton was elected a Fellow of the Royal Society in recognition of his anatomical studies, especially for the work he did relating to the origin and distribution of the superior laryngeal nerve. His career soared with prestigious professional appointments, including professor of human anatomy and surgery to the Royal College of Surgeons (1859–1862), and in 1871 Surgeon Extraordinary to Queen Victoria. He served the Royal

College of Surgeons as a life-member of the Council (1854), examiner, Vice-President (1865–1866), and as President in 1867 (Shenker & Ellis, 2007).

Rest and Pain

Hilton's delivered a series of 18 lectures entitled, *"On the influence of mechanical and physiological rest in the treatment of accidents and surgical disease and the diagnostic value of pain,"* over a period of three years (1860–1862) at the Royal College of Surgeons. The lectures were popular and, because of the demand, were quickly published as a book in1863. Given a shortened title *"On Rest and Pain,"* five editions of this classic work appeared over the years, the most recent, entitled, *"Rest and Pain,"* in 1950 with editorial comments and annotations of current concepts in relation to Hilton's views (Hilton, 1950). In Lecture XIV, Hilton recommended the value of rest as a therapeutic agent for cases of hip-joint diseases in children. A consummate anatomist, he demonstrated his vast knowledge of anatomy and confidently discussed the underlying principles regarding referred pain and the accompanying discomfort:

> I shall not dwell upon the anatomy of the hip-joint, except to remind you that its muscles perform their functions in groups, that each group has a trunk nerve of its own, and that each nerve contributes a branch to the hip-joint itself...you see a branch of the anterior crural nerve passing to the hip-joint; a branch of the obturator going to the capsular ligament and to the ligamentum teres; and a branch proceeding to the posterior aspect of the hip-joint from the sacral plexus which supplies the gemelli, the quadratus femoris, and the obturator internus. This anatomy should be borne in mind, because it explains how it happens that the remote and "sympathetic pains," associated with an inflammatory condition or chronic disease of the hip-joint, are not always found at the same part of the limb. We all know very well that, in some cases of hip-joint disease, one of the earliest symptoms is remote from the actual seat of mischief namely, pain within the knee, or on the inner side of the knee-joint; and we are familiar with the explanation of it namely, that the obturator nerve, which contributes a branch to the ligamentum teres, sends a branch to the interior of the knee-joint, to the inner side of it, and sometimes even lower down. (Hilton, 1950)

Skilled and Cautious Surgeon

Hilton was a skilled and cautious surgeon who relied on his knowledge of anatomy and experience when operating. Apparently, he was the first person to describe a case of acute osteomyelitis in a patient after examining postmortem specimens. Hilton reduced a case of obturator hernia by abdominal section, and was one of the first to carry out a lumbar colostomy. In drainage of abscesses, Hilton cautioned not to damage the underlying nerves and major blood vessels. After incising the skin and subcutaneous tissues with a scalpel, Hilton recommended using a sinus forceps to penetrate the deep fascia. The forceps is then opened up for drainage of the pus. Hilton is remembered today, especially among medical students, for a "*white line*" and a "*law*" that are eponymously associated with his name. Hilton's white line is the transitional landmark for the lymphatic drainage of the anal canal, and Hilton's law postulates that the nerve innervating a joint usually innervates the muscles that move the joint, as well as the skin overlying the joint (Shenker & Ellis, 2007).

Chapter 16

WILLIAM HUNTER AND HIS LEGACY

THE HUNTER BROTHERS

In eighteenth century Britain, the brothers John and William Hunter were among the most famous men in the field of medicine. William was already a distinguished surgeon and obstetrician, as well as an anatomist, when he invited his younger brother John to join him in London (Bynum & Porter, 1985). From William, John received his initial training in human dissection and surgical anatomy (Peachey, 1924). He became a consummate dissector and teacher at William's Great Windmill School (Moore, 2005a;b).

According to Oppenheimer (1946), the brothers were bitter rivals even though they were close at the beginning. Apparently, their relationship came to an end following a bitter dispute as to who had first described the distribution pattern of the placental blood vessels. John became famous as a surgeon and comparative anatomist (Persaud, 1996). However, his remarkable accomplishments would not have been possible without the initial training, help, and encouragement he had received from William.

John infused into the practice of surgery a scientific rationale, and William was a prominent "*man-mid-wife*" or obstetrician (Wilson, 1995), who, through his lecturing and private school at Great Windmill Street, exerted "*an even profounder effect on the whole course of medical progress in Britain*" (Oppenheimer, 1946). Indeed, what John did for surgery, William achieved the same for all of medicine through his own efforts and those of his students, including John.

LIFE OF WILLIAM HUNTER

Medical Studies and Anatomy

William Hunter (1718–1783) was, like his brother John, born in Long Calderwood, East Kilbride, on May 23, 1718 (Fig. 139). He began his studies in 1731 at Glasgow University, and in 1736 left to become a friend and apprentice to the Scottish physician William Cullen (1710–1790). Cullen was involved in Glasgow and Edinburgh medical schools. He was an outstanding teacher and attracted large numbers of students to both institutions. Cullen was one of the first to lecture in English. His textbook on medical practice dominated the era and it was translated into other languages.

In 1739, William Hunter attended the anatomy lectures given by Alexander Monro, *Primus*. Monro learnt his anatomy from William Cheselden (1688–1752) in London and after returning to Edinburgh he established the influential School of Anatomy. Shortly after, he left for London where he attended St. George's Hospital and worked with William Smellie. From August 1741 to April 1742, William Hunter became an assistant to the anatomist and obstetrician, James Douglas, who took him into his family. In addition to being a pupil of James Wilkie at St. George's Hospital, William was also a "*dissecting pupil*" under the "*leading anatomical teacher in England,*" Frank Nicholls (1699–1778), in London (Robb-Smith, 1971), and he also attended the lectures of Antoine Ferrein in Paris (Brock, 1994).

Practice of Midwifery

The surgeon Simon Foart Simmons, who presented

an account of William's life at a meeting of the Society of Physicians of London in August 1783, following his death remarked as follows (Brock, 1983): His patron, Dr. James Douglas had acquired considerable reputation in midwifery, and this probably induced Mr. Hunter to direct his views chiefly to the same line of practice. His being elected one of the surgeon-men-midwives first to the Middlesex and soon afterwards to the British Lying-In Hospital, assisted in bringing him forward in this branch of his profession in which he was recommended by several of the most eminent surgeons of that time, who respected his anatomical talents and wished to encourage him.

William brought several improvements to the practice of midwifery. For example,

> he allowed the women to deliver themselves, or at least give nature leave to exert her own powers. He even allowed to dispose of its placenta; all of which gave a new turn to midwifery; and instruments which had been to much used, whereby Dr. Hunter almost laid aside. He also allowed milk breasts to take their own, only applying poultices to them.... Although by these incidents he was established in the practice of midwifery, it is well known that in proportion as his reputation increased, his opinion was eagerly sought after in all cases where any light concerning the seat or nature of the disease could be expected from an intimate knowledge of anatomy. (Brock, 1983)

In August 1747, William Hunter was admitted to the Corporation of Surgeons of London, and a year later he was appointed surgeon accoucheur at Middlesex Hospital.

John Hunter joined William in September 1748, just two weeks before William's anatomy course began. In the same year, William was appointed physician accoucheur to the Middlesex Hospital and a year later as accoucheur at the British Lying-In-Hospital. He graduated with the Doctor of *Physic* degree from Glasgow University on October 2, 1750, and in 1751, he was admitted to the Faculty of Physicians and Surgeons of Glasgow. He gave up his membership of the Corporation of Surgeons.

William was appointed physician extraordinary to Queen Charlotte, wife of King George III, having first attended to her the year before. From a letter that was written by William's niece, Agnes Baillie, to her brother Matthew Baillie, we learn that William "*was the first man that ever attended any queen in the country*" because she was attended to by a woman in her first confinement. In 1762, Mr. Hunter attended the Queen

Figure 139. William Hunter (1718–1783). The first great teacher of human anatomy in London, surgeon-anatomist, obstetrician, and founder of the famous anatomy school, which was located at Great Windmill Street. (Courtesy of Bethesda, MD: U.S. National Library of Medicine, National Institutes of Health, Health & Human Services.)

about the middle of her pregnancy for the first time on account of "*cough and heat*" which she suffered from. Her physician recommended "*bleeding*," but first consulted with Hunter. The queen was relieved of five ounces of blood. The pregnancy progressed normally and a son, the Duke of Cornwall, was eventually born (Paget, 1897).

William Smellie

William Hunter worked with Doctor William Smellie (1697–1763), of Lanark, "*who had come to London two years before him, and was laying the foundations of great success practicing as an apothecary and accoucheur.*" As a result, he became an expert in the practice of midwifery.

Smellie was admitted to the College of Physicians and Surgeons in Glasgow in 1733, and in 1745 he received the doctor of medicine degree from Glasgow University. Following further studies in Paris and London, he established in London an obstetrical practice and a school for midwifery. Doctor Smellie had the advantage of him in experience, had been lecturing and writing for many years, but his person is said to have been coarse, and his manners awkward and unpleasing; and fortune had treated William Hunter more generously.

Smellie was the leading midwife (obstetrician) in Britain. He recommended that

Those who intend to practice Midwifery, ought first of all to make themselves masters of anatomy, and acquire a competent knowledge in surgery and physic; because of their connection with the obstetric art, if not always, at least in many cases. He ought to take the best opportunities he can find of being well instructed; and of practicing under a master, before he attempts to deliver by himself. (A Treatise on the Theory and Practice of Midwifery, 1752).

Smellie is best known for the obstetrical forceps, which he designed, and for the management of a breech presentation. In addition to his obstetrical books, Smellie also published an important anatomical "*set of anatomical tables, with explanations, and an abridgment, of the practice of midwifery.*" With Smellie as his mentor. William now had a more promising future ahead, especially after the midwives began to be more tolerant of men practicing midwifery (Paget, 1897; Smellie, 1968).

WILLIAM'S ANATOMY SCHOOL AND MUSEUM

About 1746, William Hunter began teaching at his school, which later became famous and was called the Great Windmill Street School (Thomson, 1942). William advertised in the *London Evening Post* of January 9–12, 1748, that he will be giving a course of lectures on anatomy from October thirteenth. The students were assured that they would be able to dissect as it was done in Paris. His brother, John, who had recently joined him learned quickly from William. Moreover, he was a meticulous and an enthusiastic

dissector preparing specimens for the classes, as well as tutoring the students. The first classes were held in Convent Garden, later at several other locations, and then from 1767 in a house William had purchased Great Windmill Street.

William had the house rebuilt so as to accommodate the anatomy lectures and dissection, as well as a residence for himself (Figs. 140 & 141). "*Besides a handsome amphitheatre and other convenient apartments for his lectures and dissections, there was one magnificent room*

Figure 140. William Hunter's Anatomy School on Great Windmill Street in London. The salmon-colored brick building on the right was the site of the dissecting room, museum and Hunter's home. It is now a dinner theater.

Figure 141. Commemorative plaque in honor of Doctor William Hunter (1718–1783) on the front wall of the building that was his home, anatomy school, and museum. (Great Windmill Street, London.)

fitted up with great elegance and propriety as a museum" (Brock, 1983). In 1767, William gave his first lectures at the school in Great Windmill Street, and a year later he began living there.

William Hewson

When John Hunter left in 1760 to join the army as a surgeon, his place was taken by William Hewson (1739–1774). Educated at Newcastle Infirmary and Edinburgh University, Hewson worked with William Hunter as an assistant. After leaving Hunter, he independently started his own anatomy school in 1772. He married Mary Stevenson, a friend of Benjamin Franklin. The anatomy school was located in his mother-in-law's house in London.

Benjamin Franklin lived in the house with the anatomy school for a period of 16 years between 1764 and 1775. The restored building, now known as the Benjamin Franklin House museum, is a major tourist attraction. During the restoration of the building, construction workers found over 1,200 fragments of bones and the skeletal remains of four adults and six children that were buried deep in the basement. The bones are more than 200 years ago and showed signs that they were used for anatomical studies (Kennedy, 2003). Probably, the skeletal remains are from the bodies that were obtained following public execution or stolen from graves by resurrectionists, which were not uncommon practice at that time.

Hewson was a good teacher and researcher (Fig. 142). In 1769, he was awarded the Copley Medal, and

in 1770 he was elected a Fellow of the Royal Society. Hewson discovered fibrin, a protein that is involved in the clotting of blood; a he also studied the lymphatic vessels in animals and made important discoveries on cell membrane and the shape of red blood cells.

The partnership between William and Hewson became strained, as a result of which Hewson opened his own school. It was a thriving success, but, in 1774, Hewson died at the age of 35 years from a infected wound he incurred during dissection. In the same year, his book on the lymphatic system was published (Figs. 143 & 144). Following in the footsteps of William Hewson at the Great Windmill Street School were William Cumberland Cruikshank (1745–1800), John Sheldon (1752–1808), and Matthew Baillie (1761–1823), who was William Hunter's nephew.

John Sheldon

John Sheldon (1725–1808) was an apprentice to the surgeon Henry Watson (1702–1793) who had an anatomy school in London. Sheldon learned how to inject dead bodies and embalm them. As a resident pupil, he continued his surgical training with John Hunter and he attended the lectures of William Hunter. After a year of clinical training at the Westminster and Lock

Figure 142. William Hewson (1739–74). (Department of Anatomy, University of Manitoba.)

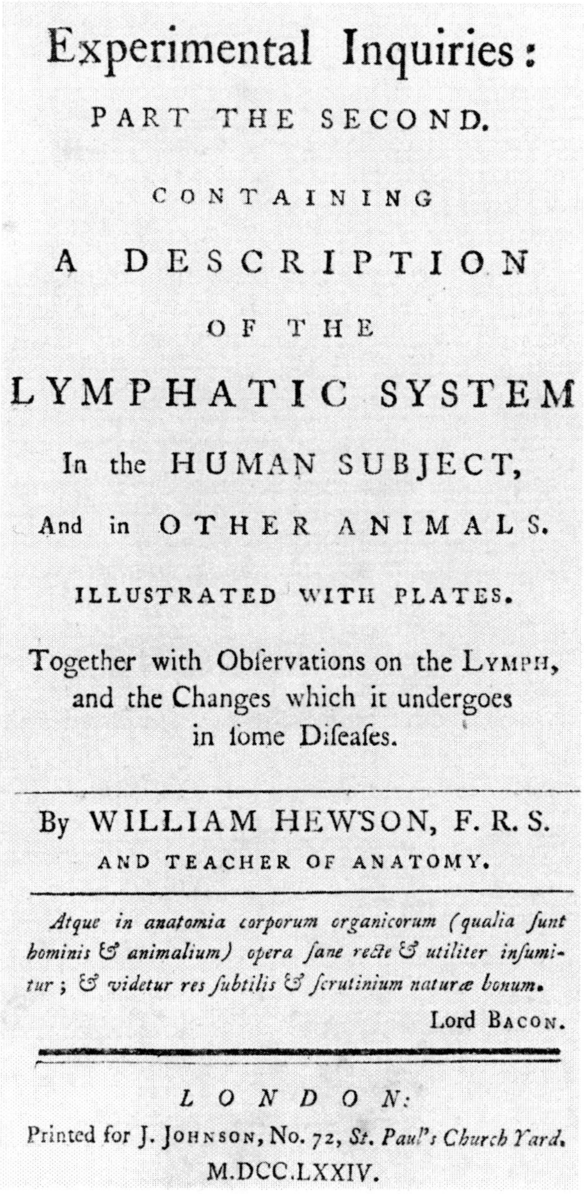

Experimental Inquiries:

PART THE SECOND.

CONTAINING

A DESCRIPTION

OF THE

LYMPHATIC SYSTEM

In the HUMAN SUBJECT,

And in OTHER ANIMALS.

ILLUSTRATED WITH PLATES.

Together with Obſervations on the LYMPH,
and the Changes which it undergoes
in ſome Diſeaſes.

By WILLIAM HEWSON, F. R. S.
AND TEACHER OF ANATOMY.

*Atque in anatomia corporum organicorum (qualia ſunt
hominis & animalium) opera ſane recte & utiliter inſumi-
tur ; & videtur res ſubtilis & ſcrutinium naturæ bonum.*

Lord BACON.

LONDON:
Printed for J. JOHNSON, No. 72, St. Paul's Church Yard.
M.DCC.LXXIV.

Figure 143. Title page of William Hewson's book on the lymphatic system. This "*essay*" was dedicated to Benjamin Franklin. Hewson also wrote a book on the "*properties of the blood.*" He was a pupil of John and William Hunter. (Courtesy of the Neil John Maclean Health Sciences Library, University of Manitoba.)

Hospitals, Sheldon obtained the diploma of the Surgeon's company, which led to an appointment at the Lock Hospital. Following the death of one of his patients, he embalmed her body, which he kept in his home. Sheldon was appointed a lecturer by William Hunter at the Great Windmill Street School in 1776.

With a high reputation and experience in teaching, Sheldon established his own school of anatomy where he taught anatomy, physiology, and surgery (Dukes, 1960). He studied the lymphatic vessels using special cannula he made for perfusion and demonstration of the vessels. His findings on the lymphatics were published as a monograph in 1784 (Sheldon, 1784). He was preparing a second volume, which remains unpublished.

In recognition of his research work, Sheldon was elected a Fellow of the Royal Society in 1784. Two years later he was appointed a surgeon at the Westminster Hospital. As professor of anatomy at the Royal Academy, Sheldon was also required to deliver an annual course of lectures at the Academy. Not surprisingly, there was much gossip regarding the mummy of the woman, which he kept in his home until his death. His wife donated the mummy to the Royal College of Surgeons. It was destroyed by bombs during the war (Capener, 1952).

Matthew Baillie

Matthew Baillie studied medicine at Oxford University, after spending several years learning anatomy and helping his uncle at the school. When William died in 1783, he left his school, house and the estate in Scotland to Matthew, who kept the house and school but transferred the estate to John. Because of the increasing demands of his clinical practice, Matthew Baillie sold in 1799 his partnership in the school to James Wilson (1765–1821). Matthew Baillie was a highly regarded physician. His book *Morbid Anatomy*, published in 1795, was a great success. This was the first published work, in the English language, on pathological anatomy.

When William Hunter started his school, there was little formal medical teaching in Britain, except for at the universities. Most medical doctors gained their knowledge by attending private lectures and through an apprenticeship system (Singer & Holloway, 1960). Without any clear understanding of the causes and nature of diseases, the doctor treated the conditions as best as he could, and this was extremely limited. Practical anatomy teaching was sadly deficient with only the occasional human dissections (Rolleston, 1939). Even at the University of Leiden and at Edinburgh, few cadavers were used for demonstrations throughout the course.

At the turn of the eighteenth century, private teachers of anatomy were prevented by the regulations of the Company of Barber-Surgeons from carrying out dissections. John Bell, who had studied anatomy with Professor Alexander Monro, *secundus*, in Edinburgh, remarked "*that in Dr. Monro's class, unless there be a fortunate succession of bloody murders, not three subjects are dissected in the year.*" In 1714, William Cheselden (1688–1752), who was one of the most famous surgeons of his era, was censored for dissecting human bodies in his home. He had to apologize to the Court of the Company of Barber-Surgeons.

When the surgeons parted from the barber-surgeons as a result of the 1745 Act of Parliament, William Hunter wasted no time in realizing his plans for opening a school of anatomy. As a result, William Hunter pioneered the changes that revolutionized the teaching of human anatomy in Britain. His method of lecturing was unique, and there were regular dissections and demonstrations; a museum was established and preserved specimens were available for the students.

William's school had a large number of anatomical specimens that were prepared by William himself, his brother John, and their assistants. William kept on increasing the collection of specimens by purchasing others that were offered for sale. There were also specimens of real diseases and of comparative anatomy, as well as of fossils. William's valuable coin collection and rare books were of special interest. A contemporary of Hunter noted that William possessed "*the most magnificent treasure of Greek and Latin books that has been accumulated by any person now living, since the days of Mead.*" Indeed, the museum was unique and "*perhaps without a rival*" in Europe (Brock, 1983). Most dignitaries and scientists, and men of learning, visiting London, requested the opportunity to see the museum.

William Hunter had put together one of the finest collections of anatomical specimens and created one of the most famous museums for teaching in Europe (Cooke, 1984). According to his will, the collections of books, manuscripts, drawings, and diverse specimens

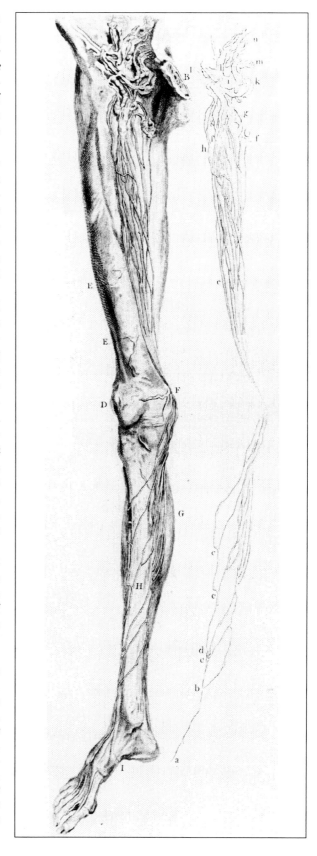

Figure 144 (*Right*). A drawing from William Hewson's book showing the lymphatic vessels of the lower limb. ("*The lymphatic vessels appear...more regularly cylindrical than they are represented by Nuck, Ruysch and others,*"). (From Hewson, W.: *Experimental Inquiries...A description of the Lymphatic System in the Human Subject and in Other Animals,* J. Johnson, London 1774. Courtesy of the Neil John Maclean Health Sciences Library, University of Manitoba.) In the preface, Hewson remarked that "*the vessels which are the subjects of this essay, having only of late been made known to anatomists, and not being easily traced by dissection, have not been completely described, nor have they ever been delineated....*"

were left to Glasgow University "*for the improvement of students and the use of the public,*" but he stipulated that the museum should remain at the school in London for 20 years so that his nephew could continue the teaching as before. Between 1807 and 1809, the collections were relocated to Glasgow. Concerning William Hunter's papers and drawings, as well as details of the catalogs available of his anatomical and pathological specimens, reference should be made to the extremely useful "*handlist*" prepared by Brock (1994).

There were several anatomical museums in Britain, but none had matched that of the Hunters. It was a model for other museums established in medical schools or owned privately by anatomists (Figs. 145 & 146). At that time, some natural scientists and surgeons privately collected anatomical specimens and formed museums. Not surprisingly, these museums had a curious appeal for the general public (Edwards & Edwards, 1959; Cope, 1966).

The Windmill Street School was the first real medical school in London, and it attracted students from throughout Britain, as well as from the British colonies in America.

The success of the school was due to the reputation of the Hunter brothers (Peachey, 1924), but the museum was also one of the great attractions (Thomson, 1942). Brock (1983) has given us a splendid account of Hunter's museum. It was unique "*in the extent and variety of his collections,*" which "*had direct relevance to current interest and was organized and catalogued in a rational manner.*" The prosected anatomical and the pathological specimens were of obvious value as demonstrations for students attending William's lectures and practical classes.

From a letter dated July 15, 1809, sent by John Burn to Benjamin Rush, we learned that the number of medical students at Glasgow University was more than doubled because of the attraction of Hunter's museum (Brock, 1983). In addition to the anatomical preparations, there were other important collections dealing with fossils, minerals, shells, coins, and paintings. William's library comprised of some 10,000 volumes. It included medical books, old and contemporary, works of the classical scholars, and almost all other areas of learning. In order to establish such an impressive museum, William must have drained his wealth over the years.

The source of William's vast income is uncertain, and it could not have been solely from the anatomy classes and his medical practice, however successful they were. His bank records revealed "*extensive dealings in government lotteries, and how much he played the market, buying and selling both government and East India Company Stock*" (Brock, 1983).

Private Anatomy Schools

The Windmill Street School continued until about the middle of the nineteenth century with shifting success. Through the years, several distinguished surgeon-anatomists, and luminaries in medicine, were involved

Figure 145. An eighteenth century anatomical museum, from F.B. Albinus's *De natura hominis libellus*, 1775. (From Edwards, J.J. & Edwards, M.J.: *Medical Museum Technology*, Oxford University Press, 1959; by permission of Oxford University Press.)

Figure 146. The early nineteenth century museum of John Heaviside. From a drawing at the Royal College of Surgeons of England. (From Edwards, J.J. & Edwards, M.J.: *Medical Museum Technology*, Oxford University Press, 1959; by permission of Oxford University Press.)

as partners and teachers in the school. These included Charles Bell (1774–1842), Herbert Mayo (1796–1852), and Caesar Henry Hawkins (1794–1884). In addition to John Hunter, his brother, John Shaw, his brother-in-law and a former pupil of the school, assisted Charles Bell from 1821 to 1826 (Thomson, 1943). Shaw died in 1827. His *Manual of Anatomy* was published in 1821. There were three English editions of the book and it was also translated into German (Cope, 1966).

In 1828, London had 11 schools of anatomy; seven were privately owned and four were located in the hospitals. The most influential of them was the Great Windmill Street School, which was in existence for over 60 years. Through the students who had studied here, it contributed profoundly to the establishment of anatomy departments and the teaching of anatomy throughout the world.

According to Desmond (1989), the private medical schools "*proliferated to meet the needs in the 1820s and 1830s, only to collapse in the 1840s.*" The demand had existed because of the massive growth of the population and also on account of students from the middle class keen on entering the medical profession. By 1832, there were 17 private schools, which were approved by the inspector of anatomy. These schools were in competition with each other, as well as with the teaching hospitals, for students (Power, 1857).

The Blenheim Street or Great Marlborough Street Anatomical School was one of the earliest private anatomy schools in London. It was started by Joshua Brookes who was born in London in 1761; not much is known of his early life. Brookes attended the anatomy and surgery lectures of William Hunter and

several other well-known anatomists in London. He also attended in Paris the anatomy and surgery demonstrations of the eminent French anatomist and surgeon Antoine Portal (1742–1832). A man of considerable reputation, Portal was personal physician to Louis XVI, a professor at the College de France for 64 years, and the author of an impressive seven-volume work on the history of anatomy and surgery (1770–1773).

Returning to London, Brookes began to teach anatomy, giving lectures and providing dissections throughout the year. In Paris, he had developed a method for preserving the body for several weeks by injecting the specimens with a nitric solution. He also injected a colored solution into the aorta and auricle, which improved the appearance of these and other structures. For this work, he was elected a Fellow of the Royal Society in 1819, but he did not receive any recognition from the Royal College of Surgeons. Brookes was regarded as one of the best anatomy teachers in London. A skillful dissector he spent most of the day in the dissecting room preparing specimens for his anatomy classes. Over a period of 40 years, it has been estimated that Brookes taught more than 5,000 students. In addition to the demands of his anatomy school, Brookes collected over three decades a large number of human and zoological specimens, including more than 6000 preparations, models, and casts...and more than 3000 specimens of the human body in both health and diseased states preserved in jars, including heads, limbs, and organs (Kell, 2004–2013). His museum occupied the upper two floors of his home and it was second only to that of John Hunter. The museum was an important part of the school for teaching and attracting students.

In 1826, Brookes retired from teaching apparently because of his poor health, but there might have been another reason. A new regulation of the Royal College of Surgeons required that candidates for examinations must now have certificates from the universities, a London hospital, or a surgeon, which Brookes was not in a position to provide. The lack of students and without any other source of income Brookes pitifully declined into poverty. Because he was unable to sell his museum, the specimens were sold piecemeal at auction in 1828 and 1830. Brookes died in 1833 and was buried in St. James Church in London (Kell, 2004–2013).

The demise of these independent institutions was largely due to financial reasons of decreasing enrollment because the Royal College of Surgeons had passed bylaws deeming that it would not recognize the summer classes given by the private medical schools. Whereas the hospital medical schools carried out the anatomy dissection courses during the winter months on account of the poor condition of the cadavers, the private medical schools were more innovative in their preservation of the cadavers. This enabled them to conduct the courses in summer effectively and with considerable financial success.

The surgeon and anatomist Thomas Cooke (1841–1899) opened in 1870 the London School of Anatomy in Brunswick Square. It was the last of the private medical schools in London. The facilities included a large dissecting room with a storage tank for the cadavers, a small cottage, and a shed where books were stored. The courses offered were anatomy, physiology, and operative surgery on cadavers. More than 100 students attended the school in its best years. Cooke died in 1899, but his son continued the school until the early part of the twentieth century (Morton, 1991).

WILLIAM'S ANATOMY OF THE HUMAN GRAVID UTERUS

William Hunter was not only a famous obstetrician, or "*man-midwife*" as he was then called (Wilson, 1995), but he was also a brilliant anatomist. His best known work is the treatise on *The Anatomy of the Human Gravid Uterus*, which was published in 1774 (Fig. 147). It was also the book, which led to the problems with his brother. William acknowledged in the preface of the book the help he received from his brother, "*whose accuracy in anatomical researches is so well known that to omit this opportunity of thanking him would be in some measure to disregard the future reputation of the work itself.*" John claimed the discovery in the very first paragraph of a short paper entitled "*On the Structure of the Placenta,*" which he communicated to the Royal Society in 1780. John referred to his brother's Atlas as that very accurate and elaborate work which Dr. Hunter has published on the gravid uterus, in which he has minutely described and accurately delineated the parts, without mentioning the mode of discovery, but John considered himself "*as having a just claim to the discovery of the structure of the placenta and its communication with the uterus together with the use arising from such structure and communication.*"

Curiously, John made his claim to the discovery of the structure of the placenta and its vasculature after several years had passed since the publication of William's atlas. John maintained his headstrong position that led to several exchanges of letters with the Royal Society, which the society wisely did not encourage. William was deeply hurt and felt betrayed by John. For the rest of their lives, the brothers did not speak to each other. Other factors of a more personal nature might have been involved (Peachey, 1924; Oppenheimer, 1946; Brock, 1983).

Despite his great fame and wealth, William remained devoted to anatomy for which he had a passion only to be matched by that of his brother John. Paget (1897) remarked that over the years, William Hunter had been an extremely hard-working person, hardly leaving the city from the time he went to Long Calderwood in 1751 to the time of his death in 1783, save to see some patient in the country. He never married, he had no country-house; he looks, in his portraits, a fastidious fine gentleman; but he worked until he dropped, and he lectured when he was dying. His school was at its zenith, and he gave his whole mind to it; he stood high above the men of his time for the charm and eloquence of his lectures; he held two hospital appointments, and his private practice was one of the largest in London. Moreover, he must be doing literary work monographs and matters of pathology and already, so early as 1751, he had made the beginning of that magnificent *Atlas of the Anatomy of the Gravid Uterus*, not published till 1755, which is perhaps the greatest book that has yet been written on this special subject.

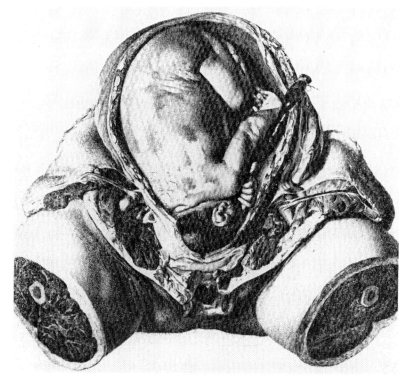

Figure 147. Fetus in the womb. This illustration is from William Hunter's treatise. *The Anatomy of the Gravid Uterus*, 1774. (With permission, from Major, 1954.)

AN INDELIBLE STAMP ON MEDICINE

William Hunter "*consoled himself for his loneliness with the fine arts, and with hard work....He had the friendship of Reynolds, Gainsborough, Hogarth, Johnson, and all the great men of his time; he stood well with the King, who appointed him in 1764 physician extraordinary to the queen, and in 1768 professor of anatomy to the new Royal Academy; and he had an immense practice. He spent his money in science and art; he gathered for himself, and he gave to the nation, the most wonderful art-collection which alone would perpetuate his name...*" (Paget, 1897). In addition, there were valuable pictures, portraits, coins, curiosities, and more than 12,000 books from the time of the earliest printed works.

William Hunter was one of the most prominent obstetricians of his time (Simmons, 1783; Peachey, 1924).

Because of his students and theirs in turn, scattered in many countries, his legacy became perpetuated. It helped to pave the way toward a more rational and modern epoch in medicine. According to his nephew, Matthew Baillie, William was probably the best teacher of anatomy that ever lived. No one possessed more enthusiasm for the art, more preserving industry, more acuteness of investigation, more perspicuity of expression, or indeed a greater share of natural eloquence....."*His scientific discoveries, books and monographs are of enduring value, but the capstone of William's monument was the Great Windmill School, which under the joint direction of Matthew Baillie and William Cruikshank, continued to turn out many of England's best medical men, and left an indelible stamp on medicine everywhere*" (Kobler, 1960).

DISCOVERIES AND CONTROVERSIES

The anatomical discoveries of John and William Hunter included: the lacrimal ducts in man, the seminiferous tubules, clinical and anatomical features of congenital hernia, and the lymphatic system, which Paget (1897) described as "*a new truth of vast importance, second only to Harvey's discovery of the circulation of the*

blood; it changed the whole character of medical science and practice."

There were counterclaims, rebuttals, and controversies with respect to these discoveries, especially from the Monros in Edinburgh and from the surgeon Percivall Pott. William responded in a highly uncommon way by publishing, in 1762, his scathing and bizarre book entitled *Medical Commentaries* Part I; a supplement followed two years later. Paget (1897) described this work as follows:

> Surely one of the strangest books that a physician or surgeon ever wrote. From beginning to end, it is an incessant attack on those who discovered what the brothers also discovered; every device of italic type, notes of exclamation, and long quotations, interrogation and interjection, heavy sarcasm, charges of stupidity, falsehood, and flagrant theft all these things make the book, and there is nothing else in it, hardly one line that is quiet. It was the method of controversy fashionable in his time, full of sound and fury.

FINAL YEARS

During the later part of his life (Brock 1985; 1994), William suffered greatly from frequent bouts of gout. About ten years before his death, his health was already suffering, and at one stage, he even contemplated returning to Scotland where he would retire. On March 20, 1783, following a brief illness of just over a week, William fainted from exhaustion while giving the introductory lecture relating to the course on the operations of surgery. He suffered a stroke and died on March 30, 1783. With a clear mind, he spoke his memorable words on his deathbed: *"If I had strength enough to hold a pen, I would write how easy and pleasant a thing it is to die."*

William did not make John the executor of his will, and he left nothing of his wealth or museum to him. Even their ancestral home at Long Calderwood was bequeathed to his nephew, Matthew Baillie, who later passed it on to his Uncle John. In any case, the property legally should have been given to John (Brock, 1983).

William acquired great wealth from his practice of obstetrics, student fees, and various investments, but according to John Hunter himself, William's attachment to money arose from prudence and not from a love of it, or a love to be rich; for whatever he was really attached to he was in the strictest sense a miser (Brock, 1983). The body of William Hunter was interred in the Rector's Vault of St. James Church in Westminster.

Chapter 17

THE TRANSCENDENT GREATNESS
OF JOHN HUNTER

John Hunter (1728-1793), the pioneer of scientific surgery (Fig. 148), "*was anatomist, biologist, naturalist, physician, surgeon, and pathologist, all at once, and all in the highest*" (Paget, 1897). He was born on February 13, 1728, the tenth and last child of John and Agnes Hunter, in East Kilbride in the county of Lanarkshire, about 12 miles south of Glasgow. His famous brother William Hunter (1718–1783), with whom his life and career were closely linked, was ten years older. Throughout his youth, John showed no interest in getting "*the same education that the sons of country gentlemen then got,*" but instead as the youngest son and spoiled by his mother, "*he would do nothing but what he liked, and neither liked to be taught reading nor writing nor any kind of learning, but rambling amongst the woods, braes, etc., looking after birds'-nests, comparing their eggs number, size, marks, and other peculiarities...*" (Letter written by John Hunter's niece Mrs. Agnes Baillie to her brother Doctor Matthew Baillie).

FOUNDATIONS IN ANATOMY AND SURGERY

In September 1748, John left home to join his brother William, who at that time was the best anatomist and "*accoucheur*" in London, where he also had a private school for the teaching of anatomy and surgery. John impressed his brother with his considerable skill in dissection. He began preparing anatomical specimens for William's classes and, before long, John was assisting with the demonstrations.

Keen on furthering John's education, William arranged for John to spend some time with the great surgeon William Cheselden (1688–1752) at Chelsea Hospital, and, following his death, John joined Percivall Pott (1714–1788) as a surgeon's pupil at St. Bartholomew's Hospital. In addition to dressing and bandaging of wounds, John probably assisted with minor operations. In 1754, he became a surgeon's pupil at St. George's Hospital and two years later he was appointed a house surgeon. He remained in the service of this institution for the next 25 years.

For the understanding of diseases, one needed a thorough knowledge of the structure of the human body. This problem John approached not only by dissecting human cadavers but also through extensive studies on almost every species of animal he could find. John "*found surgery little more than a trade...and showed that there were processes of disease which could be studied just as Harvey had studied the processes of nature, and that only by investigating the changes due to disease in the light of a knowledge of the functions of normal tissues and organs could surgery be properly applied*" (Graham, 1956). Pursuing the "*ultimate truths of natural function,*" John dissected and described more than five hundred species of animals, birds, fishes, and insects. His insightful observations showed how much the study of these lower animals can contribute to the understanding of man's structure and functional activities in health and disease. John's personal record revealed not only an expansive mind but one that was both descriptive and analytical, relentless in the pursuit of details. He was always striving to comprehend the harmony existing between structure and function in all forms of life (Persaud, 1996).

Figure 148. John Hunter (1728–1793), anatomist and surgeon. (Portrait painted by Sir Joshua Reynolds and published in January 1788 by W. Sharp, London. (Courtesy of Bethesda, MD: U.S. National Library of Medicine, National Institutes of Health, Health & Human Services.)

ANATOMICAL SPECIMENS: THE HUNTERIAN MUSEUM

The extraordinary collection of the Hunterian Museum at the Royal College of Surgeons in London, England, is an enduring legacy of Hunter's genius and prodigious labor. John Hunter's museum "*differed from others in existence during the latter part of the eighteenth century in that it was not merely a collection of exhibits but an illustration of his theories, and in particular of the constant adaptation in living things of structure to function. It was, in fact, John Hunter's unwritten book*" (Dobson, 1968). The most diverse fields are represented in the collection: pathology, geology, and paleontology. The impressive collection dealing with comparative anatomy and physiology was started by John in 1763 following his return from Portugal after the Seven Years' War (Fig. 149). He had served as a staff surgeon to the expedition for two years, and at 35 years of age "*he found himself, in point of fortune, better than nothing by his half-pay; that enabled him to pay his house rent and some other necessaries requisite ever for those who sit down in practice waiting for patients*" (Paget, 1897).

When the Company of Surgeons, later to become the Royal College of Surgeons, acquired the collection in November 1799, it consisted of about 14,000 specimens that included some 500 different species of animals. Hunter had spent at least £70,000 over the years for the specimens, which at the time of his death represented the "*bulk of his fortune.*" It was his wish that after his death the museum should be preserved as an entity and not to be sold in pieces. Following persistent efforts on the part of his executors and former students, Parliament finally voted a sum not exceeding £15,000 for the purchase of the collection. Later, additional grants were made to the Royal College of Surgeons for the building of a museum to house the collection. Since then, the collection had been increased with specimens from other sources, not taking into account the serious losses from bombing of the college in May 1941 (Dobson, 1954; 1968; Allen, 1974).

Figure 149. John Hunter is shown here standing in an oxen-drawn cart, speaking to an assistant at this estate in Earl's Court, near London. (Courtesy of Bethesda, MD: U.S. National Library of Medicine, National Institutes of Health, Health & Human Services.) Hunter purchased the property where he built a small house and kept a menagerie. In addition to the camel and monkey depicted above, there were leopards and other exotic animals. Hunter pursued here "*most of those researches which form the subject of his papers in the Philosophical Transactions, or are detailed in his work on the Animal Oeconomy*" (Ottley, 1839).

SCIENTIFIC DISCOVERIES

Few will be in a position to say what exactly were John Hunter's scientific achievements. Indeed he did not make any great or spectacular discoveries, nor has he produced any profound writing of enduring value. John's contribution to surgery is not to be found in the form of a single discovery or writing but from the many basic scientific principles he formulated through critical thinking and accurate observations (Foot, 1794; Adams, 1817; Oppenheimer, 1946).

These have led to other discoveries that benefitted the practice of medicine and surgery. Hunter's scientific approach was straightforward and is conveyed in a letter to his former pupil Edward Jenner: "*but why do you ask me a question, by the way of solving it. I think your solution is just; but why think, why not trie the Expt*" (Hunter, 1976). Jenner became one of the most famous physicians and his name is immortalized for his discovery of vaccination against smallpox. Not only did John Hunter skillfully dissect, but he designed and carried out numerous experiments from which he made critical observations that led to the formulation of some of the most fundamental principles of diseases.

Because of Hunter's expansive scientific outlook and his peculiar trait of trying to bridge in harmony different disciplines, it is extremely difficult to categorize his numerous and diverse contributions to science and medicine. To unravel the thread he has woven seems irrational. Qvist (1981) provided a practical and useful classification that is based on ten subject areas: digestive system, reproductive system, circulatory system, nervous system and special senses, musculoskeletal system, metabolism, comparative anatomy, surgery and clinical subjects, geology and paleontology, and natural history in general. In each interest area the papers are chronologically listed, but difficulties are obvious even here. For example, within the grouping "*reproduction*," one finds contributions dealing with the descent of the testes into the scrotum, placental structure, and a secretion in the crop of breeding birds. (Figs. 150–153).

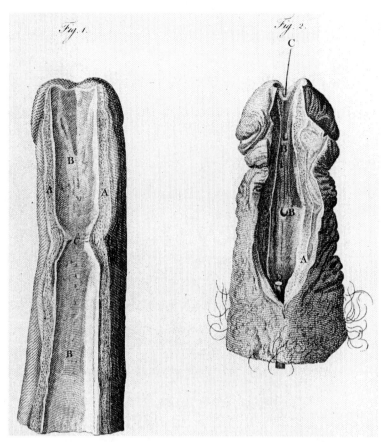

Figure 150. Longitudinal section of the penis. In the left Figure (1), observe the stricture (C) of the urethra (B); on the right (Figure 2), the lacunae are revealed as a site of obstruction when passing a catheter (D). A bristle (C) is shown introduced into a lacuna (From Palmer, J.F.: *The Works of John Hunter, F.R.S.*, Longman, Rees, Orme, Brown, Green, and Longman, London, 1837. Courtesy of the Neil John Maclean Health Sciences Library, University of Manitoba.) Hunter made these observations in relation to his investigation of "*the venereal disease* (Hunter, 1796a)."

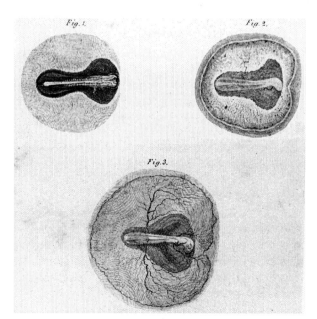

Figure 151. Three stages in the development of the chick embryo, showing the formation of blood vessels. (Plate 1 in John Hunter's *A Treatise on the Blood*....George Nicol, London, 1794. Courtesy of the Neil John Maclean Health Sciences Library, University of Manitoba.)

Figure 152. Lateral wall of the nasal cavity, medial aspect with the bony septum removed to reveal the nerves of the right nostril. The following structures are readily recognizable: the olfactory nerves (U); the frontal sinus (B); the membrane (S) lining the nasal septum, which has been removed; the hard palate (H); the sphenoidal sinus (P); the sella turcica (Q); a branch of the trigeminal nerve (T); and the opening of the pharyngotympanic tube (L). (From Palmer, J.F.: *The Works of John Hunter, ER.S.*, Longman, Rees, Orme, Brown, Green, and Longman, London, 1837. Courtesy of the Neil John Maclean Health Sciences Library, University of Manitoba.)

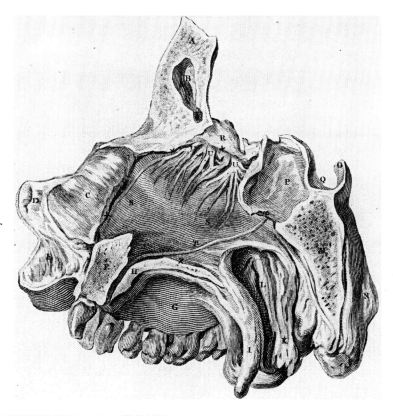

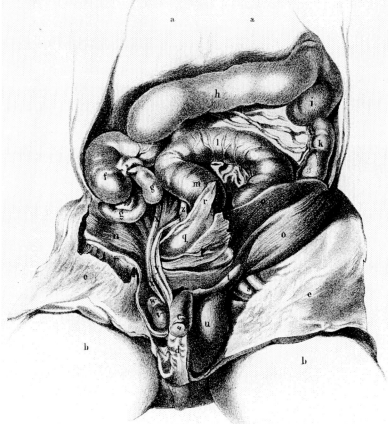

Figure 153. Dissection of a fetus about seven-months-old. Most of the intestines have been removed to show the descent of the testes from the abdomen into the scrotal sac. Note the liver (a), the transverse colon (h), sigmoid colon (1), the right testis (v), the epididymis (w), ductus deferens (y), the left testis (u), and spermatic cord (t). (From Palmer, J.F.: *The Works of John Hunter, ER.S.*, Longman, Rees, Orme, Brown, Green, and Longman, London, 1837. Courtesy of the Neil John Maclean Health Sciences Library, University of Manitoba.)

SURGEON

The most important development in the field of surgery was not any innovative technique or landmark discovery but a change in thinking "*from its traditional reliance on empiricism and toward a reliance on scientific principles...,*" which "*has brought surgery nearly everything in the way of competence and respectability*" (Buckman, 1987). This intellectual revolution, which established the practice of surgery on a scientific basis

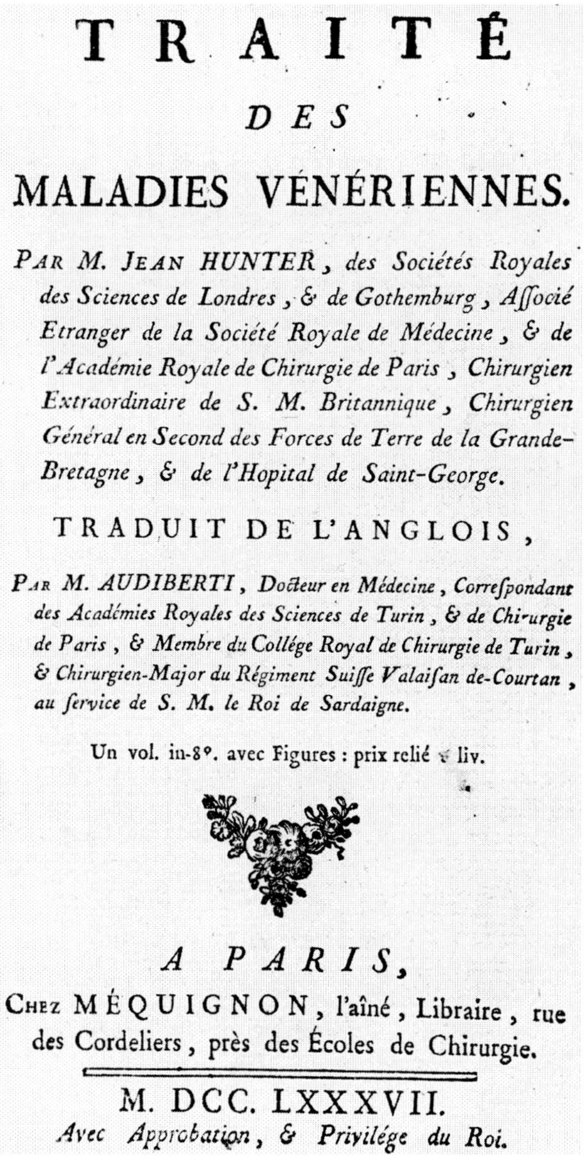

TRAITÉ

DES

MALADIES VÉNÉRIENNES.

PAR M. JEAN HUNTER, *des Sociétés Royales des Sciences de Londres , & de Gothemburg , Affocié Etranger de la Société Royale de Médecine , & de l'Académie Royale de Chirurgie de Paris , Chirurgien Extraordinaire de S. M. Britannique , Chirurgien Général en Second des Forces de Terre de la Grande-Bretagne , & de l'Hopital de Saint-George.*

TRADUIT DE L'ANGLOIS ,

PAR M. AUDIBERTI, *Docteur en Médecine , Correfpondant des Académies Royales des Sciences de Turin , & de Chirurgie de Paris , & Membre du Collége Royal de Chirurgie de Turin , & Chirurgien-Major du Régiment Suiffe Valaifan de-Courtan , au fervice de S. M. le Roi de Sardaigne.*

Un vol. in-8°. avec Figures : prix relié 4 liv.

A PARIS,

CHEZ MÉQUIGNON, l'aîné , Libraire , rue des Cordeliers , près des Écoles de Chirurgie.

M. DCC. LXXXVII.

Avec Approbation , & Privilége du Roi.

Figure 154. Title page of John Hunter's treatise on venereal diseases (French Translation, Paris, 1787. Courtesy of the Neil John Maclean Health Sciences Library, University of Manitoba.)

was forged by John Hunter. It is not necessary to discuss here the many surgical principles established by John Hunter; references should be made to recent publications (Kobler, 1960; Buckman, 1987; Perry, 1993; Allen et al., 1993; Persaud, 1996), which present these in a modern context. It is remarkable to reflect that the scientific concepts of cryopreservation with respect to the circulation of blood; hemorrhage; electrical stimulation of the heart in resuscitation; inflammation and the suppuration of wounds; organ transplantation and the nature of cancer, which he believed to be several diseases, are only some of the discoveries and procedures made by John Hunter more than two centuries ago (Perry, 1993).

Most surgeons and anatomists would know of the adductor canal but are probably not aware that it was first described by Hunter as a result of his pioneering work for the surgical treatment of popliteal aneurysm. From observations made on the growth of deer antlers and of the blood circulation in the horns, he discovered the concept of a collateral circulation. The experiments he carried out on the carotid artery of a dog convinced him that aneurysms were not due solely to a weakness in the arterial wall.

In December 1785, Hunter operated on a 45-year-old coachman for popliteal aneurysm by ligating the artery proximal to the aneurysm so as to prevent its further expansion and rupture. Because of the rich collateral circulation around the knee joint it was feasible for him to ligate the artery at a higher level away from the diseased segment. The patient died following a fever, but as the autopsy showed it was not on account of the operation. Hunter subsequently modified the procedure, which he carried out with greater success. His recommendation to operate on an aneurysm as early as possible, before it becomes larger, is still valid today.

John Hunter's brilliant studies on bone growth came about from a chance observation he had made while dining at the home of an acquaintance. He found that the bones of a pig soaked in madder were stained reddish. Madder is a dye derived from the plant *Rubia tinctorum.* Hunter's experiments on pigs and fowl revealed that bone grows by deposition of new bone on the external surface and resorption of older bone on the inside. Because of this remodeling, bone becomes larger without changing its external shape. By inserting two lead shots some distance from each other into the tibia of the pig, he also demonstrated that bone growth occurs at the epiphyseal ends of the bone (Hunter, 1835). Current treatment of certain bone diseases is based on an understanding of these fundamental concepts.

FULFILLMENT IN MARRIAGE

On July 22, 1771, John Hunter married Ann Home, sister of Everard Home, who later became a distinguished surgeon. It is not known whether his brother William approved of the matrimony, but in any case he did not attend the ceremony. John's wife is described as sociable and a woman of beauty, wit, and charm. He found fulfillment in his marriage, which allowed him the peace of mind to pursue his work. Four years after his marriage, he wrote to Rev. James Baillie, his brother-in-law, the following:

As to myself, with respect to my family, I can only yet say, that I am happy in a wife, but my children are too young to form any judgment of.... I am not anxious about my children but in their doing well in this world. I would rather make them feel one moral virtue, than read libraries of all the dead and living languages. You know I am no Scholar,... I must continue to be one of the happyst men living.... (Paget, 1897)

SCIENTIFIC STUDIES AND PUBLICATIONS

From Everard Home, we learn

that Mr. Hunter was so tardy in giving his observations to the public; but such was his turn for investigation, and so extensive the scale upon which he instituted his inquiries, that he always found something more to be accomplished, and was unwilling to publish any thing which appeared to himself unfinished. (Hunter, 1794)

Hunter presented his observations in a large number of communications to the Royal Society and these were later published in the *Philosophical Transactions*. Far more remained unpublished as notes and manuscripts.

Three of Hunter's books were published during his lifetime: *The Natural History of Human Teeth* (Parts I and II, in 1771 and 1778, respectively; King, 1994), *A Treatise on the Venereal Disease* (Hunter, 1786a), and *Observations on Certain Parts of the Animal Oeconomy* (Hunter, 1786b). After his death, three further works,

which had already been written, were published (Le Fanu, 1946): *A Treatise on the Blood, Inflammation and Gun-Shot Wounds* (1794), *Observations and Reflections on Geology* (1859), and *Memoranda on Vegetation* (1860). Some of his unpublished manuscripts were plagiarized and later burnt by his brother-in-law, Sir Everard Home (Oppenheimer, 1946; Dobson, 1954). Factual errors, misquotations, misrepresentations, and even new material can be found in John Hunter's writings that were reprinted or edited after his death (Qvist, 1981). Following Hunter's death, James Palmer in 1835 edited and published all of his writings in four volumes, entitled *The Works of John Hunter* (Hunter, 1835). Also, the great anatomist and zoologist Richard Owen (1804–1892) (Hunter, 1861; Smith, 1992) collected, arranged and revised Hunter's unpublished writings, which were published as *Essays and Observations on Natural History, Anatomy, Physiology, Psychology and Geology* (Figs. 154–156).

JOHN AND WILLIAM: HARMONY AND DISCORD

John and his brother William had a closely bonded relationship until the publication in 1774 of William's magnificent book on *The Anatomy of the Human Gravid Uterus*. In the preface, William acknowledged the help he received from John *"Whose accuracy in anatomical researches is so well known that to omit this opportunity of thanking him would be in some measure to disregard the future reputation of the work itself"* (Hunter, 1774). John, however, claimed that he

had not only carried out the dissections but also made the discovery on the structure of the placenta and its relationship with the uterus, as well as the vascular arrangements.

The brothers wrote to the Royal Society attacking each other. The letters were read but not recorded in the minutes. The society was embarrassed and did not mediate in the dispute (Oppenheimer, 1946; Kobler, 1960; Brock, 1994).

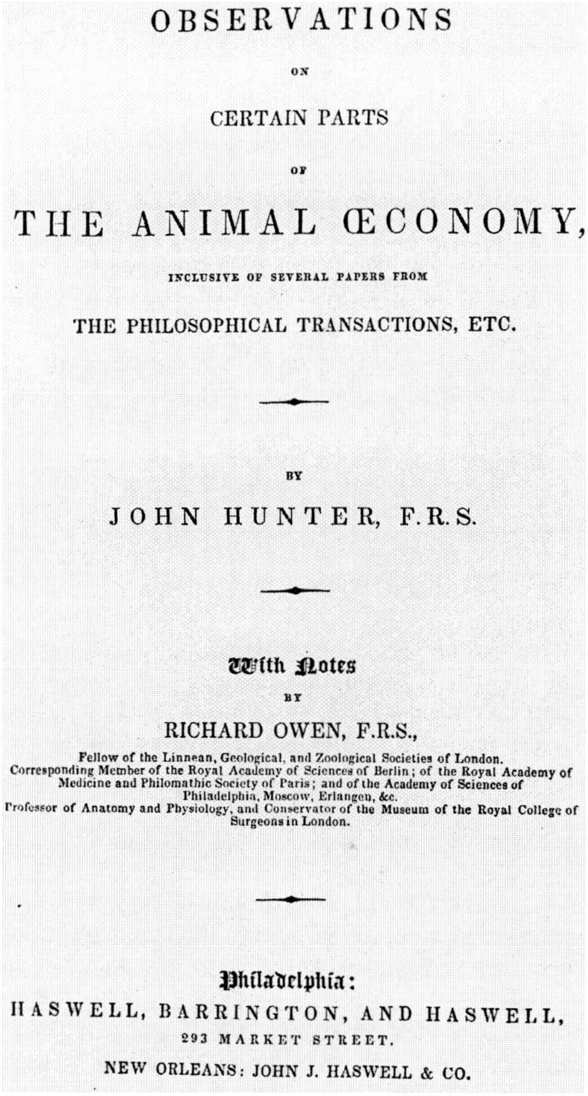

OBSERVATIONS

ON

CERTAIN PARTS

OF

THE ANIMAL ŒCONOMY,

INCLUSIVE OF SEVERAL PAPERS FROM

THE PHILOSOPHICAL TRANSACTIONS, ETC.

BY

JOHN HUNTER, F.R.S.

With Notes

BY

RICHARD OWEN, F.R.S.,

Fellow of the Linnean, Geological, and Zoological Societies of London.
Corresponding Member of the Royal Academy of Sciences of Berlin; of the Royal Academy of
Medicine and Philomathic Society of Paris; and of the Academy of Sciences of
Philadelphia, Moscow, Erlangen, &c.
Professor of Anatomy and Physiology, and Conservator of the Museum of the Royal College of
Surgeons in London.

Philadelphia:

HASWELL, BARRINGTON, AND HASWELL,

293 MARKET STREET.

NEW ORLEANS: JOHN J. HASWELL & CO.

Figure 155. Title page of Observations on *Certain Parts of the Animal Oeconamy....* by John Hunter. (American Edition: Haswell, Barrington, and Haswell, Philadelphia, 1840. Courtesy of the Neil John Maclean Health Sciences Library, University of Manitoba.)

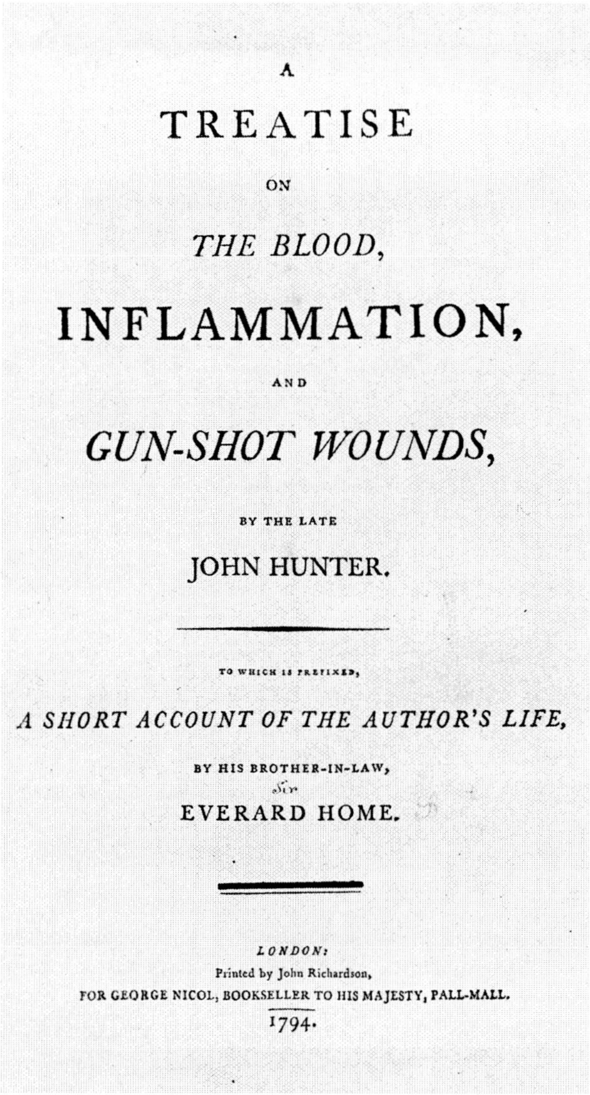

A

TREATISE

ON

THE BLOOD,

INFLAMMATION,

AND

GUN-SHOT WOUNDS,

BY THE LATE

JOHN HUNTER.

TO WHICH IS PREFIXED,

A SHORT ACCOUNT OF THE AUTHOR'S LIFE,

BY HIS BROTHER-IN-LAW,

Sir

EVERARD HOME.

LONDON:

Printed by John Richardson,

FOR GEORGE NICOL, BOOKSELLER TO HIS MAJESTY, PALL-MALL.

1794.

Figure 156. Title page of *A Treatise on the Blood, Inflammation, and Gun-Shot Wounds* by John Hunter. George Nicol, London, 1794. (Courtesy of the Neil John Maclean Health Sciences Library, University of Manitoba.)

PROFESSIONAL RECOGNITION AND HONORS

During his lifetime John received many professional recognitions and honors. He was elected a fellow of the Royal Society in 1767, and he received the membership diploma of the Company of Surgeons a year later. On December 9, 1765, he became surgeon to St. George's Hospital in London and held this position for 25 years. John was appointed surgeon extraordinary to the king in 1776 and elected to membership of the Royal Society of Gothenberg (1781), the Royal Society of Medicine and Royal Academy of Surgery of Paris 1783), and of the American Philosophical Society (1787). In 1790, John became surgeon-general of the army and inspector of hospitals. He was one of the first directors of The Royal Humane Society, which was founded in 1774 and was appointed the first vice-president of the Royal Veterinary College in 1781, which he helped to establish.

DEATH AND LEGACY

John Hunter encountered considerable opposition from his colleagues at St. George's Hospital when he presented to them his proposals for the improvement of the surgical training of the pupils and for the remuneration of the surgeons. On Wednesday, October 16, 1793, John attended a committee meeting at the hospital where

> one of his colleagues flatly contradicted something he had said. Then came the end. Angina seized him; he turned toward another room, to fight out the pain by himself, and Dr. Matthew Baillie followed him; he went a few steps, groaned and fell into Dr. Robertson's arms, and died. (Paget, 1897).

John Hunter was interred in a vault at St. Martin-in-the-Field's Church, near Trafalgar Square in London. His remains were transferred on March 28, 1859 to Westminster Abbey. The grave lies in the north nave marked by a memorial tablet with the inscription

> The Royal College of Surgeons of England has placed this tablet on the grave of Hunter to record admiration of his genius, as a gifted interpreter of the Divine power and wisdom at work in the laws of organic life, and its grateful veneration for the services to mankind as the founder of scientific surgery.

Now after more than two centuries, John Hunter's memory and work are kept alive through the Hunterian Society, the Hunterian Museum (Dobson, 1954), and the Hunterian Orations. More importantly, however, is that the work of all great teachers are embolden through their students. For John Hunter there had been a great many distinguished pupils who not only became famous through their own accomplishments but continued the inherited Hunterian values and traditions right through the years into our generation. These luminaries included Edward Jenner, Matthew Baillie, Sir William Blizzard, Sir Anthony Carlisle, Charles White, Sir Astley Cooper, Henry Cline, William Hewson, Philip Wright Post, John Morgan, William Shippen Jr., and Philip Syng Physick (Perry, 1993). The last three names were the pioneers of scientific medicine in the United States of America (Persaud, 1997).

John Morgan (1735–1789) and William Shippen Jr. (1736–1808) were the founders of the medical school of the College of Philadelphia (now the Perelman School of Medicine at the University of Pennsylvania), which was the first medical school to be established in America. William Shippen Jr. was appointed professor of anatomy and surgery. The "*Father of American Surgery,*" Philip Syng Physick (1768–1837), began medical studies in Philadelphia, which he continued in England and Scotland. Philip Syng Physick worked under John Hunter, was a surgical resident at St. George's Hospital. He received his medical degrees from the Royal College of Surgeons in England, and from the University of Edinburgh. Philip Syng Physick was a skilled surgeon, an outstanding teacher, and inventor of useful medical devices. Philip Syng Physick became professor of surgery at the University of Pennsylvania.

Twenty years after becoming a fellow of the Royal Society and six years before his death, John Hunter was presented with the Copley Medal, the highest distinction of the society. The following remarks are taken from the citation of Sir Joseph Banks, president of the society, on that occasion:

> To you, then Mr. Hunter, I most willingly deliver this testimony of the regard of the Royal Society; this regard by which she distinguishes those who are in her opinion, the most meritorious;... and be assured, Sir, that this Society will, with gratitude, bear daily testimony to the advantage which Mankind receives from the natural sagacity of your professional and the indefatigable industry of your scientific exertions. (Dobson, 1969)

Chapter 18

PROFESSIONALISM AND RECOGNITION

Similar to many countries in continental Europe, the training of doctors and the practice of medicine in England, up to the end of the sixteenth century, were poorly organized and controlled. From the knowledge of the power of healing herbs, gained through experience, and simple therapeutic measures, such as proper diet and bed rest, other expertise evolved to care for the wounded and to deal with bones that might be dislocated or broken (Majno, 1975; Haeger, 1989; Sournia, 1992; Rutkow, 1993). Soon there were a diverse group of individuals treating the sick (Figs. 157–159), ranging from the monks in the monastery (Flemming, 1957), who attended to the poor, and the "*wise women*" in the villages, to the physicians, surgeons, barber-surgeons, and apothecaries, as well as the itinerant quacks (Riesman, 1935; Keevil, 1957; Lyons and Petrucelli, 1978; Dobson & Walker, 1980; Sournia, 1992).

The physicians were considered to be more learned, having studied in Latin and Greek, the accumulated wisdom from antiquity, especially the works of Aristotle and Galen. Most would have received their medical degree from Oxford or Cambridge University, having first studied for the bachelor of arts degree there. The entire course lasted about 14 years. The physicians, therefore, looked down upon the surgeons and barber-surgeons who acquired their skill by apprenticing themselves to someone more experienced within their company or guild. The physicians carried out neither operations, nor did they dispense medicines. They may, however, instruct the surgeons and even supervise their "*cutting procedures.*" The pills and potions prescribed by the physicians were then dispensed by the apothecaries from their well-stocked shops.

In addition to the qualifications from Oxford and Cambridge, many British physicians had travelled to the continent where they obtained their medical training and degrees from the universities that were flourishing there at that time. Many foreign universities at that time gave the medical degree quite easily. For example, the M.D. degree from Rheims was obtainable without any residence, requiring only a thesis, which anyone could have written, and a modest fee. Curiously, the bishop of London, dean of St. Paul's, acting with the authority of the Church, was also empowered to grant the license to practice medicine.

At the beginning of the sixteenth century, in the reign of Henry VIII, through an Act of Parliament, a charter was granted by the King in 1518 establishing the College of Physicians (Thomas, 1971). This was largely due to the efforts of Thomas Linacre (c. 1460–1624), who was one of the most famous physicians in England and to the king. He was appointed the first president of the college (Moore, 1918). The charter conferred legal status to the English physicians and it became the prototype for physicians in other European countries as they worked towards greater professional recognition. Four years later the charter of the college was confirmed by statute. In 1540, the surgeons received similar recognition when, by an Act of Parliament (Fig. 160), the Company of Barbers and members of the Guild of Surgeons in London were united to form the Barber-Surgeons' Company (Dobson & Walker, 1980).

Even though there were two separate groups of doctors now with different rights and privileges, the surgeons were denied the exclusive right to treat all surgical patients. In practice, however, the regulations were not strictly enforced and each trespassed into the other's privileges. There were also a large number of "*healers*" and unqualified practitioners.

Thomas Vicary's (1490–1561) book *A Profitable Treatise of the Anatomie of Mans Body*... was first printed in 1548, of which no copy exists today. It was

Figure 157. In this late thirteenth century drawing (MS. Ashm, 399, fol. 34r, top half; courtesy of the Bodleian Library, Oxford), a physician is depicted standing at the rear end of the bed with a female patient lying on it. He and others are pointing to the spilled flask and its contents (urine). Urine analysis and feeling the pulse were routine diagnostic techniques of the medieval doctor.

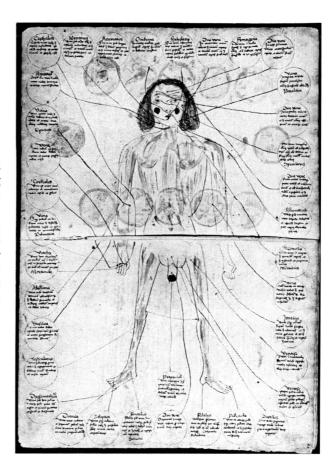

Figure 158. An anatomically realistic drawing from a fifteenth century manuscript showing bloodletting points. The nude body is boldly depicted with muscles, external genitalia, and other distinct features. (Vatican, Ms. Palat. Lat. 1709, fol. 44v-45r; Foto Biblioteca Apostolica Vaticana.) Bloodletting was carried out in order "*to reduce or to correct excessive or corrupt humours*" and based on Greek and Roman healing practices. It complemented cauterization, and specific places on the body were identified (as shown on the figure) for various ailments (MacKinney, 1965).

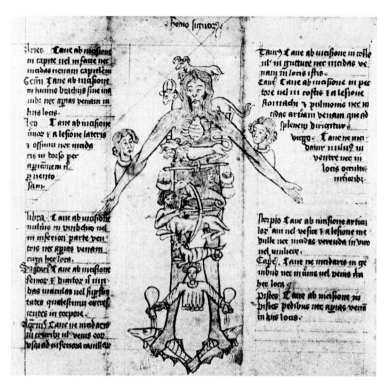

Figure 159. Astrological Man. (Wellcome Ms. 40. Calendarium 1463. Wellcome Institute Library, London.) In the Middle Ages and during the Renaissance, astrology was taught in several European universities, and many physicians were also astrologers. It was believed that the planets ruled different parts of the human body. Thus, Jupiter lungs, liver and feet; Mars left ear and genitalia; Sun right side of heart; Venus neck and abdomen; Mercury arms, hands, shoulders and hips; Moon left half of body and stomach.

Figure 160. Henry VIII is shown handing to Thomas Vicary, in 1540, the Act of Union between the Barber's Guild and the Guild of Surgeons of London to form a United Company of Barbers and Surgeons. Provisions were also made for human dissection and anatomical demonstrations. This oil painting by Hans Holbein, the Younger, 1541, is in possession of the Royal College of Surgeons, England. (Courtesy of Bethesda, MD: U.S. National Library of Medicine, National Institutes of Health, Health & Human Services.)

reprinted by his colleagues in 1577. Although the "*anatomy was archaic, completely ignoring the work of Vesalius*" (Russell, 1987; Moore, 1918), Vicary's remarks about the "*qualities of Surgeon,*" especially with respect to anatomy, are of relevance:

And I doo note foure thinges moste specially that euery Chirurgion ought for to haue: The first, that he be learned; the seconde, that he be expert; the thirde, that he be ingenious; the fourth, that he be wel manered. The first (I sayde); he ought to be learned, and that he knowe his principles, not onely in Chirurgerie, but also in Phisicke, that he may the better defende his Surgery....Also he must knowe the Anatomie; for al authors write against those Surgions that worke in mans body, not knowing the Anatomie; for they be likened to a blind man that cutteth in a vine tree, for he taketh more or lesse than he ought to doo. And here note wel the saying of Galen, the prince of Philosophers, in his Estories, that it is as possible for a Surgion not knowing the Anatomie, to worke in mans body without error, as it is for a blind man to carue an Image & make it perfyt. The .ij. I said, he must be expert; for Rasus sayth, he ought to knowe and to see other men work, and after to have vse and exercise....

King James I granted the apothecaries in 1617 a special charter with the exclusive privilege of compounding, stocking, and selling drugs that would be prescribed by the physicians. It was a privilege that the apothecaries had shared with the grocers. In 1632, the College of Physicians brought in laws that prevented the apothecaries from prescribing drugs and other medicine, which they were knowledgeable in because of their apprenticeship and experience (Keevil, 1957).

In the monasteries of England, of which there were close to 800 then, monks with medical knowledge, gained from reading the classic works and from their practical experience, provided some care and treatment for the sick poor (Flemming, 1957). With the dissolution of the monasteries by Henry VIII in 1536 and 1539, the practice of medicine was now almost entirely in the hands of the physicians and the surgeons. Many of the monks with medical experience and training probably practiced medicine or became apothecaries, in order to make a living. At that time, too, there were still many unqualified practitioners roaming the country (Riesman, 1935; Lyons & Petrucelli, 1978; Sournia, 1992).

The apothecaries were organized, requiring an extensive period of seven years of apprenticeship, knowledge of Latin, and the presence of physicians at their examinations. In 1704, the apothecaries were granted the right by the House of Lords to treat sick patients, in addition to compounding and selling medicines. In principle, they were allowed to practice medicine, but to charge only for the medicines they dispensed and not for advising the patient. Nonetheless, the standing power of the Society of Apothecaries increased in time, and through various parliamentary acts, it became a powerful influence on the future course of medical education right up to the nineteenth century.

By the beginning of the eighteenth century, there were already many impressive scientific and medical discoveries that should be a part of the training of medical doctors, but the College of Physicians and College of Surgeons provided then no regular teaching, except for a few anatomical lectures and demonstrations. Already in the sixteenth century, the College of Physicians had a room for lectures in its premises. Anatomy lectures began about 1565, and the Lumleian Lectureship in Anatomy was founded in 1581. During the seventeenth century, an anatomy theater was built in which William Harvey lectured. Later, a museum and a library were added, paid for by Harvey. In those days, men of culture followed scientific progress and even attended the anatomical sessions. For example, Charles II, who greatly admired the work of the college, attended anatomical lectures and demonstrations given in 1664 by Sir George Entis.

Private medical schools soon came into existence and became increasingly popular (Power, 1857; Desmond, 1989). Some of the students from these entrepreneurial institutions were to play an important role in influencing medical education on both sides of the Atlantic. The most important of these private schools was founded by the great surgeon and anatomist, William Hunter, around 1746 (Cope, 1966). Over a period of 70 years, the Great Windmill Street School trained several thousand students. William and his younger brother, John, were famous as teachers of anatomy, and their reputation drew students to the school from all across the country, as well as from the American colonies.

As mentioned previously, the College of Physicians in London was founded through a charter granted by King Henry VIII to the better-educated physicians who had wished to establish themselves separately, as a professional group, from the surgeons, barber-surgeons and the apothecaries (Clark, 1964; Clark, 1964–72). The Company of Barbers then united with the Guild of Surgeons by an Act of

Parliament in 1540 to form the Company of Barber-Surgeons (Cope, 1959). At that time, there was no medical school in London, and it is doubtful whether doctors then knew much about the anatomy of the human body, although there might have been the occasional human dissections. Yet, the greatest revolution in human anatomy, which also marked the beginning of modern medicine, occurred just a few years later, in 1543, with the publication of Vesalius's *De Fabrica corporis humani.*

The barber-surgeons had long recognized the importance of anatomy for their professional skill (Lett, 1943). For example, in 1656, they appointed a Doctor Christopher Terne their lecturer in anatomy. He was a graduate of Leiden University and a fellow of the College of Physicians. According to Moore (1918), Doctor Terne, whose notes of three courses of lectures have

been preserved, delivered four lectures at the Barber's Hall using a dissected body:

> The course consisted of four lectures delivered at the Barber's Hall upon a dissected body. The first dealt with the skin, the abdominal muscles, the peritoneum and the umbilical vessels: the second spoke of the omentum, stomach and intestines, and their uses, the mesentery, the pancreas, the lacteals, the spleen, the liver and the lymphatic vessels: the third, of the bile ducts, the suprarenal glands, the kidneys, the urinary and genital tract: and the fourth, of the thorax, pleurae, lungs, and trachea, of the use of the lungs, of the heart with the pericardium, and of the blood.

These courses followed a long and established tradition more than a century old. The great English physician John Caius (1510–1573) had given lectures and anatomical demonstrations to the barber-surgeons for 20 years (1544–1564). Caius had studied anatomy under Vesalius at Padua. He was president of the College of Physicians and physician to Edward VI, Mary I, and Queen Elizabeth.

In many of the continental universities, especially in Italy, France, and Germany, medical training was already organized. Students followed a curriculum of compulsory lectures and practical work over several years, which ended with formal examinations. The teaching of human anatomy, even though based on the works of Galen, as well as dissection of the human corpse (Fig. 161), was an integral part of the course (Persaud, 1984). These institutions proved an irresistible attraction for prospective medical students from many countries in continental Europe, as well as from England and Scotland. In the Netherlands, medical teaching began only in the late sixteenth century following the foundation of the University of Leiden in 1575 (O'Malley, 1970).

Private medical schools were first established in London during the latter part of the eighteenth century, and in 1828, the University of London came into existence with the University College, but still without a medical program. It was largely through the efforts of the Society of Apothecaries that medical schools were established in London and other major cities (Holloway, 1964). The 1834 regulations for the training of apothecaries, followed by an Act of Parliament in 1858, which led to the formation of the General Medical Council, established the standards and requirements for medical practice in Britain (Poynter, 1966). During this period, the practical study of the human body was not entirely neglected and it thrived

Figure 161. Anatomical dissection scene from Ketham's edition of Mondino's *Anathomia* (Venice, 1500). Mondino is shown seated on the elevated chair reading from a book and presiding over the dissection. The assistant is about to make an incision (Instituti Ortopedici Rizzoli, Bologna).

Figure 162. Anatomy theater in England at the beginning of the nineteenth century (Courtesy of Bethesda, MD: U.S. National Library of Medicine, National Institutes of Health, Health & Human Services.) The amphitheatre has three circular rows for the observers. The dissecting table is positioned in the central open space. There are three glass jars with anatomical specimens on the table; one contains a conjoined twin. Other glass containers with anatomical specimens are aligned on a shelf high up on the wall at the back.

(Rolleston, 1939), especially in the private medical schools (Fig. 162) where many original discoveries were also made. The single event, however, which immensely influenced the teaching of anatomy, not only in Britain, but also in other countries, was the Anatomy Act, which was passed in 1832. The long path that led to it actually began in Scotland and Ireland.

Chapter 19

IRELAND AND ANATOMY

For this account of anatomy in Ireland, We have relied heavily on the highly readable and comprehensive work of Fleetwood (1951); several other sources (Cameron, 1916; Ball, 1928; Lassek, 1958; Coakley, 1988; Feely, 1992) also proved invaluable.

Before the Renaissance, the practice of medicine in Ireland was the domain of "*hereditary physicians*" who served the nobility, as well as others. The healing knowledge and skill they possessed were a family tradition, passing from one generation to the next. The early waves of the European renaissance, in culture and learning, had no major influence on the practice of medicine in Ireland, although a small number of doctors did travel abroad where they distinguished themselves. For example, Nial O'Glacan, was born in Donegal, where he probably was trained as a "*hereditary*" physician, travelled to Italy and France. He even became physician to the King of France. O'Glacan published several medical books, including a *Tractatus de Peste* in 1629, and some works on physiology and pathology in 1655 at Bologna (Fleetwood, 1951).

An important medieval medical treatise, which influenced the practice of medicine in Ireland, during the latter part of the fifteenth century, was the work *Rosa Anglica*. It was compiled around 1314 by the Englishman, John of Gaddesden (1280–1361), who had studied at Oxford, where he practiced medicine. The book, which was translated into Irish, is a compilation of common medical problems (palpitations, swellings, lethargy, rupture, paralysis, smallpox, arthritis, hemorrhoids, constipation) and the recommended treatment. It was a vade mecum for the physician of the day. The opening quotation from the greatest authority Galen would apply equally today: "*...do not frequent too much the courts and halls of the great, as I never did, until you have a knowledge of your books.*"

Being a compilation from many sources, especially the classical authorities, Gaddesden's treatise was a blend of Galenic teaching with whatever knowledge he might have gained through experience. For hemorrhoids, he recommended opening them externally and then applying a poultice of boiled onions. Gaddesden believed that smallpox resulted from "*corruption of the menstrual blood,*" but his accurate description of the infection and its prognosis revealed that he was a good observer (Fleetwood, 1951). The importance of a knowledge of human anatomy for the physicians then probably did not rank too highly.

Anatomical lectures and demonstrations as an essential part of medical training arrived with the establishment of the private medical schools, especially promoted on account of the obvious needs of those who had to "*cut,*" the barber-surgeons. The Dublin Guild of Barbers received a Royal Charter from Henry VI in 1446, 15 years before the London Barber-Chirurgeons received theirs in 1461, thus making it the earliest incorporation of medical practitioners. This was followed by a charter granted by Queen Elizabeth I in 1577 to the Guild of Barber-Surgeons.

In another charter granted by James II in 1657, the "*chirurgeons*" were linked with the barbers, apothecaries, and perriwigimakers, in order that "*their Arts and Misteryes may be better exercised.*" The physicians, who were better educated and administered medicines, distanced themselves apart from the Guild of Barber-Surgeons. In 1667, the College of Physicians was granted a charter, which prevented anyone without a license from the college to practice medicine within a radius of seven miles from Dublin. This protective step separated the physicians even more from the barber-surgeons who, by now, were also uneasy with their conjoint relationship.

In 1784, with the founding of the Royal College of Surgeons in Ireland (O'Brien, 1983), the union between the barbers and the surgeons came virtually to an end (Fleetwood, 1951). The charter granted by George III stated that the reputation of the profession of surgery is of the utmost importance to the publick and highly necessary to the welfare of mankind and that the publick sustain great injury from the defects in the present system of surgical education in our Kingdom of Ireland; and the regularly educated surgeons of the City of Dublin (who are become a numerous and considerable body) find themselves incompetent (from want of a charter) to establish a liberal and extensive system of surgical education.

One of the bylaws of the newly established college stipulated that

for the better advancement of the profession it shall be lawful for the College to elect or appoint a professor or professors, who shall annually give a regular course or courses of lectures on anatomy, physiology, the practice and operation of surgery and midwifery; and that all apprentices or pupils to the members of the College, whose names shall be duly registered as set forth, may attend the said course gratis.

Facilities for the storage of cadavers and for dissections were not available as late as 1788; the college had to decline a body in October of that year.

In 1692, the College of Physicians became incorporated, and with the agreement to admit to its fellowship only medical graduates of the university. Surprisingly, aspiring students could not have received any formal teaching in medicine from any of these institutions. Moreover, there was no medical school in Ireland at that time.

In 1711, Trinity College opened its medical school with the first anatomical theater in the country. This was a small "*old red brick building*" (Lassek, 1958) with neither water supply nor drainage. Professors and lecturers were appointed for the instruction of students. Both the college and the university required that "*every candidate Batchellor of Physick be examined in all ye parts of Anatomy relating to ye Oeconomia Animalis,*" as well as in other subjects.

Sir Patrick Dun (1646–1723), who was elected president of the College of Physicians, had made representations for the establishment of a medical school in Dublin. He obtained his qualifications from Oxford and came to Ireland as physician to the Duke of Ormond, Lord Lieutenant. Dun was both an influential and wealthy man. He left the greater part of his estate for the advancement of medical teaching in Dublin. "*He made provisions for lectures, anatomical dissection and a curriculum to cover materia medica, pharmacy, medicine, surgery, and midwifery.... The Dun's (subsequently King's) Professors established in the mid 1700's played a prominent role in the education of medical students in Dublin for some 200 years*" (Feely, 1992). Because of legal problems with respect to the settling of Dun's estate, many decades went by before funds became available for the appointment of the king's professors.

A rift later developed between the college and the university, which led to the appointment of several professors, including George Cleghorn (1716–1826). Like so many Irish physicians at that time, he had graduated from Edinburgh University (Hamilton, 1987) and practiced surgery in Dublin. Cleghorn was an excellent teacher who attracted large number of students to his anatomy classes. In 1803, William Hartigan (1756–1812), who held the chairs of anatomy, physiology, and surgery at the College of Surgeons, followed Cleghorn. When William Hartigan died in 1813, he was succeeded by James Macartney (1770–1843) in the position as professor of anatomy.

Macartney was educated in Dublin and London. For 13 years, he had occupied the chair of comparative anatomy at St. Bartholomew's Hospital before returning to Dublin. Macartney was extremely popular as a teacher, attracting several hundred students to his lectures (Macalister, 1900; Coakley, 1992). Space for the department of anatomy had to be increased in order to accommodate the increasing number of students. His demands for a practical examination in English, rather than Latin, as part of the final examinations, met with strong resistance from the older professors. Quarrels with others followed; in 1837, Macartney submitted his resignation to the board and Robert Harrison (1796–1858), who was already professor of anatomy of the College of Surgeons, succeeded him in the chair. Harrison was the author of a very popular *Manual of Anatomy*, known as the "*Dublin Dissector,*" several editions of which were published in the United States of America. It should be noted that Macartney was instrumental in introducing important reforms in the medical school. His successor established the anatomical museum at Trinity College.

The College of Surgeons School appointed in 1785 professors of anatomy and physiology, surgery, and midwifery. John Halahan gave his lectures on anatomy, physiology, and bandaging in his home because there was at that time no accommodation for the classes. Halahan joined William Hartigan in 1789. Eleven years later, the anatomy department was relocated to new facilities in St. Stephen's Green. Within the next few years, the medical school was joined by two brilliant clinicians, John Cheyne (1777–1836) and William Stokes (1804–1878), who were prominent among those that brought great recognition to medicine in Ireland.

The history of anatomy in Ireland's other medical schools also warrants some consideration, but it is beyond the scope of this monograph. These institutions were established later and, as a whole, have helped to forge the modern profile of medical education in the country, which really began about the middle of the nineteenth century (Lassek, 1958; Fleetwood, 1951; 1988).

As in London and Edinburgh, there were "*several small private medical schools*" in Dublin providing instruction for those interested in medical and surgical subjects even before the School of the Royal College of Surgeons became established. Anatomy was the main subject taught, and in order to establish a school one needed only a room for lecturing and dissection, as well as a supply of bodies from the body snatchers (Coakley, 1992). In order to recruit prospective students, these schools, which were located chiefly in Dublin but also in other counties, regularly advertised in the Dublin newspapers. Fleetwood (1951) quoted several advertisements so as to give an impression of the functions of these private institutions. For example, the *Dublin Weekly Journal,* which appeared on October 19, 1928, announced:

> A course of Anatomy in all its branches vis., Osteology, Myology, Angiology, Neurology, Adenology and Enterology will be given by James Brenan, M.D., at his house on Arran Key the 18th November, 1728, at twelve of the clock and will be continued every Monday, Wednesday and Friday, until the whole is completed, the operative part by Peter Brenan, Surgeon N.B. The charge of this course is two pistoles. [less than $3,000] If any students in Physic and Chirurgery be desirous to read Anatomy and Dissect they may be instructed and accommodated at the same place on reasonable terms.

Anatomy was then a major and probably the most important part of the medical course. Because the anatomists "*showed little originality*" and "*competed among themselves,*" the large number of students entering the medical schools led to further erosion in the quality of the anatomy course. Standards "*fell to rockbottom, cramming became the fashion and the grind rooms were overflowing*" (Macalister, 1884; Lassek, 1958). The opportunity to dissect the human body and to practice various operative/surgical procedures (Doolin, 1951) must have been a special feature of the courses as the following advertisement would imply:

From the *Dublin Journal,* 28th July 1767

> Mr. Maxwell, Surgeon, of the Tyrone Hospital, being solicited by many of his friends to establish in this country an anatomical school for instruction of young gentlemen of the profession, intends on Monday 14th December, at 2 o'clock, to begin, at this house at Omagh, a course of lectures on anatomy and surgery with some practical observations in midwifery, on the following terms, viz: for attending his lectures on anatomy, three guineas; dissecting pupils provided with subjects, six guineas; for attending his lectures in general and practice of the hospital, and being taught to dissect and to perform all the different operations in surgery, twelve guineas per autumn. Such pupils as choose to come under Mr. Maxwell's more private tuition may be provided with diet and lodging in his own house at fifteen guineas per annum.

The private medical schools in Ireland flourished throughout most of the nineteenth century and were organized in a manner similar to those in Britain. They provided a service that was lacking, but at the same time the schools were run as business enterprises, which provided the income for the instructors. These were the forerunners of our present day medical schools.*

The quality of teaching varied from one private medical school to another and depended on the reputation of the teachers, as well as the facilities that were available. Some of the early anatomy teachers included Richard Hoyle, Bryan Robinson, Robert Robinson, and James Cleghorn, son of George Cleghorn, as well as others (William Hartigan and James Macartney), who have been previously mentioned (Lassek, 1958; Doolin, 1951). Anatomy schools in Britain, for example, Hunter's Great Windmill Street School in London and Knox's Anatomy School in Edinburgh, attracted large numbers of students from everywhere, including from abroad, because of their reputation.

Not unexpectedly, these many schools required a considerable number of corpses for their students to dissect. The pressing demand for human material exceeded by far the supply that was available. Bodies were almost invariably obtained clandestinely and illegally (Ball, 1928; Lassek, 1958; Richardson, 1987; Coakley, 1992). At first, students and their teachers were involved, but in no time they were joined by professional grave robbers, who kept the schools well supplied for an attractive fee (Dobson, 1952).

According to Lassek (1958), between 1500 to 2000 bodies were removed annually by grave robbers from

---

*See Fleetwood (1951) for an interesting account of the private medical schools in Ireland.

the cemetery known as Bully's Acre in Dublin. The bodies were then shipped throughout the country. Encouraged by the good payments, which they demanded, the grave robbers began a flourishing trade, which permeated throughout the entire nation (Fleetwood, 1988). Notwithstanding their motives, the grave robbers, the resurrectionists, as they were then called, played a vital and perhaps even a salutary part in the progress of human anatomy.

ABRAHAM COLLES

In 1811, the gifted Irish surgeon Abraham Colles (1773–1843) published his book *A Treatise on Surgical Anatomy*, which at that time was the first work in the English language "*to emphasize the importance of topographic anatomy…an approach championed by anatomists in Paris where aspiring surgeons were guided layer by layer through operations on cadavers in the dissecting room*" (Coakley, 1992). Colles had a lifelong interest in anatomy, which began as a boy. Apparently, he found an illustrated anatomy textbook in an abandoned house that belonged to a physician. After medical studies at Trinity College in Dublin, he was awarded the Licentiate Diploma of the Royal College of Surgeons in 1795. Two years later, he received the M.D. degree from Edinburgh University. Colles then left for London where he became an assistant to Sir Astley Cooper and is remembered for his dissections of the inguinal region.

In 1799, Colles returned to Dublin and was appointed to the staff of St. Steevens's Hospital where he remained for 42 years. In 1802, at 28 years of age he was elected President of the Royal College of Surgeons. Two years later Colles was appointed professor of anatomy, physiology, and surgery at the Royal College of Surgeons in Ireland. Colles's name is associated with fracture of the distal radius, which he described. In addition, several other anatomical structures (Colles's fascia and Colles's ligament) are named after him. A skillful surgeon, Colles was the first surgeon to successfully ligate the subclavian artery. A modest man, he declined the Baronetcy in 1839 when it was offered to him.

BENJAMIN ALCOCK

Among the Irish surgeon-anatomists, one of the best known is Benjamin Alcock (1801–?). He is remembered for the clinically important fascial canal in the lateral wall of the ischiorectal fossa through which the internal pudendal artery, veins, and nerve pass. Now called the "pudendal canal," it is eponymously known as Alcock's canal (Coakley, 1992; Oelhafen et al., 2013).

Following the tradition of his family, including his father and grandfather who were physicians, Alcock began medical studies at Trinity College in Dublin where the widely known surgeon-anatomist James McCartney (1762–1813) taught anatomy in a different manner. He stressed the importance of knowing how anatomical structures are related to each other. Alcock graduated with a B.A. degree in 1821 and became a Licentiate of the Royal College of Surgeons in 1825. Two years later, he received the M.B. degree from Dublin University.

Alcock then served as an apprentice for several years under the eminent surgeon Abraham Colles where he became a skilled anatomist. In 1825, he was appointed as a Demonstrator at the Park Street School to teach anatomy. From 1836 he continued doing the same at the Peter Street School. In the following year Alcock received an appointment as Professor of Anatomy, Physiology, and Pathology at the newly opened School of Apothecaries' Hall in Dublin. In 1849, he was appointed as the first Professor of Anatomy and Physiology at the newly established Queen's College, Cork (Oelhafen et al., 2013).

Passing of the Anatomy Act in 1832 did not improve the supply of cadavers for teaching anatomy. Grave robbers still supplied the medical schools. Alcock was reluctant to participate in an illegal scheme, proposed by the governing authorities, for the procurement of corpses from poorhouses. He was forced to resign in 1854 from his position after a number of disputes and a turbulent relationship with the College. The authorities felt that his remaining in the College "*was not beneficial, nor of good example*" (O'Rahilly, 1947). Alcock took his grievances to the public and the press for support. He even petitioned the Queen who sent it back to the Lord Lieutenant of Ireland. After spending a few years living in Dublin he emigrated in 1859 to the United States. Nothing is known of his life there (Field & Harrison, 1968; Richardson, 1987; Coakley, 1992).

Chapter 20

SCOTLAND

The history of anatomy in Scotland can be traced back to the onset of the sixteenth century, when the Town Council of Edinburgh granted to the Guild of Barbers and Surgeons a Seal of Cause in 1505. It was endorsed a year later by King James IV. This enactment permitted the barbers and surgeons to practice their crafts, and from the surgeon it was expected *"that he know anatomy, nature and complexion of every part of the man's body. And likewise know all the veins of the same that he may make phlebotomy in due time...for every man ought to know the nature and substance of everything that he works or else he is negligent."* For this purpose, they were entitled to receive *"once in the year a condemned man after he be dead to make anatomy of whereby we may have experience, each one to instruct others."* A barber was not allowed to practice surgery without these qualifications, and an apprentice in surgery must be able to read and write. From these early beginnings the incorporation of Surgeons of Edinburgh was established which, in 1778, became a Royal College (Creswell, 1926; Comrie, 1932;1972; Wright-St Clair, 1964; Hamilton, 1987).

According to Wright-St Clair (1964), the first person officially authorized to teach anatomy in the city of Edinburgh was James Borthwick. On being admitted to the Incorporation of Surgeons of Edinburgh in March 1645, he swore that *"he shall observe, keep and fulfill all points of their seal of cause...and especially that point thereof anent dissecting of anatomy for the further instruction of apprentices and servants."* From 1647, knowledge of practical anatomy was also required for the entrance qualification. The following schedule was adopted:

The first day the entrant is to begin with the introduction to surgery and make a general discourse on the whole of anatomy without demonstration, secondly he is to demonstrate by ocular inspection more particularly some parts of the anatomy which shall be appointed to him by the deacon and masters and to answer the demands of his....

The *"Surregeants and Barbouris"* had petitioned the Town Council, much earlier in 1505, *"that they might have anis in the yeir ane condampit man after he be deid, to make anatomea of, quairthrow we may heif experience, ilke ane to instruct utheris, and we sail do suffrage for the soule."* It is evident from this requirement that a knowledge of anatomy was recognized as being essential for anyone, surgeons or barbers, to practice medicine, in addition to his apprenticeship. The privilege of dissecting the human body was granted to the surgeons and barbers, and the future development of the subject, both in England and Scotland, gathered momentum largely through them.

Encouraged by the physicians, who were granted a Royal Charter in 1681 to form a College of Physicians, Alexander Monteith, a surgeon, petitioned the town council in 1694 for certain unclaimed bodies which he planned to use for dissection and teaching (Comrie, 1932;1972). Monteith received the encouragement of Doctor Archibald Pitcairn, who had been professor of practice of physic in Leiden before returning to a similar position in Edinburgh. Pitcairn wanted to see a school like Leiden established in Edinburgh. In return for the privilege, Monteith promised to provide free services to the town's poor. This was the first attempt to establish a school of anatomy, although it is not certain whether Monteith actually taught any anatomy at all. The surgeons made a similar request having learnt of Monteith's petition and its approval. The town council, in approving the request, stipulated that the surgeons should build an anatomical theater within a reasonable time period where once a year an anatomical dissection should be carried out for the benefit of the public.

Monteith obtained a grant exactly as he had asked, of "*those bodies that dye in the correction-house,*" and of

the bodies of fundlings that dye upon the breast. The Surgeons obtained the bodies of fundlings who dye betwixt the tyme that they are weaned and thir being put to schools or trades; also the dead bodies of such as are stiflet in the birth, which are exposed, and have none to owne them; as also the dead bodies of such as are felo de se, and have none to owne them; likewayes the bodies of such as are put to death by sentence of the magistrat, and have none to owne them, which includes what former pretensions of that kind the petitioners have.

Monteith also obtained a room for dissections. Other conditions attached to the grant are curious. The dissection was to be during the winter season only, from one equinox to the other; "*all the gross intestines*" were to be buried within 48 hours, and the rest of the body within ten days, at the petitioners' expense. The regular apprentices of the Surgeons were to be admitted at half fee, and the right of being present was reserved to any of the magistrates who thought fit. Any friend who desired might require the body to be buried, if he refunded

to the kirk treasurer what expenses he hath been at upon the said deceased persons. The conditions attached to the grant to the Surgeons are to the same effect, without mention of the gross intestines, and with what proved ultimately the important addition, that the petitioners shall, befor the terme of Michallmes 1697 years, build, repaire, and have in readiness, ane anatomical theatre, where they shall once a year (a subject offering) have ane public anatomicall dissection, as much as can be showen upon one body, and if the failzie thir presents to be void and null. (Struthers, 1867)

The Surgeons' Hall was completed in 1697 in which the first recorded dissection took place in November 1702. Over a period of a week, a surgeon dissected and demonstrated one system or part each day. The next public dissection occurred in April 1704. A record of these important events follows:

Completion of the Surgeon's Anatomical Theatre and Commencement of the Annual Public Anatomical Dissection. On 17th December 1697, the Surgeons' Anatomical Theatre being reported to the Town Council as completed, the Council ratified its grant (of 1694), and the same day the Surgeons chose a "*committee to appoint the methods of the public dissections and the operators.*"

Although since 1505, the Surgeons had been entitled to a body annually, we have only now the commencement of what was termed a "*public anatomical dissection.*" This did not mean open to any one, for the Town Council, much against the wish of the Surgeons, restricted the use of the theater to the regular apprentices and pupils of free-men; but this was evidently a provision to secure a formal course of anatomical instruction. It is remarkable that this injunction should have emanated from the Town Council, apparently unsought; but the Chairman of the Surgeons sat at the Council Board.

Those of their number to whom the Surgeons intrusted this duty were termed "*operators.*" An interesting minute in the Surgeons' Records, of date 18th May 1704, shows the method in which this course of anatomy was conducted. It records a vote of thanks to the operators who had conducted the course in the previous month. The names and subjects are thus enumerated: The *first* day, James Hamilton – a discourse on anatomie in general, with a dissection and demonstration of the common teguments and muscles of the abdomen. The *second* day, John Mirrie – the umbilicus, omentum, peritoneum, stomach, pancreas, intestines, vasa lactea, mesentery, receptaculum chyli, and ductus thoracicus. The *third* day, Mr. Alexander Nisbet – the liver, vesica fellis, with their vessels, spleen, kidneys, glandulae renales, ureters, and bladder. The *fourth* day, George Dundas – the organs of generation in a woman, with a discourse of hernia. The *fifth* day, Robert Swintoun – the containing and contained parts of the thorax, with the circulation of the blood and respiration. The *sixth* day, Henry Hamilton – the hair, teguments, dura and pia mater, cerebrum, cerebellum, medulla oblongata, and nerves within the head. The *seventh* day Robert Eliot – the five external senses, with a demonstration of their several organs. The *eighth* day, John Jossy – the muscles of the neck and arm, with a discourse on muscular motion. The *ninth* day, Walter Potter – the muscles of the back, thigh, and leg. The epilogue or conclusion by Dr. Archibald Pitcairn. Pitcairn, though physician, had, in 1701, entered also the Incorporation of Surgeons, and we may be sure that the epilogue would not be the least interesting part of this course of anatomy in ten lectures. It will be remembered that the terms of the grant required the body to be buried in ten days, and the labor of dissection, as well as of exposition, was lessened by dividing the duty among ten lecturers, who appear to have been appointed to the duty for the current year only (Struthers, 1867).

EARLY PROFESSORS

The Incorporation of Surgeons and the town council sanctioned the appointment of a local surgeon apothecary, Robert Eliot, as public dissector of anatomy, in 1705, following a petition from him. He received a salary of £15 from the town council and promised to carry out the duty of performing annual public dissections if the surgeons would give him their support in his private teaching of anatomy to apprentices (Struthers, 1867; Creswell, 1926; Comrie, 1932; 1972). For all practical purposes, Eliot was the first professor of anatomy in Britain, although the first person to be officially so appointed was George Rolfe in 1707 at Cambridge. Following Eliot's death in 1715, John McGill was appointed as his successor, who, together with Adam Drummond, continued the teaching of anatomy in Edinburgh. Other appointments followed, including James Crawford as professor

of medicine and chemistry in 1713, which formed the nucleus for the teaching of medicine at the university.*

Apparently, these officially approved dissections were few and infrequent. Most of the teaching in anatomy took place in private schools attended for the benefit of the apprentices. This led to grave robbing in order to provide the corpses needed for dissection and demonstration, a practice which incited concerns and serious problems, and it was condemned by the surgeons. In their minutes of May 17, 1711, one finds the following remarks: ...*of late there has been a violation of the sepulchres in the Grayfriar church yard by some who most unchristianly have been stealing or at least attempting to carry away the bodies of the dead out of their graves a practice to be abhorred by all good Christians.*...

UNIVERSITY OF EDINBURGH

By its Charter of Foundation, granted in 1583 by King James VI, the University of Edinburgh had the right to teach, examine, and confer degrees in medicine. Apparently, this did not occur until the beginning of the eighteenth century. Candidates trained in continental Europe, who presented themselves for the university's degrees, were examined by the Royal College of Physicians on behalf of the university until the beginning of the eighteenth century, when the university appointed a professor of medicine (Bower, 1817–1831).

Students' knowledge of anatomy at the time would have depended, to varying extent, on the purposeful examination of the human body gained through dissection and comparison with other species. Such was the practice during the post-Vesalian era. This has

been thoroughly covered by Russell (1987) and others (Ball, 1928; Riesman, 1935; Lassek, 1958; Persaud, 1984; Luyendijk-Elshout, 1992). The changes were gradual from a modified form of Galenism to an exact science based on dissection, observation, and meticulous description. As early as 1505, long before the establishment of a medical school, anatomical dissections were authorized in Edinburgh.

Following this period, the history of anatomy in Edinburgh is inseparable from the legacy of the three generations of Monro's, who occupied the chair of anatomy in succession from 1720 to 1846 (Struthers, 1867; Comrie, 1932;1972; Guthrie, 1956), as well as from the tragedy which brought disgrace to the famous anatomist Robert Knox (MacGregor, 1884; Comrie, 1932;1972; Ball, 1928; Thompson, 1954).

ALEXANDER MONRO *PRIMUS*

Alexander Monro (Fig. 163) was born on September 8, 1697 in London, the only surviving child of a Scottish surgeon who came from a distinguished family (Struthers, 1867; Erlam, 1954; Monro, 1996). Alexander Monro served as an apprentice to his

father, and he also attended lectures and anatomical demonstrations given by the surgeons and apothecaries in Edinburgh. In 1717, Alexander Monro continued his medical education by attending the anatomical and surgical lectures of the famous William Cheselden

---

*For a comprehensive survey of the Edinburgh Anatomical School from the beginning of the sixteenth century to the turn of the nineteenth century, see Struthers (1867). This slim volume contains brief biographical sketches of many great Scottish anatomists.

(1688–1752) in London, as well as by observing patients in the hospitals. Within a period of six months, he carried out dissections on eight cases.

Alexander spent the next two years in Paris and at Leiden where he met the celebrated physician Boerhaave. After visiting in Amsterdam the distinguished anatomist Frederik Ruysch, who had developed impressive techniques for injecting the blood vessels and in preserving anatomical specimens, Alexander returned to Edinburgh in 1719. Within a month of arriving home, he presented himself to the Incorporation of Surgeons. With the appropriate training and credentials, Alexander passed the examinations to become a surgeon, and with the help of his father, he forged to the next step in his career.

Adam Drummond with John McGill resigned from their joint professorship of anatomy in January 1720. Simultaneously, they recommended that Alexander Monro should be appointed to this position:

> Adam Drummond and Mr. John McGill two of their number and present Proffessors of Anatomy represented that the state of their health and business were such that they could not duely attend the said Proffessiorship But they and the haill Calling being perswaded of the sufficiency of Alexander Monro one of their number Did therefor unanimously recommend him to the Provost and Toun of Edinr to be Professor of Anatomy. (Town Council minutes, cited by Wright-St Clair, 1964)

Appointed professor of anatomy, Alexander Monro *Primus* pursued a most distinguished career and, at the same time, started a dynasty where his son Alexander Monro *Secundus* and his grandson Alexander Monro *Tertius* occupied the chair of anatomy at Edinburgh by succeeding each other. In March 1722, Alexander M. *Primus* petitioned the town council in a most remarkable convincing manner that his appointment as professor of anatomy should be for life.

This request was approved and Alexander Monro *Primus* was appointed "*within this Citie and College of Edinburgh... ad vitam ant culpam notwithstanding of any act of Councel formerly made....*" Personal security, continuity and posterity were expeditiously ensured. A year later, and at the age of 25 years, the new professor for life was elected a fellow of the Royal Society following a recommendation made by William Cheselden.

Alexander Monro *Primus* was an immensely popular teacher and he was the first person to lecture in English and not Latin. Monro's reputation spread quickly and he attracted large number of students to his classes, which were held then in the Surgeons'

Figure 163. Alexander Monro (1697–1767). This engraving by James Basire (1775) is form a painting done by Allan Ramsay (1713–1784). From the Works of Alexander Monro, M.D. etc. published by his son, Alexander Monro, to which is prefixed the life of the author, Edinburgh 1781. (From Wegner, R.N.: *Das Anatomenbildnis. Seine Entewicklung im Zusammenhang mit der anatomischen Abbildung.* Basel, Benno Schwalbe and Co., Verlag, 1939. Courtesy of Institut für Anatomie, Ernst-Moritz-Arndt-Universität, Greifswald, Germany.)

Hall. At first, he started with an attendance of 57 students, which later increased to an average of 67 during the first ten years. Because of the size of the classes and the fear that the public, incited by grave robbing which was then prevalent (MacGregor, 1884; Bailey, 1896), might damage his valuable collection of anatomical and pathological specimens, he received permission from the town council to relocate his classes and museum to the safety of the university (Comrie, 1932; 1972) ... *as patrons of the Universitie of Edinburgh to allow him as Professor of Anatomy therein a theatre for public dissections for teaching the students under his inspection.*

The chair of anatomy, including all dissections, was now located within the walls of the university. The size of the class increased to 109 over the next ten years, and later to 146 students. The chair of anatomy was renamed the chair of medicine and anatomy in 1557. Monro *Primus* faced an enormous challenge.

He had to do a new thing in Edinburgh; to teach anatomy, and to provide for the study of it, in a town of then only thirty thousand inhabitants, and in a half civilized and politically disturbed country; he had to gather in students, to persuade others to join with him in teaching, and to get an Infirmary built. All this he did, and at the same time established his

THE
ANATOMY
OF THE
Human Bones and Nerves:
WITH
An Account of the reciprocal Motions of the HEART,
AND
A Description of the HUMAN LACTEAL SAC and DUCT.

By ALEXANDER MONRO, *Professor of Anatomy in the University of* Edinburgh, *and* F. R. S.

The FOURTH EDITION Corrected and Enlarged.

EDINBURGH,
Printed for Messrs. HAMILTON and BALFOUR. Sold by them and other Booksellers there ;

By Messieurs Innys and Manby, Rivington, Knapton, Longman, Astley, Hitch, Millar, Davidson, Oswald, and Hodges, at LONDON. J. Smith. DUBLIN : Bryson and Aikenhead, NEWCASTLE upon Tyne, and J. Watstein at AMSTERDAM, 1746.

Figure 164. Title page of Alexander Monro's *The Anatomy of the Human Bones and Nerves....*, 1746. (Courtesy of the Neil John Maclean Health Sciences Library, University of Manitoba.)

fame not only as a man of science, and gave a name to the Edinburgh school which benefited still more the generation which followed him. This really great and good man, therefore, well earned the title, often given him, of father of the Edinburgh medical school. (Struthers, 1867)

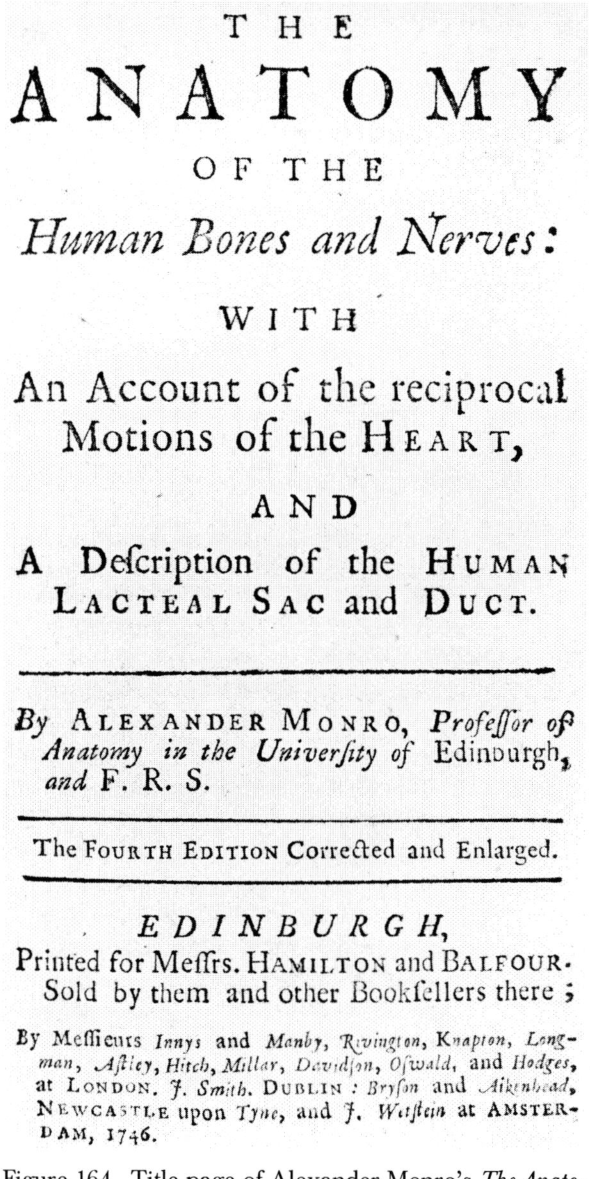

Alexander Monro *Primus* married Isabella Macdonald in 1725, and they had two sons, Donald and Alexander. Even before Alexander *Secundus* graduated in 1754, his father had him appointed the year before as his assistant. Alexander Monro *Secundus* was to become the most distinguished of the dynasty. He was the author of a large number of widely used scholarly works and he was an eloquent lecturer, attracting a large number of students from many countries to the university.

Of his several books, the most noteworthy was *The Anatomy of the Human Bones and Nerves...* (Fig. 164), which described the skeletal system accurately and in considerable detail. Even though the book did not have any illustrations, it still proved exceedingly popular. In all, there were ten editions and the work was translated in several countries.

Monro *Primus* was foremost a clinical teacher with experience in clinical medicine and operative surgery, which he constantly infused in his lectures. He was described by many of his contemporaries and students as a great teacher. Struthers (1867) has provided this account of Monro's course and lectures:

Monro's course extended from October to May, and embraced surgery as well as anatomy. His lectures were illustrated by dissections of the human body, and also, for comparison, of the bodies of quadrupeds, birds, and fishes. After giving the anatomy of each part, he treated of its diseases, especially of those parts requiring operations. He showed the operations on the dead body, and the various bandages and apparatus; and concluded the course with some lectures on physiology. He continued to give such a course uninterruptedly for thirty-eight years. He did not read his lectures. Even in giving the history of anatomy, with which he began his course, he spoke without the assistance of notes, except for the names and dates.

T. Somerville, who attended Alexander Monro *Primus's* lectures in 1757, remarked in his autobiography as follows: "*His style was fluent, elegant, and perspicuous, and his pronunciation perhaps more correct than that of any public speaker in Scotland at this time...and I think I had never before been so much capitulated with the power and beauty of eloquent discourse...."*

As a teacher, Monro *Primus* had obviously impressed William Hunter, the great surgeon, "*man-midwife*," and anatomist, when he was a student in Edinburgh. In his letter, dated September 1742 from London to his brother James at Long Calderwood, William wrote: "*I beg again and again that you will persecute your studies with resolution and never fear the event. I cannot learn why you throw Monro out of your scheme. I really think his a good course, and so fit for you that I would by no means have you neglect it, except...."*

In 1758, when he was 60 years of age, Monro *Primus* turned to the clinical aspects of medicine, as well as giving the relevant clinical lectures. A year later, Alexander Monro *Secundus* took over his teaching duties in anatomy and surgery. He eventually surpassed the distinguished reputation of his father.

Alexander Monro *Primus* died on July 10, 1767, after suffering for several years from the severe complications of rectal carcinoma, for which he needed opium. It brought to an end the life of a gifted teacher and celebrated anatomist. Monro had been referred to as a "*most admirable and lovable man; sincere, modest, without jealousy, benevolent, kind to his students; an able and active, and at the same time a calm and placid man*" (Struthers, 1867; Wright-St. Clair, 1964).

Figure 165. Alexander Monro (*Secundus*). (From Wegner, R. N.: *Das Anatomenbildins. Seine Entwicklung im Zusammenhang mit der anatomischer Abbildung*. Basel, Benno Schwalbe and Co., Verlag. 1939. Courtesy of Institut für Anatomie, Ernst-Moritz-Arndt-Universität, Greifswald, Germany.)

ALEXANDER MONRO *SECUNDUS*

Alexander Monro (*Secundus*; Fig. 165) was born in Edinburgh on March 10, 1733, the youngest son of Alexander Monro *Primus*. He entered quite early the university and in 1750 started his medical studies. Alexander Monro *Secundus* came increasingly under the influence of his father's teaching and scholarship. Soon, even as a student, he was carrying out anatomical studies, preparing scholarly papers, and eventually assisting his father with the classes.

A year before Alexander Monro *Secundus* had graduated in medicine, Monro *Primus* petitioned the town council for the appointment of his son as conjoint professor in order to assist with the classes. The supporting documentation also included a declaration signed by 20 medical students stating: "*That Alexander Monro, son of Mr. Monro, Professor of Anatomy, demonstrated a considerable part of his Fathers Course of Anatomy, and prolected in his Fathers place to us; and the other students who attended that Course last Winter entirely to all our satisfactions.*" The approval came for both Monros

to be appointed as conjoint professors for life. The younger Monro was then only twenty-one years of age.

Alexander Monro *Secundus* graduated with his M.D. degree on October 25, 1755. His doctoral thesis was entitled "*De Testibus et Semine in variis animalibus*" and it was dedicated to his father. Over a period of close to three years, Monro *Secundus* pursued further studies in anatomy from several famous anatomists, including William Hunter in London; J. F. Meckel, Berlin; B. S. Albinus, Leiden; and Pieter Camper, Amsterdam. From Meckel he benefited the most, and Monro often referred to him in his lectures. Meckel published in Berlin a paper entitled *De venis lymphatics Valvulosis*, describing the origin of lymphatic vessels from tissue spaces, which eventually led to controversy with the Monros as to who first had made the discovery.

Because of his father's illness and deteriorating health, Monro *Secundus* began to give most of the lectures in anatomy and in the summer of 1758 took up his duties as professor of anatomy. In this task, he excelled his father and large numbers of students were increasingly attending his lectures. Monro's style of lecturing was described *"by one who was present as having acted like an electric shock on the audience...he was a master of his subject and of the art of communicating knowledge to others; his style was lively, argumentative, and modern..."* (Struthers, 1867). According to Monro *Secundus*, of the 13,404 students who attended his classes in 1807, 5,831 were from outside Scotland. The reputation of Monro *Secundus* attracted many foreign students to the Edinburgh Medical School.

A new anatomy lecture theater, sufficiently large to hold 300 students, was built in 1764, on account of the size of the growing classes, and in order to house the anatomical specimens. Attendance at Monro's class ranged from 200 to 400 a year, and during his tenure more than 14,000 students attended his lectures in anatomy and surgery.

Alexander Monro *Secundus* was a practicing physician, as well as an anatomist. He was knowledgeable in surgical matters, but he himself did not carry out surgical operations. The need for a separate chair in surgery, so as to properly emphasize the clinical and practical aspects of this subject, became a persistent issue among the surgeons and students. James Rae, a surgeon, gave private lectures at Surgeons' Hall and later at the hospital. Monro was strongly against these initiatives, especially when the Incorporation of Surgeons petitioned the crown to establish a Regius Professorship in Surgery and recommended James Rae for the position.

Monro's objectives and the steps he took to prevent the establishment of a separate chair in surgery were viewed as inappropriate and even unconstitutional. This was a regrettable episode in his distinguished career. The most impressive and influential of Alexander Monro *Secundus'* many publications included the following:

(1) *Observations on Structure and Functions of the Nervous System* published in 1783, which contained magnificent copperplate engravings. Monro described in here the communication between the lateral and third ventricles of the brain; this he had already published with respect to a case of hydrocephalus in 1764. The communication which today bears Monro's name had already been described by other physicians and anatomists, including Galen and Vesalius.

(2) *The Structure...Fishes* (1785).

(3) *Bursae Mucosae...*(1788), a work in which Monro described 140 bursae in the body, including 32 in each arm and 37 in each leg (Fig. 166).

In 1758, Alexander Monro *Secundus* entered into a long protracted controversy with William Hunter over the discovery of the origin of the lymphatic vessels. According to William Hewson, Alexander Monro *Secundus* had plagiarized his work, which Monro described in a paper he had published in 1757.

Monro *Secundus* had married Katharine Inglis in 1762. They had two sons, David and Alexander. The latter eventually succeeded him in the chair of anatomy. In fact, a year before Alexander had graduated in Medicine, Monro *Secundus* had petitioned the town council, in 1798, for his son to be appointed as a conjoint professor together with him, the same his father (Monro *Primus*) had done for his career.

From 1802, Monro *Secundus* continued in the chair, but most of the teaching was done by his son Alexander Monro *Tertius*. In 1808, when he was 75 years of age, and having held an appointment for 54 years, Alexander Monro *Secundus* retired. He died on October 2, 1817 at the age of 85 years, having suffered a stroke a few years earlier. Doctor Monro *Secundus* was buried in Greyfriars churchyard next to his wife who died in 1803. In assessing the life and career of Monro *Secundus*, Struthers (1867) remarked,

To be at the same time the successful teacher of so splendid a class, the leading physician of his day, and the author of works of original research in anatomical science, formed a rare combination, the effect of which, extending and accumulating over half a century, may enable us to understand the greatness to which the reputation of the second Monro grew, both at home and abroad, and the honour in which his name is held among anatomists, and in the Edinburgh School.

ALEXANDER MONRO *TERTIUS*

Most people who know of the Monros regard Monro *Tertius* as the least talented of the three to occupy the chair of anatomy in Edinburgh.

The most common anecdote of him is that of a teacher who read verbatim from his grandfather's notes. Even though more than a century had gone by, and the dates were all wrong, Monro *Tertius* apparently never bothered to make any changes. Some of these anecdotes clearly passed from one generation of students to another, and in time they were exaggerated. Nonetheless it seems from the available sources that Monro *Tertius* was an appalling teacher who paid little attention to the content, clarity, or style of his lectures. Moreover, he had a penchant for straying away from his subject matter and remaining there. Without being too judgmental, it could have been that he tried, but his performance was uninspiring and fell short badly of his illustrious predecessors in most respects.

Alexander Monro *Tertius* was born on November 5, 1773 in Edinburgh and entered the university in 1790. He received his M.D. in 1797, and in the same year, he became a fellow of the College of Physicians. After furthering his studies in London and Paris, he returned to Edinburgh in 1800. A year earlier he was appointed conjoint professor with his father. Alexander Monro *Secundus* had petitioned the town council with such a request. The council approved their joint professorship for life in November 1798, and a month later Alexander Monro *Tertius "was admitted into the University in absentia."*

At first Monro *Tertius* gave most of the lectures, but from 1808 onwards, he was put completely in charge of all the anatomy teaching. His lectures were awful and his competence became questionable. Students and others ridiculed him, and he was the object of many of their jokes. Charles Darwin was a medical student for a brief period in Edinburgh. He did not enjoy Monro's lectures. In a letter (January 6th, 1826) to his sister Caroline, he wrote as follows: *"...after which I attend Monro on Anatomy I dislike him & his lectures so much that I cannot speak with decency about them. He is so dirty in person & actions"* (Burkhardt, 1996). As the years passed, his classes were unruly and his teaching became a concern for all within the university. Struthers (1867) remarked that *"his talent as a teacher of anatomy was not great."*

Monro *Tertius* published a large number of scientific works. These included *Observations on Crural Hernia* (1803), the *Morbid Anatomy of the Human Gullet, Stomach, and Intestines* (2nd edition, 1830), *Outlines of the Anatomy of the Human Body* (four volumes, 1813), *Observations on Aneurism of the Abdominal Aorta* (1827), *The Morbid*

Figure 166. Title page of the German edition of Alexander Monro's (Secundus, 1733–1817) monograph on *"all the bursae mucosae of the human body."* The original English edition appeared in 1788. (Courtesy of Institut für Anatomie, Ernst-Moritz-Arndt-Universität, Greifswald, Germany.)

Anatomy of the Brain (1827), *The Anatomy of the Urinary Bladder and Perineum of the Male* (1842; Fig. 167).

Monro informed the university of his intention to retire. He was 73 years old and had taught anatomy for close to 50 years. Doctor Monro was succeeded by John Goodsir, F.R.S. as professor of medicine and anatomy in 1846. Doctor Monro died on March 10th, 1859 at the age of 85 years and was buried in the Dean's Cemetery.

OTHER MEDICAL SCHOOLS

Three generations of Monros successively occupied the chair of anatomy at the University of Edinburgh, and for 126 years, they greatly influenced the teaching of anatomy (Guthrie, 1956). Many of their contemporaries and students have also made significant contributions to medical teaching, as well as in the evolution of anatomy as a scientific discipline. Some of these luminaries who, through their research and writings, achieved fame and brought distinction to Scottish medicine have been discussed elsewhere in

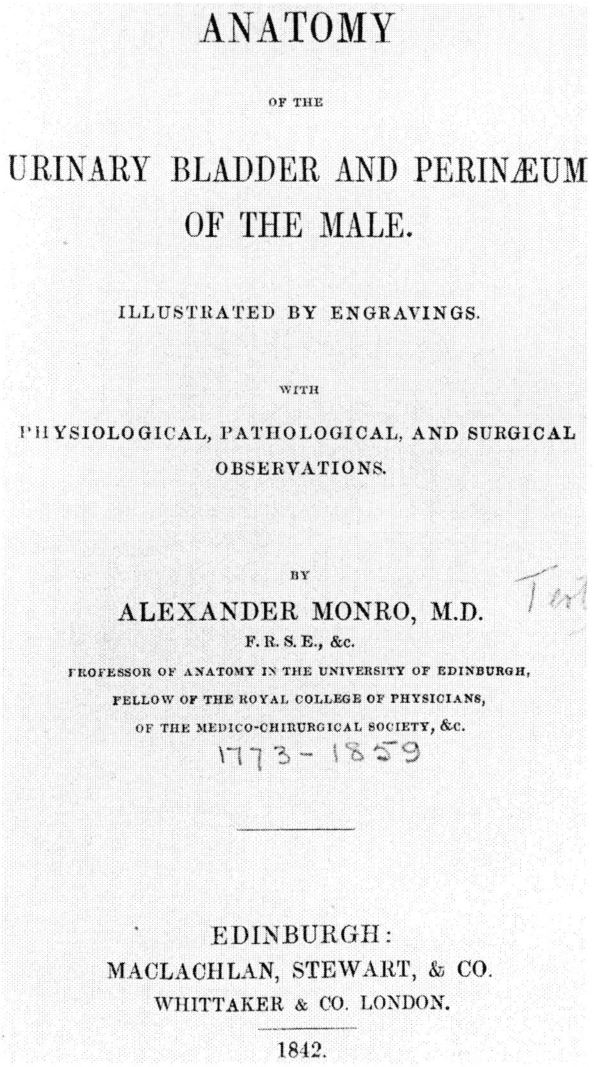

ANATOMY

OF THE

URINARY BLADDER AND PERINÆUM
OF THE MALE.

ILLUSTRATED BY ENGRAVINGS.

WITH

PHYSIOLOGICAL, PATHOLOGICAL, AND SURGICAL
OBSERVATIONS.

BY

ALEXANDER MONRO, M.D.
F. R. S. E., &c.
PROFESSOR OF ANATOMY IN THE UNIVERSITY OF EDINBURGH,
FELLOW OF THE ROYAL COLLEGE OF PHYSICIANS,
OF THE MEDICO-CHIRURGICAL SOCIETY, &c.

1773 - 1859

EDINBURGH:
MACLACHLAN, STEWART, & CO.
WHITTAKER & CO. LONDON.
1842.

Figure 167. Title page of Alexander Monro, *Tertius* (1773–1859), book on *The Anatomy of the Urinary Bladder and Perineum of the Male,* 1842. As mentioned in the introduction, the book was especially intended for the lithotomist who "*has to make his incisions through a narrow path, to deviate from which is death! The details of the anatomy of the parts of the pelvis, so essential to the safe performance of puncturing and cutting into the bladder, are, in the course of a few years, too apt to escape from the memory,...*" (Courtesy of the Neil John Maclean Health Sciences Library, University of Manitoba.)

this book; for others, reference should be made to the following valuable sources: Struthers (1867), Comrie (1972; 1932), and Hamilton (1987). In the same light, the impact Scottish medical training had on physicians in Britain and elsewhere should be appreciated. Because the first of the new medical schools in London began teaching in 1821, "*Scotland had a virtual monopoly of university medical education and in the first half of the century almost 95% of doctors in Britain with a medical degree had been educated in Scotland*" (Hamilton, 1987). Indeed, the seeds for medical teaching not only in England but also in the British American Colonies grew out of Scottish soil.

As in Edinburgh, medical teaching began quite early in other Scottish cities (Comrie, 1972; 1932; Hamilton, 1987) and with it also the teaching of anatomy. For example, from the beginning the constitution of the University of Glasgow permitted the teaching of medicine, but it did not really start until 1714 when John Johnstoun was appointed professor of medicine. Six years later, Thomas Brisbane was elected to the combined chair of botany and anatomy, even though "*he had a considerable distaste for dissection,*" so much so that "*the students appealed to the Principal to intervene and arrange for anatomical teaching*" (Hamilton, 1987). In 1790, James Jeffray was appointed to the chair of anatomy and botany. He held this position for 58 years, the longest such tenure in a Scottish University. From the beginning, Jeffray recognized the practical importance of human anatomy in medicine. He improved "*the conditions under which anatomy was taught, by obtaining increased accommodation for the dissecting room...*" (Comrie, 1972).

By the beginning of the eighteenth century, the number of medical students in Glasgow had increased considerably, especially with the establishment of the private extramural medical schools, the College Street School in 1796, and the Portland Street School in 1827. Clinical teaching was poor, and practical anatomy suffered because there was a severe shortage of cadavers for medical students to dissect. This led to grave robbing, carried out by students and often with the encouragement of their teachers. Students who were successful in supplying a corpse did not have to pay the fees for their anatomy classes. In 1814, a Glasgow anatomist was prosecuted for obtaining bodies that were stolen from their graves.

The University of Aberdeen had its origin with the establishment of a *Studium generale* in 1494. From the beginning a "*mediciner*" was appointed to teach medical subjects, including anatomy. William Gordon was one of the first "*mediciners*" who made any attempt to teach practical human anatomy. Appointed in 1636, he had petitioned the Lords of the Privy Council to provide him with four bodies, male and female, for public dissection (Comrie, 1972).

Anatomy classes at St. Andrews University also began in the eighteenth century with the establishment of a professorship of medicine in 1722; formal medical teaching, however, began much later. Attempts to initiate medical teaching in Aberdeen can be traced to the late fifteenth century (Comrie, 1972). There was not much progress until more than two centuries later. Yet, during this period and even later foreign medical graduates and experienced persons with skill in healing were able to obtain the M.D. degree from St. Andrews or Aberdeen merely by applying for it and paying the required fees. Both universities found this to be an attractive source of income. Many distinguished eighteenth century doctors (Cullen, Pitcairne, Jenner, Smellie, etc.) obtained their M.D. degree in such a manner (Hamilton, 1987).

Compared to the medical schools in Edinburgh and Glasgow, regular anatomy teaching began much later in Aberdeen and in St. Andrews.

In 1828, the College of Surgeons required of all medical students that they must attend two courses in anatomy. A year later, the practical part of the anatomy course was extended from three to six months, and to 12 months in 1838. The University of Edinburgh separated the chair of surgery from anatomy in 1831. With the establishment of a separate chair in anatomy, the scope of the anatomy lectures changed; and "*anatomical instruction and surgical teaching entered upon the modern epoch in Edinburgh*" (Comrie, 1972).

Chapter 21

ROBERT KNOX: HONOR AND TRAGEDY

ROBERT KNOX AND THE RESURRECTIONISTS

On that evening of November 29, 1827, no one could have foreseen the tragedy that would follow one of the most famous anatomy teachers when the roguish William Burke and William Hare arrived with Donald at Robert Knox's anatomical establishment. On November 27, 1827, Donald, an old-age pensioner, died of natural cause in the lodging house of William Hare at Tanner's Close in West Port, Edinburgh. In order to recover the £4 Donald owed him, Hare, with the help of another lodger, William Burke, decided to sell the body to the anatomists. At least 16 people were killed by Burke and Hare who then sold the bodies to the *"porters"* of the various anatomy rooms. The last of these bodies was discovered in Knox's rooms (Comrie, 1972).

On the way to the medical school, Burke and Hare met a student and asked for Doctor Monro's rooms. The student realizing their mission directed them to his own teacher, Doctor Knox, where they were paid £7.10s for the body. "*So big a sum and so easily got proved sadly ominous: the two Irishmen loved their labour less and their whisky vastly more from that hour of selling their friend's carcase....*" Hare, the vilest of the two old monsters suggested a fresh stroke of business, namely to inveigle the old and infirm into his den and "*do for them*" (Lonsdale, 1870). Prowling the streets, they found their victims – widows, orphans, prostitutes, and the mentally retarded, who were lured to Burke's house in the West Port "*and there dosed with whiskey and suffocated.*" The bodies were then sold to Doctor Knox's school (Lonsdale, 1870).

As pointed out by Ball (1928), "*all teachers of anatomy were accepting from the resurrectionists any bodies which were offered – no questions being asked,*" because at that time no teacher of anatomy "*in Edinburgh, in London, or in Dublin, dared to quarrel with the "sack 'em-up men." To have done so, at once would have caused the ruin of the anatomist's school.*" It was simply accidental that Knox became ensnared with Burke and Hare, because initially Hare had wanted to take the bodies to the university rooms of Alexander Monro *Tertius.* The result was Doctor Knox's fall from grace, now the victim of an indignant public, as well as the law. He was humiliated and his reputation was completely tarnished. At the same time, the sordid events surrounding the resurrectionists and Doctor Knox hastened the legislative process leading to an urgently needed reform in making cadavers available for dissection (Ladell, 1983; Richardson, 1987; Desmond, 1989).

THE ANATOMIST ROBERT KNOX

The best biographical account of Doctor Knox was written by Henry Lonsdale, his former student and colleague. Recognition should be given to Doctor Knox as an anatomist of the highest eminence, but who at the same time was the unfortunate victim of circumstances and of the era during which he had lived (Lonsdale, 1870; MacGregor, 1884; Comrie, 1972; Currie, 1933; Rae, 1964). Some of the more important aspects of Doctor Knox's life and professional career will be briefly mentioned.

Robert Knox (1791–1862) was born on September 4, 1791 in Edinburgh (Fig. 168). He attended high school, which he completed as the best student and the gold medalist. In 1810, Knox entered the medical school of Edinburgh University. His father, a teacher of mathematics and natural philosophy, as well as a prominent Freemason, might have been involved in laying the foundation stone of the university on November 16, 1789. The grandparents were tenant farmers in the estates of the Earl of Selkirk.

Alexander Monro *Tertius* was the professor of anatomy at the medical school, and "*it is said that, on his first examination for the M.D., Knox was plucked for his anatomy....As the anatomical teaching of Monro Tertius had been of poor service, Knox went to Barclay to make amends for lost time.*"

KNOX AND JOHN BARCLAY

It is now necessary to make a few remarks about Doctor John Barclay (Fig. 169) because he was not only a teacher of Knox, but later they would become involved professionally. The relationship had a profound influence on Knox's life. Doctor John Barclay (1758–1826) was a well-known anatomist. He was born in Perthshire and had studied to become a preacher of the Church of Scotland, but later entered the medical school at the University of Edinburgh from where he obtained the doctor of medicine degree in 1796. Barclay served as an assistant to the distinguished surgeon-anatomist, John Bell (Struthers, 1867); and he continued his anatomical studies in London before returning to Edinburgh in 1797.

Figure 168 (*Left*). Robert Knox (1791–1862). This portrait shows Robert Knox, 50 years of age, delivering a lecture. It was made from a calotype, which forms the frontispiece of Lonsdale's book *Sketch of the Life and Writings of Robert Knox, the Anatomist,* London 1870. (Department of Anatomy, University of Manitoba.)

Figure 169 (*Above right*). John Barclay (1758–1826). (Department of Anatomy, University of Manitoba.)

Barclay started privately to teach anatomy on a small scale, and through his skill as a teacher he attracted many dissatisfied students from the university classes. Soon, he was running a famous private anatomy school, in a house he had purchased, where he taught for 27 years (1797–1825). Barclay had also a good knowledge of comparative anatomy and he had learnt much about practical anatomy as required for the practice of surgery, which he emphasized in his lectures and demonstrations.

On June 19, 1804, Barclay's Anatomy School was recognized by the Royal College of Surgeons, which added to its importance as an institution for the instruction of medical students (Ball, 1928).

Struthers (1867) remarked that "*among the former Anatomists of Edinburgh, there is no name that I have been accustomed to hear so frequently or with so much respect as that of Dr. Barclay.*" Barclay devoted all of his time to anatomy, including lecturing, research, and the preparation of museum specimens. This partly contributed to Barclay's success because he was the first teacher of anatomy in Edinburgh to do so.

Knox greatly impressed the examiners in anatomy at the university when he appeared for the second time, because "*he had anatomy at his finger's end, and could set forth his knowledge in the choicest Latin the language in which the examinations were at that time conducted*" (Lonsdale, 1870). It would seem that Knox had benefitted from Barclay's anatomy classes.

Having graduated in Medicine in 1814, Knox received a commission as assistant surgeon in the army and in 1815 he was sent to Brussels "*to render aid to the wounded, and this was followed by similar duties in the wild*" of South Africa where he was stationed from 1817 to 1820. Returning to the United Kingdom he was granted from military headquarters a leave of absence for a year, which enabled him to further his medical studies in France.

Knox returned to Scotland with new insights in medicine, surgery, natural philosophy, ethnology, osteology, comparative anatomy, and paleontology. The love for these subjects was awakened in him through lasting friendships he forged with such eminent men as Cuvier (1769–1832), Geoffroy St. Hilaire (1772–1844), and others. Much of Knox's work later would reflect these diverse interests, especially in the areas of comparative anatomy and physical anthropology (Blake, 1870–71; Desmond, 1989).

Knox wrote numerous scientific papers on various aspects of comparative anatomy. His most influential works included scholarly monographs on *The Races of Men: A Fragment* (1850) and *Great Artists and Great Anatomists* (1852). Knox also regularly presented his scientific observations to members of the Wernerian Society, the Anatomical and Physiological Society, and the Royal Society, in Edinburgh. In December 1823, Knox became a fellow of the Royal Society of Edinburgh. A year later he was married.

Knox maintained his interest in the practice of medicine and he became a member of the Medico-Chirurgical Society. On April 2, 1824, the College of Surgeons accepted his proposals for the formation of a Museum of Comparative Anatomy, which would replace the "*relics of the Art of Chirurgerie, the fleams and cutting instruments of the barber-surgeons, and the other crude apparatus, surgical and obstetrical, stood dwarfed forms and other deformities, gather as curiosities from a superstitious past*" housed in the Museum of the Royal College of Surgeons, in the "*Old Hall*" (Lonsdale, 1870).

In his declining years, John Barclay, greatly impressed by the achievements of his former student, invited Knox to join him as a co-partner of his school. Knox gave his first course of anatomy lectures during the Winter 1825–1826 in a masterly fashion. Following Doctor Barclay's death in 1826, Knox succeeded him and was now owner of the school. Lonsdale (1870) remarked that soon "*Dr. Knox may have had some imitators but he had no rivals in the schools of anatomy. From 1826 to 1835 there was but one temple worthy of the name in Edinburgh, in which aspiring youths might worship in the spirit of Galen, and sing the hymns that the anatomic-theosophist delighted in; and that temple was "Old Surgeons' Hall," where Robert Knox presided as high priest, oracle, and philosopher.*"

Soon, Knox "*came to be designated by his class as Knox primus et incomparabilis.*" Furthermore, "*no medical lecturer in the United Kingdom ever enjoyed such popularity, or won his spurs so quickly as Robert Knox*" (Lonsdale, 1870). Struthers (1867) noted that "*Dr. Knox was able to invest human anatomy with a new interest. His forte as an anatomist was, not in detail or the relation to surgery and medicine, but in bringing comparative anatomy to the explanation of human anatomy.*" According to Comrie (1972), Knox "*had an extraordinary power of lucid exposition…and he appears to have infused an interest into the dull facts of anatomy.*" The classes were overcrowded, averaging as many as 335 students annually between 1826 and 1835; there were 504 students in the 1828–29 session. To meet this demand, Knox lectured three times daily on the same topic.

Doctor Knox was the envy of other anatomy teachers in Edinburgh. It is therefore not surprising that later when he was "*assailed by the press, attacked by mobs, and slandered by Philistines,*" except for a few, his colleagues deserted him and some even joined in the attacks (Lonsdale, 1870; Biddis, 1976). Indeed, "*by virtue of his success as a teacher, Knox from the start was placed in a dangerous situation*" (MacGregor, 1884).

For over a century, Robert Knox's name has been associated with little more than grave robbing and the atrocities committed by Burke and Hare, with disregard for the man himself, his scientific achievements, and the reputation he had then. Like all other anatomy schools in England, Scotland, and Wales, Knox's establishment relied on a steady supply of dead bodies for students to dissect (Currie, 1933) so that students may learn anatomy and practice surgery (Richardson, 1987; Desmond, 1989). Placing Knox in the era he lived in (Biddis, 1976), and with this in mind, he should be judged. "*All teachers of anatomy were accepting from the resurrectionists any bodies which were offered – no questions being asked. A century ago, no anatomical teacher in Edinburgh, in London, or in Dublin, dared to quarrel with the "sack-'em-up men." To have done so, at once would have caused the ruin of the anatomist's school – to say nothing of the loss of time and credits to the students (Ball, 1928).

It is doubtful whether a man of such eminence would *collaborate* with the depraved resurrectionists in committing the murderous acts. One must agree with Ball (1928) that much time has passed since the crimes were committed, "*All the players have disappeared: the murderers, the rival anatomists, yea, even their students are no more. Let the dead rest.*"

FALL FROM GRACE

Famous as he was, Knox, nonetheless, suffered a terrible fate of disgrace, loneliness, and poverty even though William Burke in the concluding paragraph of a signed confession from his condemned cell, dated January 21, 1829 (MacGregor, 1884), declared that:

Docter Knox never incoureged him, nither taught or incoreged him to murder any person, nether any of his asistents, that worthy gentlemen Mr. Fergeson was the only man that ever mentioned any thing about the bodies. He inquired where we got that young woman Paterson.

Knox was silent throughout this difficult period, and his first attempt to defend himself took the form of a letter, dated March 17, 1827, addressed to the *Caledonian Mercury*. I have taken the liberty to quote from Ball (1928) a personal letter that was written to him by a prominent member of the Royal College of Surgeons of Edinburgh regarding Doctor Knox and the charges against him:

I fancy that what underlay Christison's denunciation of Knox was the University jealousy of Knox's success. Everybody thought Knox was careless, everybody thinks so still; but in those far-off days no anatomist dared ask his suppliers where they got their subjects, far less how they got them. Christison overlooks the fact that Knox's men received the bodies, handled them, and injected them. Knox paid for them. If the bodies had been rotten Knox would have been warned that they were not worth the money and the suppliers told to bring them fresher. Knox's business was to get on with his lectures and demonstrations, not to get handling the bodies until they came into the dissecting-room itself. If the local committee which was formed for the express purpose of bringing home Knox's criminality failed to find reasons then to impeach him, we may be pretty certain that there was not a shred of evidence that could be a cause of complaint. Before the occurrence Knox was not a society man. That he was ruined afterwards no one denies, but far less wouldruin a man today just as certainly. The anatomical world was the poorer for the popular virulence against Knox.

Knox's reputation suffered and he left Edinburgh in 1844 for a short period in Glasgow, and he moved to London where he lectured and practiced medicine (Comrie, 1972). After decades of ridicule, isolation, a wretched existence, deteriorating health, and poverty, Doctor Knox suffered a stroke. He died on December 20, 1862 an unhappy man. His wife had preceded him in 1841.

WILLIAM BURKE AND WILLIAM HARE

The dreadful murders committed by Burke and Hare have been retold many times (Lonsdale, 1870; MacGregor, 1884; Ball, 1928). Their last victim was a poor old woman named Docherty. Her body was discovered in a secured tin chest in Doctor Knox's dissecting room by the authorities who carried out a search the following morning. That a poor old woman was murdered so that her body can be sold to the anatomists horrified everyone and caused a national furor. It transcended the robbing of graves, which was already vile. Fear for one's personal safety now became a problem. It led to their arrest (Fig. 170) and a much publicized trial that soon became a national spectacle.

The trial of Burke and Hare began on December 24, 1828 and, except for several refreshment breaks, continued until midmorning of the following day. The public's indignation was flared up on account of the wretched murders they committed over a period of a year. For this reason, it was important to secure for the prisoners the best lawyers in order to ensure a proper legal course and the administration of justice.

Hare, the most hideous of scoundrels, and his wife turned approvers, and so escaped the gallows, only to be hunted, however, from town to town like wild beasts. They returned to their native country (Ireland)

BURKE, THE MURDERER!!

Figure 170. William Burke in his prison cell, fitted with leg irons. (Courtesy of Bethesda, MD: U.S. National Library of Medicine, National Institutes of Health, Health & Human Services.)

and were no more heard of except in the pages of fiction. On Christmas morning 1828, the jury returned, after deliberating for less than an hour, the following verdict: The jury find the panel, William Burke, guilty of the third charge in the indictment; and find the indictment not proven against the panel, Helen McDougal.

This was then followed by the sentence of the Lord Justice-Clerk:

William Burke, you now stand convicted, by the verdict of a most respectable jury of your country, of the atrocious murder charged against you in this indictment, upon evidence, which carried conviction to the mind of every man that heard it, in establishing your guilt in that offence. I agree so completely with my brother on my right hand, who has so fully and eloquently described the nature of your offence, that I will not occupy the time of the court in commenting any further than by saying that one of a blacker description, more atrocious in point of cool-blooded deliberation and systematic arrangement, and where the motives were so comparatively base, never was exhibited in the annals of this or of any other court of justice. I have no intention of detaining this audience by repeating what has been so well expressed by my brother; my duty is of a different nature, for if ever it was clear beyond the possibility of a doubt that the sentence of a criminal court will be carried into execution in any case, yours is that one, and you may rest assured that you have now no other duty to perform on earth but to prepare in the most suitable manner to appear before the throne of Almighty God to answer for this crime, and for every other you have been guilty of during your life. The necessity of repressing offences of this most extraordinary and alarming description, precludes the possibility of your entertaining the slightest hope that there will be any alteration upon your sentence. In regard to your case, the only doubt which the court entertains of your offence, and which the violated laws of the country entertain respecting it, is whether your body should not be exhibited in chains, in order to deter others from the like crimes in time coming. But taking into consideration that the public eye would be offended by so dismal an exhibition, I am disposed to agree that your sentence shall be put into execution in the usual way, but unaccompanied by the statutory attendant of the punishment of the crime of murder – viz., that your body should be publicly dissected

Figure 171. Execution of the notorious murderer William Burke. (Courtesy of Bethesda, MD: U.S. National Library of Medicine, National Institutes of Health, Health & Human Services.) "*Shortly after eight o'clock on the morning of Wednesday, 28th January 1829, Burke was hanged in the presence of an enormous crowd, estimated by various writers at 20,000 to 37,000 persons. All has assembled to show their delight*" (Ball, 1928). The etching (Nimmo's Lithographic Office, 1829, Edinburgh) shows Burke hanging from the gallows built on an elevated platform. The area around the platform is crowded with people. Some are leaning from windows and others are sitting on rooftops.

and anatomised, and I trust that if it ever is customary to preserve skeletons, yours will be preserved, in order that posterity may keep in remembrance your atrocious crimes.... (MacGregor, 1884).

Burke was publicly hanged, in the Lawnmarket of Edinburgh, just after 8 o'clock on January 28, 1829 in the presence of a cheering crowd estimated to be between 20,000–37,000 people (Fig. 171). According to one report, "*Every countenance bore an expression of gladness and revenge was so near, and the whole multitude appeared more as if waiting to witness some splendid procession or agreeable exhibition*" (Ball, 1928). Together with the abuses, taunts, and curses that were directed at Burke as he stood before the scaffold, the crowd demanded that his accomplice Hare, as well as Dr. Knox, should also be hanged "*Hang Hare too!*" "*Where is Hare?*" and "*Hang Knox!*"

The body of the deceased was ordered to be delivered to Professor Alexander Monro's Department at the University, and the following morning a prominent and select group of men (Robert Liston, Surgeon; George Combe, Phrenologist (Stone, 1829); Sir William Hamilton, Philosopher; and Mr. Joseph, Sculptor) examined the body. In accordance with the sentence, Burke's body was publicly dissected by Professor Monro *Tertius* in the crowded anatomy theater the same afternoon. The anatomical theater was full to capacity.

Ball (1928) described the scene and the crowd, which required the help of the police, as the most exciting and riotous anatomical afternoon recorded in Scotland's history. The audience was mixed and impatient. A huge crowd besieged the classroom door.... The lecture concerned the brain, a portion of the anatomy of the human body on which Monro was supposed to have bestowed especial attention. After the calvaria had been removed, "*the quantity of blood that gushed out was enormous, and by the time the lecture was finished, which was not till three o'clock, the area of the class-room had the appearance of a butcher's slaughter-house, from its flowing down and being trodden upon*" (West Port Murders, Edinburgh, 1829, p. 254).

Following a long riot that started from the crowd jostling to see the remains, it was arranged that admission would be limited to 50 persons at a time with the result that a constant stream of people, calculated at a rate of 60 per minute, passed the table and viewed the dissected body of Burke. There were seven or eight who fainted or had their clothes torn. On the following day, at least 25,000 persons viewed the corpse. The crowd in the streets of the Old Town of Edinburgh *"made it appear as if the occasion were one of general holiday."* Burke's skeleton was prepared and is now located at the medical school of the University of Edinburgh. A contemporary witness recorded that "*After this exhibition Burke was cut up and put in pickle for the lecture-table. He was cut up in quarters, or rather portions, and salted, and, with a strange aptness of poetical justice, put into barrels*" (MacGregor, 1884).

Hare and his wife were spared from the gallows because of the immunity that was promised them. They turned against Burke and gave evidence for the Crown. Hare escaped to London where for 40 years he survived as a blind beggar. His wife managed to reach Belfast and not more is known about her. Another accomplice, Helen McDougal, suffered greatly and probably found her way to Australia where she died.

Chapter 22

THE ANATOMICAL LITERATURE

AFTER VESALIUS

Vesalius's Magnum Opus *De humani corporis fabrica libri septem*, published in 1543, is the best known book on human anatomy ever printed. It became an unparalleled success. As one of the most important works in the history of anatomy, it ushered in the modern era of medicine and established standards that had to be matched or surpassed. Later books combined anatomical facts observed from human dissection and aesthetics with varying degrees of success.

To compile a bibliography of textbooks and atlases, even for the period covered in this volume, would be a formidable task. For this reason, we have attempted to reflect significant trends in the evolution of major publications in anatomy and to present a mere broad sweep of *selected* works, which have marked the various phases of transition, from Vesalius to about the middle of the nineteenth century (Thomas, 1974).*

Russel (1987) observed that "*anatomy, like all sciences, has a bibliography which is international in its coverage; but to prepare such a bibliography would be a vast undertaking far beyond the powers of a single individual.*"

When examining any particular anatomical work, it is important to take into account the medical knowledge and progress of that period. An anatomical book that was published in the seventeenth century, considered then to be outstanding, might well be of lesser interest today (Choulant, 1962; Wolf-Heidegger & Cetto, 1967; Russel, 1987; Norman, 1991; Roberts & Tomlinson, 1992.

Vesalius's masterpiece *De humani corporis fabrica...* was widely copied and plagiarized (Fig. 172) because it revealed a true image of the human body based on systematic dissections new anatomical discoveries. Just two years after Vesalius's book was published in Basle, the first plagiarized copy appeared in London. One of the best known plagiarists was Thomas Geminus (c. 1510–1562) whose *Compendiosa totius anatomie delineatio, aere exarata* was published in 1545 in London (Persaud, 1984; Locy, 1911; Choulant, 1962; Sudhoff, 1964; Putscher, 1972; Lind, 1975; Persaud, 1984; Hildebrand, 1988; Roberts & Tomlinson, 1992 for anatomical writings and illustrations before Vesalius).

THOMAS GEMINUS: PLAGIARIST

Thomas Geminus was the pseudonym for Thomas Lambrit who was an engraver and printer. Not much is known about his birth, but he died in May 1562. Vesalius severely rebuked Geminus as "*an extremely inept imitator*" because Geminus had included in his book the text of the Vesalius's *Epitome*, passages from the Fabrica, and 40 woodcut plates copied from *De humani corporis fabrica*. The plagiarized book was successful after several translations and editions. Of interest, geminus had the woodcuts redrawn and engraved as copper plates, the first to be done in England (Fig. 173).

Geminus acknowledged Vesalius on the first page (Ala) of the book, and he had the woodcuts finely engraved on copper. As a result, the figures were sharper, and on the whole, the book was elegantly

---

*See Russell (1959a) for an account of plagiarism involving anatomical works and anatomists.

Figure 172. Frontispiece of a rare copy of Vesalius's work. This edition was issued in Amsterdam in 1617. (Neil John Maclean Health Sciences Library, University of Manitoba.) The illustrations can be traced to Thomas Geminus, an engraver, who copied the Vesalian plates for his own book published in 1545.

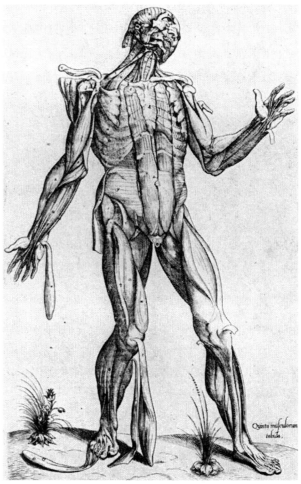

Figure 173. One of 40 engraved copper plates by Thomas Geminus appearing in his book *"Compendiosa totius Anatomie delineatio aere exarata"* after the woodcuts from Vesalius' *"Epitome,"* an abbreviated form of his 1543 work *"De humani corporis fabrica."* (With permission from: Keynes G. 1959. The anatomy of Thomas Geminus. A notable acquisition for the library. Ann R Coll Surg Engl, 25:171–175.)

produced. It was the second book to be printed in England with engraved plates, and these were considered to be the first of any artistic importance. The first English edition of Eucharius Rosslin's *Rosengarten*, based on a medieval work on midwifery, was published in 1540 with the title *The Byrth of Mankynde newly translated out of the laten into Englysshe* (O'Dowd

& Philipp, 1994). Later editions (1545, 1552, 1560) contained anatomical plates that were derived from the work of Vesalius, as well as from Geminus (Power, 1927).

CHARLES ESTIENE

The dissecting manual of Charles Estienne (1504–1564) entitled *De dissectione partium corporis humani* was published in Paris in 1545. A French translation appeared a year later. Among the 62 beautiful woodcut plates were illustrations of the entire venous and nervous system, printed for the first time (Figs. 174 & 175). It is recognized as one of the most outstanding anatomical works of the sixteenth century.

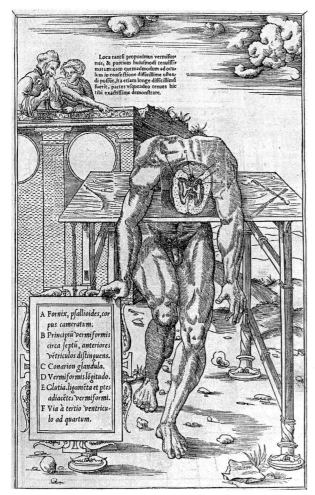

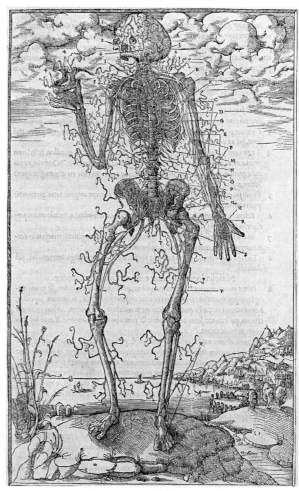

Figure 174 (*Left*). Draped over a table, an anatomical figure displays a cross-section of his brain while touching a frame that holds captions. In the background, spectators observe from atop a fanciful parapet. To cut costs, Estienne took some of his illustrations from non-anatomical books, replacing a section of the woodblock with an insert that depicted the body's interior. In this figure, the boundary of the insert is seen in a square around the head. A dissection des parties du corps humain.... Paris, 1546. Woodcut. (Courtesy of Bethesda, MD: U.S. National Library of Medicine, National Institutes of Health, Health & Human Services.)

Figure 175 (*Right*). A skeleton with a thin covering of tendons poses before an apocalyptic sky. La dissection des parties du corps humain.... Paris, 1546. Woodcut. (Courtesy of Bethesda, MD: U.S. National Library of Medicine, National Institutes of Health, Health & Human Services.)

JUAN VALVERDE DE HAMUSCO

Juan (c. 1525–1587), the Spanish anatomist who lived in Rome, introduced the work of Vesalius to Spain through his book *Historia de la composici On del cuerpo humano*, Rome, 1556. Of the 42 engravings in the book, only four new plates were made; the others were copied from Vesalius's work. Valverde's book, "*the first great original medical book in Spanish*" (Norman, 1991), contains not only his observations and discoveries but also a critique of Vesalius's errors. It was a popular work among medical students and doctors in Spain. Valverde's book might have served as the medical source of anatomical knowledge that was diffused in Spanish America (Fig. 176).

Other popular anatomical textbooks that were later available in Spain included: *Anatomia Completa del Hombre* (Madrid, 1728) by Martin Martinez (1684–1734); *el Compendio Anatomico* (Madrid, 1750–52) by

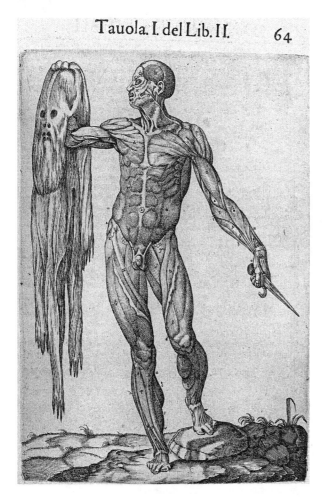

Tauola. I. del Lib. II.   64

Figure 176. One of Valverde's most striking original plates is that of a muscle figure holding his own skin in one hand and a knife in the other, which has been likened to Michelangelo's Saint Bartholomew in the Last Judgment section of the Sistine Chapel. The original illustrations were most likely drawn by Gaspar Becerra (1520?–1568?), a contemporary of Michelangelo, and the copperplate engravings are thought to have been carried out by Nicolas Beatrizet (1507?–1570?), whose initials "NB" appear on several of the plates. (Courtesy of Bethesda, MD: U.S. National Library of Medicine, National Institutes of Health, Health & Human Services.) (http://www.nlm.nih.gov/exhibition/historicalanatomies/valverde_bio.html).

Juan de Dios Lopez (1711–73); and *el Curso Completo de Anatomia del cuerpo humano* (Madrid, 1796–1800) by Jaime Bonells e Ignacio Lacaba (Febres-Cordero, 1987).

MARTIN MARTINEZ

In eighteenth century Spain, Martin Martinez (1684–1734) was the most distinguished surgeon and anatomist. He was responsible for improving medical education in Spain and also for the establishment of an anatomical theater in Madrid with the blessings of the Royal Court. For the training of students and surgeons, Martinez believed that an understanding of human anatomy gained from dissection surpassed theories.

Martinez received his medical training at Alcala de-Henares and graduated at the age of 22 years. He became professor of anatomy in Seville and was later elected a member of the Royal Society of Seville. He also served as physician to the King. He was an eloquent lecturer who attracted large number of students. Martinez's book, *Anatomia Completa del Hombre...*, published in 1728, was a popular and comprehensive textbook in human anatomy, the sixth edition of which appeared in 1775.

REALDO COLOMBO

The work of Realdo Colombo (c. 1510–1559; Fig. 177), pupil and successor of Vesalius at Padua, *De re anatomica libri XV*, published in 1559, after his death. Except for the woodcut on the front page, the book

Figure 177. This is the only illustration to Realdo Colombo *De re anatomica*, published in Venice in 1559, the year of his death. It is known, however, that he had planned an illustrated text. In his letter to Duke Cosimo de'Medici of 17 April 1548, he requests leave from his post as lecturer in anatomy at the University in Pisa in order to work on his busok, mentioning the assistance he is receiving from the "leading painter in the world" as well as how, on a previous stay in Rome, he dissected cadavers and supervised artists. This "leading painter" has usually been identified as Michelangelo, whose friendship with Colombo is documented in the 1553 biography of the artist by Ascanio Condivi. Colombo dispatched the body of "a young and very handsome moor" for Michelangelo to dissect and also treated him successfully for kidney stones. Colombus is noted for offering an early description of the pulmonary circulation and for being a proponent of vivisection, the subject of the fourteenth book of *De re anatomica*. In the title page, the dissection is being followed by several observers, two of whom are consulting books, one of which is illustrated. At the lower left a young man is seated, taking notes or sketching on a pad. (Wellcome Library, London.)

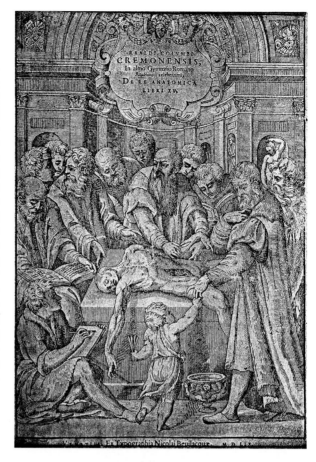

contains no illustrations and much of the text had been plagiarized from other authors. Nonetheless, the work is important on account of Colombo's description of the pulmonary circulation (translated into English by Banister in 1578) and of the action of the pulmonary, cardiac, and aortic valves. Whether Colombo was the first to describe the pulmonary circulation has been a controversial issue for centuries. The *Observationes anatomicae* of Gabriele Falloppio (1523–1562), Vesalius's other pupil, was published in 1561 in Venice.

According to Russell (1949), "*only 178 separate works which went through a total of 445 editions*" were published in English between 1548 and 1800. Of these, 139 were by British scholars and the remaining 39 books were from other parts of Europe. The earliest anatomical work was that of Thomas Vicary in 1548. No copy exists today of this compendium, which was derived from many sources, including the works of Lanfrank and Mondeville. After Vicary's death, his colleagues at St. Bartholomew's Hospital published a new edition of the book in 1557.

In 1578, a very popular book, *The Historie of Man...*, was compiled and published by John Banister (1533–1610). It contained several plates that were copied from Vesalius's work. Reference was made to the pulmonary circulation as described by Colombo. Banister was a reputable Nottingham surgeon and physician who also demonstrated practical anatomy to the barber-surgeons in London.

JULIUS CASSERIUS

Eleven years after his death, Julius Casserius, c. 1561–1616) masterpiece, *Tabulae anatomicae*, with 107 plates, was published posthumously in Venice in 1627 (Fig. 178). "The beautiful copperplates were engraved by Francesco Valesio after Odoardo Fialetti, a pupil of Titian." Casserius had commissioned these plates covering the whole field of human anatomy but died before the work was completed. The editor added 20

Figure 178. A copperplate engraving from *De formato foetu liber singularis*, commissioned by the Paduan anatomist Casserius. It shows a classic renaissance Venus filleted lotus-like to reveal her child. The plate, as well as some 95 others, represent the high point of Paudan/Venetian anatomical illustration and were published in three separate titles after Casserius' untimely death. Courtesy of Bethesda, MD : U.S. National Library of Medicine, National Institutes of Health, Health & Human Services.

plates made by the same artist/engraver (Norman, 1991). This work is considered to be one of the most artistically striking atlases ever produced with ornate landscapes in the background (Fig. 179).

The first acknowledged work on surgical anatomy, *Anatomia chirurgica...*, was authored by Bernardino Genga (1665–1734) and published in Rome in 1672. He emphasized the importance of anatomy for surgeons. Bernardino Genga's other book, Anatomia per use er intelligenza del disegno, was published in 1691.

This book is considered to be an excellent work on human anatomy for the artist on account of the quality of the plates. An English translation of Genga's book, dedicated to the fashionable physician and collector, Richard Mead (1673–1754), appeared in London in 1723 and republished in 1767. Of interest, Genga accepted Harvey's theory of blood circulation, but he believed that Colombo and Cesalpino had suggested a similar concept earlier.

Anatomists, whose works have also appealed to artists and sculptors, included Bernhard Siegfried Albinus (1697–1770), John Flaxman (1755–1826), Giuseppe del

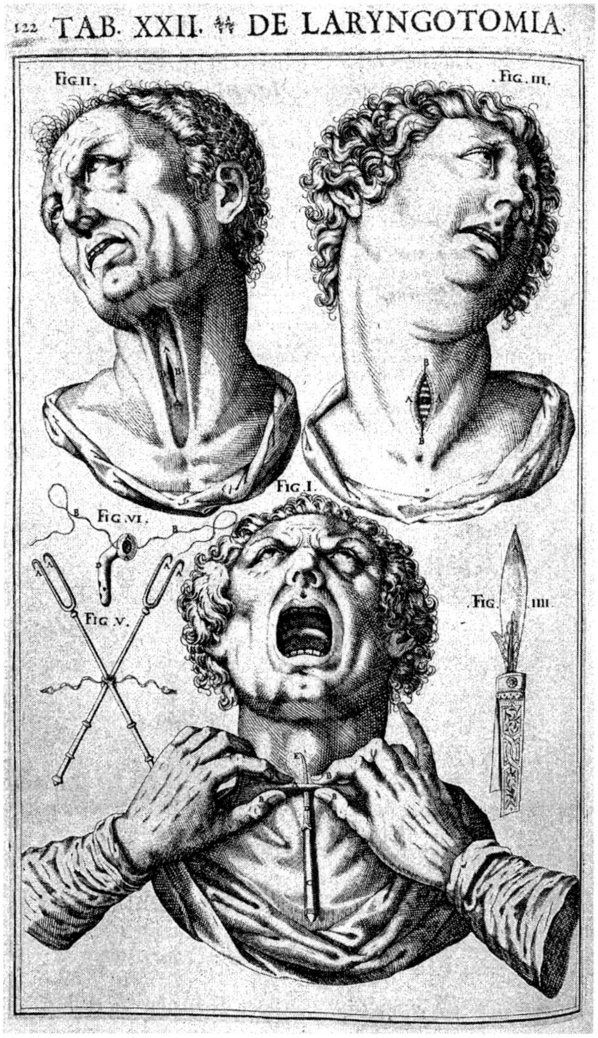

Figure 179. Casserius oldest description of tracheostomy ("laryngotomia") De laryngotomia. From *De vocis auditusque organis historia anatomica...tractatibus duobus explicata ac variis iconibus aere excusis illustrata*. Published by V. BaldinusFerrara 1600–1601. (Wellcome Institute Library, London.)

Medico, Paolo Mascagni (1752–1815), and Charles Bell (1774–1842), as well as others (Knox, 1852; Major, 1954; Kennedy & Coakley, 1992; Roberts & Tomlinson, 1992; Smith, 2006; Rifkin et al., 2006).

William Harvey's discovery of the circulation of the blood became known in continental Europe through Jan de Wale (1604–1649), professor of anatomy at Leiden. It was mentioned in his work *Epistolae duae de mortu chyli et sanguinis*, which appeared in 1641 in Leiden.

JEAN RIOLAN

Jean Riolan, the Elder (1539–1605), was a wealthy and influential person in Parisian society. He served for a year as Dean of the Medical Faculty (1585–1586), and he was the personal physician to Marie de Medici (1553–1642). His son Jean Riolan, the Younger, (Johannes Riolanus, 1580–1657) studied medicine in Paris and graduated in 1604. Both were respected physicians who also pursued anatomical studies (Fig. 180). Jean Riolan, the Younger, was a skilled dissector who believed that for a better understanding of the human body it is necessary to dissect it.

Anatomical Discoveries

Although a traditionalist and follower of Galen, Riolan established a solid reputation as an anatomist because of his discoveries and books. Most influential were his *Anthropographie* (1618; second edition 1626); and his *Encheiridium Anatomicum et Pathologicum* in 1648 which presented both normal and pathological anatomy (Fig. 181). It was one of the best anatomical textbook of the era following the addition of plates that were taken from Johannes Vesling's *Syntagma*.

Riolan described the *appendices epiploicae* and the anastomosis between the superior and inferior mesenteric arteries at the inner margin of the large intestine. The colon is supplied by short branches from this arterial circle – the marginal artery. He also described the palpebral part of the orbicularis occuli muscle. Riolan also made important contributions to our knowledge of the spermatic vessels and branchial apparatus.

Riolan and the Theory of Blood Circulation

As a Galenist, Riolan fiercely opposed Harvey's theory of blood circulation (Donley, 1946). Extracted from one of his anatomical treatises is the following: "...I shall demonstrate, however, that there is no circulation of the whole mass of the blood, but of a portion of it only, because the blood which is contained in the portal vein and in the smaller branches of the aorta and the vena cava, has normally no circulation at all. Hence it is that there is a circulation of that blood only which occupies the larger branches of the aorta and the vena cava and which seeps through the

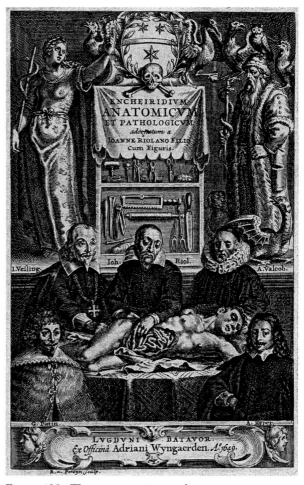

Figure 180. The present print is a later copy, in reverse, of the title page, designed and engraved by Crispijn de Passe the second for Jean Riolan the younger's *Anthropographia et osteologia*, of Paris 1626: see Wellcome Library catalogue no. 588752i. In the present print, which served as the title page to Riolan's *Encheiridium anatomicum et pathologicum* of 1649, the original French observers are replaced by named Dutchmen in different poses, particularly anatomists from Leiden. Lugduni Batavor[um] [Leiden] : Ex officina Adriani Wyngaerden, 1649. (Courtesy of Wellcome Library, London.)

middle septum of the heart from the right ventricle to the left without traversing the lungs. In this way, the blood, both arterial and venous, is supplied in abundance to all the parts twice or thrice during each day.

According to this new doctrine of the circulation of the blood, the medicine of Galen need suffer no change, as assuredly it must according to the teachings of Harvey" (*Opuscula anatomica*, 1694).

JOHANNES VESLING

Johannes (Veslingius) Vesling (1598–1649) was one of the earliest anatomists who promoted William Harvey's work on the circulation of blood (Fig. 182). He described it in his book on human anatomy, "*Syntagma anatomicum...,*" (*Greek: Συνταγμα, Constitution*) which was first published in 1651. Veslings's popular textbook was based on his own observations and meticulous dissection. The clear and concise text together

Figure 181 (*Left*). Front page of *Encheiridium Anatomicum et Pathologicum*. 1968.

Figure 182 (*Right*). In this title page to the Amsterdam 1666 edition of Johannes Vesling's *Syntagma anatomicum*, with commentary by Gerardus Blasius, Vesling is seated next to a table covered with a cloth, decorated with two crossed bones, which bears the title of the book. He is identified by his name which appears above his hat and by the cross of the Order of the Holy Sepulchre in Jerusalem. He is depicted wearing the same cross on a chain in his engraved portrait in the Padua 1647 edition of the *Syntagma anatomicum*, which is the second enlarged edition with added plates by G. Giorgio (see Garosi 1963, tav cxxxiii for Vesling's portrait and this catalog no. 25026 for the title page to the 1647 edition). Vesling indicates to four gentlemen illustrations of the heart in a book displayed by a draped skeletal corpse. Only two details of these illustrations correspond with Vesling's plate on the heart (Tab. 1, Cap. X). Suspended from two pilasters is a swag made up of surgical instruments. Through an arch topped with two angels holding a cartouche with a skull crowned with laurel leaves one sees a view of buildings. (Courtesy of Wellcome Library no. 25042i.)

with stunning figures contributed to the success of the "*Syntagma.*" It was widely used during the seventeenth century. Several Latin editions and translations (English, French and German) were published. An English translation by Nicholas Culpeper, *The Anatomy of the Body of Man – wherein is exactly described every part thereof in the same manner as it is commonly shewed in publick anatomies: and for the further help of young physicians and chyrurgions, there is added very many copper cuts...*appeared in 1653, followed by several reprints.

Johannes Vesling was born in Minden, Westphalia. There are several versions of his early education and medical studies. He went to Vienna and apparently studied medicine at several German and Italian universities. In Venice, he was a popular anatomy lecturer, and in 1627, he carried out an anatomical demonstration for the physicians of the city. In 1632, Vesling was appointed professor of anatomy and surgery at the University of Padua. He embarked on a tour of Egypt where he studied the local plants. His research resulted in his scholarly work "*De plantis Aegyptilis,*" which was published in 1638. In 1638, Vesling was appointed to the chair of botany at the University of Padua and Director of the botanical garden. Vesling continued his work in anatomy and made many original discoveries. Following a trip to Crete, he became sick and died in 1649. His remains lie in the church of St. Anthony in Padua (Porzionato et al., 2012; Vesling, 2008).

CASPAR BARTHOLIN

Caspar Bartholin, the Elder (1585–1629), was the first of a distinguished line of Danish scholars. He was born in Malmo, Sweden and became professor of medicine (1613) and theology (1624) at the University of Copenhagen. Bartholin was the first to describe the functions of the olfactory nerve, and his *Anatomicae institutiones corporis humani* (1611) was an immensely successful work (Fig. 183). In 1633, Bartholin's book

Figure 183. A corpse, with an incision to its abdomen, is the subject of an anatomy attended by a group of figures in primarily antique dress. In the niche in the right background is a skeleton in the pose of the first skeleton plate in Vesalius's *De humani corporis fabrica* (Basel 1543). On the wall to the left of this niche and above a set of surgical instruments is the relief portrait of the Danish anatomist and physician, Thomas Bartholin, framed in palms. This is comparable to the portrait of Bartholin (see this catalogue, no. 817), after H. Ditner, engraved by G. Appelmans, who also engraved this title page to the fourth edition of Bartholin's compendium of the works of his father, Caspar Bartholin snr. Wellcome Library no. 25162i T. Bartholin, Anatome ex omnium veterum recentiorumque observationibus imprimis institutionibus, b. m. parentis Caspari Bartholini, ad circulationem Harvejanam, et vasa lymphatica quartum renovata..., Leiden 1673 [i.e. 1674]. (Wellcome Library, London.)

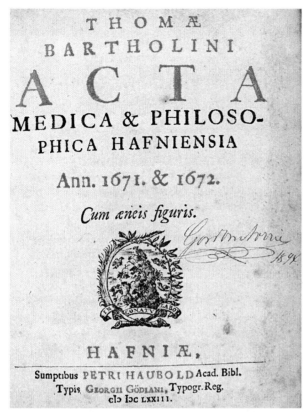

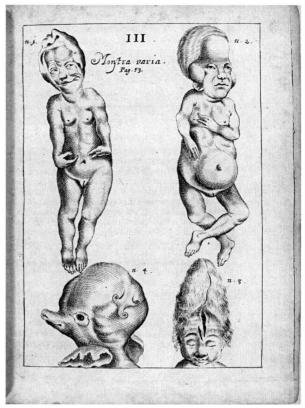

Figures 184–185. *Acta Medica et Philosophica Hafniensia,* a recent acquisition of The Lilly Library, was edited by Thomas Bartholin and was one of the first medical periodicals distributed throughout Europe. It contained information on a diverse array of subjects, including medicine, botany, and zoology. Almost immediately after the development of the newspaper in the seventeenth century, scientists and physicians began to disseminate natural knowledge through periodicals. In the earlier half of the century, knowledge was spread across geographic boundaries almost exclusively through correspondence, personal travel, and word of mouth. The periodical allowed doctors to give short observations on natural subjects – usually written in Latin – such as the birth defects presented in these images. As knowledge spread more rapidly, other physicians could more easily test the observations and experiments of their colleagues. http://www.indiana.edu/~liblilly/anatomia/.

was among the descendants of the elder Bartholin was Caspar Bartholin, the Younger (1655–1738), Danish anatomist, who described the greater vestibular glands of the female genital tract and the duct of the sublingual salivary gland. Another son, Thomas Bartholin, the Elder (1616–1680), was professor of anatomy at Copenhagen University from 1648 to 1661. He was the first person to describe the entire lymphatic system (Nielsen, 1942) in a work entitled *De lacteis thoracis in homine brutisque nuperrime*

observatis, historia anatomica; an edition was printed in London in 1653. Based on a revision of his father's anatomical treatise, Thomas Bartholin also authored a popular textbook of human anatomy. It was translated into English (*Bartholinus anatomy; made from the precepts of his father, and from the observation of all modern anatomists; together with his own…*) and published in 1662 in London. He also publish a periodical in Latin for Medical Doctors (Figs. 184 & 185).

MICHAEL LYSER

One of the best and detailed textbooks of anatomy in the seventeenth century was that compiled by the Danish anatomist, Michael Lyser (1626–1659). His

Culter anatomicus, published in 1663, was also the first work to give specific and detailed instructions on dissecting. Lyser had studied in Copenhagen and at

Padua. For many years, he was an assistant prosector to Thomas Bartholin and is supposedly the person who first discovered the lymphatic vessels.

It was not unusual for authors then to blatantly copy the work of others. For example, John Browne (1642–1702) published a treatise in 1681 with the title *"A compleat treatise of the muscles," as they appear in humane body, and arise in dissection; with diverse anatomical observations not yet discovered.* This work lacked originality and was plagiarized from the main sources. The text was *"an almost verbatim copy of Molin's description of muscles"* and the plates were taken from the 1627 edition of Julius (Giulio) Casserius's treatise *Tabulae Anatomicae.* Browne's book was extremely popular with several English editions, as well as Latin and German translations appearing. Of some importance, however, is *"that the 1687 edition of this work was the first book ever to appear in which the names were actually printed on the muscles"* (Russell, 1940; 1987).

ISBRANDO DE DIEMERBROECK

The Dutch anatomist, Isbrando de Diemerbroeck (1609–1764), published an expansive treatise entitled *Anatome corporis humani* in 1683 (Fig. 186). He was appointed professor of anatomy and medicine at Utrecht University Because of the text and its engraved plates, this work on the structure of the human body enjoyed considerable popularity. Two editions of the English translation were published in London. Diemerbroeck

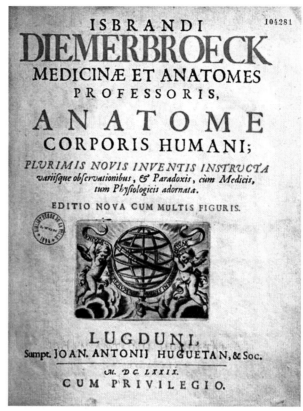

Figure 186. Front cover of *Anatome Corporis Humani*, published in 1683.

was also a well-known physician who wrote about the plague (*Tractus de Peste*) and other diseases.

GOVERT BIDLOO

In 1685, Govert Bidloo (1649–1713) (Fig. 187) published his *Anatomia humani corporis*, which contained 105 copperplate engravings of exceptional quality, drawn by the famous Belgian painter Gerard de Lairesse and engraved by Pieter van Gunst (Figs. 188–194). The text, however, was not exceptional. Bidloo was born in Amsterdam and received training in both surgery and medicine. He was a pupil of the distinguished anatomist Frederik Ruysch, in Leiden, and in 1688 he succeeded Anton Nuck (1650–1692) as professor of anatomy at Leiden. In the same year, William of Orange became King William III of England and Bidloo was appointed royal physician to accompany him. He spent the rest of his years in the service of King William III in England, as well as being professor of anatomy and surgery in Leiden (Beekman, 1935; Zimmerman, 1974).

WILLIAM COWPER

The anatomical illustrations in Bidloo's textbook, considered to be elegant *"masterpieces of Dutch baroque art"* (Norman, 1991), were plagiarized by the English surgeon and anatomist William Cowper (1666–1709). Cowper purchased sets of the copperplates and, on the title page, he pasted a printed sheet of irregular paper with his own name over Bidloo's name and title. Cowper added to the illustrations his own text, which was far better than that of Bidloo. He then published the combined work (*The anatomy of human bodies, with*

Figure 187 (*Left*). Govert Bidloo (1646–1713). Engraved by Abraham Blooteling from a painting by Gerard de Lairesse. (From Wegner, R.N.: *Das Anatomenbildnis. Seine Entwicklung im Zusammenhang mit der anatomischen Abbildung.* Basel, Benno Schwalbe and Co., Verlag, 1939. Courtesy of Institut fur Anatomie, Ernst-Moritz-Arndt-Universitat, Greifswald, Germany.)

Figure 188 (*Below left*). Bidloo's text was the first large-scale anatomical atlas to be published since Vesalius's *De Humani Corporis Fabrica* in 1543 and 1555. Whereas Vesalius's images were woodcuts, Bidloo's were created with copperplates. Anton Nuck described the work as follows: "that Mons. Bidloo, a skilful chirurgeon of Amsterdam, had newly shewed him above 100 anatomical figures of the parts of a man as big as the life, ingraven on copper, with a description of the parts, but not of their use." This illustration depics a dissection of the forearm. Govard Bidloo. *Anatomia Humani Corporis, Centum & Quinque Tabulis.* Amsterdam: Sumptibus Viduae Joannis à Someren, Haeredum Joannis à Dyk, Henrici & Viduae Theodori Boom, 1685. http://www.indiana.edu/~liblilly/anatomia/.

Figure 189 (*Below right*). The illustration shows reflected muscles of the thorax and the beginning of exposure of the anterior abdominal muscles. Govard Bidloo. *Anatomia Humani Corporis, Centum & Quinque Tabulis.* Amsterdam: Sumptibus Viduae Joannis à Someren, Haeredum Joannis à Dyk, Henrici & Viduae Theodori Boom, 1685. http://www.indiana.edu/~liblilly/anatomia/.

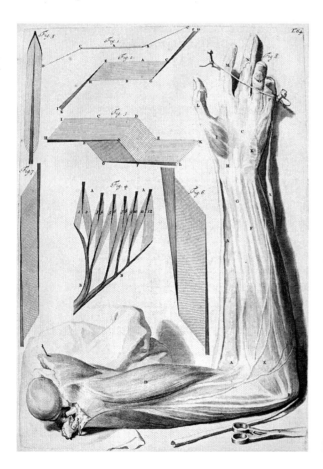

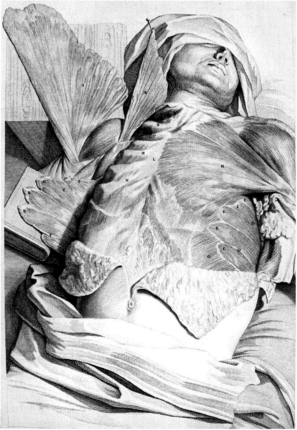

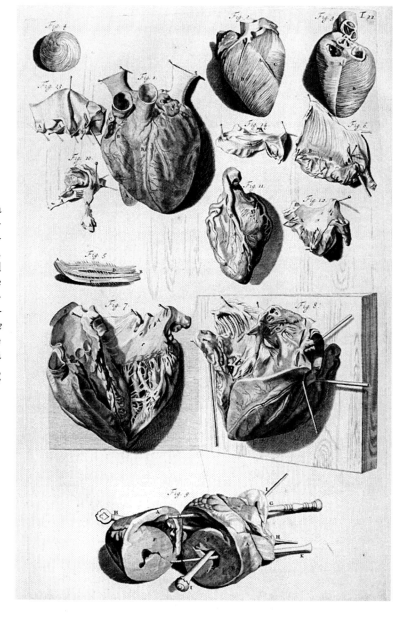

Figure 190. The illustration shows a human heart with several pins strategically placed to demonstrate the directional flow of blood through the heart. This artistic representation was not novel with Bidloo, but the quality and the size of the images are unmatched in the seventeenth century. Govard Bidloo. *Anatomia Humani Corporis, Centum & Quinque Tabulis*. Amsterdam: Sumptibus Viduae Joannis à Someren, Haeredum Joannis à Dyk, Henrici & Viduae Theodori Boom, 1685. http://www.indiana.edu/~liblilly/anatomia/.

figures drawn after the life by some of the best masters in Europe) in 1698 as his own. Remarkably, Bidloo's copperplate illustrations blended with Cowper's English text and resulted in a work that was better than the effort of each of them individually.

Bidloo denounced Cowper, accusing him of plagiarism, but Cowper responded shamelessly to Bidloo's accusation (Fig. 195). He compiled an impressive atlas (*Myotomia reformata*), published posthumously in 1724, and he also reported the *discovery* of the bulbourethral glands that bear his name. However, Jean Mery had already described the glands 15 years earlier in a paper presented to the French Academy of Medicine.

From being an outstanding and respected surgeon, as well as an anatomist, Cowper's reputation sharply diminished. "*Driven by an ambition so powerful that it blinded his scruples*" (Zimmerman, 1974), Cowper produced "*the most elaborate and beautiful of all 17th century English treatises on anatomy and also one of the most extraordinary plagiarisms in the entire history of medicine*" (Norman, 1991). Two years before the plagiarized work was published, Cowper was elected a fellow of the Royal Society of London. Long forgotten, he was not only a surgeon but also an accomplished scientist who carried out experiments in domestic animals to resolve surgical problems. He demonstrated the existence of capillaries

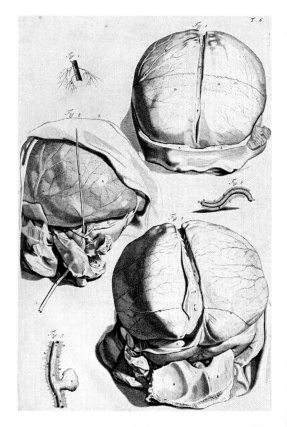

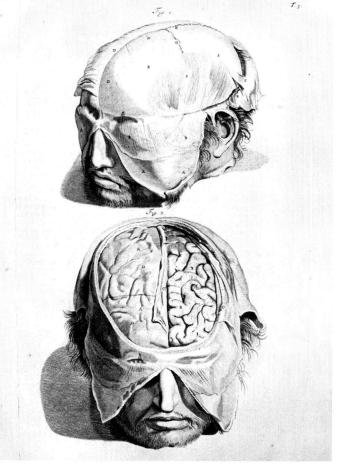

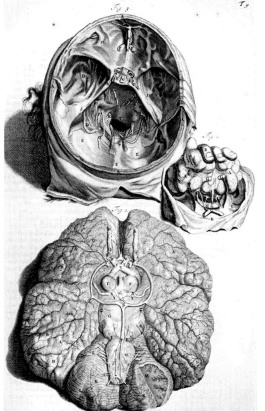

Figures 191–193. Bidloo's images of the brain and the nerves were particularly useful to the seventeenth-century researcher and physician as they showed structures in much greater detail than would have been possible in a smaller book. However, the text came under considerable criticism, as it did not contain much explanation of the figures. http://www.indiana. edu/~liblilly/anatomia/.

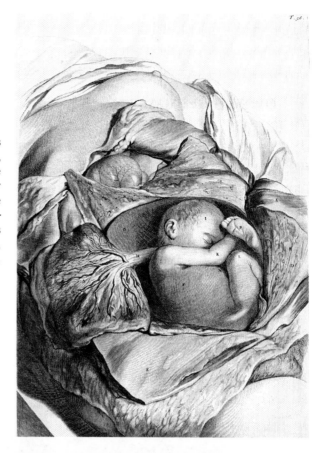

Figure 194. Like many of the other images in Bidloo's tome, this image is represented as if we, the viewers, were present to witness the cadaver. Thus, we see the almost fully–formed child in the womb, surrounded by layers of skin and muscle peeled back during the middle of a dissection. Govard Bidloo. *Anatomia Humani Corporis, Centum & Quinque Tabulis.* Amsterdam: Sumptibus Viduae Joannis à Someren, Haeredum Joannis à Dyk, Henrici & Viduae Theodori Boom, 1685. http://www. indiana.edu/~liblilly/anatomia/.

Figure 195. Best known as an anatomist, Govard Bidloo's most famous work was his monumental *Anatomia humani corporis*, published in Amsterdam in 1685, containing 107 copperplate engravings. Like so many large and expensive anatomical atlases of the time, the work was not a financial success, and in 1690 he published a Dutch translation entitled, *Ontleding des menschelyken lichaams*, using the same plates. When this edition did not sell well either, Bidloo's publisher sold 300 of the extra printed plates to William Cowper, a noted English anatomist. Cowper published the plates with his own, English language text in Oxford in 1698 under the title *Anatomy of the humane bodies*, without mentioning Bidloo or the artists of the original plates. Cowper went so far as to use Bidloo's engraved allegorical title page, amended with an irregular piece of paper lettered: "The anatomy of the humane bodies...," which fits over the Dutch title. http://www.nlm.nih.gov/exhibition/historicalanatomies/bidloo_bio.html.

and described the shunts (anastomoses) existing between arterial and venous circulation in the lungs and spleen, which he observed in a cat and dog. In 1705, Cooper described degenerative diseases of the aortic valve in an elderly man, which he presented to the Philosophical Transactions (Buckman & Futrell 1986).

A pioneer of the scientific method in surgery, Cooper spent his last years in clinical work and he began to give private lessons in anatomy. One of his pupils was William Cheselden (1688–1752), who succeeded Cowper in giving the anatomy lectures, following Cowper's death in 1709.

LORENZ HEISTER

Lorenz Heister (1638–1758; Fig. 196), who was professor of anatomy in Altdorf/Nurnberg and in Helmstedt, wrote a well-liked anatomy textbook *Compendium Anatomicum...* in 1719. Even though there were no helpful illustrations in the book, it was one of the most popular works in Europe at that time and it

went through five editions, the last appearing in 1741. Of curious interest is the book entitled *Anatomische Tabellen*, compiled by a schoolteacher, Johann Adam Kulm, and published in Danzig in 1722. This work was translated from German into Latin (1732), then into Dutch (Ontleedkundige Tafelen, Amsterdam, 1734) for ship doctors and surgeons. These anatomical tables brought new understanding and were widely used in the Far East, following translation into Japanese (1774, Tokyo) by the physician Gempaku Sugita.

WILLIAM CHESELDEN

William Cheselden was a distinguished surgeon and anatomist (Fig. 197). His book, *The Anatomy of the Humane Body*, which appeared in 1713, became one of the most popular textbooks to be published in Britain (Fig. 198). There were 18 editions over a period of 93 years with different publishers, a German translation in 1790, and two American editions (1795 and 1806). Cheselden's other major work *Osteographia*, or the anatomy of bones, was published in 1733 (Fig. 199). Like Cheselden's book on anatomy, the small and practical compendium of James Keill entitled *The Anatomy of the Humane Body Abridged* was immensely popular among students. There were 20 editions of the book between 1698 and 1774 (Russell, 1949). A leading anatomist, James Keill (1763–1719), lectured at both Oxford and Cambridge in addition to practicing medicine in Northampton.

Keill was succeeded by Frank Nicholls (1699–1778), who published in 1732 an illustrated synthesis of his lectures, entitled *Compendium Anatomicum*. Nicholls left Oxford for London where he established himself as one of the leading teachers of anatomy. Both William Hunter and his brother John were his students. Nicholls was well known on account of his corrosion and injected anatomical specimens; and as such he was among the earliest researchers in the field of microscopic anatomy (Robb-Smith, 1971).

Figure 196. Lorenz Heister. Line engraving by G. D. Heumann, 1719, after J. Kenckel. 1719 after: G. D. Heumann. (Wellcome Library no. 4108i.)

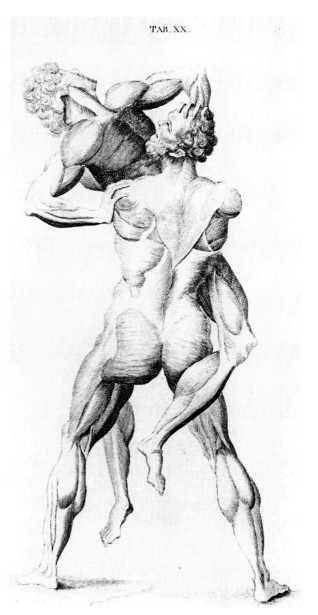

Figure 197. Portrait of William Cheselden. (Wellcome Library no. 1741i.)

Figure 198 (*Right*). Copperplate engraving from William Cheselden's *The Anatomy of the Human Body*. First American Edition. Boston, 1795. Two men are shown locked in an embrace. This beautiful and vibrant work was done "*after the famous statue of Hercules and Antaeus.*" The musculature is in full artistic display. (Courtesy of the Neil John Maclean Health Sciences Library, University of Manitoba.)

BARTHOLOMEO EUSTACHIUS

Bartholomeo Eustachius (c. 1510–1574) was an eminent anatomist of the Renaissance era (Fig. 200). The medical historian August Hirsch described him in his Biographical Dictionary "as one of the greatest anatomist that ever lived" (1884–1888). Eustachius carried out meticulous dissections and described his findings in considerable detail (Fig. 201). He was a contemporary of Vesalius, but at the same time an intense critic of some aspects of Vesalius and his work.

The *Tabulae anatomicae* of Eustachius is a remarkable work because of the elegant and accurate copper plates, engraved from the drawings made by Eustachius (Figs. 202–206), which he had completed in 1552. The plates were discovered at the beginning of the eighteenth century and published in 1714 with a commentary by Giovanni Maria Lancisi, who was the personal physician to Pope Clement XI. Eustachius discovered not only the pharyngotympanic (auditory) tube but also the thoracic duct, and the abducent nerve. Furthermore, he provided an accurate description of the uterus, the cochlea, the tensor tympani and stapedius muscles, the laryngeal

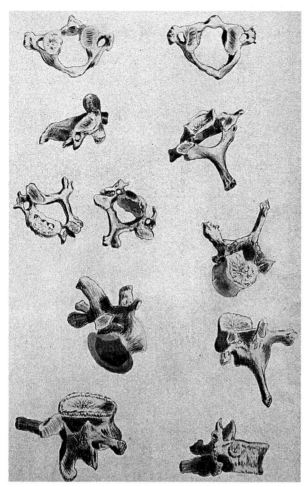

Figure 200. A portrait of Eustachius. Plate from *A History of dentistry from the most ancient times until the end of the eighteenth century*, by Vincenzo Guerini. Scanned from google books. (http://books.google.com/books?id= _CcJAAAAIAAJ&pg=PPA178-IA1).

Figure 199. Eleven vertebrae from *Osteographia*, London, s.n, 1733, tab. xi. (Wellcome Library no 562913.)

Figure 201. Bartholomaeus Eustachius (1520–1574) performing an anatomical dissection before several observers, in an anatomy theatre decorated with an articulated skeleton. Engraving, 1722, after P. L. Ghezzi, 1714. (Wellcome Library, London.)

Figure 202 (*Right*). Title page of T*abulae Anatomicae*. Amsterdam. Eustachius's work, edited and published by Massimini, Andrea (1727–1792), Martine, George (1702–1741), in 1783. (Courtesy of Bethesda, MD: U.S. National Library of Medicine, National Institutes of Health, Health & Human Services. http://www.nlm.nih.gov/exhibition/historicalanatomies.)

Figure 203 (*Bottom left*). Abdominal viscera, including the digestive tract. Tabula X from B. Eustachius's work, edited and published by Massimini, Andrea (1727–1792), Martine, George (1702–1741), in 1783. (Courtesy of Bethesda, MD: U.S. National Library of Medicine, National Institutes of Health, Health & Human Services. http://www.nlm.nih.gov/exhibition/historicalanatomies.)

Figure 204 (*Bottom right*). (Tabula XVIII) The cranial and spinal nerves. Eustachius's work, edited and published by Massimini, Andrea (1727–1792), Martine, George (1702–1741), in 1783. (Courtesy of Bethesda, MD: U.S. National Library of Medicine, National Institutes of Health, Health & Human Services. http://www.nlm.nih.gov/exhibition/historicalanatomies.)

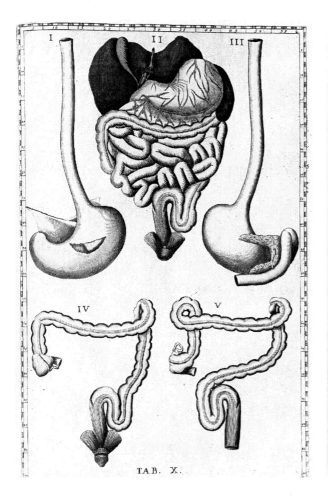

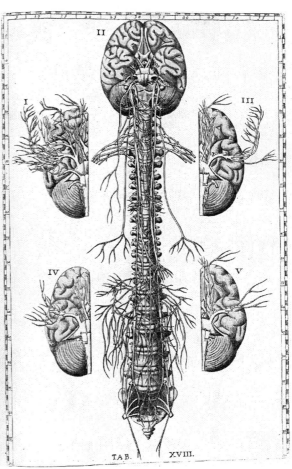

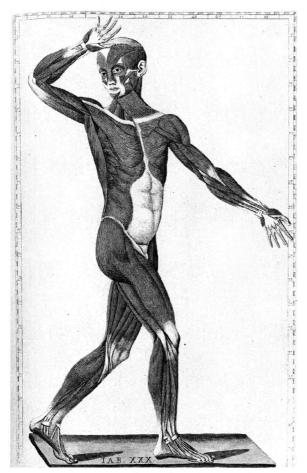

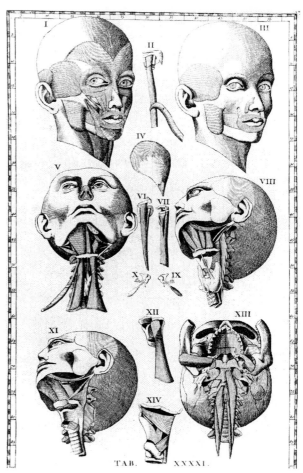

Figure 205 (*Left*). (Tabula XXX) Muscles of the anterior and lateral aspects of the body. The tendons of the arm, hand, leg, and foot are elegantly displayed. Some of the facial muscles are also revealed. Eustachius's work, edited and published by Massimini, Andrea (1727–1792), Martine, George (1702–1741), in 1783. (Courtesy of Bethesda, MD: U.S. National Library of Medicine, National Institutes of Health, Health & Human Services. http://www.nlm.nih. gov/exhibition/historicalanatomies.)

Figure 206 (*Right*). (Tabula XXXXI) Muscles of the face and neck. Observe the smaller diagrams showing the ossicles of the ear and the larynx. (From the *Tabulae Anatomicae of Eustachius.* Courtesy of the Neil John Maclean Health Sciences Library, University of Manitoba.)

muscles, and the origin of the optic nerve. Eustachius discovered the adrenal gland and he also described in detail the external and internal structure the kidney. In his treatise on the structure and function of the teeth, *Libellus de dentibus*, one finds the first account describing the pulp of the tooth and the periodontal ligament.

Eustachius began his career as physician to the Duke

of Urbino and later to Cardinal Giulio della Rovere in Rome. He was appointed Professor of Anatomy at the Archiginnasio della Sapienza where he carried out his anatomical studies. Because Eustachius's magnificent work lay forgotten in the Vatican Library for close to a century and a half, this great anatomist did not achieve the fame he deserves as a founder of modern anatomy.

JACQUES (JACOB) WINSLOW

One of the first systematic textbooks on descriptive anatomy was that of Jacques (Jacob) Benige Winslow

(1669–1760; Fig. 207), a Danish-born anatomist, who had studied with Johannes Buchwald (1658–1738) in

Denmark, Friedrich Ruysch (1638–1738) in Holland, and with Joseph Guichard Duverney (1648–1730) in Paris. In 1703, Winslow obtained the doctor of medicine degree from the University of Paris and secured an appointment as a physician to the Hospital Beneral at Bicetre. In 1721, he secured a position at the Jardin du Roi, and after Francois Joseph Hunault's death, Winslow was appointed professor of anatomy and surgery in Paris. His book, *Exposition anatomique de la structure du corps humain*, was published in 1732 in Paris (Fig. 208). Essentially, the work focused on the descriptive anatomy of the human body, but lacking the high artistic attainments of Albinus's work. An English translation (*An anatomical exposition of the structure of the human body...*translated from the French original by G. Douglas) appeared in 1732 in London. Six editions of the English translation were published over the years. It "*was a deservedly popular book, extensively used by students*" (Russel, 1987). The book was dedicated to the Scottish obstetrician and anatomist James Douglas (1675–1742).

Winslow was one of the most famous anatomists during the eighteenth century. He published 30 treatises and made many important contributions in diverse areas of anatomy. Winslow studied the muscular system and observed that muscles do not act independently. As a result, he described the synergistic and antagonistic action of several groups of muscles. Winslow provided one of the first descriptions of cranial anatomy and he introduced the term "*nervus sympathicus*." In his book, Winslow described ten cranial nerves originating from the medulla oblongata, including the first cervical spinal nerve. In addition, he described the superior, middle, and lower cervical ganglia, and also the first thoracic ganglia. Other notable contributions related to the heart, osteology, and teratology. In order to establish death and to avoid premature burial, Winslow proposed that burial should take place after signs of putrefaction are evident (Bellary et al., 2012). His name is eponymously associated with the opening (foramen of Winslow) into the lesser sac (epiploic foramen). He named the lateral sellar compartment

Figure 207. Portait of Jacques (Jacob) Benige Winslow. (Wellcome Library no. 9714i.)

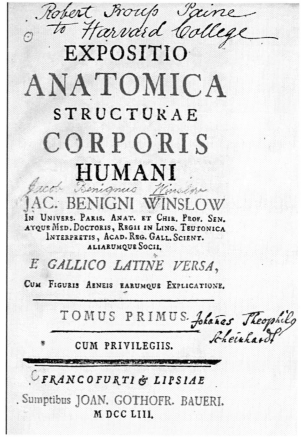

Figure 208. Front page of *Exposition anatomique de la structure du corps humain*. (Public Domain.)

the "*cavernous sinus*" because he believed it resembled the "*corpora cavernosa*" of the penis (Parkinson, 1995).

Jacques (Jacob) Benige Winslow was a student of the anatomist Guichard Joseph Duverney (1648–1730).

GUICHARD JOSEPH DUVERNEY

The famous French anatomist Guichard Joseph Duverney (1648–1730), "the best known man of science in Paris," was born in Feurs in Massif Centrale (Guerrini, 2009). At 14 years of age, he began medical studies in Avignon, which he completed five years later. In 1667, Duverney moved to Paris where he carried out dissections and anatomical studies on human cadavers and domestic animals. In 1680, he was appointed professor of anatomy at the Jardin du Roi, a position he held for 40 years. Duverney was a skilled dissector and a popular lecturer. His public lectures and exhibitions at the Jardin du Roi attracted large audiences, which included dignitaries, medical students, and the public. Overwhelmed with grief, King Louis IV requested Duverney to dissect his favorite elephant that had just died. The King, his guests, and members of the Academy of Sciences witnessed this demonstration. In recognition of his scientific work, Duverney was elected a Fellow of the Académie Royale des Sciences in 1675.

Duverney is known as the father of otology in recognition of his pioneering work on the ear. His book, "*Traité de l'organe de l'ouïe*" (Treatise of the organ of hearing), published in 1683, was one of the most complete works on otology then available. It was translated into several languages and widely used as reference source until the beginning of the nineteenth century. Duverney had formulated a theory of hearing similar to one that was proposed later by the German physician and physicist Hermann von Helmholtz (1821–1894). Duverney "was the first truly academic physician in the modern sense: He was a teacher, an investigator, and a surgeon. He published an early account of the anatomy and diseases of the ear, and he also recognized osteoporosis as a clinical problem" (Hunter et al., 2000). In his book, "*Maladies des os*" (On diseases of the bone), which was published posthumously, Duverney described the clinical condition of osteoporosis and also a new type of isolated pelvic fracture that involved one side of the pelvic bone (Duverney's fracture). In 1731, a book entitled *Human Osteogeny* explained in two lectures, read in the Anatomical Theatre of the Surgeons of London was published, with Robert Nesbitt (1700–1761) as the author. It was the first work to suggest that bone may develop in membrane as well as in cartilage, a fundamental concept that was rediscovered in the nineteenth century. The book provides an excellent description of bone growth and it is considered as one of the classics of anatomy. A German translation of this book appeared in 1743.

The anatomical copper plate engravings (Figs. 209–211) in the artistically lavish *Tabulae anatomicae* by the Italian Baroque artist Pietro da Cortona (1596–1669), published in 1741, were prepared from drawings he made. The text for the book was written by Cajetano Petrioli, the editor, who also added several small anatomical figures in the margins of the plates (Norman, 1986). Similarly, the *Essai D'Anatomie* was produced by Gautier D'Agoty in 1745 in Paris, France, on the drawings of Duverney (Fig. 212).

ALBRECHT VON HALLER

Over a period of many years, Albrecht von Haller (1708–1777), the great Swiss scholar and one of the most gifted men of science (Fig. 213), compiled his multivolume bibliographic work. It revealed the vast range of his interests and knowledge (Asher, 1902; Hintzsche, 1948; Nussbaum, 1968). In addition to many unpublished manuscripts, von Haller's writings included the following major treatises: (1) *Iconum anatomicarum,*...8 parts in 1 volume (Figs. 214–216), Gottingen (1743–1756); (2) *Bibliotheca anatomica,*...2 volumes, Zurich (1774–1777) (Reprinted, Hildesheim, 1969); (3) *De partium corporis humani praectpuum fabrica*

et functionibus. 8 volumes, Berne Lausanne (1777–1778). (4) *Elementa physiologiaecorporis humani...*(1747).

Von Haller was born in Berne and, in 1723, at the age of 15 years, he started his studies in Tubingen. At Leiden University, von Haller was a pupil of Boerhaave; and after graduation he began to practice medicine in 1729.

Von Haller was appointed professor of anatomy, surgery, and botany at the newly established University of Gottingen in Germany in 1736. In Gottingen, he organized an anatomical museum, with a lecture theater, as well as a botanical garden. Here, von Haller

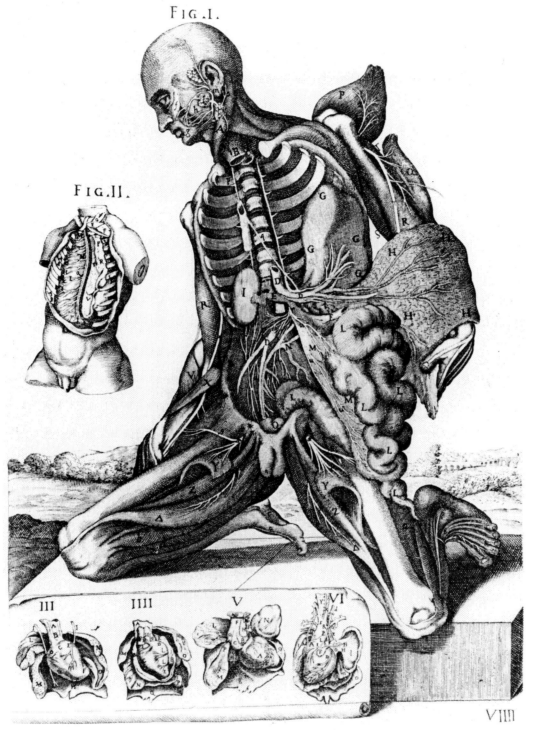

Figure 209. Illustrations from the *Tabulae anatomicae* of Pietro da Cortona (1596–1669). The figures show the deep abdominal blood vessels and nerves, various abdominal organs, the intestine, dissection of the thorax, the heart and the lungs. (From *The Anatomical Plates of Pietro da Cortona*. Dover Edition, 1986. Introduction by Norman, J.M.; by kind permission of Dover Publications, Inc., New York.)

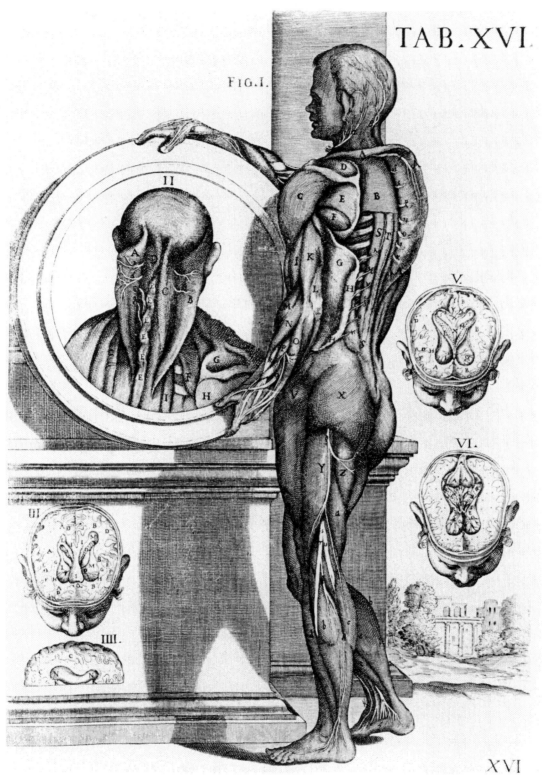

Figure 210. Illustrations from the T*abulae anatomicae of Pietro da Cortona* (1596–1669). There are four sketches of various stages in the dissection of the brain. A dissected man is shown standing and holding a mirror reflecting a prosection of the back of head and neck. (From *The Anatomical Plates of Pietro da Cortona.* Dover Edition, 1986. Introduction by Norman, J.M.; by kind permission of Dover Publications, Inc., New York.)

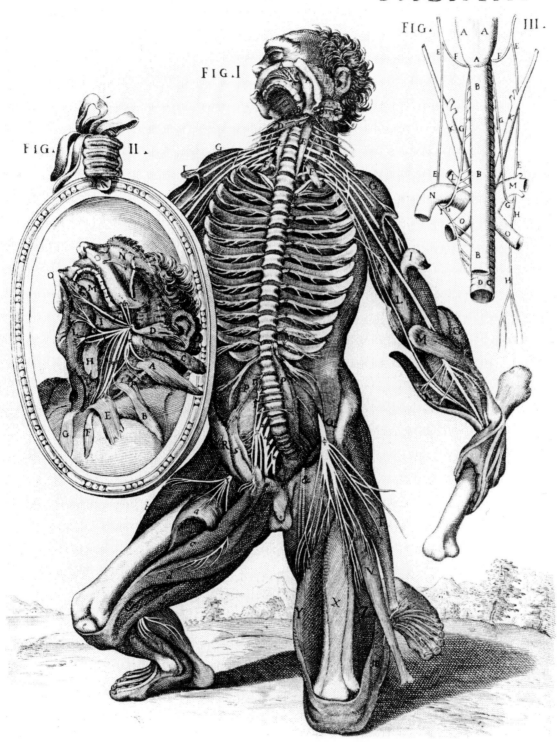

Figure 211. *Tabulae anatomicae of Pietro da Cortona* (1596–1669). This dramatic engraving shows the spinal nerves, the tongue and its nerve supply, and the nerves innervating the extremities. A separate drawing shows the larynx, trachea, and the esophagus, and the nerves relating to them. (From *The Anatomical Plates of Pietro da Cortona*. Dover Edition, 1986. Introduction by Norman, J.M.; by kind permission of Dover Publications, Inc., New York.)

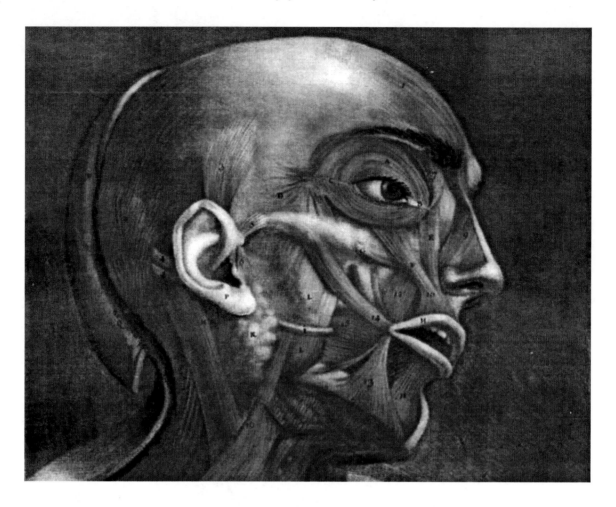

Figure 212 (*Above*). *The Essai D'Anatomie* was produced by Gautier D'Agoty in 1745 in Paris, France. It is a remarkably detailed atlas of the head, neck, and shoulder areas of the human body with explanatory text in French. The anatomical images were based on human cadavers dissected by Joseph Duverney (1) and produced using the mezzotint method of engraving and printing. (Courtesy of Tulane University Digital Library) Christiess. (2011). GAUTIER D'AGOTY, Jacques (1717–1786) and Joseph Guichard DUVERNEY. *Essai d'Anatomie*, en Tableaux imprimes, qui represent au naturel tous les muscles de la face, du col, de la tete, de la langue & du larynx…– Myologie complette en couleur et grandeur naturelle, compose de l'essai et de la suite de l'essai d'anatomie, en tableaux imprimes. Paris: Gautier [vol. I]; Gautier, Quillau pre et fils, and Lamesle [vol. II], 1745-46-[48]. | Books & Manuscripts Auction. Retrieved June 14, 2011. (From http://www.christies.com/LotFinder/lot_details.aspx?intObjectID=4959941.)

Figure 213 (*Left*). Albrecht von Haller. (Lithograph by Vigneron after the well-known engraving of Bause, 1773; with permission from Major, 1954.)

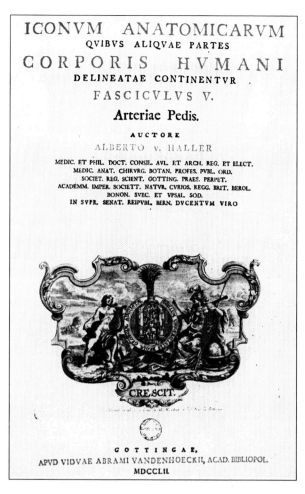

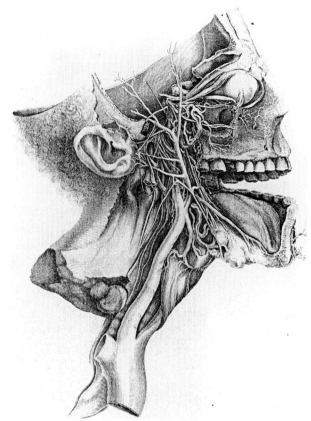

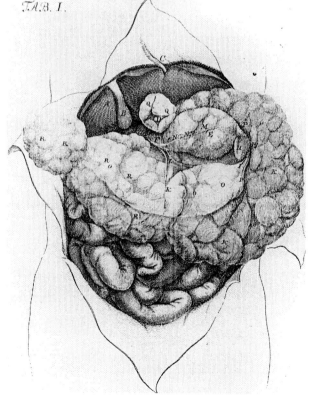

Figure 214 (*Above left*). Title page of Albrecht von Haller's *Icones anatomicae*. Fasciculus V. dealing with the arteries of the foot, was published in 1752 as part of a series. (Courtesy of Institut für Anatomie, Ernst-Moritz-Arndt-Universität, Greifswald, Germany).

Figure 215 (*Above right*). Dissection of the right side of the face, orbit, and the cervical region. This elegant engraving is from the anatomical treatise of Albrecht von Haller. (From Albrecht von Haller, 1708–1777: *Iconum Anatomicarum Parium Corporis Humani*, Gottingae, Abrami Vandenhoeck, 1745. Courtesy of the Institut fur Anatomie, Freie Universitat Berlin.)

Figure 216 (*Right*). An engraving of the abdominal viscera. (From Albrecht von Haller, 1708–1777: *Iconum Anatomicarum Parium Corporis Humani*, Gottingae, Abrami Vandenhoeck, 1745. Courtesy of the Institut fur Anatomie, Freie Universitat Berlin.) The beautiful and accurate illustrations, combined with clear descriptions, established Albinus as the founder of descriptive anatomy.

dissected more than 400 cadavers and carried out physiological research on irritability and sensibility, resonance and hearing, muscle contraction and nerve function. Of interest, von Haller also wrote three political romances and many lyrical poems. A detailed and critical account of von Haller's published works and unpublished manuscripts would demand a team of scholars from several disciplines and many years.

BERNHARD SIEGFRIED ALBINUS

Bernhard Siegfried Albinus (1697–1770; Fig. 217), professor of anatomy at Leiden, was the author of an admirable work entitled *Tabulae sceleti et musculorum corporis humani* (1747) (Figs. 218–220). It contained 40 large copperplates, showing the skeletal system and muscles, in a "*seamless combination of precision, refinement and anatomical synthesis, which resulted in images far removed from the flesh-and blood reality of dissection*" (Kemp, 2010). The drawings and engravings were done by Jan Wandelaar. "*They established a new standard in anatomical illustration, and remain unsurpassed for their artistic beauty and scientific accuracy*" (Norman, 1991). An

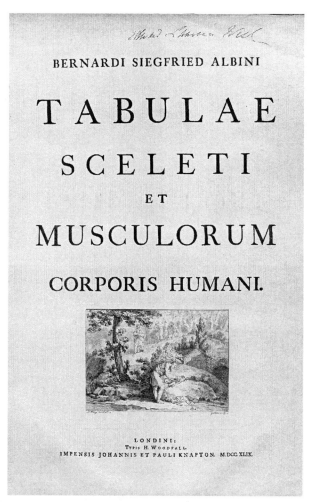

Figure 217 (*Left*). Portrait of Bernhard Siegfried Albinus. (Courtesy of Wellcome Library no. 461i.)

Figure 218 (*Right*). Title page of *Tabulae sceleti et musculorum corporis humani* by Albinus, Bernhard Siegried. Artist: Wandelaar, Jan, 1690–1759.?Engravers: Charles Grignion (1717–1810); Jean-Baptiste Scotin (b. 1678); Ludovico-Antonio Ravenet (fl. 1751); and Louis-Pierre Boitard (fl. 1750). Publication: Londini: Typis H. Woodfall, impensis Johannis et Pauli Knapton, M.DCC.XLIX. [1749]. (Courtesy of Bethesda, MD: U.S. National Library of Medicine, National Institutes of Health, Health & Human Services.)

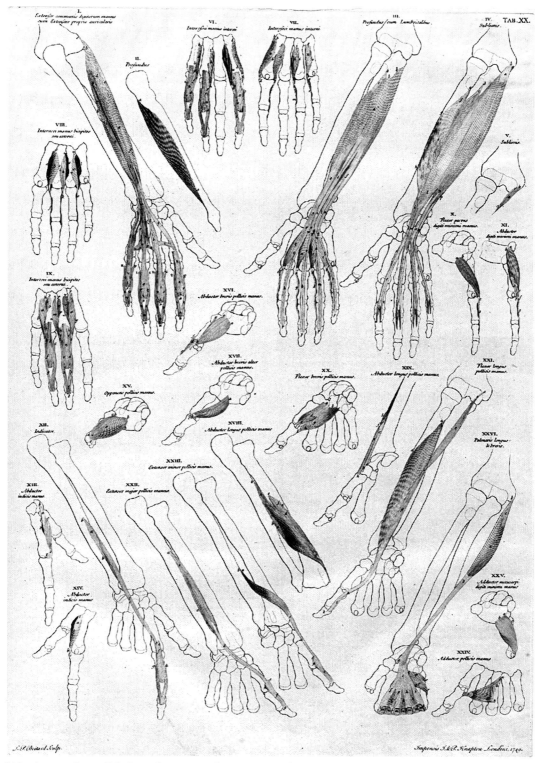

Figure 219. A page from *Tabulae sceleti et musculorum corporis humani* by Albinus, Bernhard Siegried depicting different dissections of the muscles of the forearm. Artist: Wandelaar, Jan, 1690–1759.?Engravers: Charles Grignion (1717–1810); Jean-Baptiste Scotin (b. 1678); Ludovico-Antonio Ravenet (fl. 1751); and Louis-Pierre Boitard (fl. 1750). Publication: Londini: Typis H. Woodfall, impensis Johannis et Pauli Knapton, M.DCC.XLIX. [1749]. (Courtesy of Bethesda, MD: U.S. National Library of Medicine, National Institutes of Health, Health & Human Services.)

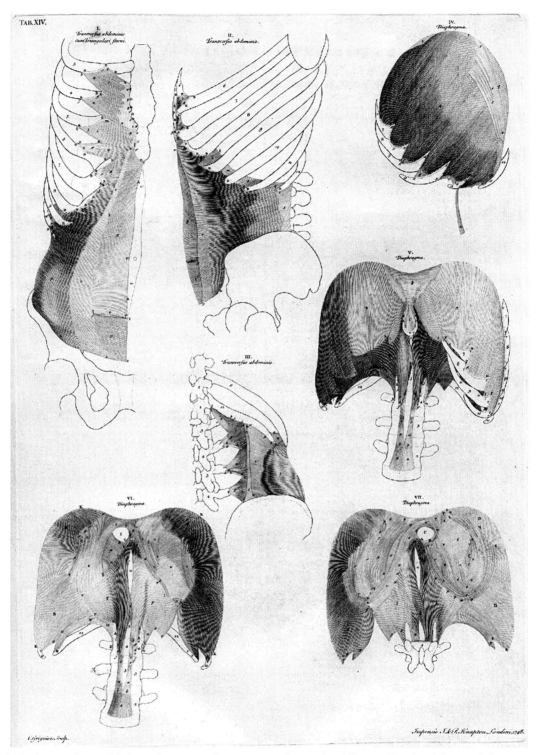

Figure 220. A page from *Tabulae sceleti et musculorum corporis humani* by Albinus, Bernhard Siegried depicting different dissections of the muscles of the anterior abdominal wall and diaphragm. Artist: Wandelaar, Jan, 1690–1759. Engravers: Charles Grignion (1717–1810); Jean-Baptiste Scotin (b. 1678); Ludovico-Antonio Ravenet (fl. 1751); and Louis-Pierre Boitard (fl. 1750). Publication: Londini: Typis H. Woodfall, impensis Johannis et Pauli Knapton, M.DCC. XLIX. [1749]. (Courtesy of Bethesda, MD: U.S. National Library of Medicine, National Institutes of Health, Health & Human Services.)

English translation of this book was published in1749 and reprinted with re-engraved plates.

Of the books published by members of the Monro dynasty (Chap. 10), the monograph on the synovial membranes by Alexander Monro *Secundus* (1733–1817), entitled "*A description of all the bursae mucosae of the human body*" (1788), was both important and original. Monro *Primus* (1697–1767) also published a useful textbook on "*The anatomy of the humane bones,*" Edinburgh 1726. This book ran four editions until 1746.

SAMUEL THOMAS VON SÖMMERRING

Samuel Thomas von Sömmerring (1755–1830; Fig. 221) was the author of an encyclopedic textbook (*Vom Baue des menschlichen Korpers*, five volumes, Frankfurt, 1791–1796) that was popular in Germany. He also produced one of the first atlases of the female skeleton entitled *Tabula sceleti feminini junta description* (1797). The former book was based on dissection, factual information, and Sömmerring's observations. Another edition of this work in eight volumes was published between 1839 and 1845 in Leipzig, and it was translated into French (eight volumes, Paris 1843–1847). The latter work contained drawings that were anatomically accurately proportioned, as well as artistically appealing.

Von Sömmerring's scientific contributions were many and diverse (Metzner, 1954; Aumtiller, 1972). For example, he published monographs on the structure and function of the lungs, the cranial nerves, smell and taste, the eyes and sight (Fig. 222), hearing, pipe smoking and cancer of the lip, and on developmental defects. Recalling his student days, Sömmerring wrote as follows: "*I spent the Christmas holidays in Gottingen and it was with the greatest difficulty that I refrained from making a dissection on Christmas Eve. The temptation was very great. I had just completed the preparation of the plexus sympatheticus....*"

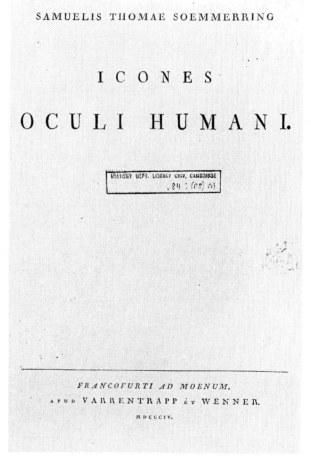

SAMUELIS THOMAE SOEMMERRING

ICONES

OCULI HUMANI.

FRANCOFURTI AD MOENUM,
APUD VARRENTRAPP ET WENNER.
MDCCCIV.

Figure 221. Samuel Thomas von Sömmerring (1755–1830). (Department of Anatomy, University of Manitoba.)

Figure 222 (Right). Title page of Samuel Thomas Sömmerring's work on the human eye *Icones Ocult Humani*, 1804. (Courtesy of the Library, Anatomy School, University of Cambridge.)

JUSTUS CHRISTIAN LODER

The German anatomist and surgeon Justus Christian Loder (1753–1832; Fig. 223) published a stunning atlas, "*Tabulae anatomicae*," comprising of a collection of large anatomical plates that showed the entire human body. For Loder, the, "*Tabulae anatomicae*" (1794–1803) was an ambitious and well-conceived project, but for the publisher, Friedrich Justin Bertuch, it became an uncertain financial undertaking. The alliance of author and publisher was punctuated with personal and financial problems because the plates were costly to produce and initial sale of the atlas was

L O D E R.

Zwickau, bei Gebr. Schumann.

Figure 223. A portrait of Justus Christian Loder. (Courtesy of Wellcome Library no. 5995i.)

lagging (Regenspurger & Heinstein, 2003). For that era, the "*Tabulae anatomicae*" (1794–1803) was the largest and most comprehensive anatomy atlas in German-speaking countries. The first collection of plates appeared in 1794 and publication of additional plates continued until 1803 when the atlas was completed.

The original plan for the "*Tabulae anatomicae*" was to make new drawings and plates from the available prosected anatomical specimens. This approach was abandoned by Loader and his publisher because of the costs involved and also the urgency to complete the "*Tabulae anatomicae*." For the new atlas, Loder "*borrowed*" heavily from the publications of other anatomists. The first collection of 15 plates, which dealt with osteology, appeared in 1794, followed by others usually after a year. The final collection of plates was published in 1803. Of interest, the complete work in six sections consisted of 182 plates with 1431 figures, but only 309 were newly drawn. The structures of the body were clearly depicted and extraneous stylistic diversions were avoided. Several skilled artists were involved in the project, but their names are not mentioned in the atlas.

Loder selected the best figures that were then available based on his perception of anatomical accuracy and even of his relationship to the original authors. He had visited William Hunter (1718–1783) in 1783, studied Hunters' specimens in the museum, and attended his lectures. As a result, he was familiar with Hunter's original specimen and depiction of the "*Anatomy of the human gravid uterus*." Loder used this impressive drawing in his work. Figures from the atlases of Bernhard Siegfried Albinus, Albrecht von Haller Frederic Ruysch, Alexander Monro, John and William Hunter, Antonio Scarpa and many others became part of Loader's ambitious project. On the other hand, he avoided using figures from the work of his rivals, especially the famous anatomist Samuel Thomas von Sömmerring. The publisher, Friedrich Justin Bertuch, had agreed to print 1000 copies of the "*Anatomische Tafeln*." The atlas was produced for a select group of users who could afford it. These included surgeons, anatomists, and others who might have an interest in the subject. It was too expensive for students to purchase.

Justus Christian Loder was born in Riga, Russia in 1753. From 1773 to 1777, he studied at the medical faculty of the University of Göttingen and graduated with the doctor of medicine degree. In 1778, he was appointed professor of surgery, obstetrics, and anatomy at the University of Jena where he remained for the next 25 years. Loder was involved in the establishment of the new anatomical theater, a teaching hospital for medical students, and a maternity house (*Accouchierhaus*) in Jena. He was well known and, as a very good lecturer, he attracted large numbers of students to his lectures. The great German intellectual Johann Wolfgang von Goethe was a friend of Loder. He attended Loder's anatomy lectures and studied the collection of skulls in the museum. The anatomical discoveries of Goethe, including a figure of the intermaxillary bone (Zwischenkieferknochen), were reported for the first time in Loder's, "*Anatomisches Handbuch, which was published in 1788*" (Pfannenstiel M, 1949; Regenspurger & Heinstein, 2003).

In 1883, Loder took up an appointment at the University of Halle because of problems with some of his colleagues. Following closure of the university by Napoleon in 1806, Loder received an appointment as the personal physician to the Prussian royal family, then residing in Konigsberg. Later, he moved to Moscow where he spent the rest of his life until his death in 1832. In Moscow, Loder received an appointment in 1810 as honorary professor of anatomy, and director of the military hospital from 1812 to 1817. He also served as the personal physician to Czar Alexander.

KARL VON BARDELEBEN

In an earlier era when communication was limited, there was an obvious need for scientists to meet and discuss their work. Since 1822, an anatomical society had existed as a section of the Gesellschaft Deutscher Naturforscher und Ärzte (German Society of Scientists and Physicians). In 1886, German anatomists founded the Anatomische Gesellschaft (Anatomical Society) for the advancement of research and education. Based in Germany, it was conceived as a multinational association for all anatomists. Soon after, the Anatomical Society of Great Britain and Ireland (1887) and the American Association of Anatomists (1888) were formed. The journal *Anatomischer Anzeiger*, renamed "Annals of Anatomy" in 1991, was founded in 1886 by the German anatomist Karl von Bardeleben (Fig. 224). It became the official journal of the Anatomische Gesellschaft (Schierhorn, 1978; Linss, 1981; Kühnel, 2011). Von Bardeleben served as the editor of the journal for the first 32 years until his death in 1918.

Karl von Bardeleben (1849–1919) was born in Giessen, the eldest son of Heinrich Adolph Bardeleben (1819–1895) who was professor of surgery in Berlin. He studied medicine in Greifswald, Heidelberg, Berlin, and Leipzig. In 1871, he graduated as a physician and obtained the doctor of medicine degree from the University of Berlin. Von Bardeleben began his career as an assistant to the anatomist Wilhelm His in Leipzig. In 1873, he was appointed as a prosector (Lecturer) at the University of Jena where he became a full professor with extensive teaching duties. He was of the old school with considerable expertise in all areas of anatomy: descriptive, topographic and comparative anatomy, embryology, and anthropology. In 1883, Karl von Bardeleben was elected a member of the German Academy of Sciences, Leopoldina.

Figure 224. Portrait of Karl von Bardeleben from Anatomischer Anzeiger 51 (1918/19) http://archive.org/stream/anatomischeranze51anat#page/n11/mode/2up.

Von Bardeleben's books and scholarly monographs span an impressive and diverse field which included anatomical atlases, dissecting manuals for medical students, anatomy and embryology textbooks, the brain and spinal cord, and Goethe's anatomical writings: *Beiträge zur Anatomie der Wirbelsäule*, 1874; *Anleitung zum Präparieren der Muskeln, Fascien und Gelenke*, 1882;

Anleitung zum Praeparieren auf dem Seciersaale, 1884; *Atlas der topographischen Anatomie des Menschen. Topographie des Gehirns und des Rückenmarks*, 1894 & 4th edition 1908; *Die Anatomie des Menschen. I. Teil: Allgemeine Anatomie und Entwicklungsgeschichte*, 1913; Editor: *Handbuch der Anatomie des Menschen* (8 Volumes), 1896–, *Goethe als Anatom* (Pagel, 1901; Linss, 1981).

JOHN BELL

The surgeon-anatomist John Bell (1763–1820; Fig. 225) wrote an important book on *the anatomy of the bones, muscles, and joints*, published in Edinburgh in 1793 (Figs. 226–229). The book ran three editions in Britain, as well as an American edition (1816). John Bell was one of four gifted sons, born in Edinburgh to the Reverend William Bell. Two of the brothers were lawyers and became university professors; John and his younger brother Charles studied medicine in Edinburgh and became eminent surgeons. John studied anatomy with the highly regarded surgeon Alexander Wood (1726–1807) and attended the lectures of the physician William Cullen (1710–1790) and the anatomist Alexander Monro, *secundus* (1733–1817). He received the M.D. degree in 1779 from Edinburgh

University and became a Fellow of the Royal College of Surgeons (F.R.C.S) in 1786.

John Bell was a pioneer of surgical anatomy. In 1790, he had built a lecture theater and dissecting room in Edinburgh where he lectured. The anatomical and surgical specimens for teaching were housed in an adjacent museum. John was a popular and innovative lecturer because of his surgical background. Many of his colleagues resented him because of the large numbers of students he attracted. Charles Bell, his younger brother, attended the classes and later assisted him with the teaching over several years (Kaufman, 2005; Chisholm, 1911; Betanny, 1885–1900).

John Bell and his brother Sir Charles Bell (1774–1842) collaborated in writing an impressive four-volume work on, "*The anatomy of the human body*," (1797–1804), which was an immediate success. It was translated into German, and seven editions of this work appeared in Britain (1797–1824), and in the United States, there were six editions (1809–1834). Russel (1987) remarked that "*This comprehensive and excellent textbook stands as a monument to the brothers…and this work is the first great text-book contributed by the British school to modern anatomy.*"

Charles Bell (Fig. 230) wrote also *A system of dissections* (1798–1803) for medical students and *Engravings of the arteries…*(1801) for surgeons. Both these works were very popular, ran into several editions, and were translated into German. In addition, he authored a classic work on the hand, which included anatomy, physiology, comparative anatomy, adaptive mechanism and biomechanics. Of great artistic merit were the meticulous engravings in Bell's…*Anatomy of the Brain* (1802) (Figs. 231 & 232). During the early part of the nineteenth century, the anatomical textbooks of the brothers were standard works (Comrie, 1972) and widely used in Britain and other countries.

For most of his career, John Bell remained one of the leading surgeons in the country. In 1800, a rabid controversy with James Gregory, an influential professor of medicine at the University of Edinburgh, led

Figure 225. John Bell (1763–1820). (Department of Anatomy, University of Manitoba.)

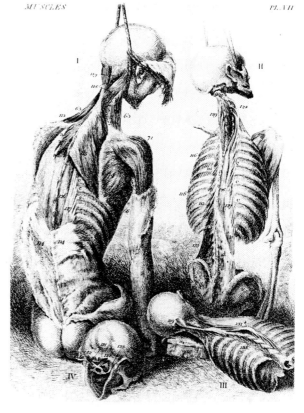

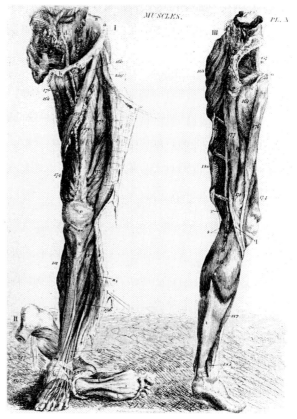

Figure 226 (*Above left*). Muscles of the face and neck. (From John Bell's *Engravings of the Bones, Muscles, and Joints....* First volume of the *Anatomy of the Human Body.* (Courtesy of the Neil John Maclean Health Sciences Library, University of Manitoba.) John Bell (1763–1820) illustrated and etched his own work. In this plate, the drawing was "from a subject that had been hanged, and the neck being broken...."

Figure 227 (*Above right*). Dissection of the trunk, showing various functional groups of muscles. (From John Bell's *Engravings of the Bones, Muscles, and Joints....* First volume of the *Anatomy of the Human Body.* (Courtesy of the Neil John Maclean Health Sciences Library, University of Manitoba.) The intercostal, trapezius, levator scapulae, deltoid, rhomboids, serratus anterior, and other muscles are clearly demonstrated.

Figure 228 (*Right*). Muscles of the lower extremity. (From John Bell's *Engravings of the Bones, Muscles, and Joints....* First volume of the *Anatomy of the Human Body.* Courtesy of the Neil John Maclean Health Sciences Library, University of Manitoba.)

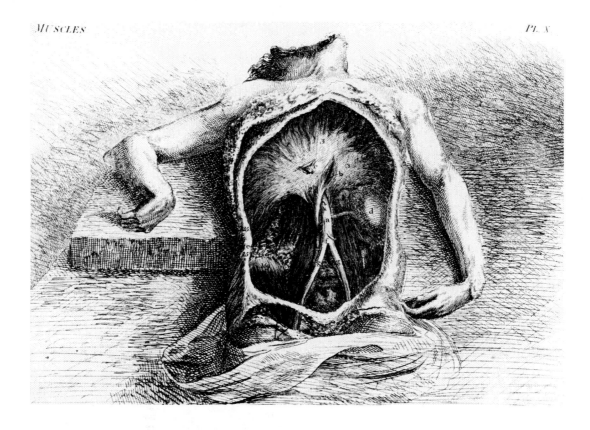

THE

ANATOMY OF THE BRAIN,

EXPLAINED IN A

SERIES OF ENGRAVINGS.

BY

CHARLES BELL,

FELLOW OF THE ROYAL COLLEGE OF SURGEONS OF EDINBURGH.

LONDON:

PRINTED BY C. WHITTINGHAM, DEAN-STREET, FETTER-LANE,

FOR T. N. LONGMAN AND O. REES, PATERNOSTER-ROW, AND T. CADELL, JUN.

AND W. DAVIES, IN THE STRAND.

1802.

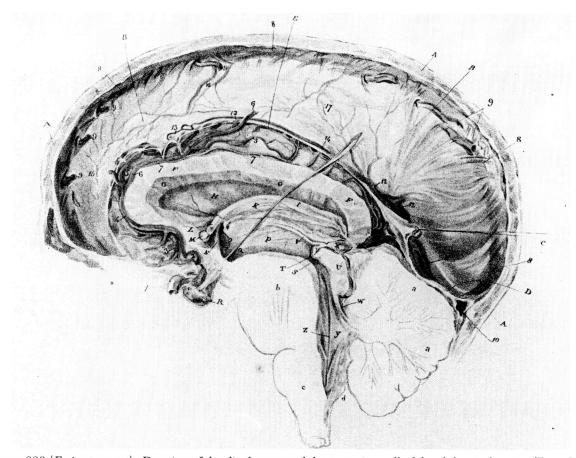

Figure 229 (*Facing page, top*). Drawing of the diaphragm and the posterior wall of the abdominal cavity. (From John Bell's *Engravings of the Bones, Muscles, and Joints....*First volume of the *Anatomy of the Human Body.*) (Courtesy of the Neil John Maclean Health Sciences Library, University of Manitoba.) The full extent of the diaphragm and the openings in it are depicted. Observe the quadratus lumborum and psoas major muscles, as well as the abdominal aorta and several branches.

Figure 230 (*Facing page, bottom left*). Sir Charles Bell (1774–1842). (Department of Anatomy, University of Manitoba.)

Figure 231 (*Facing page, bottom right*). Title page of Charles Bell's *The Anatomy of the Brain.* The engravings were done by Bell himself. (Courtesy of the Neil John Maclean Health Sciences Library, University of Manitoba.)

Figure 232 (*Above*). Medical aspect of the right cerebral hemisphere. The engraving was made by Charles Bell. It shows parts of the brain in great detail, including the falx cerebri, dural venous sinuses, the ventricles and their communications, the corpus callosum, the hypophysis, pineal gland, and several arteries and veins. (From Bell, C.: *The Anatomy of the Brain.* London, 1802. Courtesy of the Neil John Maclean Health Sciences Library, University of Manitoba.)

to his exclusion as a surgeon at the Royal Infirmary, which distressed him greatly. John Bell continued with his clinical work and writing. In 1816, he was seriously injured after a fall from his horse and because of ill health, he went to Italy where he spent the last three years of his life. He died in Rome in 1820; his tomb lies just behind that of the poet John Keats in the Protestant cemetery (Kaufman, 2005; Chisholm, 1911).

The most influential of Charles Bell's many books was the work entitled *Essays on the Anatomy of Expression*

and Painting (1806). This book immortalized his name, and it appeared in several editions, with different titles. Undoubtedly, Charles Bell's most unique and inspired work was *The Hand, Its Mechanism, and Vital Endowments, as Evincing Design* (London, 1833) (Figs. 233 & 234). As an invited contribution to the famous Bridgewater Treatises, it was meant to illustrate "*the power, wisdom and goodness of God, as manifested in the Creation,*" but it probably prepared the way for Charles Darwin's *Origin of Species,* published in 1859.

THE HAND

ITS MECHANISM AND VITAL ENDOWMENTS

AS EVINCING DESIGN

BY

SIR CHARLES BELL K.G.H.

F.R.S. L.&E.

LONDON

WILLIAM PICKERING

1833

Figure 233 (*Left*). Title page of Charles Bell's book on *The Hand...*, 1833. (Courtesy of the Neil John Maclean Health Sciences Library, University of Manitoba.)

Figure 234 (*Below*). Diagram from Bell's work on *The Hand*, demonstrating the "*mechanism of the arm*" and the "*interchange of velocity and force.*" (Courtesy of the Neil John Maclean Health Sciences Library, University of Manitoba.) The shoulder region, the scapula (A), the humerus (B), the deltoid muscle (C), and the antagonist muscle (D) are shown.

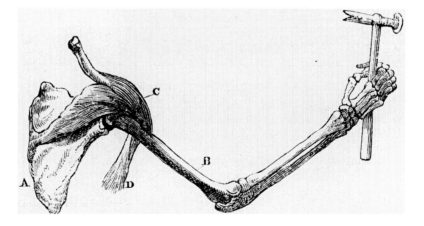

GIOVANNI BATTISTA MORGAGNI

Giovanni Battista Morgagni (1682–1771; Fig. 235), professor of anatomy at Padua, was concerned with diseases of the organs and of the whole body. His now classic treatise *De sedibus, et causis morborum per anatomen indagatis libri quinque*, published in 1761, has established him as the father of pathological anatomy. Morgagni correlated the autopsy findings with the clinical history of approximately 700 cases. Morgagni was close to 80 years of age when he completed his work.

MARIE-FRANCOIS XAVIER BICHAT

At the submacroscopic level, Marie Francois Xavier Bichat (1771–1802), professor of anatomy and surgery in Paris, studied the nature of tissues and how each individual tissue is affected in disease. Working with ordinary lenses, he identified 21 types of tissues. The two books Bichat wrote contained not a single illustration, but these seminal works revolutionized descriptive anatomy and established the foundations of modern histology.

Bichat's *Anatomie generale, appliquee a la physiologic et a la medicine,* four volumes (in two), was published in 1812 in Paris (Fig. 236). An English translation appeared in 1822 in Boston. The second work, *Traite d'anatomie descriptive,* five volumes, was published in Paris between 1801–1803; the last two volumes were unfinished at the time of his untimely death from tuberculosis at the age of 31 years. Another influential work of Bichat was his *Recherches physiologiques sur la vie et la mort,* Paris 1800.

JOHANN FRIEDRICH MECKEL

Johann Friedrich Meckel (1781–1833), the younger, was a brilliant comparative anatomist, embryologist, and pathologist. His name is associated with the cartilage of the first pharyngeal arch and with a

ANATOMICORUM PRINCEPS

Figure 235. Giovanni Battista Morgagni (1682–1771). (From Wegner, R.N.: *Das Anatomenbildnis. Seine Entwicklung im Zusammenhang mit der anatomischen Abbildung.* Basel, Benno Schwalbe and Co., Verlag, 1939. Courtesy of Institut fur Anatomie, Ernst-Moritz-Arndt-Universitat, Greifswald, Germany.)

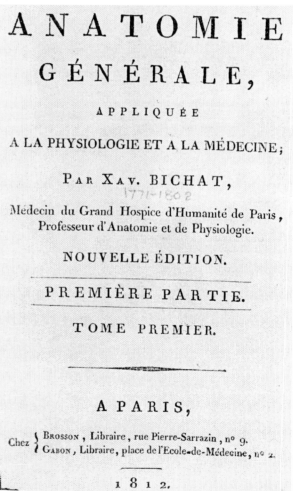

Figure 236. Title page of Marie Francois Xavier Bichat's *Anatomie Generale.* (Courtesy of the Neil John Maclean Health Sciences Library, University of Manitoba.) Bichat (1771–1802) was professor of anatomy and physiology in Paris, as well as physician at the Hotel-Dieu. He established the foundations of microscopic anatomy.

Figure 237 (*Left*). Johann Friedrich Meckel I (1724–1774). (Vorsatzbild fur den 24. Teil der Allgemeinen deutschen Bibliothek, Berlin and Stettin, 1775; (Courtesy of Professor Dr. R. Schmidt, Anatomisches Institut der Martin-Luther-Universitat, Halle (Salle), Germany.) Engraved by Johann Friedrich Schleuen from a painting by Anton Graff.

Figure 238 (*Below*). Meckel's graduation thesis, a classic description of the pterygopalatine (Meckel's) ganglion and the dural space lodging the Gasserian ganglion (Meckel's cave). (Morton's Medical Bibliography, Fifth Edition, Edited by Jeremy M. Norman)'s *Tractatus anatomico physiologicus de quinto pare nervorum cerebri*. Gottingae: Apud Abram Vandenhoeck, Acad. Typogr., 1748. (Courtesy of http://www.kumc.edu/dc/rti/.)

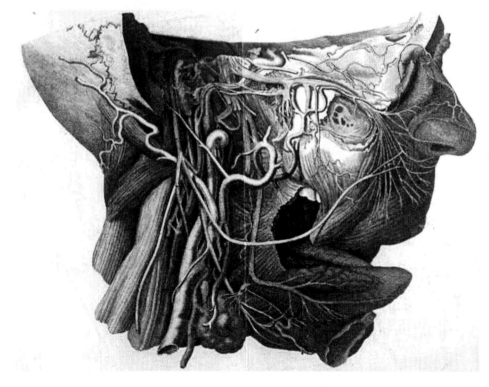

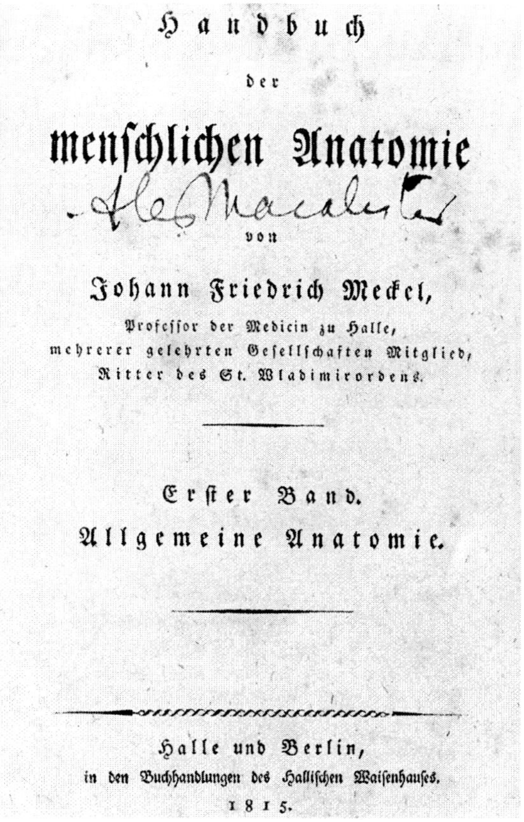

Handbuch

der

menschlichen Anatomie

von

Johann Friedrich Meckel,

Professor der Medicin zu Halle,

mehrerer gelehrten Gesellschaften Mitglied,

Ritter des St. Wladimirordens.

Erster Band.

Allgemeine Anatomie.

Halle und Berlin,

in den Buchhandlungen des Hallischen Waisenhauses.

1815.

Figure 239. Title page of the first volume of Johann Friedrich Meckel's *Handbuch der mendchlichen Anatomie.*

diverticulum of the ileum; that of his equally famous grandfather Johann Friedrich Meckel (1717–1774), who was professor of anatomy, botany, and gynecology in Berlin (Fig. 237, facing page), is eponymously linked with the trigeminal cavum in the middle cranial fossa (Fig. 238, facing page). Meckel, the younger, published between 1815 and 1820 a four-volume encyclopedic work, entitled *Handbuch der menschlichen Anatomie* (Fig. 239), which became a standard work of reference. An English translation (*Manual of general anatomy*) was published in London, 1837. His *Tabulae anatomico-pathologicae* (1817) combined normal and pathological anatomy and was considered then to be of importance. Meckel's collection in Halle consisted of 12,000 anatomical specimens. Many of the specimens were prepared by Gustav Wilhelm Münter (1804–1870). He was described as a "*diligent, skillful, reliable, and indispensable assistant of the Meckel Collections*" (Kapitza et al., 2002).

JOHANN SAMUEL EDUARD D'ALTON

Johann Friedrich Meckel (1781–1833) was succeeded by Johann Samuel Eduard d'Alton' (1803–1854). The anatomical collection in Halle was unique and of a special attraction. D'Alton kept the tradition of adding to the collection. Educated at Bonn University, he received the doctorate of medicine degree in 1824 and became professor of anatomy at the Academy of Arts in Berlin. From 1834 to 1854, d'Alton served the university in Halle as professor of anatomy and physiology, and also as the Rector on two occasions (1845/1846).

Industrious and conscientious, d'Alton was also a strict and demanding person, which made him unpopular with students and colleagues. He pursued studies in the areas of comparative anatomy, embryology, and experimental teratology (Zwiener et al., 2002; Zwiener, 2004). He authored many scientific papers and several major works on human and comparative anatomy, including: *Handbuch der menschlichen Anatomie.* 1848; *Anatomie der Bwewgungswerkzeuge oder Knochen-, Bände- und Muskel-Lehre des Menschen,* 1862; with J. S. E.; Pander: *Die Skelette der Raubvögel,* 1838); and in the field of teratology (*De Monstrorum Duplicium origine atque evolution,* 1848; *De monstris quibus extremitates superfluae suspensae sunt,* 1853).

JULES GERMAIN CLOQUET

J.⁰ⁿ CLOQUET.

Jules Germain Cloquet (1790–1883) was one of two brothers who held positions as professors of anatomy and clinical surgery in Paris (Fig. 240). Jules Cloquet's name is associated with the narrow canal in the vitreous humor of the eye, which transmits the retinal artery in fetal life. His elaborate atlas *Anatomie de l'homme* (five volumes) was published in Paris between 1821 and 1831 (Figs. 241 & 242). It contained 300 large plates and was the first anatomical work to be illustrated by lithography.

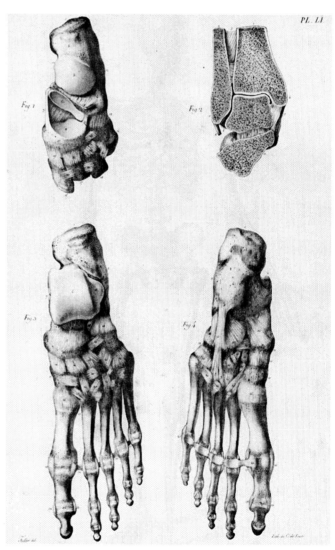

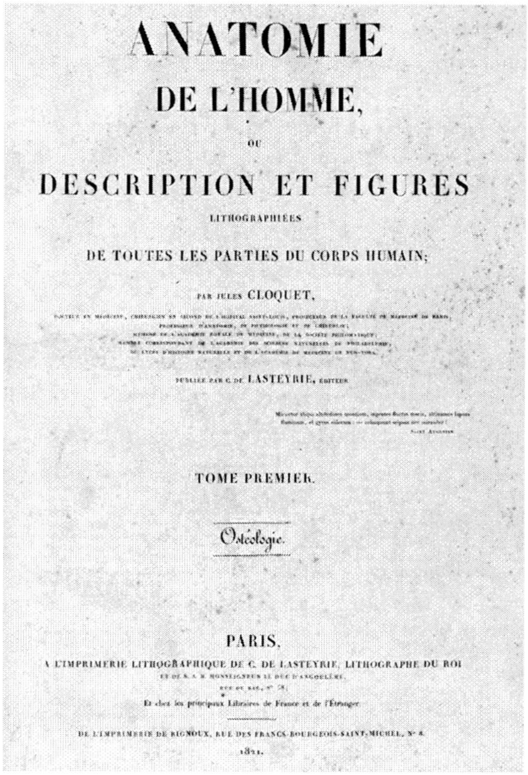

Figure 240 (*Top left*). A portrait of Jules Germain Cloquet. (Courtesy of Wellcome Library no. 1868i.)

Figure 241 (*Bottom left*). Front page of *Anatomie de l'homme*. From Google Books

Figure 242 (*Top right*). A page from *Anatomie de l'homme* depicting the bones and ligaments of the foot.

ANATOMICAL CHARTS AND PLATES

During this period, it was common practice to issue anatomical charts and plates in order to assist medical students. For example, between 1822 and 1833, the British surgeon John Lizars (1794–1860) had published *"A system of anatomical plates of the human body, accompanied with descriptions."* A similar set of anatomical plates (*A system of anatomical tables, with explanations*) for medical students was issued earlier (1786) by John Aitken (1747–1822) in London. There had been many other popular collections of anatomical illustrations, in the form of atlases, published in most European countries (Choulant, 1920; Herrlinger, 1970; Major, 1954; Roberts & Tomlinson, 1992, Jolin, 2013). These more realistic anatomical illustrations replaced the anatomical fugitive sheets of an earlier era that were still in circulation (Carlino, 1999; Lint, 1924). The two anatomical volumes of Paolo Mascagni (1752–1815; Fig. 243), professor of anatomy at Siena, Italy, entitled *Anatomia* universa, are the largest of all medical books from the standpoint of format.

The 44 life-size engraved plates are reproduced in double elephant folio size measuring 950 × 635 mm., and include an almost incredible level of detail (Norman, 1991). The work was published posthumously over a period of ten years (1823–1832); some sets of the plates were colored by hand and are breathtakingly beautiful. Mascagni's beautifully illustrated work on the lymphatics, *Vasorum lymphaticorum corporis humani...*, published in 1787, was considered then as the definitive work on the subject. It was a book inspired by the personal observations he made

Figure 243. Paolo Mascagni (1752–1815). (From Wegner, R.N.: *Das Anatomenbildnis. Seine Entwicklung im Zusammenhang mit der anatomischen Abbildung.* Basel, Benno Schwalbe and Co., Verlag, 1939. Courtesy of Institut fur Anatomic, Ernst-Moritz-Arndt-Universitat, Greifswald, Germany.)

of the amoeboid movement of leucocytes across the walls of blood vessels in cases of inflammation.

TOPOGRAPHIC ANATOMY

Noninvasive imaging of the body is commonly used as a diagnostic procedure. For this reason, a knowledge of cross-sectional anatomy is essential for reading and interpreting radiographic images, computer tomography (CT), magnetic resonance imaging (MRI), and also positron emission tomography (PET) scans. Medical students today are exposed to digital images of cross-sectional anatomy early in their training, which reinforces anatomical knowledge for clinical practice. Contemporary anatomy textbooks and atlases have CT and MRI diagnostic images coupled with labeled diagrams and photographs of the corresponding cadaveric specimens. The digitized thin transverse serial sections of a frozen male and female

cadavers from the U.S. National Library of Medicine Visible Human Project, has provided a good reference framework for learning more about the relationships of organs and other structures in the body.

The Dutch anatomist Pieter de Riemer (1760–1831) and the Russian surgeon-anatomist Nikolai Ivanovich Pirogov (1810–1881; Fig. 244) pioneered the use of frozen cadavers for anatomical studies. They sawed frozen cadavers and made cross-sectional slices to show the disposition of the organs and related structures (Fig. 245). Pirogov introduced applied topographical anatomy in Russia. He was the first anatomist to make extensive use of frozen sections of cadavers for anatomical teaching. His four-volume

Figure 244. Nikolai Ivanovitch Pirogov. Reproduction of painting, 1953, after I. E. Repin. (Wellcome Library no. 10001i.)

atlas on topographical anatomy (1851–1854), with illustrations from the frozen slices, was published over a period of four years.

Remarkable images of cross-sectional anatomy, derived from frozen cadavers, were produced by the German anatomist Christian Wilhelm Braune (1831–1892). He used a broad, fine-edged saw to slice the cadavers, both male and female, including a pregnant woman. Braune carefully studied the frozen slices and produced the most accurate atlas of topographical anatomy. His "*Topographisch-anatomischer Atlas nach Durchschnitten angefrornen Cadavern*" was published in 1872 and revolutionized the teaching of anatomy (Fig. 246). The atlas was translated into English five years later (*An Atlas of Topographical Anatomy after Plane Sections of Frozen Bodies*). It was illustrated with a series of stunningly beautiful, colored, and labeled lithographs of cross-sections of the entire body in different planes. The labels had a uniquely circular arrangement around the specimen (Figs. 247–248).

Braune studied medicine in Göttingen and Würzburg. He was appointed professor of topographical anatomy in 1872 at the University of Leipzig where he

carried out anatomical and experimental studies on movement of the body and the biomechanics of gait. Works resulting from these studies include the following books: *Die Lage des Uterus und Fötus am Ende der Schwangerschaft nach Durchschnitten an gefrorenen Kadavern* (The location of the uterus and fetus at the end of pregnancy according to a cross-section of a frozen cadaver) (Fig. 249), published in 1873 and with the physiologist Otto Fischer, "*Der Gang des Menschen*" (The human gait) in 1892. By now, anatomists were not only writing textbooks for medical students but

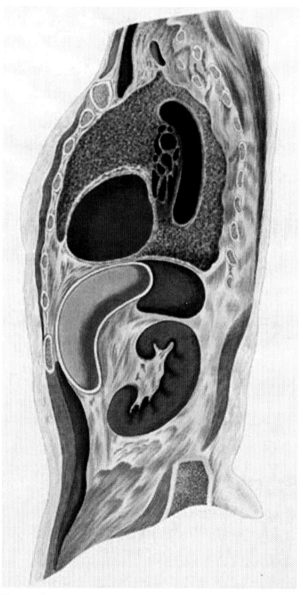

Figure 245. A midsagittal section of a specimen from Pieter de Riemer. http://digi.ub.uni-heidelberg.de/diglit/riemer1818.

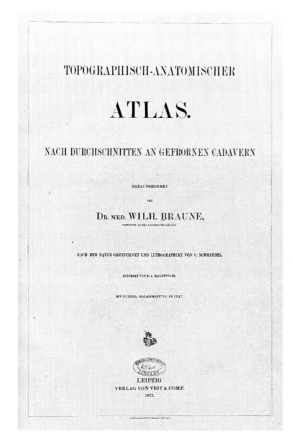

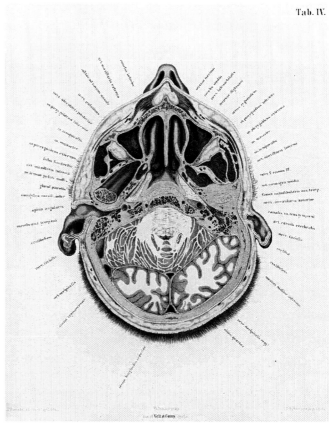

Figure 246 (*Top left*). Front page of *Topographisch-anatomischer Atlas, nach Durchschnitten an gefrornen Cadavern.* Publication: Leipzig: Verlag von Veit & Comp., 1867–1872. (Courtesy of Bethesda, MD: U.S. National Library of Medicine, National Institutes of Health, Health & Human Services. http://www.nlm.nih.gov/exhibition/historicalanatomies.)

Figure 247 (*Top right*). A trasverse section of a head, in *Topographisch-anatomischer Atlas, nach Durchschnitten an gefrornen Cadavern.* Publication: Leipzig: Verlag von Veit & Comp., 1867–1872. (Courtesy of Bethesda, MD: U.S. National Library of Medicine, National Institutes of Health, Health & Human Services. http://www.nlm.nih.gov/exhibition/historicalanatomies.)

Figure 248 (*Right*). Front page of *Die Lage des Uterus und Foetus Ende der Schwangerschaft, nach durchschnitten an Gefrornen Cadavern.* Publication: Leipzig: Verlag con Veit & Comp., 1872. (Courtesy of Bethesda, MD: U.S. National Library of Medicine, National Institutes of Health, Health & Human Services. http://www.nlm.nih.gov/exhibition/historicalanatomies.)

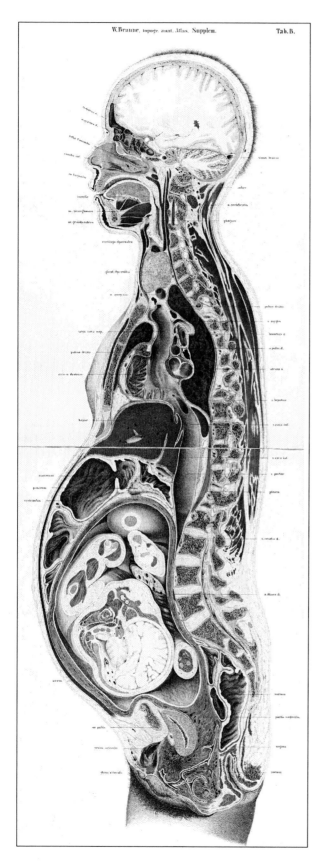

Figure 249. A midsagittal section of a pregnant woman depicted the fetus in situ, in *Die Lage des Uterus und Foetus Ende der Schwangerschaft, nach durchschnitten an Gefrornen Cadavern.* Publication: Leipzig: Verlag con Veit & Comp., 1872. (Courtesy of Bethesda, MD: U.S. National Library of Medicine, National Institutes of Health, Health & Human Services. http://www.nlm.nih.gov/exhibition/ historicalanatomies.)

also have been presenting themselves as authors of research and other scholarly treatises, based on experiments and observations. This tradition had already started a century ago with John and William Hunter.

The illustrations in Jacob Rueff's (1500–1558) book, *De conceptu e generatione hominis* (1554), which was adapted from Eucharius Roslin's work *Rosengarten* (1513) and William Raynalde's *Byrthe of Mankynde* (1545), convey the state of knowledge of human embryology during the middle of the sixteenth century. In many respects, the splendid drawings of the fetus in the womb by Leonardo da Vinci, done much earlier, were far more realistic.

GIULIO CESARE ARANZI

The Italian surgeon and anatomist Giulio Cesare Aranzi (1530–1589), also known as Julius Caesar Arantius, provided the first useful description of the fetus and the placenta in his book, *De humano foetu libellus* (Rome, 1564;) Fig. 250. He was born in Bologna to a poor family. At an early age he began his medical studies as an apprentice to his uncle Bartolomeo Maggi (1477–1552), a prominent surgeon at the University of Bologna and physician to the court of Julius III. In 1548, Aranzi joined the medical faculty of the University of Padua where, as a 19-year-old medical student, he discovered the muscle that elevates the eyelid (levator palpebrae superioris). Aranzi received the doctor of medicine degree in 1556 and was appointed as a lecturer at the University of Bologna. In 1570, he created separate professorships for anatomy and surgery and held both chairs for 33 years until his death. For the first time, anatomy was established as a major independent discipline in medicine.

Aranzi made several important anatomical discoveries. He found that during pregnancy maternal and fetal blood are separated, and he also described accurately the anatomy of the fetus and the placenta. After studying the cavities of the heart, the valves, major

Figure 250. Title page of *De Humano Foetu Liber* published in Venice, 1595. (With permission from Gurunluoglu et al., J Med Biogr, 2001;19:63-9.)

blood vessels, and the flow of blood, Aranzi discovered nodules in the semilunar valves (Aranzio nodules), the ductus arteriosus, which is usually attributed to Leonardo Botallo (1530–1600), the ductus venosus, and the foramen ovale which showed that blood flows from the right atrium (right side of the heart) to the left atrium (left side of the heart). Moreover, he described the coracobrachialis muscle, inferior cornu of the cerebral ventricles, the fourth ventricle of the brain, and the hippocampus, which he so named because it resembled a seahorse.

A skilled surgeon, Aranzi developed special surgical instruments and innovative techniques for use in surgical practice (Gurunluoglu et al., 2011). In his book, "*Observationes anatomicae*" (Basel, 1579), he described many surgical conditions, including hydrocephalus, ascites, goiter, hemorrhoids, abscesses, and fistulae.

EMBRYONIC DEVELOPMENT

One of the successors of Vesalius at Padua, Fabricius ab Aquapendente (1533–1619), wrote treatises entitled *De humano foetu, De formatione ovi et pulli,* and an embryology book, *De formato foeto,* that included many illustrations derived from the dissection of several species of embryos, as well as his theories of development.

The modern era of embryology (Meyer, 1939; Needham, 1959; Oppenheimer, 1975) began with William Harvey (1578–1657) through his experimental studies on the development of chick embryos and the publication in 1651 of his book *Exercitationes de generatione animalium* (Keynes, 1966). Harvey considered this book to be of greater significance than his more famous work, *De motu cordis,* dealing with the circulation of blood. For him this book represented the highest achievement of his life's work. The two prevailing theories in embryology were *epigenesis* and *preformation.* These were represented by the greatest of scientists at that time, William Harvey (1578–1657) and Marcello Malpighi (1628–1694).

In addition to correcting past misconceptions about embryological development, Harvey rejected the preformation theory, and he proposed that all animals originate from the egg or ovum. Furthermore, he established that the organism develops gradually through the addition of a part to already existing structures, and not from the growth of preformed *homunculus* in the egg. Harvey's concept of epigenesis remained controversial for a very long time (Meyer, 1936; Keynes, 1966).

On the other hand, Malpighi believed that the chick is already formed even before incubation; in time, the preformed chick merely grows and becomes larger. With the aid of a simple microscope, Malpighi was able to trace the formation of the chick embryo from the earliest stages. He provided the most accurate description of the chick embryo from the initial stages as it developed. Malpighi communicated his observations to the Royal Society of London in 1672 and in two books he had published: *De ovo incubato observationes* (London, 1673) and *Dissertatio epistolica de formatione pulli in ovo* (London, 1673).

Experimental manipulation of the embryo began with Walter Needham (c. 1631–1691) who carried out chemical experiments in developing embryos, the results of which he reported in his *Disquisitio anatomica de formatu foetu* (London, 1667). This pioneering work in chemical embryology also contained the first practical instructions for dissecting embryos.

Unlike his teacher Hermann Boerhaave, Albrecht von Haller (1708–1777) rejected the preformation theory in support of Harvey's theory of epigenesis. Boerhaave and von Haller were among the most influential scientists for the greater part of the eighteenth century. Von Haller correctly calculated the growth rate of chick and human embryos, and in 1758 he published his splendid work, *Sur la formation du coeur dans le poulet* (two volumes). Von Haller was one of the earliest functional anatomists, and he carried out important physiological experiments on muscle contraction and nerve conduction (Asher, 1902; Nussbaum, 1968).

One of the most important discoveries in the history of embryology first appeared in the doctoral dissertation (*Theoria generationis,* Halle, 1759) of the brilliant German anatomist Caspar Friedrich Wolff (1733–1794). It was considered to be the most outstanding doctoral work of the century. Wolff observed the embryological masses of tissues, which partly contribute to the development of the renal and genital systems (Wolffian bodies and Wolffian duct). His other important work, *De Formatione Intestinorum,* was published in 1768/1769 (Uschmann, 1955).

The concept of epigenesis through the differentiation of tissues was now firmly established. Wolff's work was in opposition to the preformation doctrine, and it contradicted the religious doctrine of creation. In

Berlin, he encountered considerable hostility and professional jealousy. Unable to secure a teaching position there, he accepted an invitation from the Academy of Sciences in St. Petersburg in Russia where he spent the rest of his life.

William Hunter's *Anatomy of the Human Gravid Uterus*, published in 1755, is, with its superb engraved plates, a magnificent scientific work on the fetus and placenta. It was supplemented later by the beautifully illustrated atlas of embryos in Thomas von Sömmerring's *Icones embryonum humanorum* (Frankfurt, 1799).

The three germ layers of the embryo were discovered by Heinrich Christian Pander (1794–1865), which he reported in his doctoral *Dissertatio sistens historiam metamorphoseos, quam ovum incubatum prioribus quinque diebus subit* (1817). Pander named the three layers "blastoderrn." Before the end of 1817, a German translation of Pander's dissertation (*Beitrage zur Entwicklungsgeschichte des Huhnchens im Ei*) was published. Sixteen copperplate engravings were added because the original Latin work did not have any illustrations.

JOHANNES PETER MÜLLER

In his eulogy for the comparative anatomist and physiologist Johannes Peter Müller (1801–1858, Fig. 251), the great German pathologist Rudolph Virchow asked: "*...how can one...adequately praise a man who resided over the whole domain of the science of animal life; or how can one depict the master-mind which extended the limits of this great kingdom, until it became too large for his own individual government?*" (website b; c). Müller was regarded as one of the finest scientific minds of his era. He inspired an entire generation of leading scientists, including Hermann von Helmholtz, Emil du Bois-Reymond, Theodor Schwann, Jacob Henle, Ernest Haeckel, Wilhelm Wundt, Albrecht von Kölliker, Wilhelm Peters, and Rudolph Virchow. Berlin quickly became a leading center for medical research because of the work of Müller and his students.

Müller introduced the most recent research tools that were then available – microscopic, chemical, and physiological – in his laboratory. His discoveries in the natural sciences and medicine have helped to establish physiology as a separate discipline. Today, Müller's immense scientific contributions are mostly forgotten, except for the "*paramesonephric ducts*" which are named eponymously after him (Müllerian ducts), even though Johann Friedrich Meckel, Jr. and Robert Remak also contributed to its discovery.

Johannes Müller was born in Koblenz on July 14, 1801 where he received an early education in mathematics and the classics. At the age of 18 years, Müller began medical studies at the University of Bonn and graduated in 1822. He then went to the University in Berlin for further studies where he became acquainted with microscopy. Returning to the University of Bonn in 1824, Müller was appointed as a Privatdozent (Lecturer) of physiology and comparative anatomy, but by 1830, he was promoted to full professor of anatomy and physiology. Following the death of the anatomist Karl Asmund Rudolphi (1771–1832), Müller was invited by the university in Berlin to succeed him.

A brilliant researcher, Müller was the author of numerous research publications and books on comparative anatomy, the nervous system, cancer,

Figure 251. Johannes Peter Müller. (Courtesy of the Library, Institut für Anatomie der Humboldt-Universität Berlin, Germany.)

development, physiology, and other areas in natural sciences (Haberling, 1924; Wendler, 1984). His major works included the following: *Zur Physiologie des Fötus* (1824); *Zur vergleichenden Physiologie des Gesichtssinns* (1826); *Über die phantastischen Gesichtserscheinungen* (1826); *Bildungsgeschichte der Genitalien* (1830); *De glandularum secernentium structura peritoni* (1830); *Beiträge zur Anatomie und Naturgeschichte der Amphibien* (1832); *Vergleichende Anatomie der Myxinoiden* (1834–1843); *Handbuch der Physiologie des Menschen*, 2 Bände. (1837–1840); *Über den feinern Bau und die Formen der krankhaften Geschwülste* (1838); *Über die Compensation der physischen Kräfte am menschlichen Stimmorgan* (1839); *Systematische Beschreibung der Plagiostomen* (1841), with Friedrich Gustav Jakob Henle; *System der Asteriden* (1842), with Franz Hermann Troschel; *Horae ichthyologicae: Beschreibung und Abbildung neuer Fische* (1845–1849); *Über Synapta digitata und über die Entstehung von Schnecken in Holothurien* (1852).

Müller's major undertaking was the *Handbuch der Physiologie des Menschen für Vorlesungen* (two volumes, 1833–1840). It was an encyclopedic treatise on human and comparative anatomy, which incorporated the most recent physiological studies and developments in the natural sciences. Physiology entered a new phase with the publication of Müller's magnum opus. It became one of the most influential scientific works of the nineteenth century and was translated into English by William Baly (Müller, J.: *Handbook of Human Physiology*, Coblenz: J. Holscher, 1834–1840, translated by W. Baly, London: Taylor & Walton, 1837.). Müller founded and edited the "*Archiv für Anatomie, Physiologie und Wissenshaftliche Medicin*," which became a highly regarded scientific journal.

In recognition of his enormous research achievements, Müller was awarded the Copley medal of the Royal Society of London, and he also received royal recognition from Prussia, Sweden, Bavaria, and Sardinia. In Koblenz, a life-size bronze monument of him was erected by the city in 1899, and the renaming of a street (Johannes-Müller-Straße) in 1894 honors his memory. Over the years, Müller had suffered from a bipolar (manic-depressive) condition, which might have contributed to his personality. He was found dead in his room on April 28, 1858, apparently due to suicide.

KARL ERNST VON BAER

Inspired by the work of Heinrich Christian Pander (1794–1865) and Jan Evangelista (Purkyne) Purkinje (1787–1869), Karl Ernst von Baer (1792–1876) carried out experimental studies that led to fundamental discoveries relating to embryonic development (Fig. 252). Von Baer found that the notochord is present in all vertebrates, and this was soon followed by his discovery of the mammalian ovum in the ovary. Moreover, von Baer formulated two important concepts: corresponding stages of embryonic development; and that general characteristics precede the special ones. The remarkable discovery of the ovum (*De ovi mammalium et hominis genesis*) was published in 1827, followed a year later by the first volume of his influential work (Baer, Karl Ernst von., 1828–1837).

Not surprisingly, von Baer is universally acclaimed as the father of modern embryology, as well as the greatest embryologist of the nineteenth century. He was born in Estonia and was educated at the University of Dorpat, from where he graduated in medicine. After further studies in Wurzburg and Vienna, von Baer was appointed professor at Konigsberg (1817–1834) and later at St. Petersburg (1834–1862).

MICROSCOPY

The discovery of the microscope provided a novel means of studying organs and tissues. Transcending the limits of the simple lens scientists were able to derive more accurate information on the nature and organization of living matter. It was the Englishman Robert Hooke (1635–1703) who first coined the term "*cell*," having observed "*microscopical pores*" and regularly shaped compartments in very thin sections of cork he had examined under the microscope (Bracegirdle, 1985). These were described as "*all cellular or porous in the manner of a honeycomb, but not so regular*" in his book, *Micrographia*, published in 1665. One of the most brilliant scientists of his time, Hooke made lasting scientific contributions in the areas of astronomy, physics, chemistry, and botany (Nichols, 1995). As an architect, he also designed several important buildings after the Great Fire of London in 1666.

Figure 252 (*Left*). Karl Ernst van Baer (1792–1876). (Department of Anatomy, University of Manitoba.)

Figure 253 (*Right*). Antoni van Leeuwenhoek (1632–1723). (Museum of the History of Science, Oxford.) Leeuwenhoek "*developed the simple microscope...and carried out a series of astonishing investigations as to the numbers of such tiny creatures in a drop of liquid, and he described Protozoa, bacteria, spermatozoa, blood corpuscles and capillaries...*" (Bracegirdle, 1986). He made hundreds of microscopes, and he communicated his observations over the years in the form of letters (more than 250) to the Royal Society of London.

One of the first persons to observe a human cell was Anton (Antoni) van Leeuwenhoek (1632–1723), a clerk in a cloth warehouse in Amsterdam, who later became an accomplished microscopist (Figs. 253 & 254).*

However, the Italian anatomist Marcello Malpighi (1628–1694) had already given the first accurate description of the red corpuscles (*De polypo cordis*, Bologna 1666) and of the renal corpuscles (Malpighian corpuscles or bodies), which he described in his work *De viscerum structura exercitatio anatomica* (Bologna 1666). Malpighi also authored a major and definitive work on plant tissues (*Anatome plantarum*, London 1675–1679). Because of his many pioneering discoveries, Malpighi is considered as the founder of microscopic anatomy. He was also personal physician to Pope Innocent XII.

Mattias Jakob Schleiden (1804–1881), who was a lawyer before turning to botany, published a paper entitled *Phytogenesis* in 1838, which established that plants too are made up of cells. It was the work of Schleiden that inspired Theodor Schwann (1810–1882), the German physiologist and professor at Louvain and Liege, to formulate his cell theory.

That the cell, with its nucleus, is the basic structure of all living tissues was in the early nineteenth century a brilliant revolutionary concept, not unlike William Harvey's discovery of the circulation of blood almost two hundred years earlier.

For the sake of completion, it should be mentioned that the term "*protoplasm*" was first used in 1839 by the great scientist Johannes Evangelista Purkinje

---

*Leeuwenhoek's description of the blood corpuscles (*particles*) is contained in letters he wrote to the Royal Society in London, starting from August 1673 (Dobell, 1932; Bracegirdle, 1985). For an interesting account of Leeuwenhoek's microscopes, see Van Zuylen (1981) and Ford (1983).

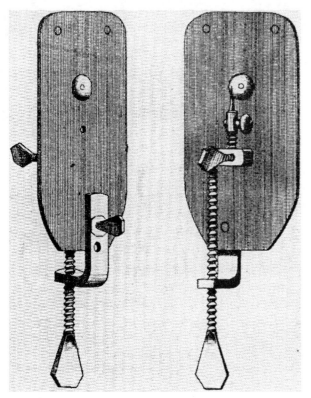

Figure 254. Antoni Van Leeuwenhoek's microscope. "*This simple microscope consisted of a single fixed lens set in a metal plate. The object, placed on the point on the plate's reverse side, was moved until in focus*" (With permission from Major, 1954.)

(1789–1869); it was present in the doctoral dissertation (*De formatione granulosa in nervis aliisque partibus organismi animalis*, Breslau, 1839) of his student Joseph Rosenthal. (It should be noted, however, that in 1867 the German anatomist Max Schultz conceived the cell as the basis of life, containing a mass of protoplasm and a nucleus. The term cytoplasm was introduced in 1862 by Rudolf Albert von Milker (1817–1905), a Swiss anatomist and embryologist).

The grouping of tissues by Marie Francois Bichat (1771–1802), founder of histopathology, progressed with other microscopic discoveries. Johannes Muller (Haberling, 1924; Wendler, 1984) described the histology of secretory glands (*De glandularum secernentium structura penitiori*, 1830) and other tissues; Friedrich Jacob Henle described the epithelial cells of the skin and intestines (*Symbolae ad anatomiam villorum intestinalium...*, 1837; *Uber die Ausbreitung des Epithelium im menschlichen Korper*, 1838) and classified further the tissues

based on more accurate histological characteristics (for more details see *Allgemeine Anatomie*, Leipzig, 1841).

In 1854, the English microscopist Lionel Smith Beale (1828–1906) published a landmark book entitled *The Microscope in Medicine*. He was professor at Kings College in London and had discovered the "*Beale's cells.*"

Galen's views on the structure and functions of the brain, spinal cord, and the nerves have prevailed at least up to the beginning of the eighteenth century.* In retrospect, Galen (A.D. 131–192) stands out as one of the greatest medical scholars not only of antiquity but of all time because of the profound influence his voluminous writings had on medical scholars through the Middle Ages (Temkin, 1973). Galen's contribution to the nervous system is considered to be his best work.

He has left us with detailed descriptions of the dissection of the brain of animals and the names of many structures he had identified are still with us, including the corpus callosum, the corpora quadrigemina, the fornix, the pineal body and the septum pellucidum. Galen knew of the ventricles, their communications, and of the choroid plexus. His description of the third and fourth ventricles reflected keen observations. In describing the brain, which he considered to be the seat of the soul, Galen identified seven pairs of cerebral nerves. Because they originated from the brain, they were considered to be nerves of sensation in contrast to the thirty pairs of spinal nerves which he recognized and designated as nerves of motion. (Persaud, 1984)

Galen failed, however, to differentiate nerves from tendons. Like Hippocrates, he believed that Nature was just because the distribution of the blood vessels and nerves throughout the body is "*in accordance with the value of each part.*" Because the structures passed safely to their destinations, he thought of Nature as being "*not only just, but also skillful and wise*" (May, 1968).

When Giovanni Alphonso Borelli (1608–1679), professor of mathematics at Pisa, published his *De Motu Animalium*, Borelli

took a distinctly physical view of the nervous action; the conception of animal spirits so dear to the older writers (and destined to persist for a further century) he replaced by that of a nervous fluid, the succus nervous. This he considered to have a

―――――――――
*See Persaud (1984) and Von Staden (1992) for other pre-Vesalian theories.

fluid consistency "like spirits of wine" and to be agitated in two directions along the nerves: centrally to generate sensations, and peripherally, from the brain to the muscles, to give rise to movements. (Gordon-Taylor & Walls, 1958)

In 1664, Thomas Willis (1621–1675) published his *Cerebri anatome*, one of the great classics in neurosciences. His work on the anatomy of the brain and diseases of the nervous system represent the foundations of modern neurology (Feindel, 1978; Hughes, 1992).

Von Sömmerring (1791–1796), in his books, provided a modern classification of the cranial nerves. Bichat (1771–1802), in his *Anatomie Generale*, published in 1801, suggested that different nerves might be responsible for different functions. The *Tabulae Neurologicae* of Antonio Maria Scarpa (1752–1832), ophthalmologist and professor of anatomy at Pavia, beautifully and accurately depicted the smallest branches of nerves.

Like other anatomists, he too was perplexed about the function of the ganglia of the spinal nerves and of the large ganglia associated with the trigeminal nerve.

The famous Scottish anatomist and surgeon Sir Charles Bell (1774–1824) distinguished the sensory from the motor nerves in 1807 and made many important contributions to our understanding of the nervous system. His classical work *The Nervous System of the Human Body*, published in 1830, contains his original papers read to the Royal Society.

At the turn of the nineteenth century, the anatomy and function of the brain was largely unknown. Although there had been several major neuroanatomical treatises published as the years advanced, the brain was an enigma then and it still continues to be so through the present day. (For an excellent account of Sir Charles Bell's life and his work, see Gordon-Taylor and Walls (1958).

Chapter 23

THE HUMAN BODY REVEALED

Many gross anatomical structures in the human body were already recognized and depicted by the middle of the sixteenth century (Knight, 1980; Persaud, 1984). Others remained to be discovered and, with the passage of time, some were even rediscovered (Persaud, 1997). For this reason alone, it is advisable to use the term "*describe*" rather than "*discover*." While the working and functions of the organs of the body might have perplexed the anatomist, their description, and the relationship of diverse structures to each other, were straightforward and fairly accurate. This is evident from the work of Vesalius and other anatomists who immediately followed him.

Knight (1980) observed that "*This ubiquitous material, which keeps the human body from falling apart, has been a glorious haven for anatomical name droppers. Anatomists and surgeons by the score are commemorated in sheets of fascia or various bands, ligaments, holes, arches, spaces, and triangles which are left here and there about the body in the complex architecture of the fibrous network*" (Field & Harrison, 1968; Dobson, 1962; Lindner, 1989; McMinn, 1990; Moore, 1992; Olry, 1995). Even though the use of eponyms is discouraged, many structures are still associated with anatomists and surgeons who discovered or first described them.

In addition, there had been occasional landmarks of important events, which contributed to the evolution of anatomy as a scientific discipline. In this section, it will not be possible to include everything that has happened. Instead, we have gathered here only *some* of the more prominent names, noteworthy events, important discoveries and descriptions in order to present a conceptus of the progress that was made over the centuries. We are conscious of the many omissions and the repetition of information given in other sections of this book, which are unavoidable in an inventory of this nature.

SIXTEENTH CENTURY

Realdus Colombus described the pulmonary circulation and stated that there was no opening in the interventricular septum. William Harvey developed his hypothesis of blood circulation from the work of Columbus, which he extensively cited in his classic book *De mortu cordis*. Columbus also carried out investigations on the eye. He established that contrary to the teachings of Galen the lens was not in the center of the eyeball, but that it was located posterior to the iris.

Gabriele Fallopius (1523–1563) described the uterine tubes, as well as the chorda tympani nerve and the facial nerve canal. In addition, he introduced the term "*cochlea*" and "*labyrinth*" for structures of the inner ear. The auditory tube and the valve of the inferior vena cava were described by Bartolomeo Eustachius (1520–1574). Hieronymus Fabricius ab Acquapendente (1533–1619) described the valves in the veins. Discovery of the circulation of blood was dependent on an understanding of the function of these valves.

For human dissection and the teaching of anatomy, special facilities were required which led to the establishment of *Theatrum anatomicum* or Schools of Anatomy in various cities: Montpellier (1556), Ferrara (1588), Basle (1589), Padua (1595), and Leiden (1597). Not only medical students and doctors attended these sessions and participated in the events but also scholars from other disciplines, as well as invited dignitaries (Persaud, 1997).

The well-known anatomy theater in Padua (Fig. 255) was designed and built by Fabricius of Aquapendente, and that in Leiden by Pieter Pauw, who was once a student of Fabricius at Padua. Over a period of 19 years Fabricius dissected close to 60 cadavers.

In England, the earliest recorded dissection of a human body occurred in Oxford. It was carried out by David Edwardes who published in 1532 a small book entitled "*De indiciis et praecognitionibus*" (Brief but Excellent Introduction to Anatomy). This book is historically important landmark. It consisted of only fifteen pages, but it was "*the first work on anatomy to be published by an English author in England. It also gives the first printed reference to anatomical dissection in England*" (Russell, 1987). The British Museum in London has the only known copy of this book. David Edwardes was a fellow of Corpus Christi College and a lecturer in Greek at Oxford. In 1528, he left for Cambridge, where he received the M.D. degree, as well as taught and examined medical students in anatomy (Robb-Smith, 1971; Pratt, 1981).

Following his return to England in 1544, John Caius (1510–1573), who was a student in Padua and once shared rooms with Vesalius, introduced annual anatomical dissections and demonstrations at the Barber-Surgeons' Hall in London. In 1557, he apparently carried out a dissection in his college, Gonville Hall, which later became Gonville and Caius College. He also made provisions for regular human dissections in his College at Cambridge, by securing from Queen Elizabeth in 1565 permission to obtain two bodies annually (Rolleston, 1932).

According to Valadez (1974) there are several references to human dissection in Cambridge, but "*the actual performance of dissections around this time is limited to a few entries in the register of Great St. Mary's Church, Cambridge...*" The earliest record of a dissection there notes that in 1566 "*John Figgen mad Anotomy at the scholes and buried here, the 12th March.*"

Some famous anatomists during this era included: Andres Laguna (1499–1560; Fig 256), Vidus Vidius also Guido Guidi (c. 1500–1569), Giambattista Canano (1515–1578; Fig. 257), Giovanni Filipio Ingrassia (1510–1580; Fig. 258), Giulio Cesare Aranzi

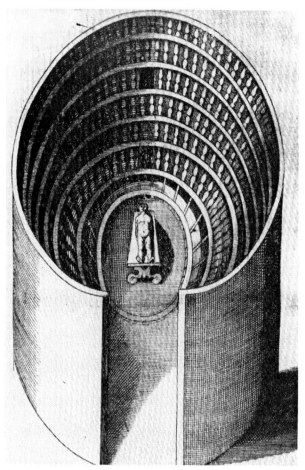

Figure 255. The *Theatrum Anatomzcum* (Anatomical Theater) in Padua. It was built in 1594 by Fabricius ab Aquapendente. (Copper engraving by Jacobus Philippus Thomasini, 1654.) The corpse is lying on the dissecting table in the middle, at ground level, of the amphitheatre. There were six elevated rows for up to 300 persons, standing.

(1530–1589), Leonardo Botalli (c. 1530–1600), Volcher Coiter (c. 1534–1600; Fig. 259), Barthelemy Cabrol, (1529–c. 1605), Felix Platter (1536–1614; Fig. 260), and Adriaan van der Spieghel (1578–1625; Fig. 261).

SEVENTEENTH CENTURY

Caspar Bauhin

The Swiss physician Caspar (Gaspard) Bauhin (1560–1624; Fig. 262) was well known for his anatomical and botanical studies. He was a student of Fabricius ab Aquapendente in Padua and of Arantius (1530–1589) in Bologna. Bauhin eventually obtained his medical degree from the University of Basel in 1580 and two years later he was appointed Professor of Greek at the University. This was followed by other appointments as Professor of Anatomy and Botany, City Physician, Dean of the Faculty of Medicine, and Rector of the University of Basel.

Figure 256 (*Top left*). A portrait of Andres Laguna. This is a faithful photographic reproduction of a two-dimensional, public domain work of art. The work of art itself is in the public domain.

Figure 257 (*Top right*). Portrait of G. Canano 1515–1579, head and shoulders. Lithograph. (Courtesy of Wellcome Library, London.)

Figure 258 (*Left*). Giovanni Filippo Ingrassia (1510–1580); Professor of Anatomy and Medicine in Naples, and later in Palermo. Ingrassia is best known for his studies on the structure of bones. His other scientific contributions were in the areas of epidemiology and infectious diseases: Ingrassia was the first physician to differentiate varicella from scarlet fever. Ingrassia's book on the bones, entitled *In Galeni librum de ossibus* (1603), described his osteological findings, including the discovery of the stapes, and that the bones mentioned by Galen were not human but from the monkey. (From Wegner, R.N.: *Das Anatomenbildnis. Seine Entwicklung im Zusammenhang mit der anatomischen Abbildung.* Basel, Benno Schwalbe and Co., Verlag, 1939. Courtesy of the Library, Institut für Anatomie der Humboldt-Universität Berlin, Germany.)

Figure 259 (*Top left*). Portrait of Volcher Coiter. (Courtesy of Wellcome Library no. 1907i.)

Figure 260 (*Top right*). Felix Platter (1536–1614), at 41 years of age. (From Platter, F. *De partium corporis humani structura et usu libri III...*, Basel 1581. From Wegner, R.N.: *Das Anatomenbildnis. Seine Entwicklung im Zusammenhang mh der anatomischen Abbildung.* Basel, Benno Schwalbe and Co., Verlag, 1939. Courtesy of Institut für Anatomie, Ernst-Moritz-Arndt-Universität, Greifswald, Germany.)

Figure 261 (*Right*). Adriaan van den Spieghel (1578–1625). The engraving was done in 1645 by Jeremias Falck. (From Wegner, R.N.: *Das Anatomenbildnis. Seine Entwicklung im Zusammenhang mh der anatomischen Abbildung.* Basel, Benno Schwalbe and Co., Verlag, 1939. Courtesy of the Library, Institut für Anatomie der Humboldt-Universität Berlin, Germany.)

Figure 262. Portait of Caspar (Gaspard) Bauhin. (Courtesy of Wellcome Library no. 954i.)

In 1579, Bauhin demonstrated the presence of the ileocecal valve during a dissection he had carried out in Paris and which he later confirmed in a dog. These findings were reported in his two major works: *Theatrum anatomi cum Caspari Bauhini...*(Basle, 1621) and *Vivae imagines partium corporis humani...*, published in 1620 (Bergmann and Wendler, 1986; Fig. 263). Bauhin also published a major anatomical treatise entitled, *Theatrum Anatomicum infinitis locis auctum* (1592). He also described and classified more than 6,000 plants, which appeared in his encyclopedic work, *Pinax theatric botanici* (1623).

Alessandro Achillini

Bauhin's name is eponymously associated with the ileocaecal valve, which was described earlier by Alessandro Achillini (1463–1512; Fig. 264). Achillini studied medicine and philosophy at the University of Bologna where he was born and had spent most of his professional life. Known as the *"second Aristotle"* because of his extensive philosophical writings, Achillini taught both philosophy and medicine in Bologna and Padua. A celebrated anatomist, Achillini wrote two

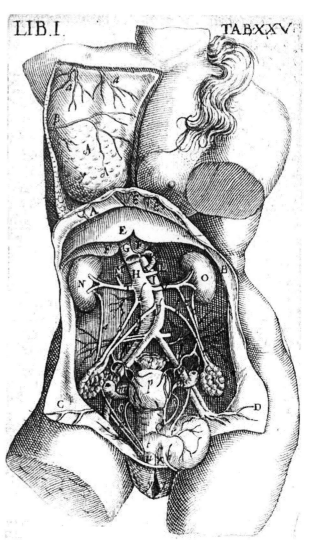

Figure 263. A picture from *Theatrum anatomicum novis figuris aeneis illustratum et in lucem emissum opera & sumptibus*. Theodori de Bry p.m. relicta viduae & filiorum Joannis Theodori & Joannis Israelis de Bry. Francofurti at [sic] Moenum, typis Matthaei Beckeri, 1605. (Courtesy of http://www.kumc.edu/dc/rti/human_body_1605_bauhin.html.)

major treatises: *De humani corporis anatomia* (Venice, 1516) and *Annotationes anatomicae* (Bologna, 1520). In addition to the ileocecal valve, he also discovered two ear ossicles (malleus and incus), the tarsal bones of the foot, the fornix of the brain, the cerebral ventricles, the infundibulum, and the trochlear nerve (Matsen, 1969; Miinster, 1933).

Antonio Maria Valsalva

The name of Antonio Maria Valsalva (1666–1723; Fig. 265), who was professor of anatomy at Bologna, is linked with the aortic sinus, but he also described several anatomical features of the ear in his outstanding book, *De Aure Humana* (1704; Fig. 266). These include the auditory tube, which he named the Eustachian tube, the parts of the ear (external, middle, and inner), glands, ligament and muscles of the external ear, and the great incisura of the pinna.

Valsalva's work was compiled in a treatise of two volumes entitled *Opera*, which was published posthumously in 1740. Valsalva was a student of Malpighi at Bologna; succeeding Malpighi, he became a teacher of Giovanni Battista Morgagni (1682–1771), who was the founder of pathological anatomy. Morgagni

Figure 264. Portrait of Alessandro Achillini. Woodcut. (Courtesy of Wellcome Library no. 392i.)

A. M. VALSALVA

Figure 265. Portrait of Antonio Maria Valsalva. (Courtesy of Wellcome Library no. 9339i.)

himself described the anal columns, the appendix testis, the foramen caecum of the tongue, and the fossa navicularis. He also wrote a biography of Valsalva, his former teacher.

Johann Georg Wirsung

The main pancreatic duct was described in 1642 by Johann Georg Wirsung (1600–1643), a German anatomist who was professor of anatomy at the University of Padua. Wirsung had studied medicine in Paris and Altdorf before moving to Padua where he received the doctor of medicine degree in 1630. He discovered the duct during dissection of a 30-year-old man who was hanged for committing a murder. Wirsung did not publish his finding but engraved a sketch of it on a copper plate from which copies were made and then distributed among well-known anatomists

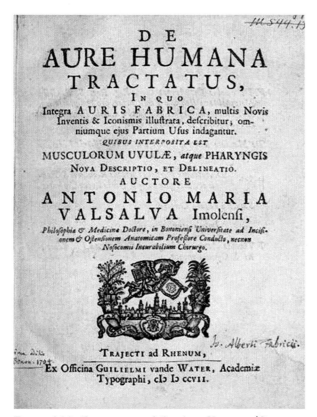

Figure 266. Front page of *De Aure Humana*. (Courtesy of Open Library. https://openlibrary.org/books/OL2669149M/De_aure_humana_tractatus.)

(Fig. 267). Moritz Hoffman (1622–1698) later claimed that he had discovered the duct in 1641 when he was a 19-year-old student at Padua. In 1643, Wirsung was murdered in front of his house, apparently resulting from the controversy surrounding the discovery of the pancreatic duct.

Heinrich Meibom

In 1666, Heinrich Meibom (1638–1700; Fig. 268) observed the tarsal glands (Meibomian glands) and cyst formation of the eyelids as a result of blockage and infection. However, these sebaceous glands were previously described by Giulio Casserius; almost seven decades later, they were rediscovered by Meibom. Heinrich Meibom was an astute physician and a very learned man. He wrote 57 medical treatises and also composed Latin poetry, which he published together with his grandfather. In 1664, Meibom was appointed Professor of Medicine at the University of Helmstedt; in addition, he became in 1678 Professor of history and poetic art in Helmstadt.

Gaspare Aselli

The bile duct and lymphatic vessels were discovered in 1622 by Gaspare Aselli (1581–1626; Fig. 269) of Milan. The first anatomical plates printed in colors

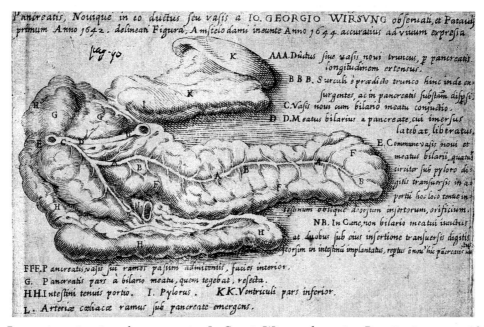

Figure 267. *Pancreatis, novique in eo ductus seu vasis a Io. Georgio Wirsung observati, et Patavii primum anno* 1642. *delineati figura, Amstelodami ineunte anno* 1644. *accuratius ad vivum expressa.* (Courtesy of Wellcome Library no. 43586i.)

Figure 268. Portrait of Heinrich Meibom. (Courtesy of Wellcome Library no. 6464i.)

were in his book *De lactibus sive lacteis venis* (Milan, 1627; Fig. 270).

Friedrich Gustav Jacob Henle

The discovery of the ducts, which collect and drain urine into the pelvis of the kidney was made by Lorenzo Bellini (1643–1704) of Pisa. Later, the convoluted tubules were discovered by Friedrich Gustav Jacob Henle (1809–1885; Fig. 271), one of the most famous anatomist and pathologist of the nineteenth century. Henle was born in Fürth in Bavaria and attended the universities in Bonn and Heidelberg. In 1832, he received the doctorate of medicine degree from the University of Bonn, and two years later he joined the well-known anatomist Johannes Müller in Berlin. Henle was appointed a prosector in the Department of

Figure 269. Portrait of Gaspare Aselli. (Courtesy of Wellcome Library no. 613i.)

Anatomy during which he became dedicated to scientific work. Over a period of six years, Henle published many original research papers and monographs in diverse areas, including cytology, epithelial tissue, the lymphatics, malignancy, mucus and pus formation, and development of hair (Talbott, 1961;Weyers, 2009).

In 1840, Jacob Henle was invited to become the first professor of anatomy, and later physiology, at the newly established University of Zurich. He was an inspiring lecturer and remained a prolific author. Henle's reputation soared with the publication in 1841 of his seminal work dealing with microscopic anatomy, "*Allgemine Anatomie*" (General Anatomy; Fig. 272). In Zurich, together with the clinician Karl von Pfeufer (1806–1869), he established an important medical journal, "*Zeitschrift für rationelle Medicin*" (Journal for Rational Medicine).

Henle accepted a second professorial position in 1844 at the University of Heidelberg and four years later he succeeded his colleague Friedrich Tiedemann (1781–1861) as the Director of the Institute of Anatomy (1848–1852). During this period, he published his

Figure 270 (*Top left*). Front page of Gaspare Aselli. *De Lactibus sive Lacteis Venis, Quarto Vasorum Mesaraicorum Genere Novo Invento*...Milan: Apud J. Baptistam Bidellium, 1627. http://www.indiana.edu/~liblilly/anatomia/viscera/aselli.html.

Figure 271 (*Bottom left*). Portrait of Friedrich Gustav Jacob Henle. (Courtesy of Wellcome Library no. 4139i.)

Figure 272. (*Top right*) Front page of Friedrich Gustav Jacob Henle *Allgemine Anatomie*, Lehre von den Mischungs – und Formbestandtheilen des menschlichen Körpers 1841. (Courtesy of Open Library. https://openlibrary.org/works/OL15705828W/Allgemeine_Anatomie.)

"*Handbuch der rationellen Pathologie*" (1846–1853-Handbook of Rational Pathology), which described diseases processes in relation to physiological functions. In 1852, Henle was appointed to the Chair of anatomy at the University of Göttingen where he remained until his death 33 years later.

Jacob Henle was the best known anatomist of his era and he had an international reputation because of his books and scientific accomplishments. Henle pioneered the use of the microscope (microscopic anatomy) in microscopic anatomy and pathology courses. He is considered to be one of the founders of modern histology and the cellular concept. Henle proposed a hypothesis for the spread of cancer, which led to Virchow's cellular pathology. He also formulated the concept of micro-organisms as the causative factor in infective diseases (Carter, 1985) with his paper on miasma and contagion, *Von den Miasmen and Kontagien*, published in 1840. Henle discovered the following: the convoluted tubules in the kidney, now known as Henle's loop, the muscular and endothelial layers of arterioles, and the inner root sheath of the hair follicle (Steffen, 2001). In recognition of his discoveries and scientific reputation, Jacob Henle was celebrated by the Prussian government, and honored with memberships from many European scientific academies, as well as honorary doctorates from the University of Breslau and Edinburgh University (Weyers, 2009).

Wilhelm His, Sr.

One of the greatest microscopist and embryologist of the nineteenth century was Wilhelm His, Sr. (1831–1904; Fig. 273) who the microtome, which led to major developments in histology and in experimental embryology. His's work was complementary to that of Jacob Henle 1809–1885), Rudolph Albert von Kölliker (1817–1905), and Rudolph Virchow (1821–1902).

The anatomist Wilhelm His, Sr. was born to an aristocratic family in Basel, Switzerland. He received his early education in Basel, and in 1849 he began medical studies at the University of Basel. After one semester, his went to the University of Bern, and from 1852 to 1853, he continued his studies at the University of Berlin and the University in Wurzburg. His, Sr. graduated in medicine in 1854 and received his doctorate degree in 1855 for studies on the cornea. In 1857, at 26 years of age, he succeeded the anatomist Georg Meissner (1829–1905) as professor of anatomy and physiology at the University of Basel. In 1872, he was invited by the Leipzig University to succeed anatomist Ernst Heinrich Weber (1795–1878) as Director

Figure 273. Wilhelm His, Sr. (Courtesy of Bethesda, MD: U.S. National Library of Medicine, National Institutes of Health, Health & Human Services.)

and Professor of anatomy.

His Sr. provided sound leadership in Basel and Leipzig where he was admired for his expertise and research. Known for his theoretical studies, he pioneered at the same time the development of experimental embryology. The microtome, which he invented in 1865, revolutionized microscopy – the study of the cellular elements of functions of tissues. At first, he studied whole chick embryos after processing (hardening) the embryos with chemicals, then cutting thin sections with the microtome for microscopic studies. His observations differed from the drawings made by the German zoologist Ernest Haeckel. As a result, he rejected Haeckel's proposal that "*ontogeny recapitulates phylogeny.*" His's collection of wax models showing the chick, fish, and human embryos at different stages of development has inspired the creation of similar work in other places. In 1886, His, Sr. discovered that each nerve fiber is attached to a nerve cell, which led to the neuron theory. In 1892, the Royal Swedish Academy of Sciences elected him as a member. Wilhelm His, Jr. (1863–1934), his son, also became a famous anatomist. His Jr. discovered the specialized bundle of fibers (Bundle of His) that transmit electrical impulse from the atrioventricular node to the ventricles.

Robert Remak

About the time Jacob Henle was working as a pro-sector of anatomy at the Friedrich Wilhelm University in Berlin, Robert Remak was a medical student at the university and attended the anatomy and physiology lectures of Johannes Müller. As a medical student he carried out research on invertebrate ganglion cells and nerve fibers, with the help of a compound microscope, and guidance of Christian Gottfried Ehrenberg (1795–1876) and Johannes Müller. Even before receiving his doctorate degree, Remak made important discoveries and had published two research papers. Remak was denied a university appointment because of his Jewish faith. According to Prussian laws, Jews were barred from teaching at the university. Remak did not receive from the university the recognition or reward he deserved despite his considerable scientific achievements (Schmiedebach, 1990; Lagunoff, 2002).

The anatomist and embryologist Robert Remak was born on July 26, 1815, in Poznan (Posen) (Fig. 274), which was returned in 1833 to the Prussian government by the Congress of Vienna. Remak pursued

Figure 274. A portrait of Robert Remak. (Courtesy of Bethesda, MD: U.S. National Library of Medicine, National Institutes of Health, Health & Human Services.)

medical studies at the Friedrich-Wilhelm-University in Berlin and graduated as a physician in 1838 at the age of 23 years. His doctoral thesis, "*Observationes anatomicae et microscopiicae de systematis nervosa structura*" contained the seminal observation that medullary nerve fibers are not hollow, but solid and flat. From ancient times, it was believed that nerve fibers were hollow for carrying air, fluids, or spirits. He also observed that the nerve fibers of the autonomic nervous system have a gray color because of the absence of a myelin sheath.

In 1838, Remak joined Müller's laboratory as an unpaid research assistant, and from 1843 to 1847 he was an assistant to Professor Johann Lukas Schönlein (1793–1864), who was professor of medicine at the Charité Hospital in Berlin. Remak continued microscopic studies in his home using earnings from private practice. In 1847, Schönlein and Alexander von Humboldt petitioned the Prussian King to grant Remak a teaching position. Only then was Remak offered a position at the university for which there was no salary. Later, promoted from his junior position to that of an Associate Professor, he became the first Jew to lecture at the university.

A gifted researcher, Remak made many important contributions to medicine. His legacy includes the following discoveries: that nerve fibers are not hollow; the unmyelinated nerve fibers (Remak's fibers) with connections to neuronal cell bodies; neurofibrils in the nerve cells; six distinct layers of cells in the cerebral cortex; that proliferation of cells (formation of new cells) occurs by a process of cell division; not four (as postulated by Karl von Baer), but three germ layers in the embryo, which he named as ectoderm, mesoderm, and endoderm; and a group of nerve cells (Remak's ganglion) which modulate the autonomic contraction of the heart (Lagunoff, 2002). Clinically, Remak wrote about galvanic electrotherapy (Fig. 275), lead poisoning, neuromuscular diseases, skin lesions, and malignant tumors. His book on the development of vertebrates, *Untersuchungen über die Entwicklung der Wirbelthiere*, published in 1850, is a classic work in developmental biology. Remak pioneered the use of electrotherapy for the treatment of patients with neuromuscular disorders, which he described in his book, *Über methodische Electrisierung gelähmter Muskeln* (1855).

In 1848, Remak devoted himself to clinical work, especially in electrotherapy (Galvanic) and neurology. Remak was a brilliant embryologist and neuroanatomist whose academic career was greatly hindered because of his Jewish faith. He experienced many hardships during his relatively short but highly

Figure 275. Front page of his book on galvanic electrotherapy. https://archive.org/stream/galvanothrapieo00remagoog#page/n9/mode/2up.

productive life. Robert Remak died in Bavaria of uncertain causes at the age of 50 years.

Wilhelm Fabry von Hilden

Wilhelm Fabry von Hilden (Fabricius Hildanus, 1560–1634; Fig. 276) was the first surgeon to operate on a patient for the removal of gallstones. He also discovered the use of the magnet for extracting metal splinters from the eye. Fabricius Hildanus is considered one of the most famous surgeons (father of German surgery) of his era. He wrote several treatises on military surgery and surgical management (Fig.

Figure 276. A portrait of Wilhelm Fabry von Hilden. (Courtesy of Wellcome Library no. 2818i.)

277). In his books, Fabricius Hildanus emphasized the practical importance of anatomy.

Regnier de Graaf

The Dutch physician and anatomist, Regnier de Graaf (1641–1673; Fig. 278), made many contributions to both anatomy and physiology, but he is best remembered for the follicles he described in the ovary. This finding was reported in his book *De Mulierum Organis Generatione Inservientibus*, published in 1672 (Figs. 279–283). It led to a bitter confrontation with Jan Swammerdam, Johann van Home, and Nicolaus Steno who accused de Graaf of plagiarism; apparently they made similar observations but had not published their work on the function of the ovary. Deeply grieved by the accusations, he wrote a rebuttal, and this incident might have contributed to his death at the early age of 32 years. In addition to experimental studies De Graaf carried out on the pancreatic juices, saliva, and bile (*De succi pancreatici natura et usu exercitatio anatomica-medica*, 1664), he also

Figure 277 (*Left*). *Observations chirvrgiqves*. (Courtesy of open library. https://openlibrary.org/works/OL15353229W/ Observations_chirvrgiqves.)

Figure 278 (*Right*). Portrait of Regnier de Graaf. (Courtesy of Museum Boijmans Van Beuningen. http://alma.boij-mans.nl/nl/object/BdH%2022806%20(PK)/.)

described the structure of the prostate gland and seminal vesicles in an earlier work (Fig. 284).

Thomas Wharton

The English physician Thomas Wharton (1616–1673; Fig. 285) described the duct of the submandibular salivary gland (submandibular duct) in his book, *Adenographia*, published in 1656 (Fig. 286). This duct had previously been described by Alessandro Achillini (1463–1512) in 1500. Wharton's name is also associated with the mucoid connective tissue surrounding the blood vessels in the umbilical cord.

Nicolaus Steno

The Danish physician and naturalist (Fig. 287)

Nicolaus Steno (Stensen) (1638–1686) was a highly accomplished anatomist who had a remarkable career that transcended the boundaries of natural sciences. He is more often remembered for his discovery of the duct of the parotid gland, but his other achievements are no less noteworthy. Steno attended the university in Copenhagen where he studied medicine, mathematics, and philosophy, which he continued in Leiden, Amsterdam, and Rostock. Following graduation, he received a hospital appointment, which he combined with anatomical studies and other scientific pursuits.

Steno explained the function of the ovaries and in a classic work *Elementorum myologiae specimen* (Florence, 1667) and he described the structure and true nature of muscle contraction (Bierbaum & Fuller, 1979; Scherz, 1964; Poulson & Snorrason, 1986; Bergmann & Wendler, 1987; Kardel et al., 1994). His

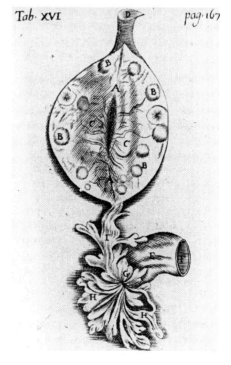

Figure 279 (*Top left*). Elegant frontispiece of R. De Graaf's *De Mulierum Organis Generationi Inservientibus*, 1672. (Courtesy of Neil John Maclean Health Sciences Library, University of Manitoba.) A woman is shown holding in her right hand a drawing of the female internal reproductive organs; her left hand is touching her breast. A rabbit lies on its back with its abdomen opened. Two of the cherubs are apparently examining the ovary of the rabbit using an optical instrument.

Figure 280 (*Top right*). Decorative title page of Regnier de Graaf's *Opera Omnia* (1678). A corpse with its abdomen opened is shown lying on a sarcophagus. A woman is elegantly pointing to it. Various medical instruments are depicted between the cadaver and the title of the book. On the floor are carcasses of animals, an opened book, and a dog. (Courtesy of Neil John Maclean Health Sciences Library, University of Manitoba.)

Figure 281 (*Right*). Section of an ovary (A) and uterine (Fallopian) tube (E). Note follicles, "*Ova diversae*" of varying sizes (B) in the ovary, as well as the suspensory ligament (D) of the ovary. (From Regnier de Graaf's *Opera Omnia*, 1678. Courtesy of the Neil John Maclean Health Sciences Library, University of Manitoba.)

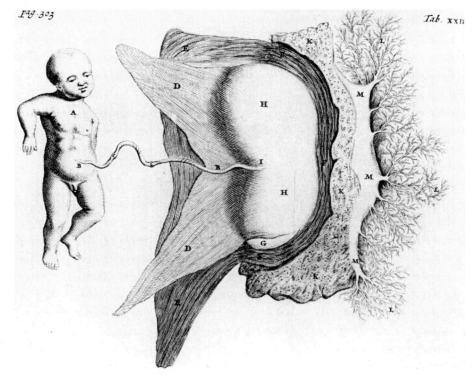

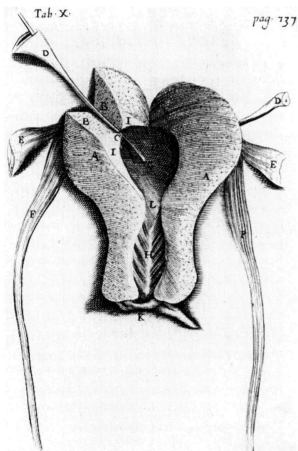

Figure 282 (*Above*). Stillborn infant attached to the placenta. The drawing shows the amnion (D), chorion (E), umbilical cord (B), false knots (C), and some structural features of the placenta. (From Regnier de Graaf s *De Mulierum Organis Generationi Inservientibus*, 1672. Courtesy of Neil John Maclean Health Sciences Library, University of Manitoba.)

Figure 283 (*Left*). Parts of the uterus, including its cavity. (From Regnier de Graaf's *Opera Omnia*, 1678. Courtesy of Neil John Maclean Health Sciences Library, University of Manitoba.) Note the round ligament of the uterus (F), the uterine (Fallopian) tube (D), and the cervical canal (H).

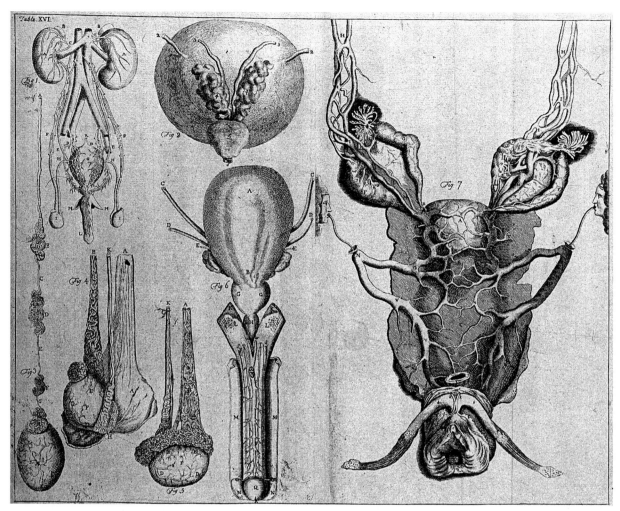

Figure 284. Several of these figures are after plates illustrating the works of Reinier de Graaf on the male and female generative organs, with engravings done after his own drawings. Figures 3, 4, 5, 6 are from De Graaf's *Tractatus de virorum organis generationi inservientibus*, Leiden and Rotterdam 1668, plates I (fig. 3); III (figs 1–2); VIII. Figure 7 is from De Graaf's *De Mulierum organis generationi inservientibus tractaus novus*, Leiden 1672, plate XIII. The bladder of figure 2 is from plate XXXI of William Cheselden's *Anatomy of the Body*, London 1740, with etchings by Gerard Vandergucht. Figure 1 is from plate XII, fig. 1, of Giovanni Maria Lancisi's edition of the sixteenth-century plates prepared fro the Italian anatomist Bartholomaeus Eustachius, first published in their entirety with Lancisi's commentary in the *Tabulae anatomicae clarissimi viri Bartholomaei Eustachii quas e tenebris tandem vidicatas*, Rome 1714. (From R. James, *A medicinal dictionary, including physic, surgery, anatomy, chymistry, and botany, in all their branches relative to medicine*, 3 vols, London 1743–1745, i, plate 16. Courtesy of Wellcome Library no. 37303i.)

theory of iso-volemic contraction of uniform muscle fibers is now of increasing significance in the field of biomechanics. Steno recognized that the heart is essentially a muscular organ. He also described muscles of the tongue, the esophagus, and ribs. His name is associated with the parotid duct, the incisive foramina of the palate, and varicose veins of the eye. Steno discovered the tarsal glands of the eyelids and found that the lacrimal glands produced tears. Moreover, he demonstrated that a ligature placed around the descending aorta resulted in paralysis of the lower limbs, but removal of the ligature restored function.

Curiously, Steno's prescient contribution to the anatomy of the brain is not widely known. He presented at the Académie Royale des Sciences a rousing lecture in which he described the brain as "*the most beautiful masterpiece of nature that is the principal organ of our mind.... Our mind thinks that nothing can set a limit to its knowledge, but when it withdraws to its own habitation it is unable to give a description of it, and no longer knows*

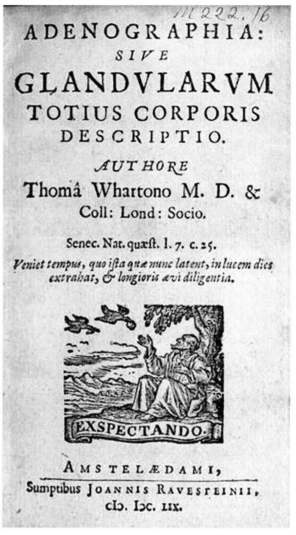

Figure 285 (*Top left*). A portrait of Thomas Wharton by W. H. Worthington. (Courtesy of Bethesda, MD: U.S. National Library of Medicine, National Institutes of Health, Health & Human Services.)

Figure 286 (*Top right*). Front page of *Adenographia* published in 1656. Courtesy of the open Library. https://openlibrary.org/works/OL15741757W/Adenographia.

Figure 287 (*Left*). Niels Steno (1638–1686). (From Wegner, R.N.: *Das Anatomenbildnis. Seine Entwicklungim Zusammenhang mit der anatomischen Abbildung.* Basel, Benno Schwalbe and Co., Verlag,1939. Courtesy of Institut fur Anatomie, Ernst-Moritz-Arndt-Universitat, Greifswald, Germany.)

itself." Steno dismissed the notion that the brain was connected with the mind through an egg-shaped gland. He felt that one should know more about the structure of the brain before speculating on how it functions. Steno stated in his dissertation that,

> It would be a great blessing to mankind if this most delicate part, and which is liable to so many dangerous diseases, were as well understood as the generality of anatomists and philosophers imagining it to be… as if they had been present at the formation of this surprising machine, and had been let into all the designs of the Great Architect. We need only view a dissection of the large mass, the brain, to have ground to bewail our ignorance.

Steno was critical of the drawings of the brain available during that era.

> The best figures of the brain are those of Willis; but even these contain a great number of important mistakes, and they want many things to perfect them. I have seen but three figures, which express in a wretched manner, the best observations that have ever been published on the brain. The principal reason why a great many anatomists have remained in their mistakes, and why they have gone no greater a length than the ancients in dissection, is because they believe that everything has been already taken notice of, and that there is nothing left for the moderns to do.

Regarding the ventricles of the brain, Steno remarked that

> the ancients were so far prepossessed about the ventricles as to take the anterior for the seat of common sense, the posterior for the seat of memory, that the judgment which they said was lodged in the middle, might more easily reflect on the ideas which came from either ventricle. Why should we believe them? Willis lodges common sense in the corpora striata, the imagination in the corpus callosum, and the memory in the cortical substance. How can he then be sure that these three operations are performed in the three bodies, which he pitches upon? Who is able to tell us whether the nervous fibers begin in the corpora striata, or if they pass through the corpus callosum all the way to the cortical substance? We know so little of the true structure of the corpus callosum that a man of tolerable genius may say about it, whatever he pleases.

Moreover, Steno did not accept Descartes's concept of the pineal gland as the link between the body and soul. He believed that "*the supposed connection of this gland (pineal) with the brain by means of arteries is likewise groundless: or the whole basis of the gland adheres to the brain, or rather the substance of the gland is continuous with that of the brain, though the contrary be affirmed by Descartes.*"

Steno made significant contributions to geology. He studied shark teeth, including those that were fossilized and embedded in terrestrial found far from the sea. Pliny the Elder (A.D. 23–79) had suggested felt that these *glossopteris* ("*tongue stones*") fell from the sky during lunar eclipses. Later, These curious objects were believed to have magical and healing properties. Steno had examined fossilized shark head and concluded that these curious stones were merely layers of fossilized material. This observation led him to formulate the principles and science of stratigraphy, which are used by geologists and paleontologist geologists.

In 1667, Steno, a Lutheran, converted to Catholicism and became a leading opponent of the Reformation. Later, he was ordained as a priest in Florence. Despite his scientific accomplishments, Steno's career suffered in Copenhagen because of his conversion to Catholicism. He left for Florence where he was appointed to the *Santa Maria Nuevo* hospital. Following an invitation from the Danish king, he returned as *anatomicus regius* to Denmark where he continued his diverse scientific studies and remained fervent in his religious convictions…. He wrote that, "*This is the true purpose of anatomy: To lead the audience by the wonderful artwork of the human body to the dignity of the soul and by the admirable structure of both to the knowledge and love of God.*" In 1677, he was appointed apostolic vicar of northern missions by Pope Innocent XI and was consecrated titular bishop of Teutopolis in Asia Minor. Now appointed Apostolic Legate for Northern Germany and Scandinavia he departed from Rome to serve the Catholic congregations in Germany, Denmark, and Norway. Steno died at a relatively young age after a serious illness. The Grand Duke Cosimo of Tuscany had requested that his body should be removed and interred in the Medici tombs in the Basilica. On October 23, 1988, Steno was beatified by Pope John Paul II and with the status and title of Blessed. Today, Steno's name is remembered by the Steno Museum in Arhus, Denmark, craters on Mars and the Moon, the Steno Diabetes Center in Denmark, and the Istituto Niels Stensen in Florence, Italy (Tubbs et al., 2011b).

Nathaniel Highmore

The English physician Nathaniel Highmore (1613–1685) (Fig. 288) described the *tunica albuginea* and the *mediastinum testis*, and his name is usually associated with the maxillary sinus. However, the earliest description and drawings of the maxillary sinus were given in 1489 by Leonardo Da Vinci. Highmore was acquainted with William Harvey and supported his theory of the circulation of the blood. Highmore was well known in England and abroad on account of his anatomical studies, especially following the publication in 1651 of his first major work, entitled *Corporis humani di squisitio anatomica...*(Wendler, 1986).

Figure 288. Nathaniel Highmore (1613–1685) at the age of 63 years. Engraving by Abraham Blooteling. (From Wegner, R.N.: *Das Anatomenbildnis. Seine Entwicklung im Zusammenhang mit der anatomischen Abbildung.* Basel, Benno Schwalbe and Co., Verlag, 1939. Courtesy of Institut fur Anatonie, Ernst-Moritz-Arndt-Universitat, Greifswald, Germany.) A famous surgeon and anatomist, Highmore was the author of two major anatomical works: *The history of generation...*(1651), and *Corporis humani disquisitio anatomica...*(1651).

The discovery of the small glands in the walls of the duodenum is attributed to the Swiss physician Johann Wepfer (1620–1695). These glands are eponymously associated with the name of his son-in-law Johann Conrad Brunner (1653–1727), professor of anatomy and rector at the University of Heidelberg, as well as in Strasburg, who also described the glands. The English surgeon and anatomist William Cowper (1666–1709) described the bulbourethral glands in 1699. Although the bulbourethral glands are known as Cowper's glands, they were discovered five years earlier by Jean Mery (1645–1722). The sublingual ducts, which open into the submandibular duct, were described by Caspar Bartholin (1655–1738) *Secundus.* He was professor of medicine, anatomy, and physics in Copenhagen; his name is more associated with the greater vestibular glands, which he also observed.

Francois Poupart

The inguinal ligament was first described by the French surgeon and naturalist Francois Poupart (1661–1709). He received his doctorate degree in Reims and worked as a surgeon at the Hôtel-Dieu. As a distinguished surgeon, he was elected a member of the Académie des Sciences for his work and also for studies in the field of natural sciences, especially entomology. Today, he is remembered for the inguinal ligament, which he recognized in 1695, and its role in inguinal hernia.

Microscopic Observations

Using injected material, Frederik Ruysch (1638–1731) made important contributions to the anatomy of the coronary blood vessels (1701). In 1665, he also described the valves of the lymphatic vessels. The lymphatics were largely discovered during the seventeenth century, and these have been mentioned elsewhere in this book [see Gaspare Aselli (1581–1626); Francis Glisson (1597–1677); Jan de Wale, Jean Pecquet (1622–1674); Olof Rudbeck (1630–1702); and Thomas Bartholin (1616–1680)].

The importance of practical anatomy for understanding the structure of the human body became increasingly obvious. Slowly, additional anatomy schools, similar to the *Theatrum anatomicum*, with dissecting rooms, museums, and impressive lecture theaters, were established in other European cities. These included Bologna (1637), Copenhagen (1643), Altdorf (1650), Uppsala (1662), Paris (1691), and Amsterdam (1691).

In the area of microscopic anatomy the landmark observations made by Marcello Malpighi, Antoni van Leeuwenhoek, Theodor Schwann, Robert Hooke, and Jan Swammerdam have already been mentioned (Ford, 1991). These fundamental findings paved the way for the great German pathologist Rudolf Virchow (1821–1902) to formulate his theory of cellular pathology. The scientific foundations of microscopic anatomy became more established at the same time.

Antonio Pacchioni

The Italian anatomist Antonio Pacchioni (1665–1726) was born in Regio Emilia. After completing his medical studies he moved to Rome and became a pupil of Marcello Malpighi. He was interested in the nature and functions of the meninges of the brain, which had been observed by earlier anatomists (Fig. 289). The brain is covered externally by the dura mater and internally by the pia mater. Lying between the dura mater and pia mater is the arachnoid mater. Pacchioni discovered that the arachnoid layer protruded through the dura mater into the venous sinuses at various sites of the brain. He believed that these arachnoid (protrusions) villi secreted cerebrospinal fluid (Olry, 1999). For this reason, they were originally called Pacchioni glands but are now known as arachnoid granulations. Cerebrospinal fluid flows from the ventricles into the bloodstream through the arachnoid granulations.

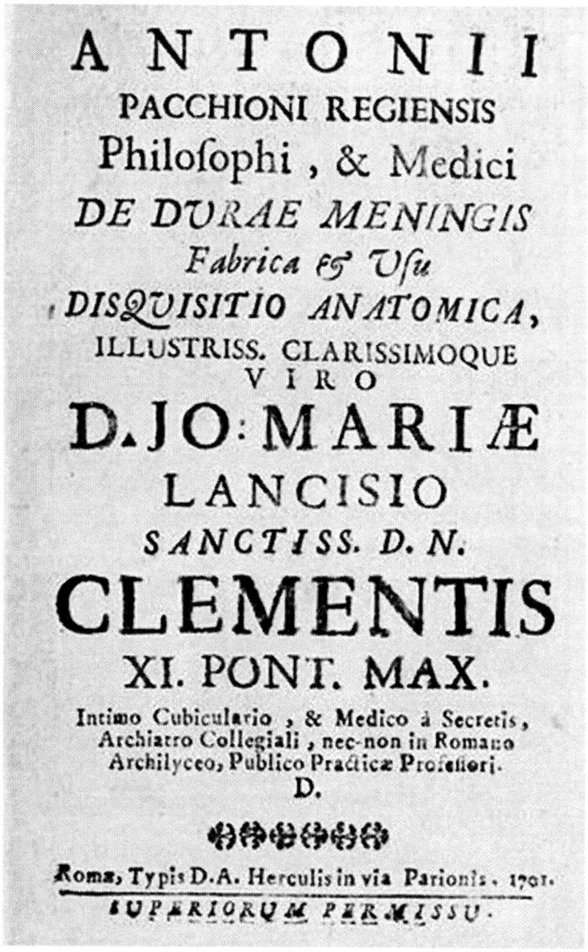
ANTONII
PACCHIONI REGIENSIS
Philoſophi, & Medici
DE DURAE MENINGIS
Fabrica & Uſu
DISQUISITIO ANATOMICA,
ILLUSTRISS. CLARISSIMOQUE
VIRO
D. JO: MARIÆ
LANCISIO
SANCTISS. D. N.
CLEMENTIS
XI. PONT. MAX.
Intimo Cubiculario, & Medico à Secretis,
Archiatro Collegiali, nec-non in Romano
Archilyceo, Publico Practicæ Profeſſori.
D.
Romæ, Typis D.A. Herculis in via Parionis. 1701.
SUPERIORUM PERMISSU.

Figure 289. Front page of Antonio Pacchioni's work on meninges of the brain, published in 1701 by Typis D.A. Herculis in via Parionis in Roma. (Courtesy of the open library. https://openlibrary.org/search?q=Antonio+Pacchioni.)

Samuel Collins

From earliest times, the brain of man was compared to that of different animals. Late seventeenth century, the physician and anatomist Samuel Collins (1618–1710; Fig. 290) published the earliest comprehensive work in comparative neuroanatomy. His two-volume folio of comparative anatomy contained the largest collection of "*handsome engravings*" depicting the brain of human and animals (Kruger, 2004, for details). Collins was an eminent anatomist and his comparative anatomical studies in two volumes; *A Systeme of Anatomy, Treating of the Body of Man, Beasts, Birds, Fish, Insects, and Plants* (London, 1685) was highly regarded by his contemporaries, and referred to by Hermann Boerhaave (1668–1738) and Albrecht von Haller (1708–1777). Largely, based on original research and dissection, Collins's work paved the way for future studies in neuroanatomy.

Samuel Collins was educated at Trinity College, Cambridge where he graduated with the B.A. (1638) and M.A. (1642) degrees. He visited other universities in Europe and received the M.D. degree from Padua University in 1654. Following his return to England, Collins was incorporated at Oxford (1652) and Cambridge (1673) universities. He was a Fellow of the College of Physicians in London and physician-in-ordinary to Charles II. In 1684, Collins was appointed Reader of anatomy and in 1694 Lumleian Lecturer to the College of Physicians. A year later, he became President of the college. He died in 1710 at the age of 93 years. According to his peers, Collins "*was an accomplished anatomist and stood foremost among his contemporaries, whether at home or abroad, in his knowledge of comparative anatomy*" (William Munk, website d).

Figure 290. A portrait of Samuel Collins. Courtesy of Bethesda, MD: U.S. National Library of Medicine, National Institutes of Health, Health & Human Services.

Raymond de Vieussens

The French physician and neuroanatomist Raymond de Vieussens (1641–1715; Fig. 291) made significant contributions to human anatomy. Vieussens studied medicine at the University of Montpellier and in 1670 obtained the M.D. degree. He was appointed physician at Saint Eloys Hospital in Montpellier where he later became the chief physician. Vieussens is remembered mostly for his seminal discoveries with respect to the cardiovascular system and the central nervous system (Fig. 292). In 1667, Willis had published his *Pathologicae Cerebri, et Nervosi Generis Specimen*. Vieussens was greatly influenced by this work, which led him to his own neuroanatomical studies. He dissected more than 500 cadavers at Saint Eloys.

Vieussens demonstrated that the spinal cord was functionally independent of the brain: he studied the

relationship between the optic nerve and the lateral geniculate nucleus; and he also described the dentate nuclei, the pyramids and the olivary bodies. Vieussens described many other anatomical structures, including the cerebellum, the white matter of the brain, the ear, and the brain's centrum semiovale, sometimes referred to as "*Vieussens' centrum*." This was later described in greater detail by Félix Vicqd'Azyr (1746–1794). Many of the anatomical structures first described by Vieussens and eponymously named after him have been replaced by a more recent terminology. These include the superior medullary velum, the fluid filled space of the septum pellucidum, subclavian loop, celiac ganglia, and limbus of the fossa ovalis. He also provided an early description of the tiny openings in the veins of the right atrium of the heart that are known as the foramina venarum minimarum, and earlier as *Thebesian foramina*, named after Adam Christian Thebesius (1686–1732) (Loukas et al., 2008a).

In his final years, Vieussens began to study the

Figure 291. A portrait of Raymond de Vieussens. (Courtesy of Bethesda, MD: U.S. National Library of Medicine, National Institutes of Health, Health & Human Services.)

Figure 292. The front page of Vieussens work on nerves *"Novum Vasorum Corporis Humani Systema."* (Courtesy of the open library. https://openlibrary.org/search?q=Vieussens.)

was elected a Member of the Academy of Sciences, Paris and he also became a Fellow of the Royal Society of London (Loukas et al., 2007b).

Werner Rolfinck

Werner Rolfinck (1599–1673; Fig. 293) was a physician, anatomist and renowned botanist.... He was a medical student in Leiden, Oxford, Paris, and finally in Padua where he received in 1625 his doctorate degree. The University of Padua was then the foremost institution known for its human dissection and anatomical discoveries.

Rolfinck returned to Germany in 1629 as professor of surgery and anatomy at the University of Jena. He influenced the establishment of anatomical theaters in other German cities during the early seventeenth century. Invariably, these were built similar to what he had experienced seen Padua. Rolfinck carried out public dissections on executed criminals in the anatomical theater. Of interest, serious criminal offences decreased in the city because of the threat of being

Figure 293. A portrait of Werner Rolfinck. (Courtesy of Wellcome Library no. 8214i.)

movement of the heart and pathological changes present in heart diseases. Using an improved fixation process he had developed, he was the first person to document the changes in patients with mitral valve stenosis and aortic insufficiency. Vieussens provided a detailed description of the pericardium, the myocardium, and the coronary vessels in his book, *Traité Nouveau de la Structure et de la Cause du Mouvement Naturel du Coeur*, published in 1715.

Vieussens remained a physician throughout his life. Not a member of the medical faculty, he carried out his anatomical studies far removed from an academic environment. Nonetheless, Vieussens was well known and highly regarded as a scientist. Vieussens dissected after execution.

Other Anatomical Discoveries

The internal structure of bone was not known until Clapton Havers (c. 1650–1702), using a microscope, observed blood vessels and regularly arranged channels in the long bones of the extremities. These channels carry blood vessels, which provide nutrition to the bone and are now known as the Haversian canals.

Knowledge of the normal anatomy of the human body paved the way for the foundation of pathological anatomy. In 1679, Theophile Bonet (1620–1689) published a treatise entitled *Sepulchretum: sive anatomia practica ex cadaveribus morbo denalis*, Bonet described the postmortem changes he observed from the 300 autopsies he carried out. It is noteworthy that publication of Bonet's work preceded by three years the birth of Giovanni Battista Morgagni, who is considered to be the father of modern pathology.

Alexis Littre (1654–1726) was a French physician

and anatomist who studied medicine in Montpellier and Paris, where he received his doctorate degree in 1691. At the University in Paris, he taught anatomy and published many scientific papers describing original anatomical and clinical findings, such as the mucous glands in the urethra and an intestinal herniation involving Meckel's diverticulum. Littre's landmark treatise, *Diverses observations anatomique* (1710), described lumbar colostomy for the management of obstruction of the large intestine, but he is more known for the hernia he first described (Littre's hernia).

Henri Francois Le Dran

The French surgeon and anatomist Henri Francois Le Dran (1685–1770; Fig. 294) wrote a practical handbook for surgeons (*Abrégé Economique, De L'Anatomie Du Corps Humain...*, Paris, 1768) based on his military experience. This manual of surgical anatomy had a useful series of 16 engraved anatomical plates. Le Dran lectured at the Royal Academy of medicine in Paris and was also the chief surgeon at the Hospital de la Charité. One of his students was Albrecht von Haller, the famous Swiss anatomist and naturalist. Le Dran was the first person to recognize that cancer was a local lesion, which progressed in stages through the lymphatics to lymph nodes in parts of the body. For this reason, he recommended to remove the tumor and the axillary lymph nodes. In a treatise on gunshot wounds, he introduced the medical term "*shock*" (choquer) resulting from a sudden severe impact or jolt (Sethi et al., 2003). In "*cutting for stones*" (lithotomy), Le Dran carried out a double bilateral procedure by dividing both segments of the prostate gland with a special scalpel. This innovative approach was reported as an anatomical discovery.

Philippe-Frédéric Blandin

The French surgeon Philippe-Frédéric Blandin (1798–1849; Fig. 295) described the anterior lingual glands, mixed glands of the tongue, which are known as "*Blandin's glands.*" He also pioneered autoplasty, grafting of skin taken from the same person. Combining his increasing knowledge of anatomy and surgery, Blandin produced an outstanding atlas in surgical anatomy. *His Traite d'anatomie topographische...* published in 1834, was a beautiful work with 20 hand-colored lithographed plates and accompanying explanatory notes. In 1826, an earlier edition of this atlas was published with 12 plates.

Blandin received his medical and surgical training

H.ᴿ F.ᵒⁱˢ LE DRAN
(Chirurgien),

Chirurgien en chef de l'Hopital de la Charité, membre de l'Académie

Figure 294. A portrait of Henri Francois Le Dran. (Courtesy of Bethesda, MD: U.S. National Library of Medicine, National Institutes of Health, Health & Human Services.)

in Paris where he pursued his professional career. From 1821, he was an anatomical assistant at the medical school, and in 1824 he was appointed as a prosector of anatomy. Two years later he received his "*aggregation*" in surgery and became a professor of surgery in the medical faculty from 1841 to 1849. In recognition of his work, Blandin was elected a member of the French Académie Nationale de Médecine.

Joseph Lieutaud

In eighteenth century France, anatomy as a scientific discipline was pioneered by the physician Joseph Lieutaud (1703–1780). He received his medicine degree from the University of Aix in 1725 at the age of 22 years. He then went to Montpellier in order to study botany. Lieutaud returned to Aix in 1731 and began to practice medicine in a different manner, free of theories and speculation. He believed in observing the patients at the bedside, recording the outcome of treatment, and learning from postmortem findings. Lieutaud's approach to medical treatment was based on facts and not books or opinions. Indeed, he was an early advocate of what is known today as "*evidence-based*" medicine.

In 1731, Lieutaud was offered the professorship of anatomy, physiology, and botany at the University of Aix where he continued his anatomical studies. In order to learn more about the human body, Lieutaud dissected more than 1200 cadavers at the Versailles Royal infirmary. He also examined autopsy specimens correlating the postmortem findings with the patient's symptoms before death. Lieutaud communicated many of his cases to the Academia des Sciences. These included diseases of the spleen, heart, stomach, gall bladder, hydrocephalus, and varied tumors, cysts, and polyps. During the smallpox epidemic of 1774, Lieutaud carried out a successful inoculation program among his patients. As a result, vaccination against smallpox became more accepted in France. Lieutaud reputation as a skilled doctor increased. A year later, he received an appointment as physician to the royal children and later to the Louis XV and following his death to his successor.

Lieutaud published works presented his clinical and practical knowledge with the existing literature. Both editions of his "*Essais anatomiques*" (1742) were widely read. Lieutaud's "*Précis de la médecine*" (1759) was a medical textbook, which presented a classification of diseases in the general population with treatment for these conditions. His *Précis de la matiere medicale* (1766) was an index in French and Latin. This

BLANDIN.

Figure 295. A portrait of Philippe-Frédéric Blandin. (Courtesy of Bethesda, MD: U.S. National Library of Medicine, National Institutes of Health, Health & Human Services.)

materia medica included compounding of drugs and indications for various illnesses. In recognition of his work, Lieutaud received an honorary doctorate from the University of Paris in 1752. He was also admitted to the Academie des Science and in 1751 became a Fellow of the Royal Society in England. Joseph Lieutaud received a royally sanctioned funeral and a street was named after him.

Giovanni Domenico Santorini

The Italian anatomist Giovanni Domenico Santorini (1681–1737; Fig. 296) was born in Venice where he died early in his career. He studied medicine at several Italian universities, including Pisa, obtaining the M.D. degree. As demonstrator of anatomy in Venice, Santorini was an assiduous and meticulous dissector. He was a good teacher and widely respected for his encyclopedic knowledge. His name is associated with several

Figure 296. A portrait of Giovanni Domenico Santorini. (Courtesy of Wellcome Library no. 8368i.)

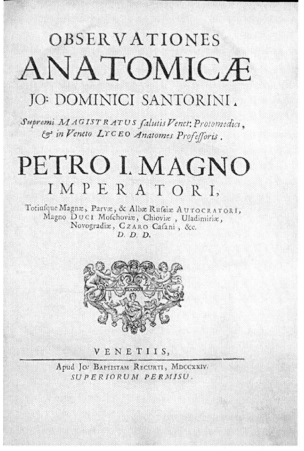

Figure 297. Front page of *Observationes anatomicae.* (Courtesy of The Manhattan Rare Book Company. http://www.manhattanrarebooks-medicine.com/santorini.htm.)

of the structures he apparently discovered. These include the accessory pancreatic duct (Santorini canal), the superior nasal concha (Santorini concha), a plexus of veins on the ventral and lateral surfaces of the prostate (Santorini plexus), the corniculate cartilage of the larynx, and a vein which passes through the parietal foramen, connecting the superior sagittal sinus with the scalp. Regarding claims of priority for discoveries,

Stern (1986) showed that the accessory pancreatic duct was already mentioned by earlier anatomists. Santorini recorded in an important work, *Observationes anatomicae* (1724), the findings he made from personal dissections (Fig. 297).

EIGHTEENTH AND EARLY NINETEENTH CENTURIES

By the beginning of the eighteenth century, most of the human body had already been mapped out, except for minute structures. The improvement and increasing use of the microscope opened new vistas and permitted the anatomists to probe deeper into the nature of living tissues. Moreover, impressive textbooks in anatomy and elegant atlases with engraved plates were published by the prominent anatomists of the day (Persaud, 1997; Jolin, 2013).

Anatomy Museums

The teaching of anatomy became more widespread with increasing emphasis on human dissection and the study of museum specimens. The establishment of magnificent museums with exquisite collections of prosected *in situ* specimens, diseased organs preserved in alcohol on display in glass cylinders (Cooke, 1984), and wax models (Schnalke, 1995), as well as other animal species for comparative studies, became fashionable (Brookes, 1828; Edwards & Edwards, 1959; Lanza et al., 1979; Bonuzzi & Ruggeri, 1988).

In some cases, the handsomely constructed museum was the focal point and a popular attraction for scholars, students, dignitaries, and the general public (Edwards & Edwards, 1959; Thomson, 1942; Cope, 1966; Brock, 1983). Anatomy theaters, private anatomy schools, and anatomical institutes were founded in Edinburgh (1697), Dublin (1711), Berlin (1713), St. Petersburg (1718), Ingolstadt (1723), Würzburg (1726), Paris (1745), London (1748), Frankfurt (1768), Vienna (1774), Dorpat (1803), Bonn (1824), and many other centers in Europe. In North America, "*anatomical schools*" were also established in many cities: Philadelphia (1750), New York (1763), Boston (1782), Dartmouth, New Hampshire (1798), Baltimore (1807), Montreal (1822), Augusta, Georgia (1828), New Orleans (1834), and Toronto (1843).

Felice Fontana

Felice Fontana (1730–1805) was a remarkable man whose scientific interests followed many paths, including the natural sciences, anatomy, toxicology, and pathology. He was born in Pomarolo, near Rovento and studied anatomy and physiology at the University of Padua. In 1755, Fontana went to Bologna where he collaborated with Marc Aurelio in investigating the irritability and sensitivity of the body with Caldani (1725–1813). He then moved to the University of Pisa where he was appointed as professor of logic (1775) and a year later professor of physics. The Grand Duke of Tuscany, Peter Leopold, appointed him court physician and with the task of establishing a museum of natural history and physics. In 1775, the museum of natural history was opened with its varied collections of scientific instruments, natural objects, and relics of Galileo. A gifted apprentice of Felice Fontana was the great Clemente Susini who also lived and worked in Florence (Ballestriero, 2010).

Clemente Susini

Clemente Susini (1754–1814) was born in Florence where he was trained at the Royal Gallery as a sculptor. In 1773, he began his training at the famous wax modeling workshop "*La Specola*" in Florence, the *Museo della Specola* (Fig. 298). Working under the guidance of Felice Fontana, he crafted a large number of remarkably accurate anatomical wax models from dissected specimens. In 1782, Susini became the chief modeler, and by 1799, he was teaching drawing and sculpturing to students at the Academy of Art.

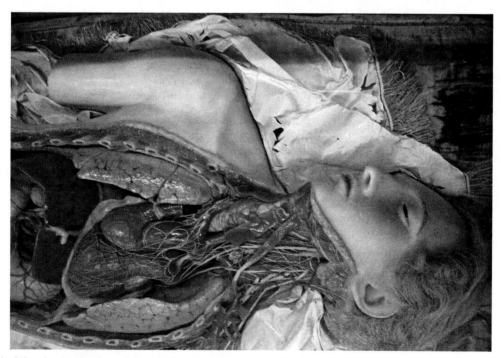

Figure 298. Wax Anatomical model made by Clemente Susini. Photograph Attribution: I, Lucarelli. Museo di Storia Naturale di Firenze, Zoologia "La Specola," Florence, Italy. Wax anatomical models.

Susini's elegant anatomical wax models were anatomically accurate and widely used for anatomical studies in Italy and other countries (Schnalke, 1995; Riva et al., 2010). During a visit to the museum, the Holy Roman Emperor, Joseph II, was so impressed that he ordered a set of the models for the benefit of medical students. Over the years, Susini and his co-workers produced more than 2000 anatomical models, which are now highly treasured. The finest of Susini's work, done later in life and which had reached a stage of perfection, is now housed in the Piazza Arsenale of Cagliari, University of Cagliari, in Sardinia.

Johann Nathanael Lieberkühn

Johann Nathanael Lieberkühn (1711–1756; Fig. 299), the Berlin physician and anatomist, was a pupil of Albrecht von Haller. His name is eponymously

Figure 299. Personification of virtue, here with Rod of Asclepius, encompasses the medalion with the image of Lieberkühn. In the foreground medical equipment and books (left) and illustrations of great doctors (Hippocrates, Galen). (Courtesy of http://www.sammlungen. hu-berlin.de/dokumente/6958/.)

linked with the tiny crypts or glands in the small intestine, but these were first described by Marcello Malpighi in 1688, and later by Brunner in 1715. Lieberkühn's dissertation (*Dissertationes quatuor. De valvula coli et usu processus vermicularis. De fabrica et actione villorum intestinorum tenuium hominis*) with his observations was published in Leiden in 1745. Caspar Friedrich Wolff (1733–1794) studied early embryonic development in the chick and formulated his theory of development from "*germ layers*" (epigenesis). He also described the mesonephros and the mesonephric duct.

Abraham Vater

The ampulla in the duodenum that is associated with both the hepatic and pancreatic ducts was described by Abraham Vater (1684–1751; Fig. 300) in 1720. The discovery of pressure receptors in the dermis and hypodermis, now called Pacinian corpuscles, is also attributed to him. Vater was born in Wittenberg, Germany, the son of a prominent physician and professor of medicine at the university. In 1706, he graduated from the University of Leipzig with the doctor of philosophy degree, and in 1710 from the University of Leipzig with the doctor of medicine degree. Vater settled in Wittenberg where he was appointed professor of anatomy and botany in the medical faculty of the University in Wittenberg, Germany; following the death of his father, he succeeded him as professor of pathology and therapeutics.

Vater visited anatomical institutes in several European countries and was particularly inspired by the work of the Dutch anatomist Fredrik Ruysch (1638–1731). His museum in Amsterdam with its unique colorful anatomical displays and special prosections was well known. Ruysch was interested in the preservation of anatomical specimens and he also developed special injection techniques, using red wax, in order to demonstrate the blood vessels. Returning to Wittenberg, Vater established a museum for training of medical students and surgeons (Wackwitz, 1985; Persaud, 1997).

James Douglas

The Scottish surgeon and anatomist James Douglas (1675–1742) described the arcuate line in the rectus sheath and the rectouterine peritoneal pouch. He was not only a respected anatomist but also an obstetrician (man-midwife) who became personal physician to the Royal family, including Queen Caroline. James Douglas was born near Edinburgh, one of 12 children, including the well-known lithotomist John Douglas

(?–1743). James Douglas studied at Edinburgh University where he received the M.A. degree in 1694. It is not known where Douglas received his medical training, but he was granted an M.D. degree from Reims University in 1699. A year later Douglas settled in London where he established himself as a skillful obstetrician. His medical practice flourished and his patients included aristocrats and members of the royal family.

Douglas carried out anatomical studies because he recognized the importance of anatomy for clinical work. He dissected diligently and prepared anatomical specimens for the anatomy classes he gave in his home. The Swiss anatomist Albrecht von Haller, who was professor of anatomy at the University of Göttingen, saw the anatomical specimens during his visit to London and he was impressed. William Hunter (1718–1783) lived in the home of Douglas and his family when he arrived in London in 1741. Douglas mentored William and both men became close friends. Douglas died a year later, but William was able to continue his anatomical studies. In 1768, William, with the assistance of his brother John, established the Great Windmill Street School of London, which became one of the leading private schools for anatomical and surgical training.

Douglas published a book on comparative myology entitled *"Descriptio Comparativa Musculorum Corporis Humani et Quadrupedis"* in London in 1707. A more expansive work, *Osteographia*, remains in manuscript form, as are many interesting clinical cases he had observed. Douglas's lasting contribution to anatomy is his description of a peritoneal deflection, forming a space of clinical importance, which is now eponymously known as the *"pouch of Douglas"* (Baskett, 1996): *"Where the peritonaeum leaves the foreside of the rectum, it makes an angle, and changes its course upwards and forwards over the bladder; and a little above this angle, there is a remarkable transverse stricture or semioval fold of the peritonaeum, which I have constantly observed for many years past, especially in women."*

Other Anatomists and Anatomical Discoveries

The adductor canal of the thigh was described by the surgeon-anatomist John Hunter (1728–1793); the maxillary antrum by the English anatomist Nathaniel Highmore (1613–1685); the long thoracic nerve to the serratus anterior muscle by Sir Charles Bell (1774–1842); the cremasteric fascia and the suspensory ligaments of the breast by the English surgeon, Sir Astley Cooper (1768–1841); the reflected inguinal ligament, perineal fascia, and the deep layer of the superficial

Figure 300. A portrait of Abraham Vater. (Courtesy of Wellcome Library no. 9362i.)

abdominal fascia by the Irish surgeon Abraham Colles (1773–1843).

Other descriptions of anatomical structures during this period include the cribriform fascia and the inguinal triangle by the German surgeon Franz Kaspar Hesselbach (1759–1816); the transverse rectal folds by the Irish surgeon John Houston (1802–1845); the lumbar triangle by the French surgeon Jean Louis Petit (1674–1750); the corniculate cartilage of the larynx and the accessory pancreatic duct by the Italian anatomist Giovanni Santorini (1681–1737).

The many classical studies carried out by Johann Gottfried Zinn (1727–1759) on the eye and orbit should also be mentioned. He was professor of medicine at the University of Gottingen and his treatise on the eye was published in 1755. Zinn's name is associated with the central artery of the retina, the annulus posterior to the eyeball for attachment of muscles, and the membrane at the edge of the lens.

The canal at the corneoscleral junction of the eye that is commonly associated with the name of Friedrich Schlemm (1795–1858), professor of anatomy in Berlin, was, in fact described a year earlier by the anatomist Ernest Lauth in 1829. Schlemm had also reported discovery of the corneal nerves. He began his medical studies in Braunschweig, which he completed at the University of Berlin in 1821. Pursuing a career as surgeon-anatomist, Schlemm became professor of anatomy at the University of Berlin in 1833. His courses in practical anatomy and operative surgery, which he carried out on cadavers, were innovative for that period and attracted students and surgeons.

William Horner (1793–1853), who was professor of anatomy at the University of Pennsylvania, described the orbicularis muscle in 1824, but the French anatomist, Joseph Du Verney, had already mentioned the muscle in 1749. Horner was the author of the first American treatise on pathological anatomy, published in Philadelphia in 1829.

The Italian anatomist, Domenico Cotugno (1736–1822), was the first person to provide a detailed description of the cerebrospinal fluid. However, he is remembered today for his outstanding studies on the theory of hearing and of the labyrinthine system, which he described.

The Heidelberg anatomist and physiologist Friedrich Tiedemann (1781–1861) published an atlas of the arteries in the body and several important research monographs. One reported in 1837 that the brains of African blacks and European whites were anatomically similar. Tiedemann is remembered for his research on comparative anatomy, blood vessels, and the brain. In addition, he carried out experimental studies on digestion, and he also wrote a book on the history of tobacco and addictive substances.

Recognition of the manubriosternal junction as an important anatomical landmark (Angle of Louis) is attributed to the French surgeon Antoine Louis (1723–1792). The Venetian anatomist Antonio Scarpa (1747–1832) described the deep layer of the superficial fascia of the lower abdomen, the femoral triangle, and the nasopalatine nerve.

A vestigial embryological remnant of the mesonephric duct, found either in the broad ligament of the uterus or in the wall of the vagina, may become infected and is of clinical interest (Gartner's cyst). It was first described (*ductus epoophori longitudinalis*) in 1822 by the Danish anatomist and surgeon Hermann Gartner (1785–1827). Gartner was born on the then Danish possession of St. Thomas in the West Indies. After medical studies in Copenhagen, he pursued postgraduate training in London and Edinburgh. Gartner became an army surgeon in 1824, but he continued his studies until his untimely death three years later.

Martin Heinrich Rathke

Martin Heinrich Rathke (1797–1860; Fig. 301), one of the founders of modern embryology, was born in Danzig (Gdansk). He studied natural sciences and medicine in Göttingen and Berlin. After graduation, Rathke returned to Danzig where he worked as a general practitioner. In 1825, he was appointed director of the city hospital. Four years later he was appointed professor of physiology and pathology at the Dorpat (Tartu) University. In 1835, Rathke succeeded the embryologist Karl Ernst von Baer (1792–1876) as professor for anatomy and zoology in Königsberg, East Prussia. He spent the rest of his life here.

Rathke discovered the pharyngeal clefts (branchial clefts) and pharyngeal arches (branchial arches) in embryos of birds and mammals (Fig. 302). He investigated the fate of these structures in embryos at different stages of development and found that the pharyngeal clefts eventually disappear. Rathke also discovered an upgrowth from the roof of the stomodeum (primordial oral cavity). This ectodermal diverticulum, which is called "*Rathke's pouch*," forms the anterior part of the pituitary gland – the adenohypophysis. In addition, Rathke pursued research in several other areas, which included the vertebrates, worms, mollusks, reptiles, the crustaceans, and the lancet fish (amphioxus), which he had discovered. Ratke published more than 125 scientific papers and several monographs and textbooks. He was elected a Fellow of the Royal Society, London and made a foreign member of the Bavarian Academy of Sciences.

Marc-Jean Bourgery

The French surgeon and anatomist Marc-Jean Bourgery (1797–1849; Fig. 303) and his collaborator, the artist Nicolas Henri Jacob (1781–1871), published an encyclopedic and influential work in eight volumes on the structure of the human body. Bourgery's treatise, *Traite complet del'antomie de l'homme comprenant la medicine operatoire*, remains one of the most remarkable and beautiful anatomical atlases ever produced (Figs. 304–306). The work consisted of eight volumes, 2108 pages, and 726 lithographic plates, of which 723 were

Figure 301 (*Top left*). A portrait of Martin Heinrich Rathke. (Courtesy of http://tartu.ester.ee/record=b1938851~S1.)

Figure 302 (*Top right*). Title page of the Heinrich Rathke's brachial apparatus and hyoid bone of vertebrates. (Courtesy of Institut fur Anatomie, Ernst-Moritz-Arndt-Universitat, Greifswald, Germany.) H. Rathke (1793–1860), German anatomist and embryologist, succeeded Karl Ernst von Baer in Konigsberg. Rathke's name is associated with the diverticulum ("pouch") from the roof of the embryonic pharynx, which gives rise to the anterior part of the hypophysis.

Figure 303 (*Right*). A portrait of Marc-Jean Bourgery. (Courtesy of Bethesda, MD: U.S. National Library of Medicine, National Institutes of Health, Health & Human Services.)

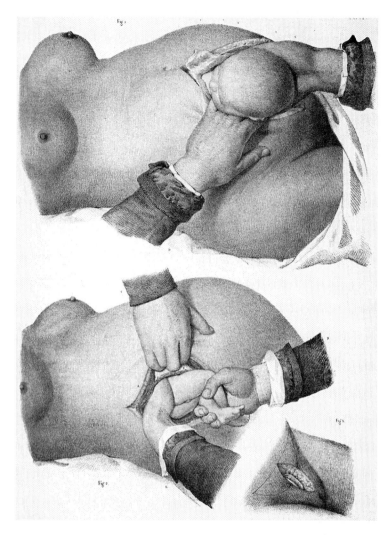

Figure 304. A Caesarean procedure. Bourgery, J. M. (Jean Marc), 1797–1849. *Traité complet de l'anatomie de l'homme.* (Courtesy of Bethesda, MD: U.S. National Library of Medicine, National Institutes of Health, Health & Human Services, 2010.)

colored by hand. The eight volumes were published over a period of more than two decades (1832–1854). It was Bourgey's life's work; the last volume was unfinished and published after his death.

Felix Vicq d'Azyr

The French physician and comparative anatomist Felix Vicq d'Azyr (1748–1794) was born in Valognes, Normandy. First, he studied philosophy in Caen and later, in 1765, moved to Paris for medical studies. Vicq d'Azyr also attended lectures in anatomy, physiology, and other subjects at the Jardin du Roi (Museum of Natural History). He was inspired by the lectures given by the celebrated anatomist Antoine Petit (1722–1794) and the naturalist Louis Daubenton (1716–1800). After receiving his doctorate degree, Vicq d'Azyr began in 1772 to teach anatomy at the Jardin du Roi for the benefit of medical students and the general public. As

a comparative anatomist, he combined structure and function in his lectures, which he felt was necessary for a deeper understanding of biomedical problems. In 1779, he was married, but a year later both his wife and child died from tuberculosis.

Vicq d'Azyr was interested not only in the anatomy of man but also in birds and large quadrupeds. For a period of eight years, he was professor of comparative anatomy at the veterinary school of Alfort. Because of his reputation, medical background, and knowledge of comparative anatomy, Vicq d'Azyr was requested by the Minister of Finance to investigate a severe cattle epidemic in the south of France which he successfully managed. He was then appointed as the Superintendent of Epidemics. More recognition and honors came with his scientific and political accomplishments. Vicq d'Azyr was named the secretary for life of the newly established Société Royale Médecine. In 1774, he was elected a member of the Académie des Sciences and of the prestigious French Academy four years later. In 1789, he was appointed as the physician to Queen Marie-Antoinette.

Vicq d'Azyr's monumental treatise, *Traité d 'Anatomie et de Physiologies*... presented his observations and concepts (Figs. 307–310). It was published in 1786 and dedicated to King Louis XVI. The work was the only volume of a planned multivolume undertaking, which Vicq d'Azyr was not able to continue because of the revolution. Vicq d'Azyr had planned "*to present a 'grand tableau' of all living creatures not in an exhaustive manner, but with sufficient examples from different species to show overall design of nature, culminating with man*" (Schmitt, 2009). He felt that it was essential to integrate structure with function in man and animals. He also proposed a new terminology for the classification of all animals that would rationalize the laws of organization in comparative anatomy.

Vicq d'Azyr recognized "*... the brain as the organ of thought and felt that the advancement of science would largely benefit from a better knowledge of the functional anatomy of this master organ*" studies (Parent, 2007). He was a pioneer among the early neuroanatomists and one of the first to make coronal sections of the brain, which he studied. The brain was first fixed (hardened) in an alcoholic solution. Not surprisingly, Vicq d'Azyr's treatise focused on the anatomy of the brain, which he copiously illustrated with large colorized original figures of his brain sections. The figures were all labeled with explanation and notes.

Vicq d'Azyr described the brain and spinal cord, the cranial nerves, and the second and third pairs of cervical nerves. The convex surface of the brain, with its convolutions of gyri and the spaces in between – sulci, were divided by him into three regions: frontal, parietal, and occipital. He identified the central sulcus,

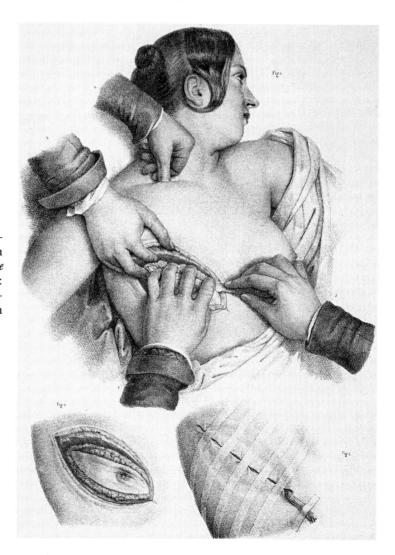

Figure 305. A picture depicting a mastectomy procedure. Bourgery, J. M. (Jean Marc), 1797–1849. *Traité complet de l'anatomie de l'homme.* (Courtesy of Bethesda, MD: U.S. National Library of Medicine, National Institutes of Health, Health & Human Services, 2010.)

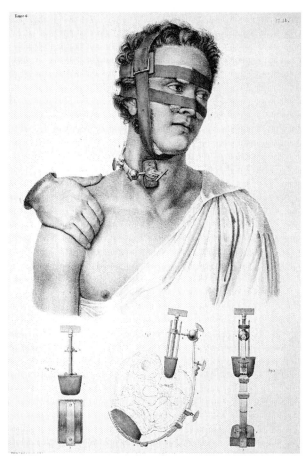

Figure 306. Carotid artery compression. Bourgery, J. M. (Jean Marc), 1797–1849. *Traité complet de l'anatomie de l'homme.* (Courtesy of Bethesda, MD: U.S. National Library of Medicine, National Institutes of Health, Health & Human Services, 2010.)

surface of the cerebral hemispheres, the caudate nucleus, the two parts of the lentiform nucleus – globus pallidus and a larger putamen, the substania nigra of the midbrain, the cerebral ventricles, and the interventricular foramen. In addition, Vicq d'Azyr also identified the cingulate gyrus, cuneus, uncus, the anterior and posterior perforated substances, the insula, spino-thalamic tract, and various cerebral and cerebellar sulci (Parent, 2007; Tubbs et al., 2011c).

Franz Joseph Gall and Phrenology

Franz Joseph Gall (1758–1828; Fig. 311), was a highly accomplished German neuroanatomist and physiologist. He studied medicine in Strasbourg and Vienna, where he settled and taught anatomy. In an era when there was considerable debate as to where mental functions might be localized, either in the ventricles or in the brain itself, Gall developed a new and controversial theory. He proposed that one can determine the development of mental and moral attributes of a person by examining the external shape of the skull (Van Wyhe, 2002) (Fig. 312).

Gall's system of craniology was taken up and popularized by his ardent follower and collaborator, the German physician Johann Christoph Spurzheim (1776–1832). Both of them carried out many neuroanatomical studies in order to support their theory, which had no anatomical basis. Gall travelled extensively across Europe giving lectures on "*phrenology*" to both scientific and lay audiences. Johann Wolfgang von Goethe, the great German poet, who attended Gall's lectures, ranked Gall's work as among the best in comparative anatomy. Not surprisingly, Gall's theory was rejected by most scientists because it completely lacked any scientific proof. As an example, the celebrated anatomist Jacob Ackermann (1765–1815) of the University of Heidelberg was highly critical of the system of phrenology and Gall's outrageous claims. The Church also condemned phrenology because it

which separated a precentral gyrus from a postcentral gyrus. Vicq d'Azyr recognized other parts of the brain, which he described in greater detail than earlier anatomists. These included the mammillo-thalamic tract, the hippocampal formation in the medial part of the temporal lobe, the basal ganglia deep below the

Fig 307 (*Facing page, top left*). A dissection exposing the brain. Felix Vicq d'Azyr. *Traité d 'Anatomie et de Physiologies.* 1786. (Courtesy of Universitats-bibliothek, Heidelberg. http://digi.ub.uni-heidelberg.de/diglit/vicqdazyr1786bd2.)

Fig. 308 (*Facing page, top right*). A dissection of the brain, exposing its vasculature and the circle of Willis. Felix Vicq d'Azyr. *Traité d 'Anatomie et de Physiologies.* 1786. (Courtesy of Universitats-bibliothek, Heidelberg. http://digi.ub.uni-heidelberg.de/diglit/vicqdazyr1786bd2.)

Fig. 309 (*Facing page, bottom left*). Several depictions of the dissection of the cerebellum. Felix Vicq d'Azyr. *Traité d 'Anatomie et de Physiologies.* 1786. (Courtesy of Universitats-bibliothek, Heidelberg. http://digi.ub.uni-heidelberg.de/diglit/vicqdazyr1786bd2.)

Fig. 310 (*Facing page, bottom right*). A dissection depicting the superior sagittal sinus and its tributaries. Felix Vicq d'Azyr. *Traité d 'Anatomie et de Physiologies.* 1786. (Courtesy of Universitats-bibliothek, Heidelberg. http://digi.ub.uni-heidelberg.de/diglit/vicqdazyr1786bd2.

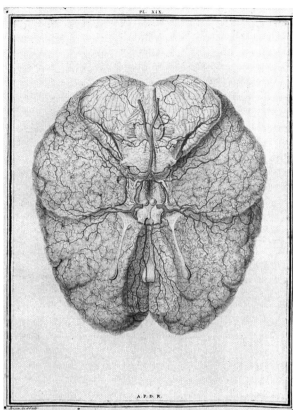

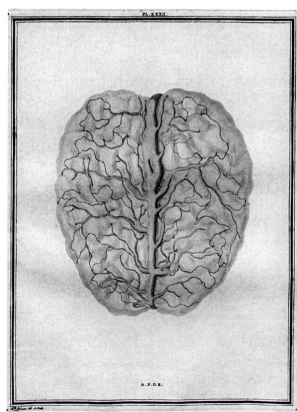

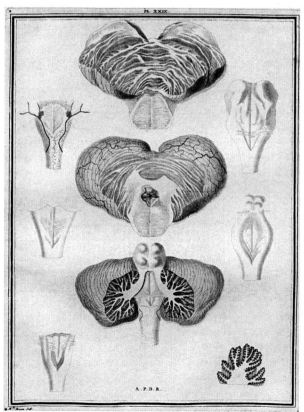

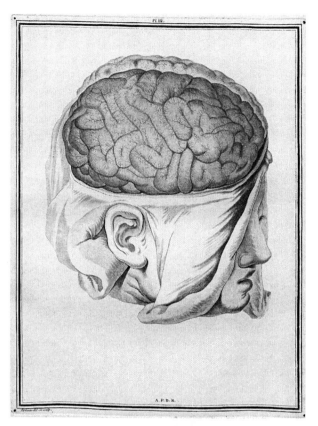

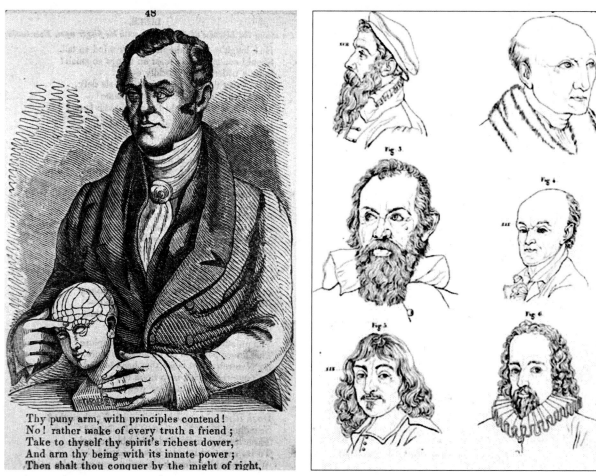

Figure 311 (*Left*). Dr. Franz Joseph Gall. *Thy puny arm, with principles contend.* (Courtesy of Bethesda, MD: U.S. National Library of Medicine, National Institutes of Health, Health & Human Services.)

Figure 312 (*Right*). A picture from the *"Anatomie et physiologie du systeme nerveux en general."* Gall & Spurzheim established the fact that the white matter of the brain consists of nerve fibers and that the gray matter of the cerebral cortex represents the organs of mental activity. They were the first to demonstrate that the trigeminal nerve was not merely attached to the pons, but that it sent root fibers as far down as the inferior olive in the medulla. In addition, they confirmed once and forever the medullary decussation of the pyramids." (Garrison and McHenry). (Courtesy of Bibliotheca Systema Naturae. The Hagströmer Medico-Historical Library Virtual Book Museum http://www. ki.se/hagstromer/bsn_detail.php?skip=0.)

contradicted religious dogma. Remaining from this era, as a legacy to the passing fad of phrenology, are many impressive collections of marked skulls, now splendidly housed in museums and medical institutions throughout Europe and in other parts of the world (Greenblatt, 1995; Simpson, 2005) .

Jan Evangelista Purkinje

One of the founders of modern physiology was the great Czech histologist, physiologist, and patriot Jan Evangelista Purkinje (Purkyne, 1787–1869; Fig. 313). He is more often remembered for the large nerve cells, with numerous dendrites, found in the cortex of the cerebellum, and for the atypical subendocardial conducting muscle fibers in the heart, which he described (Dungelova & Barinka, 1987; Tichacek, 1987; Vacek, 1987). Purkinje, however, made many important discoveries including the uniqueness of fingerprints for identification purposes, the germinal vesicle or nucleus of the ovum, sweat glands in the skin, and many other observations.

Purkinje was the first to formulate in detail the principal features of the cell theory, and he was among the first to teach microscopic anatomy as part of the university course. Already in 1819, Purkinje's

Figure 313. Jan Evangelista Purkinje (Purkyne, 1787–1869).

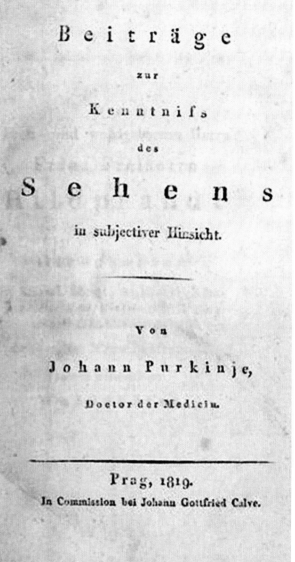

Figure 314. Purkinje's doctoral thesis, *Beitrage zur Kentniss des Sehens in subjectiver Hinsicht.* (Courtesy of the open library. https://openlibrary.org/search?q=Jan+Evangelista+Purkinje.)

doctoral thesis, *Beitrage zur Kentniss des Sehens in subjectiver Hinsicht* (Fig. 314), established his reputation as a scientist, an account of the visual phenomenon he described, which is now known as "*Purkinje effect or Purkinje's images.*"

The turn of the nineteenth century was an exciting period in European history, marked by great wars, conquests, explorations and expeditions, publication of enduring works in literature and music, as well as significant achievements in medicine. Napoleon had almost demolished the Holy Roman Empire and made himself an emperor. Catherine the Great of Russia died; the Sonata Pathetique, Fidelio, the Symphony No. 3, and the Moonlight Sonata were composed by Ludwig van Beethoven, and The Seasons by Franz Joseph Haydn. Edward Jenner had developed the technique of vaccination to prevent smallpox, and Humphry Davy produced laughing gas (nitrous oxide), which was used as a general anesthetic for minor surgery.

About this time, the anatomists in England were focusing their attention to the urgent needs of medical students for improved practical work in the dissecting room. There were more teaching institutions, especially the private anatomy schools, to cope with the increasing number of students. The demand for cadavers led to grave robbing and the emergence of a new category of tradesmen, the "*resurrectionists*" or body snatchers who, in fact, worked in collaboration with many of the prominent anatomists (Lonsdale, 1870; Rae, 1964; Cohen, 1975; Richardson, 1987; Desmond, 1989; Shultz, 1992). The public was horrified and incensed, but appropriate legislation that would facilitate human dissection was still three decades away. It was precipitated by scandals, intense lobbying of Parliament, and a series of shocking events.

Chapter 24

BODY SNATCHERS AND THE TRADE IN CORPSES

HUMAN BODIES FOR DISSECTION

In Britain, the importance of dissection for understanding the structure of the human body was recognized as early as 1542 through an Act of Parliament which permitted the Company of Barbers and Surgeons to have for this purpose the bodies of four executed "*malefactors.*" The charter granted to them by Henry VIII stipulated the following:

> And further be it enacted by thauctoritie aforesayd, that the sayd maysters or governours of the mistery and comminaltie of barbours and surgeons of London & their successours yerely for ever after their sad discrecions at their free liberte and pleasure shal and maie have and take without contradiction foure persons condemned adjudged and put to deathe for felon' by the due order of the Kynges lawe of thys realme for anatomies with out any further sute or labour to be made to the kynges highnes his heyres or successors for the same. And to make incision of the same deade bodies or otherwyse to order the same after their said discrecions at their pleasure for their further and better knowlage instruction in sight learnyng & experience in the sayd scyence or facultie of Surgery.

The modest allocation of bodies for dissection proved to be woefully inadequate for the increasing number of apprentices and students who participated in these practical anatomy teaching sessions. Legally, it was not possible to secure more bodies. Even the "*foure bodies*" proved difficult to obtain because "*despite the precautions of the Company private anatomy was, to a certain extent, carried out, and the bodies of malefactors had a market value*" (Bailey, 1896).

The Parliamentary Act of 1752 did, however, permit the judge to send to the Surgeons' Hall the body of any criminal executed in London and Middlesex for dissection. The dissection itself was perceived both by the criminal and the public as a final punishment to be inflicted on the corpse of the executed criminal. For the condemned felon this was his greatest fear and ordeal that even after death his corpse would be subjected to the anatomist's knife (Fig. 134). To quote from Bailey (1896):

> No doubt this provision much increased the dislike of the poor to any regulations by which the bodies of their friends might be given up for dissection after death. It was felt that dissection by the surgeons was part of the sentence passed on a murderer, and therefore carried with it shame and disgrace. To make provision by law, therefore, for the dissection of the bodies of any other class of persons was, not unnaturally, distasteful, in that it partly put them in the same position as murderers.

The answer to the desire for the repeal of this obnoxious clause was that nothing must be done to weaken the law; it was stated that to withdraw the part of the sentence which related to dissection would rob the punishment of its prohibitive effect. It is somewhat difficult to understand the argument; surely if the risk of suffering the extreme penalty of the law would not keep a man from crime, the extra chance of being dissected after death could hardly be expected to do so. As Sir Henry Halford said, "*I certainly think that while that law remains they [the public] will connect the crime of murder with the practice of dissection; an order to be dissected, and a permission to be dissected, seem to be too slight a distinction.*"

Another objection to the dissection of murderers came from the teachers. They stated that when the body of a notorious criminal was lying at either of the anatomical schools, the proprietor was pestered by persons of a morbid turn of mind for permission to view the body. This difficulty was also felt by the College of Surgeons, and in consequence a placard was hung up outside the place where the dissections were made, giving notice that no person could be admitted, unless accompanied by a member of the Court of Assistants. To make dissection less distasteful to the general public, and to show the advantages of anatomy, some endeavors were made to explain the structure of the human body to nonprofessional persons. In Ireland, Sir Philip Crampton lectured with open doors, and gave demonstrations in anatomy to poor people. These persons, he tells us, became interested in the subject, and often brought him bodies for dissection. A newspaper cutting of 1829 shows that this was also tried in London. A surgeon called in overseers and churchwardens of St. Clement Danes, and gave a demonstration on a body, explaining its construction, and the use of the internal organs. "*By this means,*" says the paragraph,

he so fully absorbed the self-interest of his audience as to extinguish the pre-conceived notions of horror and disgust attached to the idea of a spectacle of this description. The enlightened governors of the parish assented to the post mortem examination of the body of every unclaimed pauper, an enquiry into whose case might appear conducive to the interests of medical science.

Large crowds, as many as a hundred thousand, would sometimes gather at the site where the hanging took place. For most of the public, the hanging was turned into a festive occasion. Shops closed to allow everyone to attend; taverns and coffee houses were open and would be doing brisk business, and the drunk roamed the area in a boisterous spirit. Perhaps, it was the misunderstood perception of the anatomist, as an extension of the punishment handed down by the judge (Fig. 315) that clouded the public's understanding of the value of human dissection for the teaching of anatomy.

SHORTAGE OF CORPSES

The shortage of human bodies for dissection, as a part of medical instruction and learning, hampered the quality of teaching and the progress of anatomy in many respects. Without an adequate number of bodies the private anatomy schools would have not been able to continue.

There are many accounts of anatomists, students, and the professional body snatchers, who have disturbed the sanctity of the grave and removed the corpse (Figs. 316 & 317) for dissection in the anatomy schools (Cooper, 1843; McGregor, 1884; Ball, 1928; Cohen, 1975; Shultz, 1992). Indeed, the survival of the private schools, and the teaching of practical anatomy, depended on this illegal source of supply from the resurrectionists. The law then did not view the removal of the corpse itself as punishable, but stealing the clothing of the deceased was a criminal act.

Bailey (1896) has provided a reliable account of how the "resurrection-men in London" operated prior to the passing of the Anatomy Act, and particularly on the "doings of one gang of the resurrection-men in London." Included in this work is the incomplete "Diary of a Resurrectionist," written on 16 sheets between November 28, 1811 and December 5, 1812. The author of the diary remains unknown and his

name was probably deliberately concealed in order to protect his identity. Just a few entries from the diary will give some insight as to how the body snatchers pursued their nefarious business:

Monday 2nd. Met at St. Thomas's, Got paid for the 3 adults & settled; met and settled with Mordecei, made Him up £2 5s. 6d. and Receipt of all demands. At Home all night.

Friday 6th. Removed 1 from Barthol. to Carpue.* At night went out and got 8, Danl. at home all night. 6 Back St. Lukes & 2 Big Gates: went 5 Barthol. 1 Frampton 3 St. Thomas's, 3 Wilsont.

[C. Carpue, the founder of the Dean Street Anatomical School.]

[Dr. Frampton, of the London Hospital.]

[James Wilson, of the Great Windmill Street School.]

Saturday 7th. At night went out & got 3 at Bunhill Row. 1 St. Thomas's, 2 Brookes.*

Sunday 8th. At home all night.

Monday 9th. At night went out and got 4 at Bethnall Green.

*Joshua Brookes, founder of the Blenheim Street, or Great Marlborough Street, Anatomical School.

Figure 315. *The Reward of Cruelty.* This stark caricature of a dissection in progress was the last of a series of paintings (*The Four Stages of Cruelty*) by the renowned British artist William Hogart (1697–1764). The body of the murderer Tom Nero is dissected after he has been executed by hanging as can be seen from the rope around his neck. The dissection scene is mercilessly depicted and without any reverence at all for the deceased. One sees the eye being gorged out, intestines pulled out of the abdomen onto the floor, and an organ (heart) left for a curious dog at the front of the table. (Courtesy of Bethesda, MD: U.S. National Library of Medicine, National Institutes of Health, Health & Human Services.)

Tuesday 10th. Intoxicated all day: at night went out & got 5 Bunhill Row. Jack all most buried.

The public was outraged and strongly condemned the anatomists, as well as the resurrectionists, for these sacrilegious and odious acts which disturbed the sanctity of the grave. Desecrating the dead conflicted with societal values of respect for the dead. In the preface to the biography of Robert Knox, the famous Edinburgh anatomist, who suffered greatly from the "*injustice of the stigma cast*" upon him through the atrocities of the grave

robbers, "which happened to culminate in the rooms of his anatomical establishment," Lonsdale (1870) blamed the government with the following remarks:

The fault of the Resurrectionist system a terrible blot upon our social status as a Christian nation lay with the Executive Government of the country, who for half a century proved deaf to all remonstrance, and evinced not the slightest regard for the science of medicine till the anatomical affairs of the kingdom fell into a state of fearful chaos.

Figure 316. *Body Snatching*, Netherlands (1620–1629). (From Damhouder, J., Practycke in Criminele Saecken. Jan van Waesberghe de Jonge, Rotterdam, 1628.) The bearded man appears to be carving a human figure on the lid of the sarcophagus. The person standing next is observing him. On the right are two other men who are in the act of exhuming a body in the church cemetery. (Courtesy of Bethesda, MD: U.S. National Library of Medicine, National Institutes of Health, Health & Human Services.)

Figure 317. A grave robber flees from a corpse that has come to life! A prematurely interred person stands up in the casket after being exhumed by a grave robber who flees in terror. (From Winslow, J. B., *The Uncertainty of the Signs of Death*, M. Cooper, London, 1746; Courtesy of Bethesda, MD: U.S. National Library of Medicine, National Institutes of Health, Health & Human Services.)

TRADE WITH CORPSES

Figure 318. Sir Astley Cooper. (Cooper, *Life of Sir Astley Cooper*, London, 1843; with permission, from Major 1954.) Sir Astley Cooper (1768–1841), surgeon at Guy's and St. Thomas' Hospitals, was an avid dissector and a skilled surgeon. He was the first surgeon to ligate carotid and abdominal aorta aneurysms. Elected a fellow of the Royal Society in 1802, Sir Astley later served as president of the Royal College of Surgeons. He also attended King George IV.

Bodies from the north of Ireland were exported to Glasgow and Edinburgh for many years until this trade with corpses was exposed in 1827. Because of the large number of medical students, estimated to be about 900 in London, there was also great demand for regular shipments of bodies to London. Most of the bodies were stolen from the easily secured graveyards where destitutes were buried. In fact, it was known that the medical schools in Glasgow and Edinburgh regularly received their supply of bodies, which were shipped in barrels containing brine, from a graveyard called "*Bully's Acre.*" This was a free burial ground for the very poor and it was unguarded in the nights. A story often retold is of the shipment that went astray. As a result, when the anatomist in Edinburgh opened a barrel, he found smoked hams instead of the body he was expecting. The fate of the other barrel remains unknown (Ball, 1928).

The grave robbers often operated in gangs, which rivaled each other; the anatomy schools could not have functioned without them. The price the resurrectionists demanded for bodies kept on increasing from a few shillings to several pounds. The school had to pay them also a retainer fee (Bailey, 1896; Ball, 1928).

In 1820, the famous surgeon, Sir Astley Cooper (Fig. 318), who regularly used the services of the body snatchers, was determined to secure the body of one of his patients who died after surgery. He offered to pay the resurrectionists "*cost what it may*" for the corpse so that he could perform an autopsy on it. He boasted that "There is no person, let his situation in life be what it may, whom if I were disposed to dissect, I could not obtain" (Kobler, 1960).

HEINOUS MURDERERS

The execrable murders committed by the notorious Burke and his companion Hare remain unique in the annals of crime (MacGregor, 1884). The public was shocked and disgusted. How cold-blooded and calculated they carried out their heinous crimes, Burke himself described in his confession: *After they ceased crying and making resistance, we left them to die of themselves, but their bodies would often move afterwards, and for some time they would heave long breathings before life went out.*

Burke was hanged just after eight o'clock on January 28, 1829 in the presence of a large crowd. Early on the following day, his body was transported to the anatomical rooms of the college, where it was

dissected by Professor Alexander Monro *Tertius*. The professor gave a public demonstration of the brain and noted that it was unremarkable in appearance, except for a softness of the cerebral hemisphere, which he attributed "to the lowness of the prison diet some weeks previous to execution."

Apparently, the public was "falsely and ignorantly" informed that the lateral cerebral hemispheres were unusually developed, and that the related cranial bones were as a consequence thin. Such a finding would not have been of little interest to the phrenologists then (Stone, 1829). Burke's death mask and his skeleton have been mounted for display in the museum of the Anatomy Department at the University of Edinburgh.

These sordid atrocities brought to light "the blackest chapter in the black annals of body snatching" (Drimmer, 1981). The concerned reaction of the public, as well as the government, undoubtedly played a momentous role in Parliament, culminating with the passage of the Anatomy Act on August 1, 1832. Present-day anatomists are also the beneficiary of this important legislation (Polson & Marshall, 1975).

THE LONDON GRAVE ROBBERS

With the large number of students attending the private schools of anatomy in London during the eighteenth century, there was an obvious need for human bodies so as to demonstrate the parts. As the number of schools increased, the demand became great. Most of the bodies were obtained by illegal means, usually by the local resurrectionists or imported from Dublin and other parts of the nation. As mentioned before, without a constant supply of corpses for dissection, the private schools of anatomy would not have survived. Some schools that were started closed because of the lack of bodies for dissection. It was not uncommon during the eighteenth century for teachers of anatomy and their students to rob graves for the dissecting table.

The problems of obtaining a continual supply of bodies for teaching and dissection in London has been well documented (Bailey, 1896; Ball, 1928; Graham, 1958; Desmond, 1989; Richardson, 1989). Prior to the passing of the Anatomy Act of 1832, the grave robbers had a "seller's market" and, over a period of 15 months in 1830–1831, seven London gangs of resurrectionists were arrested (Desmond, 1989). Only a brief account can be given here and we have drawn heavily from the sources mentioned.

Graham (1958) described the case of the notorious highwayman, Dick Turpin, who was hanged in York in 1739. Even though he was buried in an unusually deep grave, his grave was opened, and his remains were later found in the garden where a surgeon lived. He was reburied, the coffin filled with lime, and his grave was guarded by watchmen. "This unsuccessful attempt at resurrection for dissection attracted a great deal of attention, as did more successful sallies, for by this time, apprentices and their masters all over the country were going forth in the dead of night to collect the anatomical material they could not obtain in no other way. Soon there was rioting, and every other medical school had its windows smashed 'and its professors of anatomy and surgery attacked by incensed mobs.'"

Alarmed by the public riots, the government legislated in 1752 that all murderers executed in London and Middlesex should be either publicly dissected or hung in chains on gibbets. A year later the following sentence was passed by the Lord Chief Justice on a young man who had murdered his wife:

Thomas Wilford, you stand convicted of the horrid and unnatural crime of murdering Sarah, your wife. This Court doth had judge that you be taken back to the place from whence you came; and there to be fed on bread and water till Wednesday next, when you are to be taken to the common place of execution, and there hanged by the neck until you are dead: after which your body is to be publicly dissected and anatomized, agreeably to an Act of Parliament in that case made and provided; and may God Almighty have mercy on your soul. (Graham, 1956)

Soon, graveyards provided a source of bodies for the surgeons and the private schools of anatomy. Often they were helped by the grave diggers who were well paid by the anatomists and the surgeons. Shady and depraved characters worked in complicity with the anatomists and surgeons for the good money that could so easily be earned.

By the end of the eighteenth century it became necessary that every grave in London had to be "*watched by angry citizens, who had usually a bell-mouthed pistol in one hand and a bottle of grog in the other.*" This did not prevent the resurrectionists from moving into other parts of the country from where bodies were collected and taken to a central location (Bailey, 1896). In 1776, more than 20 bodies were found by the police in a London home.

The professional grave robbers were now in charge of the business, and even the spring guns, trip wires, and high fences covered with broken glass did not deter the resurrectionists. Their only fear was the public and, to a lesser extent, the law. Surgeons were needed for the armies, and those caught for raiding the grave to obtain anatomical material were often given lenient sentences (Graham, 1958).

In 1536, Andreas Vesalius, the father of modern anatomy, obtained his first human skeleton by removing the remains of a thief who was hanged outside the walls of his hometown. More than two centuries later, John Hunter, the pioneer of scientific surgery, secured the body of the Irish giant, O'Brien, who died in 1783. His skeleton, which is more than eight feet in height, is now in the Hunterian Museum of the Royal College of Surgeons in London. How the body was obtained by John Hunter has been described by Ottley (1839).

JOHN AND WILLIAM HUNTER: CLIENTS OF THE GRAVE ROBBERS

William and John Hunter of the Great Windmill Street School were clients of the grave robbers for the regular supply of bodies each year on account of the large number of students. John was entrusted with the task of dealing with the resurrection men, as well as managing the affairs of the school.

When the winter course of 1749 began, he was advanced to be demonstrator to the students: thus, only a year after he left home, he held in his hands the honor of the new school. It was but a few years old, a private venture, unendowed, unsupported by any hospital, but the two young men together drove it on to success: and the younger brother bore the rough work, hobnobbing with the resurrection-men, slaving all day long in unwholesome air, dissecting, demonstrating, and putting up specimens. This giving of demonstrations is an arduous office, and in Hunter's time it was horribly unwholesome (Paget, 1897).

John seemed better suited than William for the difficult task in dealing with the resurrection men.

He was fond of company, and mixed much in the society of young men of his own standing..., and freed from restraint, are but apt to indulge in. Nor was he always very nice in the choice of his associates, but sometime sought entertainment in the coarse, broad humour to be found amid the lower ranks of society. He was employed by his brother to cater for the dissecting room, in the course of which employment he became a great favourite with the certainly not too respectable class of persons the resurrection men...." (Ottley, 1839)

The bodies supplied to John and William by the body snatchers enabled them to carry out observations that were unique at that time. With respect to the pregnant woman who was dissected, which at that time was uncommon, William stated in his masterpiece, *The Anatomy of the Gravid Uterus,* that "the body was *procured (author's italics)* before any sensible putrefaction had begun." Moreover, John's work on venereal diseases was facilitated when two condemned men, who had gonorrhea, were delivered the very next day to him for further studies (Kobler, 1960).

Irrespective of social class, safety precautions were taken by everyone to foil the intention of the grave robber. The wealthy, however, were able to do so more efficiently. Strong coffins, patent iron coffins, "mortsafes" cages around the coffins, did not deter the grave robbers. No body was safe if they wanted to exhume it. Sir Astley Cooper had his body placed in a well-secured sarcophagus at Guy's Hospital Chapel, and the remains of the Duke of Wellington were interred in four coffins (Morley, 1971).

The public was aware of the dastardly deeds of the body snatchers and the reprehensive business between them and the anatomists. They showed their anger and repulsion by repeatedly stoning the school.

SIR ASTLEY COOPER AND THE RESURRECTIONISTS

Sir Astley Cooper was not only a very skillful surgeon of his time (Cooper, 1843) but also a dedicated anatomist (Fig. 318). In fact, he had described in 1845 the fine fibrous strands of ligaments, which supports the soft glandular tissue of the breast. These are now known as Cooper's suspensory ligaments. His passion for anatomy was such that he dissected daily in a special room located in his home. Like John Hunter, his former teacher, he was also interested in comparative anatomy and dissected a wide range of animals. In Sir Astley Cooper, the resurrectionists found someone who was always willing to do business with them.

Under the encouragement of Sir Astley Cooper and other teachers, who paid high prices for anatomical material, the violation of graves in or near London became a horrible trade. The business passed into the hands of men of the most degraded character: men who, for the sake of gain, if they could not obtain their objects by the ordinary of disinterment, would adopt any means to effect their purpose. (Ball, 1928)

Yet, when Sir Astley Cooper appeared before the Select Committee of the House of Commons on Anatomy in 1828, he described the resurrectionists as "*the lowest dregs of degradation. I do not know that I can describe them better; there is no crime they would not commit, and, as to myself, if they should imagine that I would make a good subject, they really would not have the smallest scruple, if they could do the thing undiscovered, to make a subject of me.*" At that time, there were about 200 grave robbers in London. One of them, who also appeared before the committee as a witness, claimed that the largest number of bodies he obtained was 23 over four nights. He also stated that "When I go to work, I like to get those of poor people buried from the workhouses, because instead of working for one subject, you may get three or four. I do not think, during the time I have been in the habit of working for the schools, I got half a dozen of wealthier people" (Bailey, 1896).

One resurrectionist who kept a record of his business informed the committee that between 1809 and 1810 he sold the bodies of 305 adults and 44 children to the London medical schools, sent 37 other corpses to Edinburgh, and had 18 bodies in hand that were never used at all. In 1810–1811, he sold 332 adults and 47 children. The average price for an adult body was given as £4 4s. (Bailey, 1896).

Over the years; the price for a body increased. When Astley Cooper was a student with John Hunter the price was two guineas, and around 1820, eight guineas (Graham 1958). Sir Astley paid handsomely when it was necessary. "*It is no wonder, then, that of Sir Astley it might be said that no man knew so much of the habits, the crimes, and the few good qualities of the "resurrection-men"* (Ball, 1928).

Appearing before the Select Committee on Anatomy, Sir Astley Cooper was questioned about the effect of the law in preventing grave robbing, to which he replied that it only served to increase the price for subjects, which make the resurrectionists wealthy. In reply to another question, he boldly asserted that "There is no person, let his situation in life be what it may, whom if I were disposed to dissect, I could not obtain. The law only enhances the price, and does not prevent the exhumation." On the whole, the grave robbers "were a dissolute and ruffianly gang," with drunken habits, and "*they were often in pecuniary difficulties,*" despite their relatively large income derived from the sale of bodies, the retaining fee paid by the medical school at the beginning of the session, and a closing fee or "*finishing money*" which they also demanded (Bailey, 1896).

Sir Astley was a brilliant surgeon who pioneered several operative procedures, including ligations of the common carotid and the external iliac arteries for aneurysms, ligation of the abdominal aorta, amputation at the hip joint, and operating the perforated tympanic membrane in cases of deafness attributed to obstruction of the pharyngotympanic tube.

Sir Astley was a prolific author and his books included a treatise on hernia and his last work, which was on the anatomy of the breast. He became a fellow of the Royal Society in 1802 and received a knighthood in 1820. After a long and productive life, Sir Astley died in 1841.

CHARLES BELL

Charles Bell (1774–1842; Fig. 319), a contemporary of Sir Astley Cooper, revolutionized our understanding of the nervous system with his book, *A New Idea of the Anatomy of the Brain and Nervous System*, which was published in 1811. He investigated and wrote about the cerebral hemispheres and the cerebellum, the spinal nerves, and the functions of the fifth and seventh cranial nerves (Figs. 320–321). Bell discovered and demonstrated the motor function of the anterior roots of the spinal nerves, and he demonstrated that individual nerve fibers throughout the body were continuous with similar fibers in the central nervous system. For his neuroanatomical and neurophysiological studies, Bell was awarded the first medal given by the Royal Society in 1829 and two years later he was knighted.

Charles Bell was born in Edinburgh where he received his medical education and probably "*obtained*" illegally a skeleton with osteomalacia that he described in his catalogue of anatomical and pathological specimens (Ball, 1928): "*A skeleton of great value; in procuring this skeleton I lost myself for two hours, and found myself at two o'clock in the morning in the court before Pennycuick House.*"

Charles Bell distinguished himself through his research in the field of neurology, but he was also a leading surgeon and anatomist in London during his time. He was a partner in the Great Windmill Street School of Medicine, served as professor of anatomy and surgery to the Royal College of Surgeons in London, and played a part in the founding of London University. After spending 30 years in London, Charles Bell returned to Edinburgh where he became professor of surgery at the university. He died six years later at the age of 67 from a heart attack.

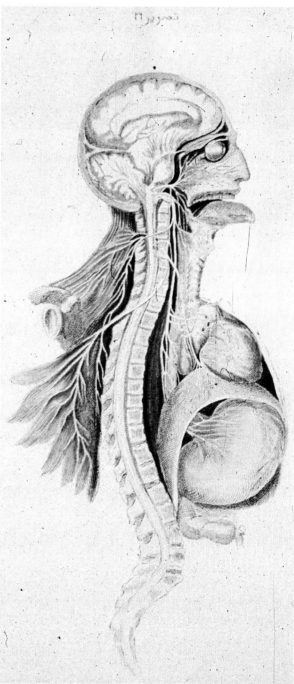

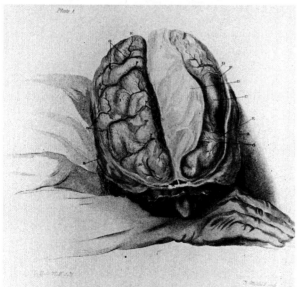

Figure 319 (*Top left*). A portrait of Charles Bell. (Courtesy of Bethesda, MD: U.S. National Library of Medicine, National Institutes of Health, Health & Human Services.)

Figure 320 (*Top right*). Nerves of respiration, according to Sir Charles Bell. (Courtesy of Bethesda, MD: U.S. National Library of Medicine, National Institutes of Health, Health & Human Services.)

Figure 321 (*Bottom left*). The brain exposed from above, in *Anatomy of the brain*, plate 1. Appears in "An atlas of anatomical plates of the human body, with descriptive letter-press in English an Hindustani, plate XXVI." (Courtesy of Bethesda, MD: U.S. National Library of Medicine, National Institutes of Health, Health & Human Services.)

BODIES FOR DISSECTION

John Hunter, Astley Cooper, and Charles Bell, three great surgeon-anatomists, would not have been able to make their valuable contributions to the progress of medicine and surgery without an understanding of the structure of the human body. The insights they have gained provided answers as to the nature of the structures involved and at the same time stimulated further research. Without the bodies, which were illegally obtained, their work would not have been possible.

The murders that were committed by Burke and Hare in Edinburgh and the trial that followed incited the public against the anatomists. They were mobbed, insulted, and dissecting rooms in all parts of the country were burnt down. It subsided when a bill was introduced in the House of Commons in the spring of 1829 to regulate the supply of dead bodies for dissection in the anatomy schools. The bill would have made provision for unclaimed bodies in hospitals and workhouses to be delivered to recognized schools of anatomy. The bill passed through the House of Commons, but it was rejected by the House of Lords on the grounds that it would be discriminating against the poor and destitute.

Perhaps the final thrust towards the legalizing of human dissection emanated from the outcry of the public following the murder in London of a young boy, Carlo Ferrari, in November 1831, just three years after Burke was hanged in Edinburgh. The 14-year-old boy was murdered by Williams and Bishop who were well-known resurrectionists. They negotiated a price with the porter at Kings College Hospital Medical School, and settled for a little less than they had anticipated. Returning late in the day with another man named Williams and a helper who carried the body in a hamper, they delivered the body to the porter who suspected foul play. This was confirmed by Mr. Partridge, the Demonstrator of Anatomy. The men were delayed on the pretext that Mr. Partridge had to change a 50-pound note. Shortly after the police arrived, the villains were arrested (Ball, 1928). The postmortem findings, described by Mr. J. F. Clarke, led to the conviction. On December 5, 1931, Bishop and Williams were executed; May was granted a reprieve.

Chapter 25

LEGISLATION OF HUMAN DISSECTION

Many factors have contributed to the discussions in Parliament and the passage of the Anatomy Act in 1832 (Ball, 1928; Goodman, 1944; Desmond, 1989; Richardson, 1991). The government was forced to take action on account of public pressure following the outrageous murders that were committed, as well as the persistent lobbying of the anatomists for a legal source of bodies for dissection. This relatively unknown law is one of the most important landmarks in the history of medicine, because it recognized the need for human dissection in the training of physicians. The act served as a prototype for other countries, even though the earliest legislation in the United States of America was passed in 1831 for the state of Massachusetts (Shultz, 1992). In all countries now, legal provisions have been made for medical schools to receive unclaimed and donated bodies for dissection, but in all these institutions, strict measures are taken to ensure that the bodies are treated with dignity and that the remains will be cremated or buried with appropriate respect.

As stated by Richardson (1991), the Anatomy Act of 1832 "is no more than a footnote in medical history," but "it was the instrument by which dissection for anatomy was at last put on a secure legal footing in Britain." The act was passed by Parliament in the summer of 1832, enacting recommendations laid down by a House of Commons Select Committee, whose findings had been published four years earlier, in 1828. The chronological events leading to the passage of the Anatomy Act are straightforward (Ball 1928; Richardson, 1989), but there were many misapprehensions relating to the act (Richardson, 1991).

Essentially, the Anatomy Act brought to the end the illegal trade with corpses, because it made provision for an adequate supply of bodies for the proper teaching of anatomy (Comrie, 1972). With this piece of legislation, resurrection men were gradually put out of business (Polson & Marshall, 1975), although grave robbing and the trafficking with bodies did continue for a while after.

Regional inspectors of anatomy were appointed to oversee the act. The dissecting room had to be approved and supervised by the inspector. The Anatomy Act required that the wishes of the deceased person and his or her next of kin be respected as to whether the remains should be dissected or not. Dissection was permissible only if there was a death certificate, and at least 48 hours must have elapsed before the dissection was performed. Moreover, the remains of the deceased must be properly interred in an appropriate manner. These are some of the important considerations in the Anatomy Act, which today characterize the laws that have been passed in other countries.

The anatomists were the beneficiaries of the Anatomy Act, which "was as much a political as a medical measure" (Richardson, 1991). Jeremy Bentham prepared an earlier draft of the Anatomy Act, which proposed that those dying in institutions, such as workhouses, should be made available to the anatomists for dissection; as such, the poor were now singled out for dissection rather than the criminal.

Jeremy Bentham had a personal interest in the matter, and his skeleton with his own clothes on and seated in his chair can be viewed today at University College in London. Not much emanated from the draft Bill of Bentham until March 1828 when two doctors were convicted of body snatching. The judge ruled that the only legal source of bodies for dissection was the condemned murderers who were hung in the gallows; obtaining bodies from other sources was illegal. This important judgment caused Parliament to act quickly because the anatomists could now be prosecuted for a criminal offense on account of the precedence that was established by this case.

Within 40 days, the House of Commons Select Committee was formed under the Chairmanship of Mr. Henry Warburton (1784–1858). He was educated at Trinity College, Cambridge, and elected F.R.S. in 1809. As one of the founders of London University, he served on its first council in 1827. Ball (1928) stated that "The name of Mr. Henry Warburton constantly appeared as the courageous and practical champion of the true interest of the medical profession in Parliament."

The task of the Select Committee was "to enquire into the manner of obtaining subjects for dissection in the Schools of Anatomy, and into the state of the law affecting the persons employed in obtaining and dissecting bodies." The committee met for the first time on April 28, 1828 and one of its first witnesses was Sir Astley Cooper, as well as other prominent members of the medical profession, and several public figures. One of these was Thomas Wakley who had founded *The Lancet*, which became one of the leading and influential medical journals in the world (Sprigge,1899). Bentham's proposal for obtaining bodies was taken into consideration by the House of Commons Select Committee. In July 1828, the report of the committee was finally published. Regarding the report, Bailey (1896) made the following comments:

> Amongst those who gave evidence before the Committee were the principal teachers of anatomy, and three of the resurrection-men. The tone of the Report was decidedly in sympathy with the teachers, but it strongly condemned the way in which they were compelled to obtain the bodies for dissection. After showing how badly off English students were for opportunities of learning anatomy, as compared with those students who really wished to master their art were compelled to go abroad, the Report proceeds: "These disadvantages affecting the teachers are such, that except in the most frequented schools, attached to the greater hospitals, few have been able to continue teaching with profit, and some private teachers have been compelled to give up their schools. To the evils enumerated it may be added, that it is distressing to men of good education and character to be compelled to resort, for their means of teaching, to a constant infraction of the laws of their country, and to be made dependent, for their professional existence, on the mercenary caprices of the most abandoned class in the community."

The first Anatomy Bill, "A Bill for preventing the unlawful disinterment of human bodies, and for regulating Schools of Anatomy," was presented to Parliament in March 1829. The essential aspects of Warburton's Bill extracted from Bailey (1896) are given below:

> …it was enacted that persons found guilty of disinterring any human body from any churchyard, burial-ground or vault, or assisting at any such disinterment, should be imprisoned for a term not exceeding six months for the first offence, and two years for the second offence. Seven Commissioners were to be appointed; the majority of these were not to be either physicians, surgeons, or apothecaries. All unclaimed bodies of persons dying in workhouses or hospitals, were, seventy-two hours after death, to be given over for purposes of dissection; but if within this specified time a relative appeared and requested that the body might not be used for anatomical purposes, such request was to be granted. Another proposed change in the law was that a person might legally bequeath his body for dissection; in such cases the executors, administrators, or next-of-kin had the option of carrying out the wishes of the testator, or declining to do so, as they thought fit. A heavy penalty was laid on persons who were found carrying on human anatomy in an unlicensed building, and it was made an offence to move a body from one place to another, without a license for so doing. All bodies used for dissection were to be buried; the penalty for failing to do this was fifty pounds. One great blot on this Bill was the neglecting to repeal the clause which ordered the bodies of murderers to be given up for dissection…this was one of the great reasons which made dissection so hateful to the poor. During the debate, a motion was made by Sir R. Inglis "to repeal so much of the Act 9 Geo. IV. cap 31, as empowers judges to order the bodies of murderers to be given over for dissection." This, however, was lost, eight members only voting for the amendment, and forty against. There was strong opposition to the Bill outside the House. Some of the private teachers were very uneasy as regarded the effect of the Bill on themselves. The measure spoke of "recognized teachers" and "hospital schools," and all those who were to be entitled to the benefits of the Act were to have licenses from one of the Medical Corporations. The proprietors of the smaller schools felt that this would result in their extinction, and that the teaching would all pass to the large schools. In the country, too, there was strong opposition to the Bill, as practitioners there felt that they were excluded from any benefit.

About this time there was a national disgust on account of the murders committed by Burke and Hare in order to sell the bodies for dissection (Comrie, 1972; Ball, 1928; Richardson, 1987). In June 1829, the Anatomy Bill was withdrawn "*on the advice of the Duke of Wellington. The run up to what was to become the Reform Crises was on the way by this time, and the Duke probably feared that the Bill would be contentious and might perhaps provoke trouble*" (Richardson, 1991).

It was the London murders committed by Bishop and Williams that shook the public and provided the momentum for the submission on December 15, 1831 of a new Anatomy Bill to Parliament. On August 1, 1832, Mr. Warburton's Anatomy Bill was passed through both houses of Parliament and finally became law (Goodman, 1944).

Richardson (1991) is correct in emphasizing that Bentham's draft Anatomy Bill and the publication of the Report from the House of Commons Select Committee occurred before the discovery of the murders committed by Burke and Hare. "Although the first Bill's *parliamentary* progress was undoubtedly affected by Burke and Hare, Parliament's reaction to their crimes was not sufficient in itself to insure the first Bill's passage through the Lords." It was the murders that were decisive for Parliament to act quickly in passing the bill.

There are several other misapprehensions about the Anatomy Act. Regarding this, reference should be made to the insightful paper by Richardson (1991). The following discussion and extracts are largely derived from this valuable source. Richardson (1991) argued that factors other than humanitarian measures to prevent the murders, as well as to assist medical teaching, were involved. It was suggested "that differentiation must be made between influential events and strategies by which the Act finally gained its passage and the aims and intentions of those promoting it."

It was the conviction of the two doctors who were involved in body snatching that the Select Committee of the House of Commons was appointed in 1828. The committee's mission was to "*endorse Bentham's project by the adoption of a formula which would not only decriminalize dissection, but would also initiate the contentious project of dissecting the 'unclaimed' poor from hospitals and workhouses.*"

Another consideration was that "*The Act was in fact very significant on a number of non-medical fronts. Most crucially, the Anatomy Act established the first centrally funded inspectorate of the 19th-century Benthamite calendar of government reform.*" This sets the place for the establishment of other inspectorates such as for immigration, prisons, railways, and others.

Until the passing of the Anatomy Act in 1832, murderers were hanged and offered for dissection, but

with the new law the interpretation was that the same punishment awaited those who are poor and unable to pay for their funeral. "At the time the Anatomy Act was proposed, news of its provisions provoked terror among workhouse inmates" and "dissection served to promote the stigma of the pauper's funeral."

Other than unclaimed bodies, there were alternative suggestions for those who committed suicide, killed in duel, horse thieves, and all criminals who died in prison should be given for dissection rather than the poor. In order to reduce the stigma of being dissected after death, it was even suggested "that the King should set a fashion by bequeathing his body for dissection, rather than the customary extensive embalmment, and hence promote the willing donation of corpses from his subjects in all walks of life of society."

It was even suggested that the Royal College of Surgeons should maintain a register of potential donors who can be paid a lump sum during their lifetime for donating their body. Yet, another proposal was for the government to abolish death duties for those who donated their bodies for anatomical studies. None of these proposals were ever considered by the Select Committee or by Parliament. This led to the view that "the proposal to dissect the poor was primarily a political one" (Richardson, 1991).

Through the Anatomy Act, the poor who inhabited the hospitals and workhouses were now more vulnerable, whereas the upper and middle classes of society were better able to protect their remains from the body snatchers. "Such a change hardly represents a triumph of secularism. It serves rather to indicate the strength of the repugnance towards dissection and concern about the repose of their own remains among many of those who promoted the Act" (Richardson, 1991).

Another misapprehension relates to "*unclaimed*" bodies. Provisions were made for inmates in hospitals and workhouses to make a written declaration in the presence of two witnesses that they did not wish to have their body dissected. The inmates were often illiterate and their wishes were often not recorded or not followed. Of importance is the observation made by Richardson (1991) that "there is little doubt that, had the Act featured an opting-in system, or had paupers' wishes to opt out been routinely be recorded and observed, the Anatomy Act would have failed miserably from the outset." Furthermore, under the Anatomy Act, "*claiming*" had to occur within 48 hours of death, but the act failed to specify whether or when relatives were to be informed."

Richardson (1991) suggested that dissection was still perceived as a punishment according to British law even after the passing of the Anatomy Act. "The Anatomy Act merely transferred what had since Tudor times been a punishment for murder to poverty." Moreover, the Anatomy Act was far from a complete success because the bodies that were available for dissection were no more than those provided by the resurrection men before the passing of the Act. True, the supply from the workhouses and hospitals met the needs of the anatomists, but this was soon followed by shortages, which needed again the assistance of the resurrection men.

The problem of the shortage of corpses for dissection forced the anatomists to develop methods for preserving the body and dissected parts (Edwards & Edwards, 1959; Polson and Marshall, 1975). More students were assigned to a cadaver on which more operative procedures were carried out. It was the intent of the act that the Inspector of Anatomy would arrange for a fair allocation of the available bodies to the schools of anatomy depending on the number of students at each institution. The result was a lack of cooperation from the authorities of the schools and strong competition between the schools. The Anatomy Act "really marked the beginning of a new phase in a major struggle, particularly in the London area,

between the anatomy schools operating within hospitals and those attached entrepreneurial schools outside, a struggle which was eventually won by the hospital schools.... The Inspection for of Anatomy was powerless to enforce a fair distribution system, and the Act contributed instead to the demise of the unattached anatomy schools" (Richardson, 1991).

Passing of the Anatomy Act was of benefit to the hospital and the medical schools, but not to the apothecaries. Only the physicians and surgeons were permitted to obtain bodies and to carry out dissections. Even though it was the apothecaries who provided medical care for the poor because they charged a lower fee, the Anatomy Act did place between them, and the rest of the medical profession, a wedge.

Finally, the passage of the Anatomy Act through Parliament was not entirely smooth sailing (Bailey, 1896; Ball, 1928; Richardson, 1987; Desmond, 1989). There was substantial opposition to it. The Act was criticized because it infringed on human rights, and the public vented their anger through riots and by attacking the anatomists and their schools. The Anatomy Act with all its modifications over the years has controlled the supply of bodies to universities up to the present time (Polson & Marshall, 1975). A well-informed public now supports a donation program, and, on the whole, public attitude to dissection has changed considerably.

Chapter 26

GERMANY

EARLY UNIVERSITIES

The first German-speaking university was founded in 1348 by Emperor Charles IV in Prague. By the end of the fifteenth century, other universities were established in the German Nations of the Holy Roman Empire, despite the political turbulence of the period. These included Vienna (1365), Heidelberg (1386), Cologne (1388), Erfurt (1392), Köln (1388), Heidelberg (1389), Wurzburg (1402), Leipzig (1409), Rostock (1419), Freiburg (1455), Greifswald (1428), Ingolstadt (1472), Trier (1473), Tubingen (1476), and Mainz (1476). Long before, during the twelfth and the thirteenth centuries, universities were thriving in Italy, England, and France. Because of their reputation, these institutions attracted students and scholars from the rest of Europe (Steiger & Flaschendrager, 1981; Cardini & Beonio-Brochieri, 1991).

In Italy alone, 15 universities were established between A.D. 1200 and 1350. The first center of learning to be established was the medical school at the University of Salerno during the tenth century, followed by Bologna where a medical faculty was organized in 1156, even though theology, law, and philosophy were the commanding disciplines. Early medieval universities included Paris (1150), Sorbonne (1252), Oxford (1170), Montpellier (1200), Cambridge (1209), Padua (1222), Salamanca (1230), Rome (1244), Salamanca (1218/1254), Lisbon (1290), Leiden (1300), Pisa (1343), Florence (1349), Prag (1347), and others. As many as 80 universities were established in Europe before the end of the Middle Ages, but nowhere were medical studies, especially in the area of human anatomy, pursued with such vigor as in the universities of Italy. Bologna and Padua were the most celebrated of them (Persaud, 1984), and it is from these seminal sources that anatomical enquiry in the German medical schools took root.

ANATOMICAL DEMONSTRATIONS

Since the early Middle Ages, human bodies were dissected in many German cities, for legal purposes or whenever a corpse was available. Until the seventeenth century, anatomical dissections and demonstrations were less frequently carried out than in France, Holland, or Italy. These were special civic events attended by gay festivities. An exception might have been the dissections that were done by Johann Konrad Brunner, professor of anatomy at the University of Heidelberg, who was able to secure the bodies of all soldiers dying at the garrison.

Professor Brunner dissected the bodies for the benefit of the medical students. "*In general such an event was attended by professors, doctors, students, noblemen, civic administrators and anyone who might be interested*"

(Lassek, 1958). The human body was compared to a machine, and it was felt that every person could benefit from knowledge of anatomy. Also, an understanding of how the body works would bring one closer to nature and God.

The importance of human dissection in medical training was clearly recognized quite early in many German cities, but long after the practice was established in Italy. Schumacher and Wischhusen (1970) have given some of the locations and years for the earliest recorded anatomical demonstrations (Fig. 322) in Germany: Prague (1348), Vienna (1404), Cologne (1479), Tubingen (1482), Leipzig (c. 1500), Rostock (1514), Wittenberg (1526), Frankfurt Oder (c. 1530), Marburg (1535), Heidelberg (1574), Wurzburg (1585),

Figure 322. An anatomical demonstration in Rostock at the beginning of the sixteenth century. This woodcut illustration appeared in the Rostock edition (1514) of the *Anatomia Mundi* by Nicolaus Thurius Marschalk. (Universitatsbibliothek Rostock; from Wischhusen and Schumacher, 1970.)

Gießen(1609), Greifswald (1624), Altdorf (1657), and Erfurt (1675).

Anatomical theaters were built in order to facilitate demonstrations of the dissected cadaver and part of the body (Schumacher, 2007). For example, "anatomical theatres" (*Theatrum anatomicum*) were built in Altdorf (1650), Rostock (1696), Berlin (1720), and in Halle (1724) that were replicas of the anatomical theater in Padua; that in Greifswald (1750) was based on the theater in Bologna. Great discoveries in human anatomy were made in these institutions by teachers who also produced magnificent treatises. Some of these works have been of enduring value and influence and remain so even up to our time.

We have abstained from discussing in any detail individual anatomists or of the scientific contributions that have emanated from various anatomical institutes (for example, Kopsch, 1913; Wegner, 1956; Stürzbecher, 1958; Nauck, 1959; Schumacher & Wischhusen, 1970; Aümuller, 1972; Hein, 1976). The task of compiling from original sources a comprehensive history of human anatomy in Germany still remains outstanding. Of all the anatomical institutions in Germany, that in Berlin is of special interest because of its uncommon beginnings and the extraordinary influence it has exerted for centuries. For this reason, some remarks about the origin of anatomical studies in the Prussian capital (Dorwart, 1958) is presented.

ANATOMY FOR MILITARY DOCTORS

The history of anatomy in Berlin is unique. Anatomical instruction began because the King wanted it, even before there was a medical school in the city, and almost a century before a university came into existence. Furthermore, all instruction was in German, the common language of the people, and not in Latin as was the practice at that time.

Friedrich Wilhelm I, who succeeded his father Friedrich I in 1712 as king, was foremost a military man (*Soldatenkönig*). He recognized the obvious benefits for his army doctors, especially the field surgeons. These surgeons were responsible for treating the wounded soldiers, and the king felt that they should have a good understanding of anatomy. Probably influenced by his personal physician Gundelsheimer, and after a visit to the renowned medical institution of Boerhaave in Leiden, the king ordered the establishment of a *Theatrum anatomicum* in the Prussian city for the training of military doctors (Dorwart, 1958; Stürzbecher, 1958; 1959; Laitko, 1987; Winau, 1987). In passing, it should be mentioned that there had been some discussions before for the establishment of an anatomical theater in the city.

Figure 323. Christian Maximilian Spener (1678–1714); first professor of anatomy in Berlin. (Courtesy of Institut fur Anatomie, Ernst-Moritz-Arndt-Universitat, Greifswald, Germany.)

It was the practice in many European cities at that time to have on display anatomical specimens in their *Theatrum anatomicum* for the public, as well as the curious, to view as a matter of general interest, not unlike our present-day museums (Schumacher, 2007). On a more scientific level, Leibniz, the natural philosopher and statesman, petitioned the Royal Court in 1700 that a "*Konigliche Societat*" (Royal Society) should be established for the study of diverse subjects which are of practical and useful value, such as astronomy, mechanics, architecture, chemistry, botany, and anatomy. Anatomy was not perceived as being a useful subject, and there was opposition to include it.

Following the establishment of the *Collegium medico-chirurgicum* in 1724, a medical institution largely for the practical training of military doctors, the *Theatrum anatomicum* was transferred to it. Almost a century later, with the foundation of the Friedrich-Wilhelm-Universitat in 1810, the *Theatrum anatomicum* evolved into the Anatomical Institute. The university was renamed the Humboldt University in honor of the great German statesman and scholar (linguist) Wilhelm von Humboldt. For this reason, the Anatomy Institute is one of the oldest scientific institutions in Berlin.

Leibniz had proposed to the Royal Court in 1701 that the distinguished anatomist Bernhard Siegfried Albinus (1653–1721) should be appointed as the first anatomist in Berlin. Since 1681, Albinus was professor of anatomy in Frankfurt Oder. Apparently he did not follow up on the invitation he received from Berlin. Instead, in 1702, Albinus went to Leiden where he pursued a brilliant career and helped to expand the reputation and great traditions of the medical faculty.

For the newly established *Theatrum anatomicum*, the king provided both the space and money. In spring 1713, Christian Maximilian Spener (Fig. 323) was appointed professor of anatomy, and teaching began after much pomp and ceremony on November 29, 1713 (Stürzbecher, 1963).

Until the eighteenth century, physicians, i.e., the so-called *medici purl* had a higher social standing than the military surgeons (*Wundarzte*). Whereas physicians received their scientific training at the university, surgeons were usually trained in an apprenticeship system with barbers or other surgeons. The distinction between them became more pronounced with time, with the surgeons pursuing their craft on a more rigorous scientific basis. Many of the names that are familiar to us from the field of surgery who received such a training; for example, Guy de Chauliac, Ambroise Pare (France), John and William Hunter

Figure 324. The Theatrum anatomicum in Berlin, built in 1713. (Copper engraving by Anton Balthasar König from a sketch by Ferdinand Gottfried Leygebe. (Courtesy of Institut fur Anatomie, Ernst-Moritz-Arndt-Universitat, Greifswald, Germany.)

(England), Hieronymus Brunschwig (Strassburg), Fabricius Hildanus (Cologne/ Berne), and Lorenz Heister (Altdorf/Helmstedt). A knowledge of anatomy was considered essential then, not unlike the present time, for the training of *Wundarzte* or military doctors. In the *Theatrum anatomicum* (Fig. 324), there was a dedication stone from the king (Friedrich Wilhelm I) giving his full support and the assurance that there will be enough cadavers for instruction (Dorwart, 1958; Stürzbecher, 1959; 1963).

The Theatrum anatomicum was erected in a corner of the upper level of what was then the royal stables (Koniglicher "*Neuer*" Marstall). The stables were first built in 1691 and over the years greatly modified to accommodate other institutions and enterprises. These included the Societat der Wissenschaften (Scientific Society), Academia der Kiinste (Academy of Arts), the observatory (das Observatorium), eventually

the Collegium medicochirurgicum, and several manufacturing concerns, in addition to the horses. The anatomical theater was located near the area occupied now by the university and State Library (Staatsbibliothek) in Berlin.

In 1827, a house was purchased from a general practitioner and remodeled for the anatomy classes. This new Anatomical Institute was situated behind the military church (*Garnisonskirche*). It had circular rows of seats for the students and for the invited and distinguished guests who sat in the first two rows. Here dissections were carried out and lectures in descriptive and regional anatomy were given to the students. Lectures and practical classes in histology, embryology, and comparative anatomy took place in the west wing of the university itself, which also housed a valuable collection of zoological and anatomical specimens (Kopsch, 1913; Dorwart, 1958).

Chapter 27

ANATOMY IN THE NEW WORLD

The arrival of the Spanish conquerors to the New World in the late fifteenth century and early sixteenth century brought them to civilizations that were established and flourishing (Cohen, 1992). These indigenous people were culturally advanced in many respects as those that thrived in Egypt, Asia, and China. They had a social fabric with an organized hierarchy that made provisions for the population. In addition to food and shelter, these also included proper sanitation and health care. Paramount were the enlightened priests who, in elaborately constructed temples (Figs. 325 & 326), invoked the deities with their "*magical*" and religious practices, as well as with human sacrifices, for the good of all.

AZTEC MEDICAL PRACTICES

Schendel (1968), in his highly readable book on medicine in Mexico, traces the development of medical practices and health-related matters from the Aztec civilization to modern times. He remarked that Cortes and his conquistadors, as well as other Spaniards, who arrived in the capital of the Aztecs during the early sixteenth century, marveled at the magnificent pyramids the Aztecs had built, the planned layout of their homes, canals with aqueducts, floating gardens, and exotic markets. They were also deeply impressed with the skills of the Aztec physicians, who had learnt from experience the curative properties of certain plants (Cruz, 1940; Guerra, 1979; Stuart, 1987) and knew how to treat a wide variety of medical and surgical problems.

As in Europe, the Aztec physicians attended to the wounded, carried out bloodletting, administered enemas, but they were far more familiar with the use of herbs for medicinal purposes. The Aztec physicians knew about asepsis and had also developed simple surgical instruments. For surgical procedures, religious practices, and human sacrifices, narcotics derived from medicinal plants were used. Like other earlier indigenous cultures, including the Toltec, Mayan, and Incas, the Aztecs also performed trepanation of the skull (Clendinnen, 1983; Sharer, 1994).

This operation might have been carried out to reduce intracranial pressure as a result of head injuries, or to release the "*evil spirits*" of insanity (Cabieses, 1979).

Schendel (1968) also cited from the work [Fray Bernardino de Sahagun (1905): *Historia General de las Casas de Nueval Espana* (The Universal History of New Spain)] of Bernardino de Sahagun, a sixteenth century Spanish Franciscan Friar, the following with respect to fractures:

The broken bones were carefully set and the limb placed between splints of wood, tied tightly with cord. A plaster then was applied to the break, composed of gum of the ocozot/tree and resin and feathers. The limb and the splints together then were encased in a second covering of rubber-like gum.

Before aligning the bones and placing the cast, the Aztec physicians reduced any swelling and pressure by bleeding. In addition to the suturing of wounds, cauterization, performing trepanation, and setting bones in cases of fractures, there were many other surgical procedures, which the Aztec physician carried out, including some dental work, circumcision, castration, and embryotomy when normal delivery was not possible.

Figure 325. The magnificent pyramid-temple at Palenque (Yucatan Peninsula), one of many Mayan sites in Central America. The Mayan civilization flourished from A.D. 300 to A.D. 900 during which they built massive ceremonial temples, pyramids, and palaces. The Mayans developed a complex system of writing (hieroglyphs) in the New World for recording important events, and they created stunning works of art, including sculptures and brilliant paintings. (Courtesy of Dr. A. Zarian-Herzberg, National University of Mexico.)

Figure 326. The astronomical observatory at Palenque. From here, the Mayans observed the sky and the movements of the heavenly bodies, especially the moon, from which they determined the cycles of the seasons. Their priests made astronomical calculations from which the weather was predicted and the dates for ceremonial events. (Courtesy of Dr. A. Zarian-Herzberg, National University of Mexico.)

There is no record of the Aztec physician performing any thoracic or abdominal operation, but they were remarkably skilled in cutting through the thoracic wall quickly with stone knives and tearing the still-palpitating heart from the helpless sacrificial victim. There is also no evidence that the Aztec physicians carried out anatomical dissections for the purpose of increasing their knowledge of the human body, even though from the sacrificial altars there were unlimited cadavers for dissection (Schendel, 1968). Yet, the Aztecs had identified many parts of the body for which there were close to 4,000 anatomical terms (Guerra, 1979). Hancock (1995) in his fascinating book cited the following from the records of an observer during the sixteenth century:

> If the victim's heart was to be taken out they conducted him with great display...and placed him on the sacrificial stone. Four of them took hold of his arms and legs, spreading them out. Then the executioner came, with a flint knife in his hand, and with great skill made an incision between the ribs on the left side, below the nipple; then he plunged in his hand and like a ravenous tiger tore out the living heart, which he laid on the plate.

FIRST UNIVERSITY AND MEDICAL SCHOOL

The scientific study of the human body in the New World began with the National University of Mexico, the first university to be founded in the Americas, long before the English had established colonies in Virginia and Massachusetts. The National University of Mexico was established on September 21, 1551 by a royal decree from Charles V of Spain to serve the kingdoms of New Spain (Los Reynos de Nueva Espagria), as the territories conquered by Cortes and his conquistadors were called. At first, religion, literature, and philosophy were taught at the university, and in 1553, the teaching of medicine began (Schendel, 1968; Amezquita et al., 1960). It is to be expected that anatomical instruction would have been a part of the course, and most likely the content of the course was similar to the standards then prevalent in Spain. Medical schools with chairs of anatomy had been established in Barcelona (1450), Valencia (1549), Valladolid (1550), Salamanca (1551), and in other major cities.

According to Schendel (1968), the teaching of anatomy was compulsory at the university in 1621. Detailed instructions came from Viceroy Palafox in 1645 that the medical teachers and their students had to regularly carry out anatomical dissections as part of the medical course and for the training of future surgeons. On October 8, 1646 Juan de Correa and Andreas Martinez de Villaviciosa carried out the first human dissection in Mexico on an executed criminal (Febres-Cordero,

1987). A year later, the early anatomical dissections were conducted in the Hospital of Jesus, the oldest hospital on the American continent.

The Hospital de Jesus Nazareno (Hospital of Jesus) was built by Cortes in 1524, initially for the benefit of critically wounded soldiers. He made financial provisions for its maintenance of the hospital, extending even after his death. In Cortes's last will and testament, it is stated that he did this "*for the discharge and satisfaction of whatever fault or burden might grieve my conscience,*" obviously in a moment of guilt and reflecting on untold acts of cruelty, torture, and great injustices he and his conquistadors had inflicted on the now completely subjugated natives. The Hospital of Jesus was built attached to an already established church (Purisima Conception). Now extensively rebuilt and modern, the hospital has been in continuous operation since its beginning. Cortes tomb lies in the church, but allegedly his remains are in a secured location.

Fray Augustin Farfan, who in 1567 graduated in medicine from the University of Mexico, wrote a small book on medical subjects: It was published in 1579 in Mexico City and reprinted several times with revisions over the next three decades. In the work, entitled *Tractado Breue de Chirurgia y del Conocimiento y ovra de algunas enfermedades q. en esta tierra mas comunmente suelen auer*, Casa de Antonio Ricardo, described the "*anatomy, medicine, pharmacology and surgery, as practiced by the Aztecs*" (Schendel, 1968).

ANATOMICAL TEACHING

Moll's encyclopedic book *Aesculapius in Latin America* is an almost inexhaustible source of information about medicine in the countries of the New World (Moll, 1944). Much of the historical and biographical data in this section have been extracted from this valuable work.

One can only speculate about the medical skill and knowledge of the indigenous population before the arrival of the Spaniards in 1492 and the establishment of the first permanent settlement four years later. The ritual of human sacrifice apparently did not serve any other purpose than to appease the gods. Various structures of the body had been named (Guerra, 1979), but these were not based on anatomical studies or observations. The teaching of anatomy began with the medical courses and schools that were organized by the colonialists in order to train physicians for the conquered territories. Teaching followed the European pattern, largely a theoretical subject with almost no patient contact.

The city council of Mexico in 1525 paid a barber-surgeon to train others. In 1538, a medical course was inaugurated at Santo Tomas University of Santo Domingo, and later in Mexico (1580), Lima (1621), Caracas (1721), Habana (1726 or 1728), Bogota (1758 or 1760), Chile (1769), Quito (1787), Guadalajara (1792), Nicaragua (1799), Buenos Aires (1801), Guatemala (1805), Merida (1805), Bahia and Rio de Janiero (1808), Puerta Rico (1816), Cuzco (1825), Oaxaca (1827), Morelia and Michoacan (1829), Yucatan (1833), La Paz (1834), San Salvador (1847), Haiti (1852), and others followed.

In Mexico, a special chair of anatomy was established in 1621. The first autopsy was carried out in 1646, and the only practical work medical students did was to examine a skeleton once every month (Gomez, 1979). Also at the University of Lima, a chair of anatomy was founded in 1711, but the position was filled only in 1729. Guatemalan students complained in 1798 of a lack of teaching in anatomy and other subjects, which was not uncommon in other countries. Descriptive anatomy, to which surgical anatomy was sometimes added, was an important part of the medical curriculum, even though the course might have been lacking in some respects.

Moll (1944) remarked "*while anatomy was a major subject, it was taught only in a restricted manner. When Esparragosa applied for his Chair in Guatemala, it was argued that such teaching was useless. In Buenos Aires the subject had to be put off in 1800 for one year as no dissection was available.*" The situation was no different elsewhere

and reflected the prevailing attitude, in Spain, to the dissection of the human body. By the latter part of the eighteenth century, permission was granted by the Spanish government for the establishment of chairs of anatomy in Mexico (1621), Chile (1773), Cuba (1797), Buenos Aires (1801), Lima (1711), Guatemala (1809), Caracas (1811), and in Quito (1837).

On the whole there was little opportunity for the teaching of practical anatomy on the cadaver, especially with respect to the female body. According to Moll (1944), "*as late as the XIXth century only one female body was sent to the Lima amphitheatre, and that one chosen among the oldest and most deformed.*" This was due to "*false modesty and religious opposition.*" As in Europe, cadavers were stolen for teaching and examination purposes.

Medical students and doctors obtained their knowledge of anatomy up to the end of the nineteenth century largely from books, atlases (Jolin, 2013), skeletons, models, and manikins. The conditions for human dissection were far from conducive. An instance in 1825 of "*being forced to make necropsies on the graves,...parched by the sun in summer, and wading through mud in winter...*" and also at the beginning of the nineteenth century, a professor in Cuba, who was afraid of the cadavers, which were subsequently replaced by manikins, have been mentioned.

Cadavers were available for dissection in the nineteenth century, but "*bodies kept for necropsies were actually loathsome...and dissection became a bugbear to students as it was done (Chile) in a corner of the cemetery in the sun on bodies several days old with no equipment but a razor, a hammer, and a saw. Of Moran's first class of three, two died, and every year several students fell victims to infection*" (Moll 1944). Until the middle of the nineteenth century, the quality of anatomical instruction varied from one country to another. It remained sadly deficient especially with respect to human dissection.

In Brazil, until the beginning of the nineteenth century, the medical professionals were physicians, barber-surgeons, apothecaries, and an assortment of other "*healers.*" Some had received their training in Europe, others in Brazil following a period of apprenticeship. The first medical school in Brazil was founded in Bahia in 1808. About the early history of human dissection and anatomy teaching in this country, the information available is sparse (de Castro Santos Filho, 1979).

UNITED STATES OF AMERICA

When medical teaching began in Philadelphia in 1765, more than 150 years had already elapsed since

the first colony was established in Jamestown. The expeditions that brought the early settlers to the new world recognized the importance of institutions that will foster learning,

> but with them the welfare of the Church and their political economy were always first in mind. The peculiar circumstances surrounding their immigration to this country, the desperate situation in which, only too often, they found themselves after their arrival, and their dependence upon the physicians and surgeons whom they had brought from Europe, combined to make the early establishment of a medical school unnecessary, if not impossible, during the first century of American life. (Ball, 1928)

Epidemics and Other Medical Problems

One of the worst feared scourges during the seventeenth century in England and the New World was smallpox (Rothstein, 1972; 1987; Savitt, 1990). In England it was endemic with epidemic outbursts regularly between 1667 to 1676. Earlier epidemics had prompted Thomas Sydenham to give an account of the disease in his book *Methodus Curandi Febres*, published in 1666. The first epidemic known in New England occurred among the Indians of Massachusetts in 1633, just 13 years after the Mayflower sailed from England with the Pilgrims.

A similar epidemic probably occurred in Plymouth in 1620, and in 1634 the Connecticut Indians were afflicted. This terrifying disease affected more the nonimmune native population than the English settlers, but they too were not spared. The epidemics were so severe that some tribes were reduced to extinction, and in Massachusetts, the natives were reduced from a population of 3000 to 300 fighting men. With regard to the Connecticut Indians, it was noted that they "*fell sick of ye small poxe, and dyed most miserably; for a sorer disease cannot befall them; they fear it more than ye plague....*" (Viets, 1937).

In the *Almanack* for 1679: "*July 10 1677. The Vessel arrived at Nantasket which brought that contagious Distemper the Small Pox, which was soon taken by some of Charlestown going aboard since which time many thousands have taken the infection, and more than seven hundred already cut off by it*" (John Foster 1679; cited in Viets 1937). As more settlers arrived from England, the problem became increasingly serious in the colonies, with frequent occurrences of epidemics and high mortality in the population.

Those with some medical skills tried to assist drawing from the scanty information contained in books and tracts brought from England. Of interest is the broadside, a single printed sheet, entitled "*A Brief Rule to guide the common-people of New England how to order themselves and theirs in the small pocks, or measles.*" It was written by the preacher-physician Thomas Thacher and first published in 1677/1678. This historical document, now preserved in the Massachusetts Historical Society, is the earliest known medical document (dated 21.11.1677/1678, from the press of John Foster in Boston) to be printed in North America (Viets, 1937). It was then a useful tract for the ordinary man with information relating to all aspects of the disease, including signs, treatment, and prognosis.

The foregoing account does not describe all the medical problems which the early settlers faced (Rothstein, 1987; Savitt, 1990), but it outlines briefly the fearful consequences of a dreadful disease during that period. It is not surprising, therefore, that the need for the training of medical personnel to deal with these terrifying epidemics, as well as other serious conditions, became obvious (Marti-Ibanez, 1958; Packard, 1973; Bordley & Harvey, 1976).

Medical Studies and Doctors

Before 1765, there were no medical schools in the colonies. Bright young men with an interest in medicine either served as apprentices to a medical practitioner or had to make the long ocean journey to study at the great medical schools of Europe, mainly in Edinburgh and London. Usually, those who studied abroad were the sons of wealthy colonists (Rothstein, 1987). Many returned and became the pioneers of modern medicine in British America (Guthrie, 1959; Michels, 1955; 1987). Some went to other European cities. Between 1765 and 1779, 112 Americans graduated in medicine from the University of Edinburgh. Not surprisingly, the only two medical schools (College of Philadelphia and King's College in New York City) that were established during the colonial era followed the program and traditions of that in Edinburgh (Packard, 1931).

In contrast, no more than about 20 students studied medicine in German universities before 1850. John Foulke of Philadelphia was the first American who went to a medical school outside of Britain. Instead, in 1781, he went to Paris and later Leipzig. According to records, Benjamin Smith Barton was the first American to graduate in medicine from a German university. He received the M.D. degree from Gottingen University in 1789, and later he became professor of botany at the University of Pennsylvania. After 1850, there was a surge in foreign students attending German universities. This was due to the remarkable advances that had occurred there and the reputation of many German professors (Bonner, 1963).

In addition to Britain and Germany, Austria, Switzerland, and France attracted a fair number of medical students from the United States even up to the beginning of the present century. Guides were published advising prospective medical students on the suitability of universities, clinics, the courses available, and names of professors (Hun, 1883).

Anatomical Studies and Medical Schools

John Eliot (1604–1690) lamented in 1647 that the *"young Students in Physick were forced to fall to practice before ever they saw an Anatomy made,"* and that up to then there had been *"but one Anatomy in the Countrey, which Mr. Giles Firmin (Firmin) (now in England) did make and read upon very well"* (Krumbhaar, 1922). According to Hartwell (1881a), this was the *"earliest utterance in America, in recognition of the importance of anatomical studies."*

On September 22, 1676, Judge Samuel Sewall of Boston made the following entry in his diary: *"Spent the day from 9 in the M. with Mr. (Dr.) Brakenbury, Mr. Thomson, Butter, Hooper, Cragg, Pemberton) dissecting the middle-most of the Indians executed the day before"* (Ball, 1928).

Six autopsies were reported in New England between 1674 and 1678. In 1691, Doctor Johannes Kerfbyle, a graduate of Leiden, performed an autopsy on the body of Governor Slaughter of New York, who had died suddenly. He was suspected of being poisoned. Doctor Kerfbyle and the five doctors who assisted him were paid £8.8s for the job.

By the middle of the eighteenth century courses in practical anatomy were available for those interested. Doctor Thomas Cadwalader (1708–1779) completed his medical training in Rheims, France, and under William Cheselden in London. In 1750, he began the first course on human dissection in Philadelphia. Also in 1750, Doctor John Bard and Doctor Peter Middleton, in New York City, *"injected and dissected the body of Hermanus Carroll, an executed criminal, for the instruction of the young men then engaged in the study of medicine."* In addition, Thomas Wood, a surgeon, informed the public in the *New York Weekly Postboy* on January 17, 1752, that he will be giving *"Course on Osteology and Myology...in the City of New-Brunswick,...in which course, all the human Bones will be separately examined, and their connections and dependencies on each other*

Figure 327. The first medical school building in the United States. (The University of Pennsylvania, Philadelphia; with permission from Major, 1954.)

demonstrated; all the Muscles of a body dissected..." At that time, there were no entrance requirements for prospective medical students, and commonly only a few certificates of courses taken for a year or two were necessary in order to obtain the medical degree.

In Newport, Rhode Island, William Hunter (1720–1777), a relative of the famous brothers, John and William Hunter, gave a course from 1754 to 1756 on the *"History of Anatomy and Comparative Anatomy."* Because of the demand, there had been many other efforts to provide practical anatomy courses as part of medical training, which paved the way for the establishment of medical schools in the colonies (Krumbhaar, 1922; Ball, 1928; Packard, 1931; Sigerist, 1933; Rothstein, 1987).

The first medical school in the United States was established in Philadelphia in 1765 by William Shippen and John Morgan as the School of Medicine of Pennsylvania (Michels, 1955; Bordley & Harvey, 1976). At that time, Philadelphia was the largest city in British America. This institution later became the University of Pennsylvania (Fig. 327). Following William Shippen's death in 1808, Caspar Wistar (1761–1818) was appointed professor of anatomy at the University of Pennsylvania.

Caspar Wistar

Doctor Caspar Wistar (1761–1818; Fig. 328) began his medical studies with two of the most prominent

CASPAR WISTAR M.D.

Pendletons Lithog.?

Figure 328. A portrait of Caspar Wistar. (Courtesy of Bethesda, MD: U.S. National Library of Medicine, National Institutes of Health, Health & Human Services.)

physicians in Philadelphia, Dr. John Redman (1722–1808) and Dr. John Jones (1729–1791). Redman was born in Philadelphia where he began his medical studies, which he completed in Europe. Dr. John Jones too had studied medicine in Europe and settled at first in New York before moving to Philadelphia. A skilled surgeon, he was one of the founders of King's College Medical School which became later the Columbia University College of Physicians and Surgeons. John Morgan and Benjamin Rush were also students of John Redman.

Wistar entered the University of the State of Pennsylvania (University of Pennsylvania) as a medical student and graduated three years later with the Bachelor of Medicine (M.B.) degree. He continued his medical studies in England and Scotland, and in 1786 he received the doctor of medicine degree from Edinburgh University. Wistar had also studied anatomy with John Hunter at the Great Windmill School in London.

Returning to Philadelphia in 1787, Wistar received an appointment to the Philadelphia Dispensary and

was elected to the College of Physicians. He established a successful medical practice and five years later was on the staff of the Pennsylvania Hospital. In 1787, he was elected a Fellow of the American College of Physicians. In 1791, Wistar joined the medical school of the University of Pennsylvania which he served until his death. He served as an adjunct professor of anatomy, midwifery, and surgery (1791–1808); professor of anatomy and midwifery (1808–1810); and as professor of surgery (1810–1818).

Because of his reputation as an excellent anatomy lecturer, Wistar attracted a large number of students to the medical school. He used large anatomical drawings and prepared anatomical specimens and models for teaching the students. Until Wistar's day, anatomical textbooks came from overseas. Wistar was the author of a two-volume anatomy textbook, *"A System of Anatomy for the Use of Students of Medicine,"* Volume 1, 1811, Volume 2, 1814. These were the first anatomy textbooks ever printed in what is now the United States. In 1815, he became President of the American Philosophical Society following the resignation of Thomas Jefferson, and succeeded Benjamin Rush as President of the Society for the Abolition of Slavery (University of Pennsylvania University Archives, Penn Biographies: Caspar Wistar, 1761–1818).

John Morgan

Morgan (1735–1789) attended college in Philadelphia. After graduating in 1757 he was a resident apothecary to the Pennsylvania Hospital, and later he served as a surgeon during the wars against France and the Indians. In 1760, Morgan left for London, England, where he was a pupil of William and John Hunter in 1762, and Alexander Monro *Primus* in Edinburgh; later, he moved to Paris and Italy (Fig. 329).

Morgan learnt the most recent advances in medical practice from the most prominent medical teachers in Europe. Apparently, William Hunter was angry with Morgan for demonstrating to the Royal Medical Society in Paris the technique of making corroded anatomical specimens without any acknowledgment to William from whom he learned the technique. William, himself, had learned the technique from Frank Nicholl (Brock, 1983).

After five years, Morgan returned to America with the most impressive credentials, including the distinction of being a Fellow of the Royal Society of London; Correspondent of the Royal Academy of Surgery at Paris, member of the Arcadian *Belles Lettres* Society at

Rome, and a Licentiate of the Royal Colleges of Physicians in London and in Edinburgh.

Morgan's "*creative mind was soon occupied with the problem of establishing in this country (United States of America) a school of medicine of the type in which he had worked abroad...He realized, for example, the importance of specialization...*

He realized the necessity of a sound preliminary education in literature and science....*He saw the fundamental importance of training in anatomy, physiology, chemistry, and physics, on the basis of which alone clinical medicine could be cultivated*" (Flexner, 1937).

Morgan had a clear vision and high ideals for "*the institution of medical schools in America*" which he conveyed, almost immediately following his appointment as professor of the theory and practice of medicine, in a public lecture he delivered at an anniversary commencement, held in the College of Philadelphia at the end of May, 1765.

Morgan had a good understanding of "*the various branches of knowledge which compose the science of medicine.*" Included among his writing is a monograph on the preservation of anatomical material. Morgan's inspired remarks regarding anatomy, although more than two centuries old, are valid today even more as medical schools attempt to refashion the anatomy course (Morgan, 1765).

William Shippen Jr.

William Shippen Jr. (1736–1808; Fig. 330), who had also studied with the Hunters in London, and Monro *Primus* in Edinburgh, was by then already teaching privately in Philadelphia for the past three years, following his return in 1762 from Europe. William Hunter was one of his teachers, and in admiration of William, Shippen named his son after him. In September 1765, he was appointed professor of

Figure 329 (*Left*). John Morgan (1735–1789). (Courtesy of Bethesda, MD: U.S. National Library of Medicine, National Institutes of Health, Health & Human Services.) Together with William Shippen, Jr., Morgan founded in 1765 the Medical College of Philadelphia. He was elected to the first professorship offered, the Theory and Practice of Physic. Regarding anatomy, Morgan remarked: "*It is anatomy that guides the doubtful steps of the young votary of medicine through an obscure labyrinth, where a variety of minute objects present themselves in such a group as, at first to perplex his imagination...*" (Lassek, 1958).

Figure 330 (*Right*). William Shippen, Jr. (1736–1808). (Portrait by Gilbert Stuart; Courtesy of Bethesda, MD: U.S. National Library of Medicine, National Institutes of Health, Health & Human Services.)

anatomy and surgery. Shippen was, therefore, the first professor of anatomy in America. For his classes, he made extensive use of a set of elegant anatomical charts and casts, which John Fothergill (1712–1780) had presented to the Pennsylvania Hospital. These were made in London, and the charts were drawn in 1755 by the well-known artist J. Van Riemsdyck, who was responsible for the illustrations in William Hunter's magnificent atlas on the pregnant uterus (Krumbhaar, 1922; Corner, 1951).

First Medical School

During this period, Philadelphia became the medical capital in America, and the teaching of anatomy was firmly established under Shippen and his successors, including the brilliant Caspar Wistar. Wistar, who became a full professor of anatomy at the University of Pennsylvania in 1808, founded a museum and wrote a widely used textbook of anatomy.

In the autumn of 1796, a young Virginian, James T. Hubard, went to study at the newly established medical school in Philadelphia. Three weeks later he wrote to his father of his experiences and early impressions. About Doctor Wistar, James Hubard made the following remarks:

> Dr. Wistar is certainly an eminent anatomist His lectures on the Bones are equal it is supposed to those of Monro. There is a large Collection of anatomical preparations in the Theatre. Last Monday he began the Lectures on the Fresh Subject. He began first by dissecting the parietes of the abdomen. Here every muscle, was most beautifully dissected and Demonstrated. He is now demonstrating the abdominal viscera. Since I have attended the Lectures on Anatomy, I am fully convinced that a person can hardly acquire any other knowledge of the human Body by mere reading those names. (Hoyt Jr., 1942)

New York City

The second medical school, established in 1767 in New York City, also began as a private school in 1763 with Doctor Samuel Clossy giving lectures in anatomy to students. He was a graduate of Trinity College in Dublin, Ireland and held the chair of natural philosophy at King's College, which later became Columbia University. When the "*medical department*" was established at King's College, Clossy became its first professor of anatomy. One of the founders of King's

College Medical School in New York was John Jones (1729–1791). He had studied medicine in London and France, and obtained the doctor of medicine degree from the University of Rheims in 1751. Jones was appointed professor of surgery, which he pioneered in America. He was also the author of a medical book dealing with fractures and wounds, the first medical textbook to be published in colonial America. The first medical class graduated in 1769; and Doctor Samuel Bard (1742–1821), who gave the convocation address, noted that there was no hospital in the city even though such an institution was necessary for the training of the medical students.

Medical Schools in Other Cities

With the increase in population and the establishment of cities, Departments of Anatomy and medical schools were founded in Boston Harvard University (1783), New Hampshire Dartmouth College (1797), Baltimore University of Maryland (1807), New Haven Yale (1812), Philadelphia Jefferson (1824), Augusta, Georgia (1828), New Orleans Tulane (1834), and in many other places (Bordley & Harvey, 1976). The teaching of anatomy was now firmly rooted in the fledgling medical colleges of the new land. By 1910, dissection was a requirement for the medical course as recommended by Flexner in his report (Flexner, 1910).

In 1810, there were five medical schools for a population of about seven million. With an increase in the population to 17 million over the next three decades, the number of medical schools increased to more than 30. This rapid increase led to a deterioration in standards and quality of training. Lamenting the deplorable state of medical education in the country, Daniel Drake (1785–1852) observed "*in most medical schools the teachers who held such titles as Professor of Anatomy, or Chemistry, or Physiology, or Pathology had no real claim to such titles.*" Improvements and changes in medical education began to occur slowly by the end of the nineteenth century (Bordley & Harvey, 1976).

Corpses for the Dissecting Room

Similar to Ireland, Scotland, and England, there had been the obvious need for human corpses in order to demonstrate the parts to students. Because of opposition from the public, anatomists experienced difficulties in obtaining bodies for dissection (Shryock, 1966). According to Rothstein (1972), human dissection or practical anatomy was seldom required for graduation because of the limited material that was available due to anti-dissection laws and public opposition to dissection. The dissecting room was usually hidden in a remote corner of the building, and the cadavers were obtained legally or stolen from the cemeteries.

Medical students had to pay a "*demonstrator*" to assist them in the dissecting room. In antebellum Virginia, medical students were also required to dissect a human cadaver, but few people donated their bodies to the medical schools for this purpose. As a result, medical students and their professors hired resurrectionists or grave robbers to supply them with bodies from the pauper and black cemeteries. There was fierce competition between the medical schools for the bodies which was eventually settled with a signed agreement in 1851 (Savitt, 1990).

Grave robbing was not uncommon, and such occurrences incited the public. In Philadelphia,

> the opening of an anatomical theatre created great alarm among some of the citizens…and on several occasions Dr. Shippen's labours were interrupted by rioters…more than once Dr. Shippen had to desert his home and conceal himself, in order to avoid bodily harm. Several times Dr. Shippen addressed the populace through the public papers, assuring them that he had not disturbed private burial grounds; and stating that the subjects anatomised by him were either persons who had committed suicide, or such as had been publicly executed except, he naively adds, now and then one from the Potter's Field. (Ball, 1928)

In New York City, a riotous mob stormed the New York Hospital because of rumors that the doctors had obtained bodies from the graveyards for dissection. Several people gained entry into the dissecting room where they found cadavers and "*mutilated*" parts of the body.

> Enraged at this discovery, they seized upon the fragments, as heads, legs and arms, and exposed them from the windows and doors to public view, with horrid imprecations. The rioters had now become so outrageous, that both the civil and military authorities were summoned to quell the tumult, and the medical students were confined in the common prison for security against the wild passions of the populace. (Thacher, 1828; cited in Ball 1928)

The riot, which lasted over two days, was quelled by the army under General Steuben's command. The confrontation between the military and the crowd resulted in several deaths as well as many with severe wounds (Hartwell, 1881a;b; Packard, 1931).

Legislation for Anatomical Dissection

One of the outcomes of the anatomy riot in New York City was the hastening of legislation that would make some provision for anatomical dissection in medical schools. On January 6, 1789, appeared "*An Act to prevent the odious practice of digging up and removing, for the purpose of dissection, dead bodies interred in cemeteries or burial places*" Chapter III of the Laws of the Twelfth Session of New York. However, the first statutory provision regarding anatomy in America appears to be the Massachusetts Act of 1784 which permit "*the bodies of persons killed in duels, and of those executed for killing another in a duel might be given up to the surgeons*" for dissection and to be "*anatomised.*" In 1831, the State of Massachusetts legalized the "*study of Anatomy in certain cases,*" probably influenced by the debate and impending laws in other states, as well as in Britain, that would make anatomy legal. This was the first Anatomy Act passed in North America (Spector, 1955).

The demand for cadavers to dissect, in order to teach anatomy, exceeded by far the supply. Body snatching continued right up until the beginning of this century. As many as 5,000 cadavers were dissected in the medical schools of the United States during the late 1870s; most of these were stolen or unlawfully obtained. "In the 1920s, reports Dr. D. C. Humphrey, stolen cadavers black ones were being shipped from state to state, just as they had been for a hundred years" and "*in 1922 the New York State Board of Medical Examiners charged that cadavers were being illegally snatched from New York City mortuary and turned over to students of naturopathy*" (Drimmer, 1981).

Undoubtedly, one of the most sensational cases of grave robbing was "*the disappearance of the body of John Scott Harrison from its supposedly secure grave and of its discovery in a chute in the Medical College of Ohio*" (Edwards, 1957). He was the father of Benjamin Harrison who was the twenty-third president of the United States. John Scott Harrison died on May 25, 1879 at the age of 74. Because of the large number of grave robberies at that time, every precaution had been taken by his relatives to protect his remains in a grave that was reinforced by a cemented brick vault and large, flat stones. A watchman was even employed to keep an eye on the grave nightly for a week.

The Medical Curriculum

Before 1825, the medical curriculum in general comprised of both basic scientific and clinical subjects, of which anatomy was especially important. For an account of the anatomy course, the following has been taken from the work of Rothstein (1987):

> The keystone of the curriculum was anatomy, the most popular and important course in the medical school. Anatomy was the only medical science with demonstrable value in the practice of medicine. It could be taught effectively only in the medical school, which had the necessary skilled teachers and equipment. The teaching of anatomy was divided into two courses. Lectures on anatomy described the human skeleton and organs, illustrated by cadavers when possible and by skeletons, drawings, models made of wax, plaster, wood, or papier-mdché, and both dry and wet specimens. The use of models had just been developed in Europe, indicating that American medical evaluation was abreast of European medical education in this regard. The lectures also included morbid or pathological anatomy, the anatomical changes that occurred in disease. The significance of anatomy for surgery was a frequent topic of discussion.

> The second part of the teaching of anatomy was *"practical anatomy"* or the dissection of cadavers. Most medical schools in the early nineteenth century offered practical anatomy as an optional course, unwilling to require it because of the difficulty of obtaining cadavers and unsympathetic public opinion. By 1848, 25 schools required a course in practical anatomy, although many of them permitted students to take it elsewhere and some did not even teach the course. Practical anatomy provided actual experience in dissection under the supervision of a demonstrator and was the only laboratory course in the medical school curriculum. Many cadavers had been preserved in barrels and were in poor condition, which reduced their education value. Nevertheless, most students regarded practical anatomy, as one recalled, *"the chief event of the course."*

CANADA

In 1759, Canada became a British colony. Less than three decades later, after a bitter and fierce revolution, the 13 British Colonies in America secured their independence from the British Government in 1783. As disbanded soldiers, Empire Loyalists, and other settlers entered Upper Canada, the population increased and the demand for medical services with it. There were some medical men, especially army and navy surgeons, in the colony, but many others had dubious credentials and made claims that bordered on quackery. At that time, there was no formal medical training available in the country, and professional standards to regulate the practice of medicine was also lacking.

Medical Practice in Upper Canada

The first step in regulating medical practice occurred in 1788 with the passing in the British Parliament of *"an Act or ordinance to prevent persons practising physic and surgery within the Province of Quebec, or midwifery in the towns of Quebec and Montreal, without license"* (Canniff, 1894). Upper Canada was then a part of the province of Quebec. During these early years in the history of the country, other bills were passed, but with shifting success, in controlling the practice of qualified physicians, as well as barring unskilled persons from engaging in the practice of medicine. As late as in 1817, the following advertisement was posted at a public place:

> Richmond, Oct. 17th 1817.
> ADVERTISEMENT. This is to certify that I, Solomon Albert, is good to cure any sore in word Complaint or any Pains, Rheumaticks Pains, or any Complaint what so ever the Subscriber doctors with yerbs or Roots. Any person wishing to employ him will find him at Dick Bells.
> "Solomon Albert"

"The population had to be protected from unqualified persons and quacks. The act to license Practitioners in Physic and Surgery was passed on March 14, 1815 because many inconveniences have arisen to His Majesty's subjects in this province from unskillful persons practicing physic and surgery therein...." Eventually, the act was repealed in 1818 because it was found impractical to enforce. The act was amended to make more provisions for the licensing of doctors:

> That it shall and may be lawful for the Governor, Lieutenant-Governor, or person administering the government to constitute and appoint, under his hand and seal at arms, five or more persons legally authorized to practise physic, surgery or midwifery in this province, to be a board whereof any three to be a quorum, to hear and examine all persons

desirous to apply for a license, to practice physic, surgery and midwifery, or either of them, within this province, and being satisfied by such examination that any person is duly qualified to practice physic, surgery and midwifery, or either, to certify the same under the hands and seals of two or more such board, where upon the Governor, Lieutenant-Governor, or person administering the Government, being satisfied of the loyalty, integrity and good morals of such applicant may, under his hand and seal at arms, grant to him a license to practice physic, surgery and midwifery, or either, comfortable to such certificate; provided always, that nothing in this Act shall extend to prevent any female from practising midwifery in this province, or to require such female to take out a license as aforesaid.

The act barred unqualified persons from practicing physic, surgery, or midwifery within the province with a fine of "*the sum of £100*" for "*every such offence.*" Through this Act of 1818, the foundation for the regulation of the medical profession was then established, with the formation of the medical board in 1819. It met regularly and one of its main functions was the examination of medical candidates. The board met about once every three months, and particular emphasis was given to the candidate's knowledge of anatomy, a lack of which would be sufficient reason to reject him.

The July 1828 minutes of the board reported that of three candidates examined, one was found "*deficient in Chemistry, Anatomy, etc.*" for which further study and a course of lectures were recommended. Also, the minutes of the April 1830 meeting of the medical board reported that four candidates were examined; two were rejected, one because he was "*deficient in Anatomy.*" The candidate was "*recommended to attend a course of lectures at some medical university*" (Canniff, 1894). Indeed, the importance of anatomy as an integral part of the candidate's medical training was constantly emphasized in the minutes of the board meetings, and passing this subject was an absolute requirement for the granting of a license to practice the medical profession.

When compared to the European medical traditions, doctors practicing in New France, and in other parts of Canada, had a scientific foundation that was essentially derived from the *mother country*. In the British Colonies of America, such as Virginia and New England, medical practice would have reflected that in England because of the well-trained and university educated physicians, as well as the experienced barber-surgeons. In most cases, however, the conditions were the same. "*There were the same halting beginnings in a hostile environment, the same surge, because of the scarcity of trained men, of quacks and charlatans, leading to the sort of pressure first to regulate the profession, and then to establish decent standards of medical education*" (Jack, 1981).

For an interesting account of the development of medical education in Canada, reference should be made to the works of Canniff (1894), Heagerty (1928), Jack (1981), and McPhedran (1993). The last work gives an excellent survey of the development of medical education in all the existing medical schools in Canada. It spans the period from 1822 to 1992, and for the earlier phase, the other two works are particularly recommended.

Medical Schools

The first recognized medical school in Canada was established in Lower Canada. It began in 1822 as the Montreal Medical Institution in affiliation with the Montreal General Hospital, and in 1829, it "*was accepted*" as the Faculty of Medicine of Montreal's McGill University (Frost, 1979). This was followed by medical schools at the University of Montreal (1843), the University of Toronto (1843), Laval University (1852), Queen's University (1854), Dalhousie University (1868), University of Western Ontario (1882), and at the University of Manitoba in 1883 (Heagerty, 1928; McPhedran, 1993). By the end of the nineteenth century, eight of the present 16 medical faculties in Canada were already established.

The early history of anatomy in the medical schools of Canada during the early part of the last century have been sparsely documented. Some of this information will now be presented, taking into consideration the period that this book covers. In fact, only the medical school at McGill University, which opened its doors to students in 1822, was functioning before the British Anatomy Act was passed. With respect to the origins of medical education in Upper Canada, the following extract from the October 1826 minutes of the Medical Board is of interest:

In regard to the profession of Medicine, now becoming of great importance in the Province, it is melancholy to think that more than three-fourths of the present practitioners have been educated or attended lectures in the United States; and it is to be presumed that many of them are inclined towards that country. But in this colony there is no provision whatever for attaining medical knowledge, and those who make choice of that profession must go to a foreign country to acquire it.

About this time, efforts were being made to establish a medical school with a program similar to that at the University of Edinburgh. Thus, in 1822, the Montreal Medical Institution began teaching with four professors and five students. Doctor John Stephenson gave the lectures on anatomy, physiology, and surgery. This institution eventually evolved into the medical school of McGill University. Doctor Stephenson received his M.D. degree from the University of Edinburgh for a thesis entitled *de Volesynthesis*, written in Latin (McPhedran, 1993).

Anatomy Classes

With respect to the teaching of human anatomy, Doctor D. C. MacCallum, who was a medical student in the year 1850, informed us of the following:

... special country excursions to secure material for practical anatomy, were of frequent occurrence. The last involving as it did a certain amount of danger, commended itself particularly to the daring spirits of the class, who were always ready to organize and lead an excursion having that object in view. These excursions were not at all times successful, and the participators in them were sometimes thwarted in their attempts and had to beat a precipitate retreat to save themselves from serious threatened injury. They contributed, moreover, to the unpopularity of the medical students.... Dissections and demonstrations were made only at stated times during the morning and afternoon of the day. There evidently existed a marked disinclination on the part of both demonstrator and student to work at night in the highest story of the lonely building, far removed from other dwellings, imperfectly heated, and lighted by candles, the light being barely sufficient to render the surrounding darkness visible. Having occupied for two seasons the position of Prosector to the Professor of Anatomy, I had to prepare, during the greater part of the session, the dissections of the parts which were to be the subject of the Professor's lecture on the following day. This necessitated my passing several hours, usually from nine to twelve o'clock at night, in the dismal, foul-smelling dissecting room, my only company being several partially dissected subjects, and numerous rats

which kept up a lively racket coursing over and below the floor and within the walls of the room. Their piercing and vicious shrieks as they fought together, the thumping caused by their bodies coming into forcible contact with the floor and walls, and the rattling produced by their rush over loose bones, furnished a variety of sounds that would have been highly creditable to any old-fashioned haunted house. I must acknowledge that the eeriness of my surroundings was such that I sometimes contemplated a retreat, and was prevented from carrying it into effect only by a sense of duty and a keen dislike to being chaffed by my fellow-students for having cowardly deserted my work....

Because of conflict with McGill University, a second Faculty of Medicine (l'Ecole de medecine et de chirurgie de Montreal) was established at the University of Montreal in 1843. It was organized by six physicians, and the first class had six students, four English-speaking and two French. Doctor Jean-Gaspard Bibaud, of the first graduating class at McGill, joined the new medical school as the anatomy professor. Lectures were given in both English and French, and this bilingual institution quickly became a serious competition for McGill on account of the quality of its medical course.

In the same year, the Toronto School of Medicine was established following the trail of several proprietary medical schools in Toronto. Toronto was then the capital of Upper Canada. Other "medical schools" were started in Ontario; some went into oblivion and others became amalgamated with the University of Toronto, except for Queen's, Trinity, and McMaster which remained independent. In 1848, a medical school was established in Quebec City. A two-year course of lectures and anatomy demonstrations was offered to the first batch of students (McPhedran, 1993). By the middle of the nineteenth century, there were less than 200 medical students in the various medical schools in Canada (Heagerty, 1928).

When the medical department of King's College, the first medical school in Upper Canada, was established in 1843, a professor of practical anatomy was appointed. The Medical Committee was of the "*opinion that it will conduce to the interests of the Medical School, and be more in accordance with the usages of some British universities if the designation of Demonstrator of Anatomy be altered to that of Professor of Practical Anatomy.*" In addition, a course was given in anatomy and physiology over a period of six months. For the practical part of the course, the students were expected to pay a fee of

£4. The proposed salary for the professors of anatomy and physiology, and of practical anatomy, was £200 and £250, respectively.

Regulating Human Dissection and the Anatomy Act

Obtaining cadavers for dissection proved to be a great problem throughout this era (MacGillivray, 1988) even though the Legislative Assembly of the Province of Canada passed on December 6, 1843 "*An Act to Regulate and Facilitate the Study of Anatomy*" (Canniff, 1894). The bill received royal sanction from the governor, Sir Charles Metcalfe, three days later. It followed the principles of the first Anatomy Act in North America, which was passed by the state of Massachusetts in 1831 and the more familiar British Anatomy Act that was passed in 1832. The first Canadian legislation with respect to the supply of bodies for medical studies began with the following remarks (Lawrence, 1958):

Whereas it is impossible to acquire a proper or sufficient knowledge of Surgery or Medicine, without a minute and practical acquaintance with the structure and uses of every portion of the human economy, which requires long and diligently prosected courses of dissections; And whereas the difficulties which now impede the acquisition of such knowledge amount almost to a prohibition of the same, and it has become necessary, in consideration of the rising importance of Medical Schools in this Province, and for the relief of suffering humanity, to make some legislative provision, by which duly authorized teachers of Anatomy or Surgery may be provided with bodies necessary for the purpose of instructing the pupils under their charge; Be it therefore enacted by the Queen's Most Excellent Majesty, by and with the advice of the Legislative Assembly of the Province of Canada, constituted and assembled by virtue of and under the authority of an Act passed in the Parliament of Great Britain and Ireland, intituled, An Act to Re-unite the Provinces of Upper and Lower Canada, and for the Government of Canada, and it is hereby enacted by the authority of the same, that the bodies of persons found dead publicly exposed, or who immediately before their death shall have been supported in and by an Public Institution receiving pecuniary aid from the Provincial Government, shall be delivered to persons qualified as hereinafter mentioned,

unless the person so dying shall otherwise direct: provided always, that if such bodies be claimed within the usual period for interment, by bona fide friends or relatives, or the persons shall have otherwise directed as aforesaid before their death, they shall be delivered to them or decently interred. (Statutes of the Province of Canada. 7 Vict., Cap. 5, ss. 1–9)

In this regard, at the January 10, 1844 meeting of the Council of the Faculty, it was resolved: That Profs. Gwynne, Beaumont and Sullivan be authorized to enter into engagement for procuring subjects; Bursar to advance requisite funds. Dr. Paget to be written to and authorized to expend the sum of money remaining in his hands, as to him may seem most advantageous for the medical department.

Cadavers for Dissection

The problems of procuring cadavers for the practical anatomy classes must have been of pressing urgency because appropriate legislation was passed by the Legislative Councils during the middle of the nineteenth century in both Upper and Lower Canada, "*making it legal and relatively easy to obtain cadavers for medical education.*" Before the act was passed, body snatching was necessary to obtain anatomical specimens. Even later the situation remained relatively unchanged.*

Jack (1981) described the "*dead-house*" in the newly established medical school in London, Ontario as

"makeshift quarters, a cottage on the grounds of the Hellmuth Boys' College which had been taken over by the University. The dissecting room was in the formal dining room....It contained two tables, a few chairs, a pile of sawdust, a shovel in a corner, old coats and aprons on hooks along the walls. A trapdoor in the floor led to the cellar where two large vats, filled with ancient wood alcohol and other things, permeated the whole building with their odours." The cottage "inspired dread and fascination throughout the neighbourhood, to corner sinister appearance, as if the ghastly activities inside were affecting the very foundations."

The indefatigable Francis J. Shepherd (1851–1929) was one of Canada's most distinguished anatomists and surgeons, as well as a dermatologist and an art

---

*See Bensley (1958), Leblond (1966), MacGillivray (1988) for a review of the problem.

connoisseur. He also served as dean of the Faculty of Medicine at McGill's University, where he had been a medical student from 1869 to 1873. Shepherd's observations of the anatomy course, extracted from his "*Reminiscences*" (Shepherd, 1919), provide some insight as to the state of anatomy teaching at that time:

> ...nearly every subject for dissection was obtained illegally, by the old method of "body snatching." Even though there was an Anatomy Act in the Province of Quebec at that time, it was hardly implemented. All of the basic subjects were taught through lectures, except for anatomy which had a laboratory (dissecting) component. Regarding the Professor of Anatomy, William Scott, Shepherd noted that he "was a handsome man, very bluff in his manner, who lectured on anatomy word for word from Wilson (Erasmus); he never entered the dissecting room, or used the blackboard, though he did pass bones about the class and a dissected subject was shown at his lectures. He was always a great friend of the students and championed their cause when in trouble...."

With respect to the anatomy course itself and the examinations, Shepherd offered the following:

> I remember I was never obliged to dissect the brain, the thorax, or abdomen. I never saw a posterior view of the pharynx until I went to London. In fact, the dissecting consisted in exposing the muscles, chiefly of the arm, leg, and neck, with the accompanying blood-vessels and nerves. One could do other things if one wished, but was never compelled to. I remember it was quite a common thing for the men on the abdomen to toss up to see who should be obliged to clean out its contents in order to get at the prevertebral muscles. We had no demonstrating as we afterwards knew it; the chief thing was to pay for our extremities and get through the work quickly. Subjects were usually plentiful and all obtained from adjacent cemeteries by French students, who paid their fees in that way. The examinations in anatomy at the end of the third year were written and oral, the oral consisting of questioning by the professor for a few minutes; no actual dissections or any specimens were shown; it was purely a test of memory of anatomical facts.

Following his return from Europe in 1875, Shepherd joined the McGill Medical Faculty as Demonstrator of Anatomy, the only such position existing then at the university. He, too, had to rely on bodies for dissections that were provided by the resurrectionists.

The body snatchers were usually medical students who paid their tuition fees from the sale of the subjects. Shepherd insisted that medical students should personally dissect the cadaver in order to gain more practical experience in anatomy. He revolutionized the teaching of anatomy at McGill's University and set high standards that were admired and emulated in other medical schools.

Shepherd was determined to change the teaching of anatomy at McGill's University from its uninspiring past he himself had experienced, just a few years earlier as a medical student, to a more dynamic discipline that was of greater relevance to medical practice. In order to do so, a regular supply of bodies for dissection was required. As late as during the latter part of the nineteenth century there were many obstacles in the way, resulting in body snatching and other sordid events:

> For some years I obtained subjects from the Cote des Neiges Catholic Cemetery, to the west of the Montreal Mountain. Two Irish students made a compact with the guardian of the cemetery, and aided and abetted by him obtained many subjects (so I learned afterwards). The dead poor, not being able to pay expenses of the vaults, were buried in winter in very shallow graves in a certain corner of the cemetery, and those freshly made graves were marked by the guardian and the students went up at night, disinterred the bodies, buried usually the previous morning, removed all clothing, wrapped them in blankets and tobogganed them down Cote des Neiges Hill....

Sometimes these bodies were missed by relatives and the dissecting rooms of the city were searched, and if the body was identified, it was confiscated by the detectives. Occasionally, they prosecuted me for receiving the body. Now, as there is no property in a dead body and no clothes were taken, the only count on which they could summon me was "*Offence against decency*," and I was usually fined $50. The judge, a Mr. Coursol, recognized the necessity of obtaining material for dissection, always fined me and nothing more was said. I seldom knew who brought the bodies, and the janitor, strange to say, was never summoned as it was supposed to be all done by medical students for the love of anatomy and in the interests of their profession, and it was thought that they had the entrée to the dead house.

Naturally, there were other sources of supply, such as country cemeteries at a distance, and sometimes I received a subject (frozen, of course) from the railway in a Saratoga trunk. Occasionally, if there had been soft weather, the smell from the trunk attracted attention and excited suspicion so that box was often opened and the body found. Needless to say, it was never claimed and no one knew who sent it.

At one time, the scandal of the "body snatching" enterprise became so great that public opinion was aroused. It is the custom in Eastern Canada in country places in winter, on account of the frozen state of the ground, to place the dead in vaults or dead houses in place of burying them, and in the spring when the ground thawed out, they were interred in the ordinary way. Well, at one time, the students in search of subjects broke open these vaults and removed all the bodies from them without disturbing the coffins, leaving the clothes behind. I have seen the French students bring in as many as 10 or 12 bodies at one time, obtained in this way. Of course, when the relatives came in the spring to bury their dead, there were no bodies and an outcry was justifiably enough raised; but it was too late to trace the subjects that had long since disappeared, and the only remedy was to guard the dead more carefully in the future. This affair so scandalized the community and the Catholic hierarchy that the Archbishops approached the Anatomical Departments and asked them what kind of law they wanted in order to obtain subjects legally. We replied that a law with a penalty attached was necessary and also that the body must be claimed by relatives and not friends. So they went to the Provincial Legislature and requested the Government to pass such a law, which they promptly did; they were opposed only by a few English members. This law put an end to "*body snatching*" and provided an ample supply of subjects for the dissecting rooms. The law runs somewhat thus: "*All persons dying in institutions (such as hospitals, jails, lunatic asylums) receiving aid from the Provincial Government, if not claimed by a relation nearer than the third degree in 24 hours, must be handed over to the Inspector of Anatomy for distribution in proper order to the medical schools.*" The penalty was, of course, the withdrawal of the aid furnished by the Provincial Government if the law was not carried out. It was proved most satisfactory, and all the subjects are obtained chiefly from the large lunatic asylums and without difficulty. The clause in the act which requires the claiming of the body by relatives, not further removed than first cousins, and this relationship to be sworn to before a magistrate, is a most important addition, for any friend or society could claim them heretofore.

Another feature of the law, which was introduced to satisfy the prejudices of the many, was the burial of the remains. The Protestant remains had to be separated from the Catholic, and the certificate of burial had to be handed over in due time to the Inspector of Anatomy. For many years now, this law has been in force and it has worked well. In summer the College collects subjects, and after preparing them with preservative injections and filling the arteries with tallow or wax, the subjects are kept in hermetically sealed safes, placed on shelves, and exposed to a continuous vapor of pure methylated spirits. In this way, I was enabled to start the session with 30 to 40 subjects, and when the cold weather came, fresh subjects were obtained in sufficient quantities to go on with our work.

Indeed, the pace of changes in facilitating human dissection throughout the nineteenth century was sluggish even though the importance of the subject for medical training was well appreciated. Nonetheless, grave robbing and other illegal practices continued well into the present century in order to provide cadavers for medical students to dissect.* In order to deal with the disgraceful and odious practice of body snatching and to make adequate provision of dissecting material for medical students, it became necessary for changes to be introduced in the statute. A revised statute, entitled "*An Act to Amend and Consolidate the Various Acts Respecting the Study of Anatomy,*" was introduced and passed in 1883 (Revised Statutes of Quebec, 1888, Cap. 4, S. 1, Clause 3958 et seq.). This bill was supported by the anatomy professors, the Archbishop of Quebec, the Bishop of Montreal, and other religious authorities. It required that the Lieutenant-Governor-in-Council should appoint an Inspector of Anatomy for each of the two sections (The "Montreal Section" and the "Quebec Section," author's note) and a Sub-Inspector of Anatomy for each judicial district. The bill further provided that the bodies of all persons dying in a public institution receiving a grant from the Provincial Government, unless claimed for burial within 24 hours by relatives within the degree of first cousin, were to be handed over, through the Inspector or Sub-Inspector of Anatomy, to the schools of medicine for use in the study of anatomy and surgery.

The Superintendent of any such public institution was to notify the Inspector or Sub-Inspector of Anatomy of the decease of patients in his charge within 48 hours of the death. Similarly, every coroner in possession of a body found publicly exposed was to notify

---

*For more details on this subject, see Lawrence (1958), Leblond (1966), and MacGillivray (1988).

the Inspector or Sub-Inspector of the finding thereof (Lawrence, 1958). Moreover, the third section of the legislation stipulated the following:

> That every superintendent or director of a public institution receiving a grant from the Government, and every coroner who shall knowingly omit or who shall neglect or refuse to comply with the provisions of the Act to be based on these resolutions, and every University of School of Medicine which shall receive corpses in its dissecting rooms, or allow the dissection within its establishment of corpses which have not been supplied to it by the Inspector of Anatomy, shall, upon a complaint to that effect before a Justice of the Peace by the Inspector or Sub-Inspector of Anatomy, be liable to a penalty of not less than one hundred nor more than two hundred dollars for each offence. (Journal of the Legislative Assembly, Quebec, Vol. 17, 1883)

From an editorial in the March 1884 issue of the *Canadian Medical and Surgical Journal,* it would appear that the new Anatomy Act helped in bringing to an end grave robbing and at the same time made it possible for the dissecting room to have an adequate supply of cadavers (Lawrence, 1958):

> The session of 1883–84 has now come to a close, and as it is the first since the introduction of our New Provincial Anatomy Act, we may be permitted to enquire what has been accomplished by it. The first noticeable fact is that, from last October until the present moment, not one single paragraph has been found in the daily papers having reference to the desecration of graves.
>
> We have, moreover, been informed by the teachers of anatomy that it is their belief that, for the first time in the history of the country, grave robbing has been entirely unknown. The main object, therefore, for which the Act was passed viz., the suppression of the resurrectionists has been completely fulfilled. At the same time the requirements of the Medical Schools have been amply met…(No authors listed, 1884).

Chapter 28

FRAGMENTS FROM THE EAST

Surgery is an art acquired through practice, but it is based on a knowledge of the structure of the human body that is gained from experience. So it must have been until the beginning of the eighteenth century when scientific thinking and concepts established its modern course.

One of the first pioneers in the development of modern scientific surgery in Europe was the surgeon-anatomist John Hunter (1728–1793). There was now no turning back, in contrast to the great heights in the practice of surgery that were once reached in Egypt, India (Majno, 1975; Lyons & Petrucelli, 1978; Estes, 1989; Rutkow, 1993), and China, only to decline with the passing of time. The reasons were twofold: over-specialization and the blending of religion with medicine. The gods were invoked by priests who kept on increasing their influence. The practice of surgery regressed into a menial trade, no longer a profession or an art (Graham 1956).

Neither ancient China nor Japan "*had any real system of surgery,*" unlike the fairly advanced state in other civilizations. From a description of the itinerant "*cutters for stone,*" in medieval Europe, given by Graham (1956), one cannot but admire the deftness and special skill they possessed, which clearly was based on a knowledge of living and practical anatomy.

These peripatetic specialists make the same meager living as their forefathers and in exactly the same way. They are "*experts*" in this one operation of cutting for stone in the bladder, and know no other. The operator puts his finger in the patient's anus and hooks down the stone, which can be felt in the bladder, so that it presses hard against the perineal tissues. With an ordinary razor an incision is made over the flinty protrusion and deepened till the bladder is opened. The stone is extracted with a scoop, and that is all. There is no preoperative preparation and no question of serious postoperative care. The wandering stonecutter moves on to the next village and his next batch of patients.

According to the teachings of the great Chinese philosopher Confucius (551–479 B.C.), the human body was sacred and not to be defiled. This belief prevented any human dissection, and doctors were taught the practical aspect of anatomy from models and diagrams. The *Nei Ching*, an ancient treatise compiled from the teachings of Huang di, who lived about 2600 B.C. and practiced preventative medicine, divided the internal organs into five *zangs* and six *fus*.

The *zangs* were solid organs and included the heart, liver, spleen, lungs, and kidneys. It was believed that the zangs did not eliminate but stored, unlike the hollow gallbladder, stomach, intestines, and urinary bladder which eliminated. The *San jiao* existed as "*three burning spaces*" (Fig. 331); the function and location of these purely imaginary organs were debated among the scholars (Graham, 1956). Much of these philosophical concepts permeated modern thinking and interpretation of human structure. Dominci Parrenin attempted to introduce modern teachings in anatomy to China but was unsuccessful.

At that time, the *Huangdi Neijing* was the chief source of information. This ancient work contained many erroneous statements, such as the 24 holes present in the lungs, and that urine is derived from feces. In his work the *Yi Lin Gai Cuo*, Wang Qingren (1728–1831) tried to correct some of these misconceptions. He observed the abdominal aorta, the vena cava, the pyloric sphincter, hepatic duct, pancreas, and the diaphragm in corpses that were decaying in the graveyard. The diaphragm was described as being "*thin as paper.*" Wang Qingren stated that it is the brain and not the heart, which controls memory (Hoizey & Hoizey, 1993).

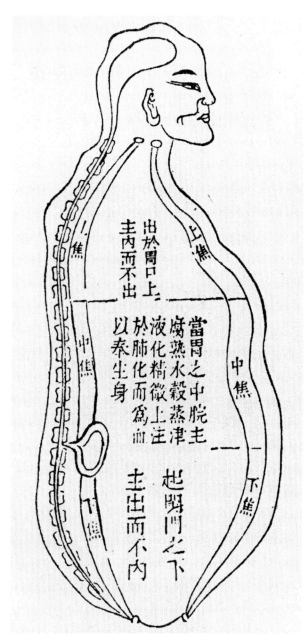

Figure 331. The *"three burning spaces"* (2697 B.C.). These were considered as viscera, implying that they *"eliminate"* but do not *"store up."* (From Hsieh, E.T.: A review of ancient Chinese anatomy. *Anat. Rec.* 20:97, 1920; reprinted by permission of John Wiley & Sons, Inc.)

Unlike Europe, scientific anatomy was not a dominant feature in the medical practice of China until the beginning of the twentieth century (Cowdry, 1920). Apparently, *"the first time human dissection was attempted in interior China was on April 22, 1915"* (Graham, 1956).

Needham (1970) claimed that "the poor condition of Chinese anatomy in the last century gave people the

idea that it had always been backward in China, but that is not the case." Representation of the human body in Chinese anatomical pictures from the seventh to the ninth century A.D. was fairly advanced and might have been the source of the well-known series of five schematic pictures (Funfbilderserie) that were widely circulated in early medieval Europe (Persaud, 1984).

According to Needham (1970), drawings of the human body are to be found in many editions of *The Clearing of the Innocent; or, The Washing Away of Wrongs.* This was the first treatise ever written on forensic medicine and it was compiled by Song Ci in A.D. 1247. Nonetheless, the assumption is that anatomy as well as all other *"sciences basic to medicine were much more advanced throughout the nineteenth century than what was known in China...."* From the drawings of Yang Jie (c. 1068–1140) we know that a human body was dissected in 1106.

William Harvey is almost universally recognized for his discovery of the circulation of blood, but at the same time, one should reflect on an account that was given by the Chinese physician Huang di about 2697 B.C. He stated that *"all the blood in the body is under the control of the heart. The blood current flows continuously in a circle and never stops."* How did Huang di arrive at such a profound and surprisingly accurate conclusion? Is it rational to assume that the physicians in ancient China knew relatively little about the structure of the human body, despite the abundant contrary evidence (Hsieh, 1920)?

The monograph of Huard and Wong (1968) is a well-documented source on Chinese medicine and most of the following account is derived from this excellent work. The legendary surgeon Hua Tuo, who was born during the Yong he period (A.D. 136–141) and died in A.D. 208, is credited with having performed complicated operations, which he carried out on patients who were under the anesthetic influence of Indian hemp. These operative procedures included thoracoplasty, laparotomy, intestinal resection, trepanation, lithotomy, and others. Huard and Wong (1968) stated that this *"practitioner in acupuncture,... anatomist and a therapeutist... is thought to have published, under the title Neichaot'u, anatomical charts showing the inside of the human body."* These charts have been transmitted from one generation to another over the ages and enjoyed a great reputation.

Perception of the human body was strongly influenced by Daoism as is evident in the work Sing-ming Kuei-che (c. 1622). The body was divided into three anatomical regions: *"The upper or cephalic region was the source of the spirits which dwelt in the body.... The middle region was represented by the spine, which was not regarded as a functional column, but as a canal linking the cerebral cavities with the genitalia.... The lower region...was the seat of genital activity represented by the two kidneys"* (Huard & Wong, 1968). Obviously, there were considerable philosophical and religious speculations, which clouded any realistic concept of the human body during this period.

Chinese medicine was also influenced by the Buddhist missionaries from India, and from these dual sources there had been an exchange of medical knowledge with neighboring countries. We know that at one time Japanese students were sent to China for training in medicine and they brought back to Japan the techniques of acupuncture. This traditional form of medical practice relates hundreds of points on the skin with the internal organs, and diseases of various organs are cured by inserting long, thin needles into specific sites on the surface of the body. The organs are related to 12 hypothetical and undissectable channels (Bowers, 1972).

During the Azuchi-Momoyama period (1559–1615), the Portuguese introduced Western science to Japan and established a medical school (Namban igaku). Later, Japanese medicine, especially in the areas of anatomy and surgery, would also benefit from the Dutch and French who by then had established trading companies in China. Kambara (1974) described the use of wooden models of the human skeleton, which were used for teaching functional anatomy of the skeletal system, with respect to the treatment of fractures and dislocations. These were made towards the end of the eighteenth century and the beginning of the nineteenth century based on *"the results of human dissection under the influences of European Anatomy."*

In 1771, Mayeno had dissected a body confirming the position of the organs as shown in European anatomical charts. The Chinese drawings of these structures and their position were different from what was revealed by the dissection (Preda, 1990). As a result, the works of Kulmus was immediately translated into Japanese and printed in 1773. Thus began the medical reforms in Japan, which led to the opening of a Dutch medical school (later Tokyo University) in 1857.

According to Akihito (1992), Yamawaki Toyo, who was a court physician in Kyoto, published in 1759 an illustrated work entitled *Zoshi* (Record of *Internal Organs*). It was based on his personal observations of the *"first officially approved dissection of the human body"* carried out in Japan. Johan Adam Kulmus's *Anatomical Tables*, written in German and published in 1734, was translated into Dutch (*Ontleedkundige Tafelen*). In

1774, a Japanese translation of this work entitled *Kaitai Shinsho (A New Book of Anatomy)* was published. The drawings were accurate and consistent with the observations made during dissection of the human body but differed from the account found in Chinese medical texts.

The publication of *A New Book of Anatomy* held great significance for the subsequent development of science in Japan. First, it revealed errors in Chinese medical books that had previously been the sole source of information for Japanese physicians and illustrated the importance of learning by direct observation and of having an open mind.... Second, it served as a focal point for gathering physicians in Edo who shared a common interest in European Science.... (Akihito, 1992).

In 1857, the first Japanese edition of Benjamin Hobson's textbook on anatomy, entitled *Zentai shinron* (New Discourse on the Whole Body) was published. Doctor Hobson was an American medical missionary in China where his book, with Chinese text, was published in 1851. This two-volume work contained illustrations showing the eye, muscles, various organs, blood vessels, and a developing fetus in the womb. Hobson's textbook was widely used by medical students and physicians in Japan and China. By the end of the nineteenth century, the European traditions in medical education, especially from Germany, were adopted in Japanese medical reforms.

Stimulated by Western medicine, there was renewed interest in human anatomy especially in the external form of the human body, which was important for locating acupuncture points. For this purpose, dolls of bronze, wood, or paper were made with the acupuncture meridians and locations clearly marked. Close to 400 acupuncture points were mapped out along the 12 meridians of the body. It was believed that the meridians of the body transmitted an active force or *qi*. Through long experience, the paths of the major arteries and nerves were avoided when the acupuncture needles were inserted into the body (Preda, 1990).

One area of related interest is Tibet-Mongolian medicine on account of the impressive anatomical charts that were made. At first these territories came under the influence of scholars and physicians who travelled with the caravans and missionaries along the Silk Road, and later from the West.

It was not until the seventeenth and eighteenth centuries, with the arrival of European physicians and surgeons, that Western medicine had any significant influence on Chinese medical sciences and practice (Bowers, 1972). Huard and Wong (1968) reported that in the late seventeenth century, Emperor Kang Xi commanded Pere Bouvet to collect anatomical charts for him. Pere Parennin (1665–1741) continued the assignment and also compiled in Manchu a medical work in eight volumes, four of which dealt with anatomy.

It is of interest that Parennin's manuscript and the anatomical drawings were largely based on the works of two contemporary European anatomists, Pierre Dionis (Paris, 1690) for the text and to Thomas Bartholin (Leiden, 1677) for 118 of the 175 diagrams. Even though the emperor was impressed with this new anatomy of the human body, he kept it a secret from his people because it conflicted with traditional Chinese doctrine.

According to Huard and Wong (1968), Emperor Kang Xi had decided on the advantage of dissecting condemned felons sentenced to death, because by so doing, they would compensate society for their wrongdoings. Later, he changed his mind on the matter. The *Work on Manchu Anatomy*, which he

considered to be highly unusual, was read only by a selected few in high positions who were not even permitted to borrow it or make any notes. The work was of interest to many Jesuits, including Pere Pierre-Martial Cibot (1727–1780) and Pere Jean Joseph Marie Amiot (1718–1793), who were permitted to examine it. As a result, Chinese anatomy did not benefit from this compiled treatise that presented a far more advanced concept of human anatomy.

Hsieh (1920) described a terrible epidemic, which occurred in the town of Zhong Li Xie during the Qing dynasty. As a result, *"hungry dogs were uncovering the bodies hastily buried in shallow graves, and devouring them."* A certain magistrate with medical interests, named Wang Qing ren, visited the cemetery daily and studied over 30 dismembered bodies. He wrote a book entitled *A Correction of Faults in Medicine*, which tested old theories and presented new observations. Wang's book, based on 24 drawings of the organs and viscera, included essays on such topics as the

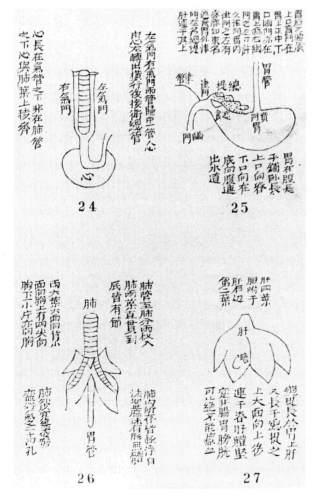

Figure 332. Drawing of *"organs and viscera"* from Wang Qing ren's book entitled Corrections of Faults in Medicine. The book was based on Wang Qing ren's thorough study of the human musculature and 24 *"corrected pictures of the organs and viscera."* (28: bladder, urethra, and vas deferens; 29: tongue and epiglottis; 30: omentum; 31: vena cava and aorta. From Hsieh, E.T.: A review of Ancient Chinese Anatomy. *Anat Rec 20:97*, 1920; reprinted by permission of John Wiley Sons, Inc.)

epiglottis, the respiratory tract, the digestive and excretory systems, the brain, and the heart (Fig. 332).

However, this direct method of observation "*was soon replaced by a rule of authority.*" Wang's followers, such as Peng Zheng Hai, who was "*able to secure an atlas of western medicine for comparison in testing out the theories of Neiching,*" indulged themselves in abstract speculations. For example, Peng Zheng Hai held on to the old theories and described the "*gate of life*" as corresponding to the cords of the kidney and the fat lying below the cords as lower burning spaces. The fat in the region of the omentum corresponds in his mind to the middle burning space and that above the diaphragm to the upper burning space...that part of the bladder, which connects with the fat as the place where the water enters. And finally...the brain carries out the action of the heart (Hsieh, 1920).

Johannes Schreck (Terenz or Terentius, c. 1576–1630) compiled the first work in Chinese that dealt with Western anatomy. This illustrated book was entitled "*Tai-si jen-shen shuo-Kai*" (Western treatise on the structure of the human body).

Following the opening of the treaty ports, and with the arrival of British and American Protestant missionaries, as well as surgeons from the East India Company, medical schools and hospitals were established. Thus, began the dissemination of Western science not only in China but also to Japan.

For the first documented anatomical dissections, demonstrations and lectures on postmortem techniques in China, one has to go beyond the chronological span of this book. The credit for this is given to James Henderson who died in 1865. Part of the *Anatomical Tables* of William Cheselden (1688–1752) was adapted for use in both China and Japan. However, more elaborate anatomical textbooks with Western ideas appeared towards the end of the nineteenth century. Huard and Wong (1968) listed among these "*an Anatomical Atlas and a monumental Anatomy (twenty volumes 1887),*" published by the Edinburgh graduate John Dudgeon, who taught medicine at the Imperial College in Peking.

Zhongshan Medical University was established in 1866, followed by Peking Medical University in 1912. In 1866, seven apprentices were enrolled in a three-year program which included anatomy, medicine, surgery, and materia medica. However, "*the barriers against dissection posed a major problem in anatomy, but occasionally the extremities of the body of a patient unclaimed by friends were superficially dissected*" (Bowers, 1972). In Japan, there were four medical schools before the end of the nineteenth century teaching

modern anatomy: Tokyo (1868), Nagasaki (1868), Chiba (1876), and Nagoya (1878).

More than five decades later, medical teaching was still hampered by the lack of human material for students to dissect. Judging from the account given by Cowdry (1920) of anatomy teaching at the Peking Union Medical College, China's most prestigious medical school and with the "*most elaborate laboratory*" in China, it would seem that there had been little progress even here in obtaining bodies for dissection. For the dissection course, only four cadavers were available over a period of 18 months. Indeed, the greatest obstacle in the teaching of anatomy to medical students then was the difficulty of obtaining bodies for dissection. The police and the authorities generally were not sympathetic, and even the "*servants*" in the building were not cooperative.

In 1913, the Chinese government had passed appropriate regulations (Order of the Board of Interior No. 51, November 22, 1913) that should have facilitated human dissection:

Article I. A physician, in case of death from disease, may dissect the body and inspect the diseased part to determine (examine) the origin of the disease, but he must first obtain the consent of the relatives of the dead person and clearly inform the local magistrate before proceeding to dissection.

Article II. The police and inspectors, in case of mysterious death, the cause and origin of which cannot be accurately ascertained without dissection, may appoint a physician to dissect said corpse.

Article III. The bodies of all those meeting death by punishment or dying in prison from disease, without relatives and friends to claim their bodies, may be given by the local magistrate to a physician for dissection, to be used for the purpose of experimentation in medical science, but after dissection the body must be sewed up and buried.

Article IV. If any are willing for the benefit of science to offer their bodies for dissection and leave word to that effect before death, they may do so, but the whole body must be sewed up and returned to his or her family after dissection.

However, other supplementary regulations created problems in the interpretation of the order, which suggested that the written consent of the individual given before death must be further approved by the relatives. The relatives rarely granted permission.

This led to the use of anatomical models, as described by Cowdry (1920):

> Owing to the pressure of other duties, the teachers were often unable to properly adapt their methods of teaching to the lack of human material. They soon come to rely upon the use of anatomical models made in Europe or Japan and give up their efforts to obtain bodies for dissection; for it takes a lot of time and energy to cultivate the authorities, to drink endless cups of tea, to make petitions for favorable legislation and arrangements for executions.

INDIA

Like the Chinese, the Indian civilization is one of the oldest known. From the archaeological excavations and findings of Mohenjo-Daro and Harappa, it is now known that there were earlier inhabitants of the Indus Valley between the Himalaya and the Vindhya ranges, long before the Aryan conquest from the northwest in 1500 B.C. Abundant evidence suggests an advanced social organization and sanitation practice. The settlements were orderly laid out and there were wells, sewers, and public baths.

With the cultural development in this ancient Indo-Aryan civilization, healing became entwined with religious practices. Health and diseases were attributed to the gods, with Dhanvantari as the patron. Three "*Doshas*" or humors (wind, bile, and phlegm) were thought to permeate the entire organism and when in harmony lead to good health. Not withstanding their devotion to spiritual pursuits, the Aryans surprisingly developed a rational and secular approach to the practice of healing based on keen observations, the judicious use of herbs and surgery (Majno, 1975).

Several phases of development within this ancient culture have been identified: The Vedic (about 1500–500 B.C.), Brahmanic (600 B.C.–A.D. 1000), and the Mughal (from A.D. 1000 until eighteenth century). The Vedas or books of knowledge were compiled during the Vedic period. It has been suggested that these revealed works of the universal spirit or creator, which embodied religion and philosophy, were formulated some 4000 years prior to the Christian era. Only four of these books have survived: The *Rig Veda*, *Sama Veda*, *Yajur Veda*, and the *Atharva Veda*. Many diseases and treatments, as well as surgical procedures, are recorded in two of these works (Zimmer, 1948).

The *Rig Veda* was essentially a medical treatise, whereas the *Atharva Veda* was a surgical work. Thus evolved traditional healing methods (Ayurvedic) together with practical skills, which were remarkably advanced for that period in the history of man. Much of this knowledge spread into Asia and later reached Europe during the Middle Ages as a result of translations that were made by Persian and Arab scholars in the eleventh century.

Apart from the use of medicinal plants for the treatment of a wide spectrum of diseases, the practice of surgery evolved to become one of the outstanding achievements of Indian medicine (Majno, 1975; Haeger, 1989; Rutkow, 1993). The Laws of Manu, which probably were formulated at about 3000 B.C. and compiled between 200 B.C. and A.D. 200, formed the basis of the social fabric of daily life in ancient India. The nose was cut off as a punishment for adultery and it is therefore not surprising that rhinoplastic procedures were well advanced. Other surgical operations included the repair of torn earlobes and cleft lip, suturing of the intestine by applying large ants and followed by decapitation of the ants after they had bitten into the edges of the wound, removal of stones from the bladder, hernia repair, caesarean section, and cataract extraction (Lyons & Petrucelli, 1978). As many as 101 surgical instruments, including a variety of forceps, scalpels, needles, and suturing material, have been described. Aspiration of fluid for the treatment of both ascites and hydrocele and the use of a magnet for the extraction of foreign bodies were mentioned for the first time.

The famous physicians of Hindu medicine were Susruta, Charaka, and Vagbhata. There is still some controversy as to exactly when these scholars lived. Susruta was a surgeon who most likely lived during the sixth century B.C. and taught at the University of Kasi or Banaras. He was a younger contemporary of Atreya who taught at Taksasila or Taxila, a famous seat of learning in the West (Persaud, 1984).

The medical wisdom of Atreya was compiled in the form of a compendium or the Samhita. This work was essentially a classification of diseases with some remarks on the skeleton. Susruta also produced a similar Samhita but with more emphasis on surgical matters, including surgical instruments and surgical operations. It is in this work that one finds significant anatomical considerations of the ancient Hindu.

Because Susruta referred to Atreya's system of describing the bones, it is generally agreed that both these men lived and compiled their work during the sixth century B.C. In regard to Charaka, it would appear that he flourished during the reign of King Kanishka, about the middle of the second century. For a deeper appreciation of the many remarkable accomplishments of ancient India in the field of medicine, reference should be made to the work of Keswani (1970), Jee (1978), and to the series edited by Singhal and Guru (1973).

Vagbhata referred to both Susruta and Charaka by names and quoted their works. Vagbhata might have lived during the early part of the seventh century about A.D. 625. The work produced was a summary of Samgraha of the eight branches of medicine.

There is compelling evidence to believe that the knowledge of human anatomy revealed at the time of Susruta was gained not only by inspecting the surface of the human body but also through dissection of human subjects (Hoernle, 1907; Keswani, 1973). He recommended to those aspiring to a career in surgery that they should acquire a good knowledge of the structure of the human body and described the method how the body should be prepared for this purpose. Regarding the importance of dissection, he stated, "*therefore the surgeon, who wishes to possess the exact knowledge of the science of surgery, should thoroughly examine all parts of the dead body after its proper preparation*" and about the method for dissecting, the following is recommended (Singhal & Guru, 1973): Therefore for dissecting purposes, a cadaver should be selected which has all parts of the body present, of a person who had not died due to poisoning, but not suffered from a chronic disease (before death), had not attained a 100 years of age and from which the fecal contents of the intestines have been removed. Such a cadaver whose all parts are wrapped by any one of "*munja*" (bush or grass), bark, "*kusa*" and flax, etc. and kept inside a cage, should be put in a slowly flowing river and allowed to decompose in an unlighted area. After proper decomposition for seven nights, the cadaver should be removed (from the cage) and then dissected slowly by rubbing with the brushes made out of any of usira (fragrant root of a plant), hair, bamboo or "*balvaja*" (coarse grass). In this way, as previously described, skin, etc. and all the internal and external parts with their subdivision should be visually examined.

Susruta was able to achieve his remarkable knowledge of human anatomy in spite of religious laws, which prohibited contact with the deceased other

Figure 333. Picture of Susruta. (Courtesy of Wellcome Library no. 10023.)

than for the purpose of cremation. Using a brush-type broom, he was able to scrape off skin and flesh from the macerated remains in a systematic manner without actually touching the body.

Susruta's *Samhita* contains a fair amount of speculations and philosophical concepts organized in a system of classifications. He stated that from surgical experience he knows of 300 bones, although 360 are recorded in the *Vedas*. He ascribed 120 bones to the extremities; 117 to the pelvis, flanks, back, and the chest; and 63 to the region above the neck.

Susruta further described the types of bones; the importance of the bony skeleton; the number of joints; the types of joints, ligaments, and muscles in different parts of the body. He assigned 20 additional muscles to the female on account of the breast and genital tract.

It has been suggested that Susruta might have arrived at the relatively large number of bones in the human skeleton because of the many dissections he carried out on children under two years of age. Despite his erroneous account of the skeleton and other speculations, e.g., 700 veins originating from the umbilicus and distributed to all parts of the body, Susruta's knowledge of human anatomy as revealed in his account of the muscles, joints, ligaments, and even blood vessels and nerves was remarkable for the period in which he flourished.

Jee (1978) remarked that Indian medicine began to show "*signs of decay*" during the rule of the conqueror because there was little "*moral and material support of the government of the day.*" When the *Peshwas* (A.D. 1715 to 1818) came to power, medicine in India experienced a revival. Medical treatises were compiled, based on rational and practical principles, and these served as authoritative sources for a very long time. For an account of the development of medical education in India during ancient times, and later, reference should be made to Keswani (1970). The information given here is largely derived from this valuable and comprehensive source.

Under the personal influence of the Emperor Akbar the Great (1555–1605), the arts and sciences flourished. But in the field of medicine, there was little progress compared to the rapid changes in Europe. Medical books were translated from Arabic into Persian, the language of the Court, and these translated works were for essentially the training of indigenous physicians. Thus,

> during the early Muslim period the indigenous schools of Hindu medicine languished, and due to a lack of patronage from the rulers, and the devastation caused to the medical and education institutions by the successive onslaughts of the earlier Muslim invaders, the cause of Hindu medicine suffered an irreparable and permanent blow. (Keswani, 1970)

Western medicine was first introduced to India by the Portuguese conquerors in the sixteenth century. The Royal Hospital that was founded then passed on to the Jesuits in 1591, and in 1703 a simple program of medical training was inaugurated. This institution later became the School of Medicine and Surgery (*Neelameghan*). However, it was the British and French

who "*later established and consolidated the modern medical system in India*" (Keswani, 1970). Indeed, this did not occur until about two centuries later, with the establishment of medical schools along European traditions, in Madras (1835), Calcutta (1836), and Bombay (1845). The records are scanty with respect to anatomy instruction during this period.

In 1827, Doctor John Tytler, who was an oriental scholar, gave a series of lectures in anatomy at the Calcutta Sanskrit College. Even though the students were introduced to the "*Western*" system of anatomy, the medical and surgical works of the ancient Indian doctors, including Charaka and Sushruta, were not neglected. This Native Medical Institution, "*which was the first institution of its kind in British India,*" combined both the Ayurvedic and Western systems of medical teaching. Pandit Madhusudan Gupta, who was a student of this school, "*later became famous for being the first Indian in modern times to dissect a cadaver.*"

The abolition of the Native Medical Institution in 1835, by an Order in Council from the Governor General of India, paved the way for "*a new medical college,*" where a certain number of Indian students will be exposed to various "*branches of medical science, in strict accordance with the mode adopted in Europe...*" (Keswani, 1970). Doctor H. H. Goodeve, a surgeon and the only full-time assistant at the new medical college, was appointed professor of medicine and anatomy. Keswani (1970) has given us an account of this period:

> The students started their work with no museum of anatomic preparations, no library, and no hospital. Besides, the teachers were confronted with the task of combating the strong national prejudice against dissecting cadavers or handling anatomic material. To overcome this, parts of the human body were first introduced in illustration at the time of teaching, and these were gradually replaced by the organs from domestic animals, wooden models and metallic representations. It took Goodeve six months before he could place a cadaver on the table, which created much interest and excitement among his students, who probably "*went through the same kind of mental stress that a Westerner might experience who contemplated a change of religion which would involve social outlawry*" (Spear). But soon the students became familiar with the sight of a cadaver from daily repetition, and on 10 January 1836, Pandit Madhusudan Gupta, accompanied by his pupils, Uma Charan Set, Dwarka Nath Gupta, Raj Krist Dey and one other whose

name is not recorded, joined Goodeve in an outhouse of the College building, and began dissection on a cadaver. Raj Krist Dey is believed to have been the first Hindu student who actually dissected the human body on that day. That day will be marked in the annals of the history of Western medical education in India, when Indians rose superior to the prejudices of their traditional education and flung open the gates of modern medical science to their countrymen. Two years later Goodeve, who had carefully watched the reaction of the students toward dissection, was pleased to address his new batch of students thus: "*in less than two years from the foundation of the College, practical anatomy has completely become a portion of the necessary studies of the Hindu medical students as amongst their brethren in Europe and America.*"

In June 1835, the Madras Medical School was established. At first, this school was intended for the training of medical "*subordinates*" who would work in the army medical services. Of interest, three years later, the school was open to civilians, but for eight years no civilian student enrolled because of the prejudices at that time against Western medical practices.

In many distant countries, anatomical knowledge was lacking or deficient. Uragoda (1987) traced the medical history of Sri Lanka, which is over two thousand years old. Long before the advent of the European medical tradition, local medical practitioners showed some understanding of the structure of the body in dealing with a wide variety of injuries and ailments. In fact, during the colonial rule, the Dutch and Portuguese both admired some aspects of local medicine. Other areas, however, were less advanced.

In 1661, the Dutch surgeon Schouten visited Sri Lanka and commented as follows: "*There are also intelligent lawyers, doctors, surgeons, and barbers. The medical men, however, have very little knowledge of anatomy, of things natural, unnatural, and contrary to nature, which ought to be the basis of their science....*"

Military hospitals were established by the Dutch and British for the benefit of the garrisons, and some early attempts were made to train local pupils about the middle of the nineteenth century. A private medical school in Sri Lanka was founded by Doctor S. F. Green in 1848 following a curriculum of the medical schools in the United States where Doctor Green was educated. The Colombo Medical School was not opened until 1870, although proposals had been made for such an institution, including anatomy teaching: "*...the time is not, I hope, very distant when I shall be enabled to propose to you, in furtherance of these views, the establishment of an Anatomical School.*" When the school was finally opened the practical aspects of anatomy were given much attention.

Chapter 29

LATE NINETEENTH CENTURY

In 1868, a Swiss national by the name of Wilhem His, Sr. (1831–1904) identified a structure he called the ganglionic rest, now know as the neural crest cell. This structure consists of a multipotent migratory population of cells, which give rise to an incredible diversity of cell and tissue types. These cells are essential for craniofacial and jaw formation, peripheral nervous system development, melanocytes, and the development of numerous other tissues. Born in Basel on July 9, 1831 and trained in the Universities of Basel, Berlin, Bern, Wurzburg, Prague, and Vienna, His's work is regarded as the very foundation of our knowledge of the embryonic history of man. After his work with the neural crest cell, His went on to demonstrate that nerve fibers originate as outgrowths of nerve cell. This theory was essential to the development of the Neuron Doctrine in 1886, which states that the nerve cell is the basic unit of the nervous system (Dixon, 1904). His stated, "*I defend the following postulate as an indisputable principle: that each nerve fiber originates as a process from a single cell. This is its genetic, nutritive, and functional center; all other connections of the fiber are either indirect or secondary.*"

Born at Corento near Brescia on July 7, 1843 and trained at the University of Pavia, an Italian physician and scientist Camillo Golgi (1843–1926; Fig. 334) went on to develop a novel technique for staining the nervous tissue in 1873. He initial called it the "*black reaction,*" but later it was coined the Golgi stain or Golgi method. This method uses a weak solution of silver nitrate and is capable of staining the most delicate ramifications of a cell. The stain enabled Golgi to demonstrate the existence of a nerve cell with many short branching extensions that served to connect other nerve cells, now called a Golgi cell. This staining technique and his subsequent investigations provided a substantial contribution to the advancement of the knowledge on the structural organization of

nervous tissue. Additionally, this technique allowed Golgi to discover the presence of an irregular network of fibrils, vesicles, and granules within the nerve cell that he called the "*internal reticular apparatus*" in 1898, which was later named the Golgi apparatus. The Golgi method, coupled with his work on the nervous system, is what earned Golgi the highest honors when he was awarded the Nobel Prize in 1906 (Bentivoglio, 2013).

Franklin Paine Mall (1862–1917; Fig. 335), born in Belle Plaine, Iowa, was trained at the University of Michigan and is most known for his role in establishing anatomy as a science in the United States. Before Dr. Mall, anatomy had a minimal scientific standing in the United States. He first began his research in Europe in 1885, working in the laboratory of Wilhelm His and then Carl Ludwig (1816–1895), the German anatomist and physiologist. While studying under His, Mall came to the conclusion that the thymus arises form the endoderm of the pharynx and not the ectoderm as previously believed. It was during his work with Carl Ludwig that he demonstrated the vascular patterns of organs and discovered the vasomotor nerves of the hepatic portal vein. After a few years of study, Mall returned to the United States and continued his research. He went on to clarify the structure of organs with his concept of the structural units, delineated the law of growth of the nervous system, laid the foundations for the study of the pathology of embryos, and followed the development of certain organs to adult state. In 1893, at Johns Hopkins University, Mall was able to organize a department of anatomy that would reflect his view of the science, ("*The study of anatomy begins with the cell, ends with the entire individual, and includes man*"), and he realized what he saw as the field's potential ("*A subject like anatomy, taught for many centuries, has recently been made a new science through the studies in embryology and histology*").

Figure 334. A photograph of Camillo Golgi. (Courtesy of Bethesda, MD: U.S. National Library of Medicine, National Institutes of Health, Health & Human Services.)

Figure 335. A photography of Franklin Paine Mall. (Courtesy of Bethesda, MD: U.S. National Library of Medicine, National Institutes of Health, Health & Human Services.)

Through his accomplishments, Mall will continue to be associated with the strongly physiological bent of modern anatomy, in laying the foundation of organogenesis in embryology, and changing the perception of anatomy in America (Sabin, 1914).

G. Carl Huber was born in Hoobly, India August 30, 1865 to German Swiss missionary parents and trained at the University of Michigan. Dr. Huber published his first paper on the nervous system in 1891, *Physiology of the communicating branch between the superior and the inferior laryngeal nerves.* He went on to publish many more papers including, *"Directions for work in the histological laboratory, for the use of medical classes in the University of Michigan,"* the first edition in 1882. Additionally, for each phase of each topic he studied, Dr. Huber did vigorous research and checked and rechecked the validity of existing theories. Dr. Huber always regarded anatomy as the truly basic science of all medicine and his work emphasized morphology rather then function. Huber's major contributions are to the knowledge of the finer structures of the sympathetic nervous system, the structure of sensory nerve endings, the degeneration and regeneration of peripheral nerves and nerve endings, the development and structure of the uriniferous tubule, the blood supply of the mammalian kidney, the structure of the seminiferous tubule, the development of the albino rat, the notochord in mammalian embryos, and the comparative neurology of vertebrates (American Association of Anatomists, 2013; Guild 1935).

The Spanish physician and scientist, Santiago Ramon y Cajal (Fig. 336) made great advances in the understanding of the nervous system with his scientific discoveries between 1888 and 1894. Born in Petilla de Aragon, Spain, May 1, 1852 and educated in Madrid, he masterfully used and modified the Golgi method of staining. These modifications allowed him to demonstrate that the nervous system was made up of individual cells. This discovery challenged the widely accepted reticular theory, which stated that the nervous system was a continuous network of interconnected fibers. It eventually led to the creation of the neuron doctrine, a theory that constitutes the fundamental principles of neuroscience. During this time,

Figure 336. A photograph of Santiago Ramon y Cajal. (Courtesy of Bethesda, MD: U.S. National Library of Medicine, National Institutes of Health, Health & Human Services.)

Cajal also reported that axons terminate freely in the cerebellum and retina (Fig. 337). He also discovered dendritic spines, identified the axonal growth cone, and summarized his views on the conduction of electric impulse by nerve cells in his law of dynamic polarization. Cajal organized all of these discoveries in his magnum opus, *Textura del Sistema Nervioso del hombre y de los vertebrados*, published in 1909. It was his research and subsequent advancements in our knowledge about the nervous system that would earn Cajal the Nobel Prize in Physiology or Medicine in 1906 (Bullock et al., 2005; Iturbe et al, 2008; Nemri, 2010).

Born in Yonkers, New York May 12, 1857, Henry Hubert Donaldson (Fig. 338) is known mostly for his study of the brain and subsequent book entitled *The Growth of the Brain: A Study of The Nervous System in Relation to Education*, published in 1895. His interests and knowledge of the brain started with the study of Laura Bridgman's brain, a blind deaf mute who had been taught to speak and had attained marked mental ability. This study has been characterized as "*probably the most thorough study of a single human brain that has ever been carried out.*" It was through this study that Donaldson became invested in the long, accurate, quantitative study of growth, which was the main theme of his work. Additionally, Donaldson is the

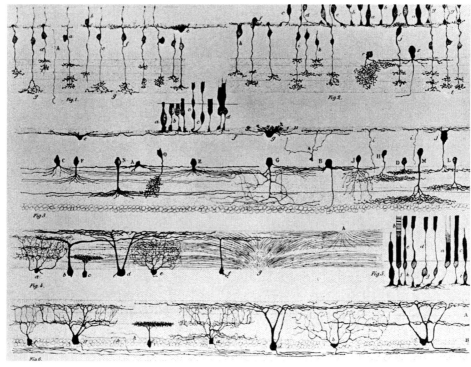

Figure 337. Santiago Ramon y Cajal nerve structure of the retina. Kgl. Univers – Druckerei v. H. Stürtz, Wurzburg. (Courtesy of Bethesda, MD: U.S. National Library of Medicine, National Institutes of Health, Health & Human Services.)

individual who is most responsible for the institutionalization of the albino rat as the dominant experimental organism in neurological research. He believed the rat was the best available mammal for laboratory work on the problems of growth (Conklin 1939; Logan 1999). Donaldson said,

> I selected the albino rat as the animal with which to work. It was found that the nervous system of the rat grows in the same manner as that of man – only some thirty times as fast. Further, the rat of three years may be regarded as equivalent in age to a man of ninety years, and this equivalence holds through all portions of the span of life, from birth to maturity. By the use of the equivalent ages observations on the nervous system of the rat can be transferred to man and tested. The results so obtained show a satisfactory agreement and indicate that the rat may be used for further studies in this field.

Another major advancement in 1895 was the publication of Oliver Smith Strong's dissertation on *The Cranial Nerves of the Amphibia*. Born in Red Bank, New Jersey, December 30, 1864 and educated at Princeton, Strong's work provided a rational basis for a classification of the cranial nerves and their central connections in all vertebrates and for their treatment in clinical practice. Some believed that it was one of the most important contributions to neurology in 50 years. Although he never qualified for the practice of medicine, he took an active part in the neurological clinic where his mastery of neuroanatomy was invaluable to the physicians. He published a highly esteemed textbook, *Human Neuroanatomy*, in collaboration with Adolph Elwyn (American Association of Anatomists, 2013; No author, 1951).

Great advances in anatomy are not always at the hands of anatomists, as proven by German physics professor Wilhelm Röentgen (1845–1923), when he accidentally discovered X-rays on November 8, 1895. Born in Lennep, Germany and trained at the Polytechnic Institute in Zurich, Rontgen discovered the X-ray while experimenting at the University of Würzburg with Lenard and Crookes tubes. While working with the tubes, he noticed a fluorescence of a plate coated with barium platinocyanide where none was expected, and in realizing its significance, placed his hand between the tube and the plate. Rontgen submitted his initial report on December 28, 1895 to the *Würzburg's Physical-Medical Society* journal, titled "*On a new kind of ray: a preliminary communication.*" He referred to the radiation as "X" to indicate that it is an unknown type of radiation, a name that sticks despite

Figure 338. A photograph of Henry Hubert Donaldson. Picture taken by Scott, Julian P. Photographer. (Courtesy of Bethesda, MD: U.S. National Library of Medicine, National Institutes of Health, Health & Human Services.)

his objections. He was awarded the first Nobel Prize in Physics in 1901 for his discovery. Röentgen understood the medical importance of his discovery when he took a picture of his wife's right hand on a photographic plate with X-rays. This was the first ever photograph of a human body part using X-rays. As word spread of this discovery, scientists through out the world constructed their own X-ray apparatus. By early January 1896, in Queens Hospital, a radiograph of a woman's hand was used as a guide to aid a surgeon in the removal of an embedded needle. By May 1896, medical radiography was sufficiently well established and became a permanent medical and diagnostic tool (Cohen & Trott 1973).

The final key contribution to anatomy in the nineteenth century was the publishing of the book entitled *The Nervous System and Its Constitutional Neurons* in 1899 by Lewellys Franklin Barker (1867–1943). This textbook, which was a compilation of his early research in the field of anatomy broadened our understanding of the nervous system and is an important reference book today. Barker was born in Norwich, Ontario and studied at the University of Toronto, graduating with the M.D. degree in 1890. Two years later he went on to Johns Hopkins where he conducted much of his research and published many outstanding works (no authors 1943). Trained in clinical medicine and in pathology, Barker was appointed professor of anatomy at the University of Chicago. Shortly after, he was appointed professor and physician-in-chief at the Johns Hopkins Hospital where he succeeded the distinguished physician William Osler. Barker received many honorary degrees in recognition of his many contributions to scientific medicine.

Chapter 30

ANATOMY IN THE THIRD REICH

It is impossible to discuss the advances in anatomy during the twentieth century and not mention the impact of the Third Reich. The dawn of World War II and the rise of Nazi Germany brought a new set of ethical dilemmas into the field of anatomical research. During the Third Reich, prisoners and inmates of concentration camps were executed and their bodies were used for teaching and research in various anatomy departments (special issue of *Annals of Anatomy* 194:225, 2012 for symposium contributions on "Anatomy in the Third Reich").

After the war, many infamous anatomists of the Nazi era brazenly continued their careers as prominent academic leaders in the field of anatomy. For example, the anatomist Hermann Voss (1894–1987) joined the Nazi party in 1937 for his career advancement. He was appointed professor and director of the anatomical institute at the newly established Reich University of Posen. Voss, who hated the Jews, communists, and the Poles, was responsible for experiments carried out on them in his institute. He also sold the skulls and death masks of Jews and Poles. From 1952 to 1962, Voss was professor of Anatomy at the University of Jena, and later became an emeritus professor at the medical faculty in Greifswald. He was the coauthor of an extremely popular three-volume anatomy textbook (Voss-Herrlinger: "*Taschenbuch der Anatomie*," Gustav Fischer Verlag, Stuttgart/New York, 1986) which now as the 17th edition is still available. Voss was also appointed as the editor of two important anatomical journals, the *Anatomischer Anzeiger* and *Acta Histochemica*.

There are the well-known accounts of the infamous anatomist August Hirt (1898–1945), who was appointed professor and director of the institute of anatomy at the Reich University in Strasbourg. He had concentration camp inmates killed specifically for his anatomical collection of Jewish skeletons. Hirt personally selected the inmates, mostly Jews, who were to be gassed. He committed suicide before his trial as a war criminal for mass murder. The equally depraved physician Paul Kremer (1883–1965) was professor of anatomy at the University of Münster and later became a Nazi doctor in Auschwitz concentration camp. Kremer was interested in the effects of starvation on the body, especially the liver. For this, he took the liver and other organs from murdered inmates he had selected for his research. Kremer was convicted for war crimes and mass murder. He was sentenced to death, but his sentence was later changed to life imprisonment. In 1958, Kremer was released from prison.

However, there are also the less publicized anatomical advances attached to unethical acquisition of cadavers for dissection during this period. Although the war occurred over 60 years ago, it is only recently that these ethical issues are being discussed in detail (Mitscherlich & Mielke, 1960; Aümuller & Grundmann, 2002; Winkelmann & Schagen, 2009; Hildebrandt, 2008; Hilderbrant & Redies, 2012).

In Germany and other European countries, the traditional method of retrieving bodies for dissection and research was through unclaimed bodies from hospitals, prisons, executions, and psychiatric institutions. Before 1933, the rate of execution in Germany was very low and so was the percentage of unclaimed bodies. This made it very difficult for universities to obtain bodies for dissection. However, with the beginning of World War II and the rise of the Nazi regime, the number of executions rose exponentially from approximately 478 between 1907–1932 to approximately 15,000 civilian plus another 15,000 military executions from 1933–1945. This, coupled with the 1943 decree that families of the executed could not claim the bodies of Jews, Poles, and those executed for high treason, explains why universities received so many bodies that they had to begin turning them away due to the lack of storage (Hildebrandt, 2009a,b; Waltenbacher, 2008).

The universities that received bodies under the Nazi regime were the 24 German universities with anatomical departments; the Austrian universities of Vienna, Graz, and Innsbruck; the Czech university of Prague and Strassburg, Posen, and Dorpat. Among these universities, the total number of bodies per institute used for dissection and research during this period ranged from 336 in Strassburg to 5,341 in Vienna (Hildebrandt, 2009a,b; Winkelmann & Schagen, 2009). Of these bodies, only some were those of European Jews and other so-called "*non-Aryan*" minorities. The majority of the bodies were of German, Austrian, Polish, Russian, and other European descent (Druml, 1999; Hildebrandt, 2009a,b). Therefore, most if not all of the anatomical advances that came from the anatomist at these universities are cloaked in ethical controversy.

Figure 339. Eduard Pernkopf (1888–1955) as a young anatomist at the Anatomy Institute in Vienna. (Photograph courtesy of the Institut für Geschichte der Medizin der Universität Wien [Inst. Hist. Med. Univ. Vienna]). (Courtesy of http://www.vet.purdue.edu/brad/medill/pernkopf.html.)

As mentioned previously, the two most infamous anatomists of this period were August Hirt and Johann Paul Kremer. Hirt was responsible for lethal medical experiments with mustard poison gas on prisoners at the Natzweiler concentration camp. Additionally, he attempted to construct a collection of "*Jewish skeletons*" from the bodies of 86 Jewish prisoners who were murdered using cyanide gas at Auschwitz (Hildebrandt, 2009a,b; Lachman, 1977; Mitscherlich & Mielke, 1960; Pressac, 1985). Kremer was the only anatomist who worked at an extermination camp. He was involved in the selection at the ramp at Auschwitz. He also was involved in researching the effects of starvation on the human body. He would select the prisoners he wished to study before their execution by intracardiac phenol-injection in order to harvest their tissues immediately postmortem (Hildebrandt, 2009a; Hoss et al., 1984).

Another ethical controversy surrounds Eduard Pernkopf's textbook, *The Atlas of Topographical and Applied Human Anatomy* (Figs. 339–341). This atlas, considered by many to be an artistic masterpiece, is a four-volume set that was begun in 1933 and spanned two decades. Eduard Pernkopf (1883–1955), author of this atlas, was working as an anatomy professor at the University of Vienna when the Nazi regime came into power in 1933. In the same year, he became a member of the party. After the start of the war, he quickly rose through the academic ranks in Vienna, all the way to become Dean of the Faculty of Medicine and Rector of the University in 1943. His academic accomplishments were paralleled by his increased involvement in the Nazi party. The combination of his extensive involvement in the Nazi movement and the possibility that the cadavers in the textbook could be Nazi victims resulted in many arguing to ban the book. On the other hand, at present, there is no evidence that Pernkopf played a role in causing the death of Nazi victims or that he used their bodies for his atlas. However, some have argued that it is impossible to view the atlas without taking into consideration the history of both the author and the period in which it was created (Angetter, 2000; Hubbard, 2001; Pringle, 2010; Williams, 1988).

The research conducted by Berlin anatomist Hermann Stieve (1886–1952) on female reproduction is another anatomical advancement that has recently come under scrutiny. In 1986, Stieve was revered for his research and described as "a great anatomist, who revolutionized the fundamentals of gynecology with his excellent clinical-anatomical research" (Gütz, 1986). However, in 2003, he was labeled the "leading

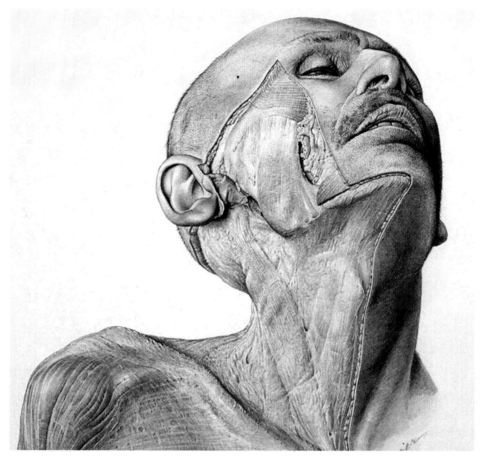

Figure 340. The superficial layer of the cervical fascia. The skin, subcutaneous tissue, and platysma have been removed from the face and neck on the right side. Watercolor illustration by Erich Lepier. (Reproduced from Pernkopf: *Atlas of Topographical and Applied Human Anatomy*, Second Edition © 1980, Urban & Schwarzenberg, Baltimore-Munich). Courtesy of http://www.vet.purdue.edu/brad/medill/pernkopf.html.)

Figure 341. Karl Endtresser, academic painter; Ludwig Schrott, Jr.; Univ. Prof. Dr. med. Eduard Pernkopf; Wener Platzer; Franz Batke, academic painter. Vienna, March, 1952. (Photograph courtesy of Univ. Prof. Dr. W. Platzer, Institut für Anatomie der Universität Innsbruck [Anatomy Inst. Univ. Innsbruck]). (Courtesy of http://www.vet.purdue.edu/brad/medill/pernkopf.html.)

anatomist of National Socialist times and the chief exploiter of human cadavers within the Nazi justice system" (Klee, 2003; Pringle, 2010). His research involved the examination of the effects of stress on the timing of human ovulation. Stieve worked closely with prison authorities in order to collect histological samples from 200 female prisoners after their death. He chose women who were recently stressed by the knowledge of their execution date. In order to interpret his findings, he would collect information on the medical history, menstrual cycle, and mental status of each woman from a third party. However, it has not been proven that he ever had any contact with the women prior to their deaths or that he had any influence on their execution dates. It has been stated that Stieve's research reduced these women to mere psychobiological mechanisms and was a callous exploitation of the anxiety of these women for research purposes (Hartman, 1962; Hildebrandt, 2009c; Winkelmann & Schagen, 2009).

The "*Clara*" cell, a nonciliated, secretory cell in the respiratory epithelium of the distal airways, named after the Austrian anatomist Max Clara (1899–1966) is an additional example of an anatomical finding tainted by the Nazi movement. In 1935, he was appointed professor and chair of anatomy at Leipzig University. During the Nazi era, he wrote two popular medical textbooks on the human nervous system and on human embryology.

Clara was a known member of the Nazi party and owed his career advancement to the Nazi regime. It has been shown that he actively sought to increase his cadaver supply by soliciting the justice administration. Additionally, he even requested specimens for anatomical research be obtained from the executed without the knowledge of their families before the 1943 decree, making his actions illegal at that time. Clara writes in his own paper in 1937 when discussing the "*Clara*" cell that he did not know whether this was an "ordinary" criminal or a victim of persecution and torture. However, there is reason to believe that on at least on one occasion, he experimented on a live prisoner prior to his/her execution. Clara has been accused of "*moral complicity*" and his work was called into question (Hildebrandt, 2008; Winkelmann & Noack, 2010; Woywodt et al., 2010). Following his release from prison, Max Clara went to Turkey where he became a professor in the medical faculty of Istanbul University.

It has been stated that the scientists of this period failed to "*examine the circumstances and implications of their work*" (Cohen & Werner, 2009). These anatomists failed to see the atrocities they were committing and that they were desecrating the bodies of victims. The actions taken by the researcher stripped the bodies of their humanity and dignity, boiling them down to mere tools of research (Hildebrandt, 2009c). In an effort to correct this wrong, policy has been established to identify and properly dispose of all the specimens that were collected unethically during this period. The identified items have been removed from the collections and interred in graves of honor. Additionally, places of remembrance have been created and the public informed about the process (Hildebrandt, 2009c; Malina & Spann, 1999).

Chapter 31

THE TWENTIETH CENTURY

In 1901, while still only a medical student, Edward Anthony Spitzka was chosen to perform the autopsy on Leon Czolgosz, the assassin of President McKinley. He was to inspect the murderer's brain for signs of degeneration. Spitzka did not find any, but his published report was widely praised. Born June 17, 1876 to the famous American neurologist Edward Charles Spitzka, it was no surprise that he chose the brain as his main focus of study. After medical school, Spitzka began to collect and examine the brains of accomplished men such as the explorer John Wesley Powell, looking for differences that set them apart from the average brain. He went on to publish over 40 papers on brain anatomy and is recognized as one of the world's leading brain anatomists (Burrell, 2003).

Florence Rena Sabin (1871–1953; Fig. 342) made the first of her many contributions to anatomy when she constructed a three-dimensional model of the medulla, pons, and midbrain in 1901. In connection with the model, she wrote a laboratory manual, *An Atlas of the Medulla and Midbrain,* which became a popular textbook. Born in Central City, Colorado on November 9, 1871 and educated at Smith College and Johns Hopkins Medical School, Sabin became a pioneer for woman in science. She was the first woman faculty member at Johns Hopkins School of medicine in 1902 and later the first woman to hold a full professorship at Johns Hopkins in 1917.

Sabin was elected the first female president of the American Association of Anatomists in 1924, the first woman elected to the National Academy of Sciences in 1925, and the first woman to head a department at the Rockefeller Institute for Medical Research. In addition to her pioneering firsts, Florence Rena Sabin's research career was long and varied. Her research and ideas on the lymphatics were original, controversial, and eventually proven correct. She believed the lymphatics represented a one-way system closed at the collecting ends, where the fluid entered by seepage from preexisting veins. Sabin's study of the origin of blood vessels led to her to the discovery that blood plasma is developed by liquefaction of the cells that form the walls of the first blood vessels. In her retirement years, Sabin pursued a second career as a public health activist in Colorado, and in 1951, she received a Lasker Award for this work (Ogilvie & Harvey, 2000).

"I hope my studies may be an encouragement to other women, especially to young women, to devote their lives to the larger interests of the mind. It matters little whether men or women have the more brains; all we women need to do to exert our proper influence is just to use all the brains we have."

Florence Sabin speaking at Hobart and William Smith College, after receiving the Elizabeth Blackwell Award.

Robert Russell Bensley (1867–1956) made a name for himself in 1907 when he devised a staining technique that allowed him to better visualize the anatomy of the pancreas. Born in Hamilton, Ontario, he was educated in Hamilton and at the University of Toronto. Bensley went on to accept a post in the Department of Anatomy at the University of Chicago in 1891; six years later he became the director. While there, he became interested in the pancreas and began his research with his student, M. A. Lane. The staining method they devised with Janus green and other dyes allowed Bensley to do three different things. First, he was able to stain all the islets of an entire pancreas of a guinea pig. Second, he was able to demonstrate the islets were loosely connected to the duct system and therefore had no exocrine functions. Third, he observed that the endocrine islets of Langerhans contained at least two different *"granular cell types"* alpha- and beta cells. Bensley's granule stain is still widely used to study the pathology and

Figure 342. Photograph of Florence R. Sabin. This image is available from the United States Library of Congress.

Figure 343. Photograph of Robert Russell Bensley. (Courtesy of Bethesda, MD: U.S. National Library of Medicine, National Institutes of Health, Health & Human Services.)

physiology of the islets cells. He published these classical studies in the *American Journal of Anatomy* in 1911 and firmly established the islet system as an independent component of the pancreas designed for internal secretion (Ricketts, 1955). Over his 26 years in Chicago, Bensley strengthened the Department of Anatomy and attracted an outstanding faculty, including C. Judson Herrick, George W. Bartelmez, Alexander Maximov, Charles Swift, and William Bloom.

Edwin Grant Conklin (Fig. 343) provided his first reports on cell size and nuclear size in 1909. Conklin was born November 24, 1863 in Waldo, Ohio, and educated at Ohio Wesleyan University and Johns Hopkins University. It was during his studies at Johns Hopkins that Conklin became interested in embryology and began his lifelong study of cell lineage. After receiving his Ph.D., he went on to become a professor of biology at Princeton University. It was during his professorship that Conklin produced his most important book entitled, *Hereditary and Environment*, which first appeared in 1914. It has passed through six editions and was translated into Russian and Japanese (Harvey, 1953).

In 1910, Ross Granville Harrison (Fig. 344) published his first major paper on nerve outgrowth, *The Outgrowth of the Nerve Fiber as a Mode of Protoplasmic Movement*, which changed the whole line of thought in neurology. Born January 13, 1870 in Germantown, Pennsylvania and trained at Johns Hopkins University, Harrison first introduced a method of cultivating tissue outside the body in 1907. He took fragments of tissue from different parts of the body and placed them in a drop of clotted frog's lymph. To his surprise, the tissue not only remained alive but continued to grow. Harrison observed that the shoots, which grew from the edge of a medullary tube, were outgrowths of nerve cells. These shoots grew rapidly, branched out, and formed a typical growth cone. Harrison had discovered that the neuroblast is able to form a nerve fiber outside of the body when removed from all sources of contamination. This discovery demonstrated the origin of the nerve, solved the problem of nerve outgrowth, and formed the basis for the present functional treatment of neurology.

Harrison's tissue culture method is still used today in the study of cancer, disease, and preventive medicine. His work exploring the growth of nerve fibers in tissue culture earned him a nomination for the Nobel Prize in 1917. He was the first American zoologist to be nominated, but because the Karolinska Institute ruled that no awards be made in Physiology and Medicine during World War I (1914–18), there was no prize given that year (American Association of Anatomists, 2013; Nicholas, 1961).

In 1915, Leslie B. Arey began one of the longest tenures of any medical school professor when he joined the faculty of Northwestern University Medical School as an instructor of anatomy. He was born in Camden, Maine, in 1891 and trained at Harvard University, where he received his Ph.D. degree in 1915. Arey began working at Northwestern University immediately after graduation and in 1925, he was appointed Chairman of the Department of Anatomy and the Robert Laughlin Rea Professor of Anatomy. Arey formally retired in 1956 but still maintained long hours in the department and continued, on a voluntary basis, teaching embryology and histology to medical and graduate students.

Arey not only molded the minds of numerous generations of physicians, but he also was an early proponent of women in medicine and graduate education. In 1917, he published his *Textbook of Embryology*, which was followed by *Development of Anatomy* in 1924. Arey taught at Northwestern for 72 years (American Association of Anatomists, 2013).

The 1918 monograph on the perilymphatic spaces of the labyrinth, published by George Linius Streeter (1873–1948; Fig. 345), was the first of a series of papers on the subject, of which previously little had been written. Moreover, students have found this monograph invaluable in clarifying the relations within the petrous bone. Born in January 12, 1873 in Johnstown,

Figure 344. Photograph Ross Granville Harrison. (Courtesy of Arizona State University. http://hpsrepository. asu.edu/handle/10776/1805?show=full.)

Figure 345. Photograph of George Linius Streeter. (Courtesy The Faculty History Project documents faculty members who have been associated with the University of Michigan http://um2017.org/faculty-history/faculty/george-linius-streeter.)

New York and educated at Columbia University, Streeter first became interested in embryology under the guidance of Wilhelm His at the University of Leipzig in Germany. He then took a position in the anatomy department of Johns Hopkins under Franklin P. Mall. While at Johns Hopkins University, Streeter discovered his interest in the development of the human auditory apparatus. His research on the topic included a full account of the embryology of the external ear, of the labyrinth and cochlea, the whole ganglionic apparatus, and the cranial nerves linked to the organs of equilibration and hearing.

In 1920, Streeter published his paper on *Weight, sitting height, head size, foot length and menstrual age of*

Figure 346. Photograph of Elizabeth Caroline Crosby. (Courtesy of University of Michigan, History of Medicine project. http://um2017.org/faculty-history/faculty/elizabeth-caroline-crosby.)

human embryo. This classical account of human prenatal growth is the go-to-guide for every embryologist when it is necessary to establish the age of a fetus or late embryo from its dimensions (Corner, 1954).

Elizabeth Caroline Crosby (1888–1983; Fig. 346) published her first book entitled *Laboratory Outline of Neurology,* in 1918 with C. J. Herrick. Born October 25, 1888 in Petersburg, Michigan and educated at Adrian College and the University of Chicago, where she obtained the M.Sc. (1912) and Ph.D. (1915) degrees for research work supervised by C. Judson Herrick, Crosby is perhaps the most famous female American neuroanatomist of the twentieth century. She received her first academic appointment at the University of Michigan in 1920 where she spent her entire academic career. Crosby's clinical and experimental approach to neuroanatomical methodology was sometimes controversial. Her final book, *Comparative Correlative Neuroanatomy of the Vertebrate Telencephalon,* was published in 1982. Crosby's collective works on the anatomy of the forebrain are classics in the field. In 1979, she received the National Medal of Science from President Jimmy Carter *"for outstanding contributions to comparative and human neuroanatomy and for the synthesis and transmission of knowledge of the entire nervous system of the vertebrate phylum"* (Ogilvie & Harvey, 2000). Crosby died at the age of 95 years, having lived a highly productive academic career that was rewarded with many honors and recognition for her work. In 1936, she became the first woman to be granted a full professorship at the University of Michigan Medical School. There she became Professor Emerita of Anatomy before leaving for Alabama in 1963, where she again became Professor Emerita of Anatomy.

In 1919, Charles Bardeen (Fig. 347) demonstrated a simple method of injecting blood vessels so as to reveal them in roentgenograms. The transformative technique produces stereoscopic roentgenograms of the vascular supply of some of the joints. Bardeen was born in Kalamazoo, Michigan, in 1871 and educated at Harvard and Johns Hopkins University. While studying at Hopkins under Franklin P. Mall, Bardeen pursued the development of the vascular and lymphatic system. This work culminated in a section in Keibel and Mall's classic *Manual of Human Embryology.* Barden went on to the University of Wisconsin-Madison where he played an intricate part in developing their medical school (American Association of Anatomists, 2013).

More enhancements in the knowledge of embryology came when Bradley M. Patten (1889–1971) who published his book entitled *The Early Embryology of the Chick* in 1920. Patten said, "*If (the book) helps the student to grasp the structure of the embryos, and the sequence and significance of the processes he encounters in his work on the chick, and thereby conserves the time of the instructor for interpretation of the broader principles of embryology it will have served the purpose for which it was written.*" Born in Milwaukee, Wisconsin, and educated at Dartmouth and Harvard, Patten is especially noted for his pioneer work using time-lapse cinematography to study the early development of the heart and great vessels. He made notable contributions to the understanding of human, pig, and chick embryology, especially in regard to the anatomy of the heart and its congenital defects. In addition to the book above, Patten published three more authoritative textbooks: *Embryology of the Pig* in 1925, *Human Embryology* in 1946, and *Foundations of Embryology* in 1958. All editions of his published texts are beautifully written, comprehensive, and classics in their respective fields (no author, 1971).

Nineteen twenty was also the year when Olaf Larsell began his studies that eventually resulted in a unified cerebellar terminology. He was born in Rattvik, Sweden, in 1886 and moved to the United States in 1891. He trained at McMinnville College (now called Linfield College) in Oregon where he graduated in 1910 with the B.Sc. degree, and in 1918 he completed his doctorate at Northwestern University. However, it was Larsell's work with C. Judson Herrick, a neurologist at the University of Chicago that sparked his interest in cerebellar anatomy. Larsell's first paper on the cerebellum of ambystoma, published in 1923, was the first in a series of studies that made him one of the world's authorities on the anatomy and function of the outgrowth of the hindbrain of the central nervous system. At the time of his death in 1964, he was completing the last chapter of his definitive, three-volume set, titled *The Comparative Anatomy and Histology of the Cerebellum*. In this series, he develops the definitive clarification and standard terminology for the lobes and lobules of the cerebellum (no author, 1964a).

Another advancement in neuroanatomy occurred in 1920 when Stephen W. Ranson released the first edition of his textbook entitled *The Anatomy of the Nervous Systems*. This textbook has been a crucial factor in the development of neuroanatomy in the United States. Born August 28, 1880 in Dodge Center, Minnesota, and educated at the University of Minnesota and the University of Chicago, Ranson first took interest in the nervous system under the influence of Professor J. B. Johnston of Minnesota. In addition to

Figure 347. A photograph of Charles Bardeen, in public domain.

the above textbook, Ranson made some influential contributions in the experimental field. He introduced his pyridine silver technique in 1914, a modification of one of Cajal's methods. With this new staining technique, he was able to appreciate a large number of unmyelinated fibers in sensory nerves. His subsequent studies went on to show that these fibers mediated pain sensation. Moreover, since Ranson could recognize unmyelinated fibers, his histological analyses of the peripheral visceral nervous system were more complete as compared to previous ones (Hinsey, 1943).

In 1921, Marion Hines published *The Embryonic Cerebral Hemisphere in Man*. Born June 11, 1889 in Carthage, Missouri, and trained at Smith College and the University of Chicago, Hines is widely known for her studies of the brain's control of movement. Her observations helped explain the nature of muscle paralysis and the stiffness associated with strokes in the cerebral cortex. She was able to demonstrate in cats and rhesus monkeys that extrapyramidal impulses are mediated exclusively by the corticospinal tracts (Magoun & Marshall, 2003; Ogilvie & Harvey, 2000).

We have been able then to separate the phenomena of spasticity from its companion paralysis and to demonstrate...a new localization for the inhibition of that substrate on movement, tone.

– Marion Hines

Edward Allen Boyden (Fig. 348) was born in 1886 in Bridgewater, Massachusetts, and was educated at Massachusetts State Teacher College and then Harvard University. His groundbreaking work on the extrahepatic biliary tract began in 1922 with the finding of a pancreatic bladder in one of 25 cats that had developed from an accessory pancreas beside the gallbladder. Wishing to compare sections of the two organs in the same physiologic state, he devised a way to empty the gallbladder. He fed the cats egg-yolk and cream and discovered that this would completely empty the gallbladder. Boyden is also known for his original studies of the urogenital system and the segmental anatomy of the lungs. In his work published in

1924, Boyden deprived chick embryos of the growing tips of the Wolffian ducts and noted that the mesonepheros begins to function as early as the fourth day of incubation. These studies eventually lead to his sound embryological explanation for agenesis of the human kidney in 1932. Boyden's studies of the lung were sparked in 1945 when he could not find any comprehensive figures of dissection of the external and interlobar surface of the lung. This research led to his book in 1955 titled, *Segmental Anatomy of the Lungs, A study of the Patterns of the Segmental Bronchi and Related Pulmonary Vessels*. In 1928, he became editor of *The Anatomical Record* and edited the next 62 Volumes (Myers, 1955; No author, 1954).

Dr. Boyden has been, as he continues to be, a potent influence in shaping the quality and the direction of American anatomy.

– Charles H. Denforth, concerning Boyden's work with *The Anatomical Record*.

Harold Cummins (1894–1976; Fig. 349) gave the science of dermatoglyphics its name when he coined the term in 1926 to describe the scientific study of the palmar and plantar ridges of the hands and feet. Born in Kalamazoo, Michigan, and educated at the University of Michigan and Tulane University, Cummins carefully preserved the skin from one sole of a cadaver between two sheets of glass and used it for future observations. Cummins published numerous studies in the field, including a 1929 paper that remains one of the most widely referenced papers on dermatoglyphic methodology to date. Many people consider his greatest contribution to the field was his discovery that patients with Down syndrome have characteristic dermatoglyphic features on the palms and fingers.

Cummins published a book in 1943 entitled, *Finger Prints, Palms and Soles*, which he dedicated to the dermatoglyphics pioneer Harris Hawthorne Wilder. This book will become a highly regarded reference work in the field of dermatoglyphics. He has also served as an expert consultant in numerous legal cases, including the Charlie Chaplin paternity case, the Oakes murder in Nassau, and the shooting of John Dillinger (Holt, 1976).

In 1935, William F. Windle began his experiments on the initiation of respiration in the fetus while visiting a scientist in Cambridge, England. This started his long-running major program in fetal physiology, with examination of *asphyxia neonatorum* and its effects on the central nervous system. Born in Huntington, Indiana, on October 10, 1898 and educated at Illinois

Figure 348. A photograph of Edward Boyden, A (Edward Allen) 1886–1976, University of Minnesota Institute of Anatomy. (Courtesy of Acc. 90–105 – Science Service, Records, 1920s–1970s, Smithsonian Institution Archives. http://siarchives.si.edu/collections/siris_arc_290371.)

Figure 349. A photograph of Harold Cummins, 1893–1976, Tulane University. (Courtesy of Acc. 90-105 – Science Service, Records, 1920s–1970s, Smithsonian Institution Archives. http://siarchives.si.edu/collections/siris_arc_296592?back=%2Fcollections%2Fsearch%3Fquery%3D Harold%2520Cummins%26page%3D1%26perpage%3D10%26sort%3Drelevancy%26view%3Dlist.)

Figure 350. A photograph of John Charles Boileau Grant. (Courtesy of http://www.anthropology.utoronto.ca/about/history.)

Wesleyan University and Northwestern University, Dr. Windle went on to become an assistant professor at Northwestern. In addition to his lung research, Dr. Windle worked many years on spinal cord regeneration, an idea that was sparked by a colleague's observation of intraspinal fiber regeneration after injections of bacterial pyrogens in animals. His research efforts in neonatal asphyxiation and spinal cord regeneration proved to have great clinical relevance for cerebral palsy and paralysis. In 1940, Dr. Windle published his classic book *Physiology of the Fetus*, a treatise that summarized much of the research done by Windle and his students from 1936 to 1940. His next book, entitled *Asphyxia Neonatorum*, was published in 1950. In 1959, as the explosive growth in neuroscience research and knowledge was beginning, Dr. Windle realized the need for additional outlets for articles in experimental neuroscience and became the founding editor of the journal "*Experimental Neurology*" (Clemente, 1985).

The Edinburgh born anatomist John Charles Boileau Grant (1886–1973; Fig. 350) published his first influential book in 1937, entitled *Methods of Anatomy, Descriptive and Deductive*. Grant studied medicine at the University of Edinburgh and graduated in 1909. Following training as a surgeon, he served as a medical officer during the First World War. For outstanding bravery he was awarded the Military Cross. Grant then decided to pursue a career in anatomy and in 1919, he accepted an appointment as Professor of Anatomy at the University of Manitoba. In 1930, the University of Toronto invited Dr. Grant to head its anatomy department (Persaud, 1983).

Because of his surgical background, Grant's approach to anatomy effectively changed how gross anatomy was taught. His *Method of Anatomy*, conceived and partly written in Manitoba, brought out the superb logic of the body's structure at the same time relying heavily on clinical significance as the major justification for the inclusion or exclusion of details. Grant's "*Method*" is the basis for clinical practice relying on logic, explanation, and rational thought. Not surprisingly, that in a span of 38 years, this book has been printed 26 times over many editions. Grant published two other influential books, *Handbook for Dissectors* in 1940 and Grant's *Atlas of Anatomy* in 1943 all while serving as an anatomy professor at the University of Toronto (Tobias, 1993).

For many years after his retirement from the University of Toronto in 1956, Grant was a Visiting Professor at the University of California in Los Angeles. He received wide recognition and many honors. He served as Vice-President of the American Association of Anatomists (1950–1952), and he was unanimously elected Honorary President of the Canadian Association of Anatomists when it was founded in 1956. The University of Manitoba conferred on him the honorary degree of Doctor of Sciences in 1956 (Persaud, 1983).

In 1940, Oscar V. Batson published a seminal article, "*The function of the vertebral veins and their role in the spread of metastases.*" This network of veins in the human body became known as "*Batson Plexus.*" Born in 1894 in Sedalia, Missouri, he received his medical doctor degree at St. Louis University in 1920. Batson was a classical anatomist, but he used modern techniques to demonstrate anatomical structures. Curious about how blood was drained from the brain when main drainage channels were blocked, he decided to investigate this problem further. Batson injected plastic material into the venous system and corroded away the tissue with alkali. The extensive network of veins that was created convinced him of the significance of the vertebral veins in drainage of blood from the head and neck. (Nemir, 1980).

The pathologist and immunologist Albert H. Coons (1912–1978) first introduced immunofluorescence in 1941, using specific antibodies labeled with a fluorescent dye to localize substances in tissues. Educated at Williams College and Harvard Medical School, he received his M.D. degree in 1937. Coons first had his idea for immunofluorescences while visiting Berlin. During his visit, Dr. Coons was thinking about the possible immunological bases for the pathogenesis of various diseases and was struck with the idea that it might be easier to find the antigen rather than the antibody. Once he returned to Harvard, he began the work that led to his 1941 publication. In the following years, the controls for immunocytochemistry were taken from this method and other nonmicroscopic techniques that use antibodies to identify specific proteins (American Association of Anatomists, 2013; Karnovsky, 1979). A modest and friendly man, Coons received many honors and recognition for his achievements. He was awarded the Albert Lasker Award in Basic Research in 1959, and in 1962, he was elected a member of the National Academy of Sciences. Coons received a named professorship at Harvard University and this was followed with many honorary academic degrees. In 1962, Coons was the recipient of the prestigious Canadian Gairdner International Award.

In recognition of his contributions to the knowledge of pathology and immunology, and especially for his development of the fluorescent antibody technique which has cast new light on hypersensitivity reactions, and which has opened important new avenues for the study of diseases, such as rheumatoid arthritis, glomerulonephritis, rheumatic fever and disseminated lupus erythematosus, in which these reactions may be involved.

Also in 1941, Roger Wolcott Sperry made the first of what was to become a number of successful challenges to existing concepts related to neuronal specificity and brain circuitry. In a series of well-written papers from 1941–1946, Sperry conclusively demonstrated that the rat's motor system was "*hard wired*" and unmodifiable by training following transplants. This finding drastically altered the surgical management of nerve injury. Sperry was born August 20, 1913 in Hartford Connecticut and was educated at Oberlin College and the University of Chicago. During his postdoctoral years, Sperry continued to work on neuronal specificity and provided strong evidence for nerve guidance by "*intricate chemical codes under genetic control.*" These experiments laid the foundation of many of our current views on neuronal specificity in brain development. It was also during his postdoctoral years that Sperry became intrigued with the corpus callosum and he began his experiments that would eventually lead to him receiving the Nobel Prize in 1981.

Sperry won the Nobel Prize for his research on the functional specialization and lateralization of the cerebral hemispheres. He was able to show that the left-brain is superior to the right in abstract thinking, interpretation of symbolic relationships, and in carrying out detailed analysis. Furthermore, it is the leading hemisphere in control of the motor system and communication (Voneida, 1997).

> When the brain is whole, the unified consciousness of the left and right hemispheres adds up to more than the individual properties of the separate hemispheres.
>
> – Roger Wolcott Sperry

In 1942, Wendell J. S. Kreig published his *Functional Neuroanatomy*. Born in Lincoln, Nebraska, in 1906 and educated at the New York University School of Medicine, Krieg went on to publish 17 books, most of them with his own publishing house called Brain Books that he created in 1955. Krieg's publications were known for their beautiful and unique illustrations. He was immensely talented and would fill his books with stunning original images. His precise and detailed artwork of the human brain in slices was the first attempt to display the brain's structure and pathways in three dimensions. Krieg was a unique neuroscientist who personally illustrated his research findings and interpretations through his meticulous drawings (Whitlock, 2007).

The anatomist Murray Llewellyn Barr (1908–1995; Fig. 351) and his graduate student Edwart G. Bertram discovered the inactive X chromosome found in female somatic cells in 1948. This important cell structure is eponymously known as the "*Barr body*." In men and women with more than one X chromosome, the number of Barr bodies visible at interphase is always one less than the total number of X chromosomes. Barr was born in Belmont, Ontario, and received his M.D. degree from the University of Western Ontario in 1933. Bertram was working on a project to produce electrical stimulation of the hypoglossal nerve of cats to deplete the Nissl material of motor neurons. During examination of the tissue, he discovered that some cats did not have a nuclear satellite, a small dot adjacent to the nucleus. Upon careful checking of the records, Barr and Bertram found that these cats were all male and postulated that the nucleolar satellite may represent the second X chromosome in female cats. Further studies demonstrated that human females also exhibited what is now called the Barr body. Barr helped in establishing human cytogenetics as a clinical discipline (Potter & Soltan, 1997). Murray Barr was appointed Professor of Anatomy at the University of Western Ontario and was the recipient of many honors. Edwart Bertram subsequently pursued a very successful career as Professor of Anatomy at the University of Toronto.

Figure 351. A photograph of Dr. Murray Llewellyn Barr. (Photo credit the Canadian medical hall of fame. The Canadian Medical Hall of Fame.)

In 1950, Barry Joseph Anson (Fig. 352) published an *Atlas of Human Anatomy*. Anson aimed to publish an anatomy "*whose pictorial content would be based upon new dissections, serially prepared, and upon variable morphological features statistically presented.*" Anson was born on March 21, 1894 in Muscatine, Iowa, and he studied at the University of Wisconsin and Harvard Medical School, graduating with a Ph.D. degree in medical sciences. He went on to teach at Northwestern University Medical School where he became Professor of Anatomy. It was during his time as a gross anatomy teacher that Anson would get his students to collect data on anatomical variations and on the dissection of special areas. Much of these findings made their way into his *Atlas of Human Anatomy*, making it especially popular amongst clinicians. He also coauthored *The Temporal Bone and Ear, The Surgical Anatomy of the Temporal Bone and Ear, Surgical Anatomy*, and *Anatomy and Surgery of Hernia* (American Association of Anatomists, 2013; Arey, 1962).

Nineteen fifty also marked the beginning of Malcolm Breckenridge Carpenter's years of study on the

Figure 352. A photograph of Barry Joseph Anson. Contributor: Northwestern University. Medical School, Chicago. (Courtesy of Bethesda, MD: U.S. National Library of Medicine, National Institutes of Health, Health & Human Services.)

Figure 353. A photograph of William Henry Hollinshead. (Courtesy of http://www.clinical-anatomy.org/honored/hollinshead.html.)

structure, function, and connections of forebrain motor centers. Dr. Carpenter, who was born in Montrose, Colorado in 1921, graduated from Columbia College and went on to receive his medical degree from Long Island College of Medicine. His work shed new light on motor disorders, particularly those caused by diseases of the basal ganglia. Moreover, his contributions furthered research into defects in the neural networks responsible for ailments like Parkinson's disease. He was also noted for his work with vascular patterns, eye movement, vestibular systems, the diencephalon, and the subthalamus. He authored the textbook entitled, *Core Text of Neuroanatomy* and coauthored *Human Neuroanatomy and Cerebellum of the Rhesus Monkey* (American Association of Anatomists, 2013).

The clinical anatomist William Henry Hollinshead (Fig. 353) published his first influential anatomy textbook entitled *Functional Anatomy of Limbs and Back*, in 1951. Born in Winchester, Tennessee, June 17, 1906, Hollinshead received his B.S., M.S., and Ph.D. from Vanderbilt University. He went on to head the anatomy department, which he established, at the Mayo

Graduate School of Medicine. There he trained and influenced countless future surgeons. He also authored two more anatomical textbooks, *Anatomy of Surgeons* (1954) and *Hollinshead's Textbook of Anatomy* (1962). All of his books are classics in the discipline of anatomy. His *Anatomy of Surgeons* remains a standard reference for clinical anatomists and surgeons (American Association of Anatomists, 2013).

In 1952, Keith L. Moore (Fig. 354) developed the buccal smear sex chromatin test for detecting sex in intersexuality. Born October 5, 1925 in Bethel, Ontario, Moore attended the University of Western Ontario where he earned his B.A., M.S., and Ph.D. He began his study of the sex chromatin during his years as a doctoral student under Dr. Murray Barr.

Moore first developed a skin biopsy method for sex chromatin testing where a small piece of skin had to be surgically removed for testing. It was less than a year later that he developed the buccal smear test, completely eliminating the need for minor surgery. This technique is simple and it is widely used. Moore has authored many influential textbooks that have

Figure 354. A portrait of Keith L Moore.

impacted anatomy education of medical and dental students worldwide. These textbooks include: *Clinically Oriented Anatomy, Essential Clinical Anatomy, The Developing Human,* and *Before We Are Born* (American Association of Anatomists, 2013; Moore, 2012). Dr. Moore was President of the Canadian Association of Anatomists, and, in 1984, he was awarded the J.C.B. Grant Award. He is a Honored Member of the American Association of Clinical Anatomists and was recipient in 2007 of the inaugural Henry Gray/Elsevier Distinguished Educator Award of the American Association of Anatomists.

Dr. Frank N. Low (1911–1998) has been described as a pioneer in electron microscopy, a teacher's teacher in the anatomical sciences, and an inspired genius of morphological interpretation. Dr. Low was born in Brooklyn, New York, in 1911 and earned his undergraduate degree from Cornell University in Ithaca, New York in 1932. He attended medical school at the University of Buffalo immediately after graduation but returned to Cornell University where he studied with Benjamin F. Kingsbury and completed his Ph.D. in histology and embryology in 1936. He published more then 100 papers and was an expert on the use of

the electron microscope. Low became internationally known for the superb quality of his micrographs (Carlson, 1999; Low, 1953). His research career began with studies aimed at peripheral visual acuity but with electron microscopy, broadened to include descriptions of the pulmonary epithelium in rats, blood cells, and microfibrils (Carlson, 1999) He was the first to describe the anatomical structure of the blood-air barrier in the lung in 1953. In a seminal paper, he published a TEM description of the pulmonary epithelium in rats, which showed for the first time that the blood-air barrier in mammals was not syncitial, and that all alveoli were lined by a continuous pulmonary epithelium subtended by a basal lamina.

Nineteen fifty-three was also the year the DNA double-helix structure was identified. It came to fruition after countless years of critical discoveries and observations when James D. Watson and Francis Crick described what is now accepted as the first correct double-helix model of DNA structure. Their double-helix molecular model of DNA is based on a single X-ray diffraction image taken by Rosalind Franklin and Raymond Gosling in May 1952, as well as their knowledge that the DNA bases are paired. James Dewey Watson was born in Chicago, Illinois, on April 6, 1928 He received his B.S. from University of Chicago and the Ph.D. degree from Indiana University. Francis Harry Compton Crick was born in Northampton, England, in 1916 and studied at University College of London. The two met in 1951 and discovered their common interest in solving the DNA structure. In 1953, they proposed the complementary double-helical configuration. In 1962, after Franklin's death, Watson, Crick, and Wilkins jointly received the Nobel Prize in Physiology (no author, 1964b).

In 1957, Oxford University Press first published Russell T. Woodburne's *Essentials of Human Anatomy.* This is one of the early textbooks to describe anatomy using the regional approach rather than a systematic method. Woodburne was born November 2, 1904 (Fig. 355) and was educated at the University of Michigan where he spent his entire academic career. He was known as an excellent teacher who influenced several generations of medical students. Woodburne published a second textbook entitled, *Guide to Dissection in Gross Anatomy* (Brahce, 1967).

Raymond C. Truex became the editor of *Strong and Elwyn's Human Neuroanatomy* in 1959, a book that was widely used for generations. Truex was born December 11, 1911 in Norfolk, Nebraska. He received his B.A. from Nebraska Western University, his M.S. from St. Louis University, and his Ph.D. from the

Figure 355. Russell T. Woodburne. (Courtesy of University of Michigan, Faculty History Projects. http://um2017.org/faculty-history/faculty/russell-thomas-woodburne.)

University of Minnesota. Truex made enduring contributions to anatomy through research and his textbooks. He authored the book *Detailed Atlas of the Head and Neck and Human Cross Section of Anatomy*. Although most people regarded Truex as a neuroantomtist, his research was mainly on the heart. His research of the cardiovascular system included seminal observations on the anatomy of the sinu-atrial and atrioventricular nodes, cardiac circulation, and visceromotor innervation of the heart (Ladman, 1981).

In 1959, Chester B. McVay authored *Inguinal Hernioplasty*, which broadens our clinical knowledge of the inguinal canal. Dr. McVay was born August 1, 1911 in Yankton, South Dakota, and received his M.D. and Ph.D. degree from Northwestern University under the eminent anatomist Barry J. Anson. He went on to coauthor, with Anson, one of the first textbooks on clinical anatomy entitled *Surgical Anatomy*. After his Ph.D., Dr. McVay went to the University of Michigan to complete his surgical training. He became one of the leading herniologists and authored *Femoral Hernioplasty* in 1969 (Herzog, 2007).

The embryologist Richard J. Blandau (Fig. 356), renowned for his basic studies of ovulation,

fertilization, and embryonic development, produced a research film, *Ovulation and Egg Transport in the Rat*, which received a Vienna Film Festival Award in 1959. He was born in Erie, Pennsylvania, on August 5, 1911 and in 1939 achieved his doctorate from Brown University in reproductive physiology and anatomy. Blandau's research thrived while he was at the University of Rochester. During this period, he expanded his reproductive-biology research in animal models but also developed a lifelong interest in human embryology and clinical studies. Blandau earned international recognition for his cinemagraphically recorded ovulation process and the transport of the ova into and through the oviduct (Slonecker, 1998).

In 1960, Alan Peters became one of the first to interpret the structure of myelin sheaths in the central nervous system in his paper *The structure of myelin sheaths in the central nervous system of Xenopus Laevis*. Peters was born in Nottingham, England in 1929 and received his doctorate in Zoology from Bristol University. After serving on the faculty of the Department of Anatomy at Edinburgh University in Scotland, he

Figure 356. Richard J. Blandau. Contributor: University of Washington School of Medicine 1955. Seattle, Washington. (Courtesy of Bethesda, MD: U.S. National Library of Medicine, National Institutes of Health, Health & Human Services.)

came to Boston as Chairman of the Department of Anatomy in 1966. When Peters began his work with the electron microscope, there was still a great deal of controversy about the myelin sheath structure and the relationship between oligodendrogliocytes and myelin in the central nervous system. Peter's work contributed significantly to solving these controversies. He provided evidence of the continuity of oligodendrocyte processes to the myelin sheath and showed that a spiral wrapping of membranes also forms the central myelin. He published numerous papers and authored the textbook *The Fine Structure of the Nervous System* (1970), which has had an immense influence on the field of neurocytology and has instructed generations of young neuroscientists (Palay, 1995).

In 1962, Leonard Hayflick discovered that most types of human cells have a natural limit to the number of times that they can divide, or reproduce, after which they become senescent – a phenomenon now known as the "*Hayflick Limit.*" Dr. Hayflick was born on May 20, 1928 in Philadelphia, Pennsylvania, and he obtained his higher education at the University of Pennsylvania. His discovery of the "*Hayflick Limit*" overturned the belief that focused attention on the cell as the fundamental location of age changes. Dr. Hayflick showed for the first time that mortal and immortal mammalian cells existed. This distinction is the basis for much of modern cancer research. Dr. Hayflick also developed a normal human diploid cell strain called W1-38 that could be used for any research throughout the world that requires a normal human cell. His strain is the most widely used and highly characterized normal human cell population in the world. Dr. Hayflick is also author of the popular book, *How and Why We Age* published in 1994 (no author, 2003).

In 1963, Jan Langman published the first edition of his *Medical Embryology*. This book introduced a clinical approach to teaching embryology and changed how the subject is taught in medical school. Langman was born October 21, 1923 and educated at the University of Amsterdam. After graduating with his M.D. degree, he started his anatomy career under the guidance of M.W. Woerdeman in Amsterdam. Langman moved to Canada where he built himself a reputation at McGill University with his well-known dramatic anatomy lectures. At McGill, he went on to establish a course in embryology. He was also involved in research and studied neuron differentiation during embryogenesis using autoradiography. After the publication of the first edition of *Medical Embryology*, Langman became Professor and Chairman of the Anatomy Department at the University of Virginia in 1964 (American As-

sociation of Anatomists, 2013).

In 1970, Raymond Damadian (Fig. 357) a medical doctor and research scientist, developed magnetic resonance imaging (MRI) as a tool for medical diagnosis. Damadian was born in New York on March 16, 1936. He earned his bachelor's degree in mathematics from the University of Wisconsin-Madison in 1956 and an M.D. degree from the Albert Einstein College of Medicine in New York City in 1960. Physicists had been using NMR, but it was Damadian who reasoned that hydrogen (in water) might prove responsive within the cells of living tissue. He discovered that different kinds of animal tissue emit response signals that vary in length, and that cancerous tissue emits response signals that last much longer than those from noncancerous tissue. Damadian published a seminal paper in 1971 on his preliminary findings and then focused on building his machine. His machine, named "*Indomitable*" was finally used in 1977. Damadian would prove on his own body that the intense magnetic fields produced no harm, but the machine was too small for him. Instead, he imaged a smaller graduate student and produced the first NMR image of a human torso (American Association of Anatomists, 2013; Damadian, 1936; Schneider, 1997).

Only two years after the MRI was introduced, Sir Godfrey Hounsfield invented the first commercial

Figure 357. Raymond Damadian. (Courtesy of http://www.inc.com/magazine/20110401/how-i-did-it-raymond-damadian.html.)

computed tomography (CT) scanner at the EMI Central Research Laboratories in 1972. Hounsfield was born in Nottinghamshire, England, on 28 August 1919 and attended Faraday House Electrical Engineering College in London. Hounsfield was working on pattern recognition of letters in the mid 1960s when he began to contemplate whether he could reconstruct a three-dimensional image from readings taken through a box at randomly selected directions. It was this work on pattern recognition and the use of computers to analyze readings that made the CT scanner possible. The CT is considered to be the greatest innovation in the field of radiology since the discovery of X-rays. This cross-sectional imaging technique makes it possible to visualize phenomenon including disease processes. In 1979, Hounsfield and A. M. Cormack were awarded the Nobel Prize in medicine for the invention of CT (Beckmann, 2006; Oransky, 2004).

Dr. John Elias Skandalakis (Fig. 358) was born in 1920 in Sparta, Greece. In 1946, he earned a medical degree from the University of Athens and he received his certification in General Surgery from the University of Athens in 1950. His training was at the First

Figure 358. John Elias Skandalakis. (With permission from Loukas and Tubbs, *Clinical Anatomy*, 2012.)

Surgical Clinic, Athens University and at the Naval Hospital in Piraeus, Greece. He moved to the United States to complete a surgical residency and also completed a Ph.D. in anatomy at Emory University School of Medicine in 1962. Skandalakis's love for the art and science of human anatomy is demonstrated in his well-known books: In 1972: *Embryology for Surgeons*, in 1983: *Anatomical Complications in General Surgery*, in 1994: *Surgical Anatomy and Technique: A Pocket Manual*, and in 2004: the *Surgical Anatomy*. His two volume text, *Surgical Anatomy, the Embryologic and Anatomic Basis of Modern Medicine*, could be considered the magnum opus of his contributions to the art of surgery. He rose quickly through the academic ranks and became Professor of Anatomy and Professor of Surgery. During his time at Emory University, Dr. Skandalakis served as Professor of Anatomy, Professor of Surgery, and Professor of Surgical Anatomy and Technique, and he established the Thalia and Michael Carlos Centers for Surgical Anatomy and Technique and the Alfred A. Davis Research Center for Surgical Anatomy and Technique.

In 1981, Dr. Skandalakis was awarded the Aven Cup by the Medical Association of Georgia for outstanding service to the community by a doctor. In 1999, he received the distinguished Medical Achievement Award from the Emory University School of Medicine. Dr. Skandalakis authored more than 300 publications, including journal articles, books, book chapters, and monographs. He was a distinguished member of the American College of Surgeons, a founding member of the American Association of Clinical Anatomists (AACA), and of the American Hernia Society (AHS), which served as the impetus for his founding of The Thalia and Michael Carlos Center for Surgical Anatomy and Technique, along with The Alfred A. Davis Research Center for Surgical Anatomy and Technique (Loukas et al., 2010c).

Gene L. Colborn (Fig. 359) was born in Springfield, Illinois, and received a B.A. degree with Honors in Religious Education from Kentucky Christian University, followed by graduate studies in Religion and Education at the University of Pittsburgh. He received a B.S. degree with Honors in Biology and Chemistry from Milligan College in 1962 and continued with graduate work in Medical Sciences at Wake Forest University on a National Sciences Foundation sponsored Cardiovascular Training Grant. Colborn received M.S. and Ph.D. degrees in Anatomy and Physiology from the Bowman Gray School of Medicine at Wake Forest University in 1967. He completed a Post-Doctoral Fellowship at the University of New

Figure 359. Gene L Colborn. (With permission from Loukas et al., Clinical Anatomy, 2010.)

Mexico School of Medicine and then accepted a position as Assistant Professor of Anatomy at the University of Texas Health Science Center at San Antonio, Texas (UTHSA). He also taught Surgical Anatomy at Fort Sam Houston. Dr. Colborn went to the Georgia Health Sciences University, formerly Medical College of Georgia (MCG) in Augusta as an Associate Professor in 1975 to direct the program in medical gross anatomy. He was promoted to Full Professor of Anatomy and Surgery, and was Founder and Director of the Center for Clinical Anatomy in 1988. At the same time, he was appointed as Clinical Professor of Surgery at Emory University School of Medicine in Atlanta, Georgia. He taught many programs in clinical anatomy to residents in surgery, OB/GYN, ophthalmology, orthopedics, emergency medicine, and general surgery at MCG and special programs in urology and gynecology at Rush University Medical School in Chicago. He is noted professionally most for his contributions to the field of abdominal and pelvic surgical anatomy together with friend and long-life collaborator Dr. John Skandalakis. Dr. Colborn was a member of many years of the American Association of

Anatomists and was a Founding Member of the American Association of Clinical Anatomists (Loukas & Tubbs, 2012).

In 1974, T. V. N. (Vid) Persaud (Fig. 360) coauthored with Keith L. Moore the first of many editions of the embryological textbooks, *The Developing Human* and *Before We Are Born: Essentials of Embryology*. The concise and richly illustrated information coupled with its highly efficient, reader-friendly manner, made these books a popular choice for learning embryology in the medical field. Dr. Persaud was born in Port Mourant, Guyana, in 1940. He pursued medical studies at London University in England and Rostock University in Germany. After internship in Berlin, he served as a Government Medical Officer in Guyana. Persaud began his academic career as an anatomist at the University of the West Indies in Jamaica. In 1972, he was offered an academic position as Associate Professor at the University of Manitoba where he worked with Keith Moore.

Dr. Persaud was Professor and Head of the Department from 1977 to 1992. He has dedicated his professional career to the study and teaching of normal and abnormal human development. Embryogenesis and teratogenesis have become the main focus of Persaud's books, monographs, and research papers. In 1991, Dr. Persaud was awarded the Award of the Canadian Association of Anatomists. He is a Honored Member of the American Association of Clinical Anatomists (2008) and recipient of the Henry

Figure 360. A photograph of Vid Persaud.

Gray/Elsevier Distinguished Educator Award of the American Association of Anatomists (2010).

Professor Robert McMinn (Fig. 361) died on July 11, 2012 at the age of 88 years after a long and distinguished career as one of the foremost human anatomists of his generation. Known to all as Bob, his major work was the publication in 1977 of *A Colour Atlas of Human Anatomy*, in collaboration with the photographer Ralph Hutchings. A totally new concept in anatomy teaching, this book showed life-sized color photographs of actual dissections of real bodies, rather than the usual line drawings with red arteries, blue veins, and yellow nerves, which had been the somewhat dull, unrealistic fare for medical students until then. Revised over the years to include more clinical emphasis, the book is now in its seventh edition and has been translated into over 30 languages.

Dr. McMinn was born on September 23, 1923 in Auchinleck. He was educated at Brighton College embarking on and at Glasgow University, where he graduated in medicine with the M.B. Ch.B. degrees in 1947. In 1950 he began his anatomical career as Demonstrator in Anatomy at Glasgow University before

Figure 361. A photograph of Robert McMinn. (With permission from Marion Philip, Clinical Anatomy 2012.)

being appointed Lecturer at Sheffield University in 1953. In 1956, he gained his Ph.D., followed in 1958 by an M.D. degree. In 1960, he was appointed Reader and then Professor of Anatomy at King's College, London University. Ten years later, he was appointed as Sir William Collins Professor of Human and Comparative Anatomy, and Conservator of the Anatomical Museum at the Royal College of Surgeons of England; the major work of the atlases was achieved during his time at the College. Throughout his career, McMinn's main research interests lay in the field of tissue repair and wound healing, and on the association between skin disease and the alimentary tract. He published numerous papers, and his first major book on these subjects, *Tissue Repair*, appeared in 1969, followed in 1974 by *The Digestive System and the Human Gut.*

Much of McMinn's time was spent in lecturing at both undergraduate and postgraduate levels. He was greatly revered in this role, both on his home turf and in numerous institutions around the world. He attributed at least part of his success to his habit of destroying his notes after each lecture, thus ensuring that next year he would have to rewrite and never become stale and predictable. For many years, he was an examiner for trainee surgeons sitting for the Primary FRCS examination, both in the United Kingdom and overseas. McMinn was an active member of the anatomical community, serving as program secretary and later, treasurer of the Anatomical Society of Great Britain and Ireland, as well as being a founder member and first honorary secretary of the British Association of Clinical Anatomists. He was editor of later editions of *Last's Anatomy Regional and Applied*, which remains one of the standard works for surgical trainees (Philip, 2012).

In 1977, the German anatomist Gunther von Hagens (Fig. 362) created "*plastination*," his process of preserving cadavers with a reactive plastic material like silicone rubber, polyester resin, or epoxy resin. This process replaces the fluid components of the human body with plastic and enables each cell of the body to remain identical to their condition prior to preservation. Additionally, the specimens are dry, odorless, durable, and completely touchable. Von Hagens was born in 1945 and began his medical studies in 1965 at the University of Jena, which he completed at the University of Lubeck. Von Hagens eventually became a resident and lecturer at Heidelberg University's Institute of Pathology and Anatomy where he invented plastination

Plastination made it possible for the first time to preserve natural anatomical specimens in a durable, research focused on the microanatomic and histologic appearance of the vessels of the peritoneum, pleura, and cerebrum. He also researched possible topographical issues concerning the cranial and cervical regions. From this research, Dr. Lang was able to publish more then 200 original papers (von Lüdinghausen, 2004).

In 1979, John V. Basmajian (Fig. 363) published his book *Muscles Alive: Their Functions Revealed by Electromyography* and became internationally known for his pioneering work in electromyography and biofeedback, the study of electrical discharges from muscles. Basmajian was born on June 21, 1921 in Constantinople, Turkey and emigrated to Canada as an infant with his Armenian parents. He graduated from the University of Toronto's School of Medicine in 1945.

After serving as a Captain in the Canadian Army Medical Corp during World War II, he joined the Department of Anatomy at the University of Toronto as a Lecturer in 1949 and in 1957, Basmajian became

Figure 362. Photograph of Gunther von Hagens. (Photo, courtesy of Wikimedia-Commons User Túrelio, Creative Commons BY-SA 2.0-de.)

realistic and aesthetic manner for educational and research purposes and allows for specimens to be frozen in time at a point between death and decay. It allows for bodies to be displayed in artistic fashion that attracts the layperson and brings anatomy to the general public. Plastinated specimens are now widely used for teaching practical anatomy to students in the health professions (Von Hagens, 2002).

Johannes Lang published the first of two textbooks that are a culmination of his life's work in 1979 entitled *Praktische Anatomie, Kopf: Gehirn und Augenschädel*. The second book, *Praktische Anatomie Kopf: Übergeordnete Systeme*, was published in 2003. These two books cover Lang's research from 1955 to 1993. Lang was born in 1923 in Germany and received his doctorate and other degrees from the University of Munich. Dr. Lang's

Figure 363. Photograph of John Basmajian. (Courtesy of American Association of Anatomists. http://www.anatomy.org/sites/default/files/images/John_Basmajian.jpg.)

Professor and Head of the Anatomy Department at Queen's University. He moved to Atlanta in 1969 as Director of Neurophysiology at the Georgia Mental Health Institute in Atlanta. He also was appointed Professor of Anatomy, Physical Medicine, and Psychiatry at Emory University. Returning to Canada in 1977, Basmajian was Professor of Medicine and Anatomy at McMaster University. He also served as the Director of Chedoke Centre for Rehabilitation Medicine.

Basmajian was fascinated with anatomy, but he particularly loved to examine the structure and function of muscle. He realized well before most scientists that an understanding of muscle activity patterns under a variety of movement situations could serve as a foundation for the specification and assessment of therapeutic interventions. It was this mode of thinking and dedication to research that led to major advances in electromyography (EMG) and its implications. EMG applications quickly led to the discovery that given appropriate visual and auditory feedback of amplified, raw muscle signals, individuals could isolate and control single motor units within the recording radius of fine-wire electrodes. From Basmajian's seminal studies emerged what is now known as *"EMG Biofeedback,"* a rapidly advancing scientific field (Wolf, 2008). Basmajian received many honors, including the Order of Ontario (1991) and the Order of Canada (1994). He served as President of the American Association of Anatomists and was made an honored member of the American Association of Clinical Anatomists.

In 1985, Ronald Bergman (Fig. 364) published the *Arterial Variations in Man* and in 1988, his landmark book, the *Compendium of Human Anatomic Variation.* Dr. Ronald A. Bergman is Professor Emeritus of Anatomy and Cell Biology at the University of Iowa College of Medicine. He grew up in Chicago and earned a B.S. (biology), M.S. (physiology and zoology), and Ph.D. (Marine Biology, 1955) degrees from the University of Illinois. He did postgraduate work as a National Foundation for Infantile Paralysis Fellow at the Karolinska Institute in Stockholm, Sweden. Following his work in Sweden, he was offered a position in the Poliomyelitis Laboratory in the School of Hygiene at The Johns Hopkins Medical Center by David Bodian, a distinguished polio researcher. When Dr. Bodian was appointed Head of the Department of Anatomy at Johns Hopkins, Dr. Bergman was asked to join him. There, Dr. Bergman continued his bench research in muscle biology and was assigned to teach gross anatomy. By Dr. Bergman's own admission, he did not know enough anatomy *"to stuff a pocket"* and that his knowledge of the subject he was to teach was *"pitiful."* Despite his early humiliation and frustration as an instructor, he exemplified the contemporary attribute of *"lifelong learning"* and developed preeminence in the discipline. In 1974 Dr. Bergman went to the American University of Beirut where he was Professor in the Department of Human Morphology until 1980. During that time, he and his colleagues held the medical school together while civil war raged in Lebanon. For his valiant work, he was awarded the Gold Order of Merit medal by the Lebanese Government. In 1980, Dr. Bergman moved to the Department of Anatomy at the University of Iowa where he taught gross anatomy and histology to hundreds of medical students before retiring in 1997. He is the author of 66 papers and more than a dozen books and chapters, including texts on MRI and CT sectional anatomy. He directed the thesis work of several graduate

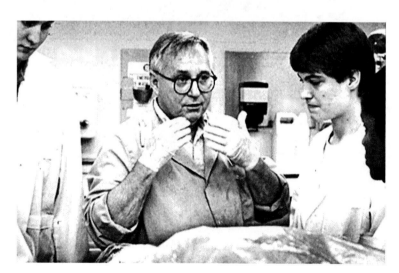

Figure 364. Ronald Bergman teaching students next to a cadaver. (Courtesy of http://www.anatomyatlases.org/Anatomic Variants/Images/Bergman.shtml.)

Figure 365. Frank H Netter. (Courtesy of Netter Images, Elsevier. http://www.netterimages. com/artist/netter.htm.)

students and supervised postdoctoral fellows and was the recipient of numerous teaching awards, and research and education grants.

Over the course of more than 35 years, Dr. Bergman methodically perused a total of 884 research journals (every English journal and a multitude of foreign journals as well) and each page from their inception and cataloged human anatomic variations. From this work, he published the *Compendium of Human Anatomic Variation: Text, Atlas, and World Literature* (1988) and *Illustrated Encyclopedia of Human Anatomic Variation* (1996). In 2006, he and his associates made these and other resources available as an online anatomy health sciences library, the *Anatomy Atlases* (http://www.anatomyatlases.org). In 2013, the AACA recognized Dr. Bergman as an Honored Member (website h).

Frank H. Netter (Fig. 365), in 1989, published his *Atlas of Human Anatomy*, often referred to as Netter's "personal Sistine Chapel." Netter was born on April 25, 1906 in Brooklyn, New York, and at an early age displayed "*an uncanny knack for making pictures.*" He attended the City College of New York and studied art at the National Academy of Design and the Art Students' League. He became a successful artist by 1920. However, with the death of his mother, Netter decided to honor her memory by enrolling in medical school at New York University Medical School. Netter started his collection of sketches as a young medical student. While in school, his personal sketches of anatomy and embryology on his notes came to the attention of his instructors and they asked him to illustrate articles and books authored by the faculty.

After graduation, Netter joined a private surgical practice but due to the Great Depression, could not afford to stay. He decided to leave the practice and become a full-time medical illustrator. As a formally trained physician and artist, Netter had a distinct advantage over most medical illustrators in that he approached each picture from the viewpoint of the clinician first and then as an artist. He drew over 4,000 illustrations and after the urging of many physicians and students, Netter dedicated his time to producing his *Atlas of Human Anatomy*. The textbook quickly became the best-selling anatomy atlas in North American medical schools and now has a following around the world based on its publication in 16 languages (Brass, 1984; Hansen, 2006). Netter was the recipient of numerous awards, including Honored member of the American Association of Clinical Anatomists. He died on September 17, 1991 (Netter, 2013).

In 1996, Dolly the sheep was born on July 5 to three mothers (one providing the egg, another the DNA, and a third carrying the cloned embryo to term). Ian Wilmut goes down in history as the first man to clone an animal. Wilmut was born in Hampton Lucey, England, July 7, 1944. He received his doctorate from the University of Cambridge where he concentrated on animal genetic engineering. During his postdoctoral years, Wilmut continued his pursuit of genetic research and was part of the 1973 team that produced the first calf from a frozen embryo. The course of his studies took a drastic change after an interesting conversation in a bar, when he was told that someone had already produced a lamb from the cells of an already-developing lamb embryo. Although this claim turned out to be false, it put Wilmut on the path to creating Dolly from adult sheep cells. Dolly was the first clone produced from a cell taken from an adult mammal. She was created using the technique of somatic cell nuclear transfer, where the nucleus from an adult cell was transferred into an unfertilized oocyte (developing egg cell) that had its own nucleus removed. The hybrid cell is then stimulated to divide by an electrical stimulus and when it develops into a blastocyst, it is implanted into a surrogate mother. The production of Dolly demonstrated that genes in the nucleus of a mature, differentiated, somatic cell are still capable of reverting back to an embryonic totipotent state, creating a cell that can then go on to develop into any part of an animal. Wilmut's accomplishment opened the door to the field of cloning and all of its possibilities (no author, 2010).

Tatsuo Sato has been instrumental in medical education by promoting basic gross anatomy education from the viewpoints of both scientific research and clinical application. Since the 1991 Joint Meeting of the American Association of Clinical Anatomists and British Associaiton of Clinical Anatomists in Norwich, England, Sato has presented videos of his dissection demonstrations of lymphatics, autonomic nerves, and visceral fasciae. These videos are widely used for the education of medical students and surgeons. In addition to numerous anatomy textbooks in Japanese, in 1992, he published *A Color Atlas of Surgical Anatomy for Esophageal Cancer* with T. Iizuka. Dr. Sato was born May 5, 1937 in Sendai, Japan. After receiving his M.D. from Tokyo Medical and Dental University (TMDU) in 1963, and following a year of residency, he passed his national board exam to become a licensed physician in 1964. Continuing his studies, he received his Ph.D. in anatomy from the TMDU Graduate School. Following his first academic position as an Assistant Professor of Anatomy at Fukushima Medical College from 1968-1970, he became an Associate Professor of Anatomy at the Faculty of Medicine of Tohoku University in Sendai (1970–1973), then was promoted to Professor and Chairman of the 2nd Department of Anatomy of TMDU in 1974, serving in this capacity until 2003. During that period, he facilitated education as the Dean of Students (1982-1984), and the Dean of the Faculty of Medicine (1995-2001). In 2009, he became the President of Tokyo Ariake University of Medicine and Heath Sciences. Dr. Sato was instrumental in the development of the body donation foundation in Japan, and facilitated the practice of the Ministry of Education to send a

letter of appreciation to the family of all body donors (1980), and key in implementing legislation for the new body was donation law. His research interests first focused on the morphological analysis of trunk musculature based upon nerve supply. From there, he entered clinical anatomy research focusing on surgical anatomy for the preservation of the lymphatic system during cancer surgery. Japanese guidelines for cancer treatment are based on his research studies. In addition to the field of anatomy, he also is an honored member of the Japan Esophageal Society, the Japanese Gastric Cancer Association, and the Japanese Breast Cancer Society. In 1997, Dr. Sato was named Honored Member of the American Association of Clinical Anatomists (website h).

In 1996, Dr. Robert Acland released his Video Atlas of Human Anatomy. Dr. Acland received his MBBS from London University in 1969, completed internships in the UK and Tanzania and residencies in plastic surgery in London and Glasgow. He began his career at the University of Louisville in 1976 as Director of the Microsurgery Teaching Laboratory and continued to hold that position until 1998. During that

Figure 366. Robert Acland. (Courtesy of http://acland anatomy.com/images/photo_acland.jpg.)

period he ascended through the professional ranks in Plastic and Reconstructive Surgery becoming Professor in 1986. He also holds Associate appointments in the Departments of Physiology and Biophysics and of Anatomical Sciences and Neurobiology. Dr. Acland is a founding member of the International Society for Reconstructive Microsurgery and the American Society for Reconstructive Microsurgery. His presentations have been highlights at the Annual Scientific Sessions of the American Association of Clinical Anatomists. Dr. Acland has been a pioneer in the field of fresh tissue dissection (website h).

Dr. Peter Abrahams (Fig. 367) trained in London as a physician. He intended to become a surgeon but was sidetracked into anatomy after writing *Clinical Anatomy of Practical Procedures* in 1973. This led to a British Fulbright Scholarship (1975-76) to the University of Iowa Medical School. He held a position as a Clinical Anatomist at University College, London (UCL) for 15 years and then held the Clinical Anatomist post at the University of Cambridge (UK) succeeding Prof. Harold Ellis before moving to the new Chair of Clinical Anatomy at Warwick Post-Graduate Medical School in 2006. He was awarded the BMA electronic publishing prize and the IAMS prize for "Interactive Skeleton" CD-ROM (1993).

In 2005, Abrahams, Craven and Lumley won the Richard Asher Prize of the Royal Society of Medicine, London for the best new medical textbook. Major published works include the McMinn and Abrahams *Clinical Atlas of Human Anatomy*, now in its 7th edition, which won the BMA book award in 2008 and Weir and Abrahams *Imaging Atlas of Human Anatomy* now in its 4th edition. In 2006, the American Association of Clinical Anatomists recognized his contributions as its Honored Member. He also won the National Teaching Fellowship from HEA in the UK-and in the same

Figure 367. Peter Abrahams, from http://www2.warwick .ac.uk.

year made a FHEA (Fellow of Higher Education Academy). Finally Abrahams was consultant to the Buckingham Palace Exhibition "Leonardo-Anatomist" and in 2013 was appointed to co-curate the "Mechanics of Man" exhibition, in Holyroode Palace for the Edinburgh Festival.

THE EARLY TWENTY-FIRST CENTURY

In 2003, the human genome had been successfully completed through The Human Genome Project (HGP), a 15-year project aimed at mapping and sequencing all human DNA. The program was started in 1988 by James D. Watson and coordinated by the U.S. Department of Energy and the National Institutes of Health (NIH). Watson was the head of the National Center for Human Genome Research at NIH. He was forced to resign in 1992, largely over his disagreement regarding the issue of patenting genes. Francis Collins was chosen to replace him in April 1993. With Collins now in control, the name of the Center was changed to the National Human Genome Research Institute (NHGRI). Collins was raised on a small farm in Virginia and obtained his B.S. at the University of Virginia, his Ph.D. degree from Yale University, and his M.D. from the University of North Carolina. Before becoming head of NHGRI, Dr. Collins was a member of the research team that identified the gene for cystic fibrosis in 1989, the gene for neurofibromatosis type I in 1990, and the gene for Huntington's disease in 1993 (Collins, 2003).

Susan Standring (Fig. 368) became the editor of the thirty-ninth edition of *Gray's Anatomy* in 2005 and also edited the fortieth and forty-first editions (no author, 2005). She was born on February 22, 1947 in Penzance (a small town at the extreme southwest tip of England) in the county of Cornwall. Professor Standring is Emeritus Professor of Anatomy at King's College London and Anatomy Development Tutor at the Royal College of Surgeons of England (RCSEng). She arrived at Guy's Hospital Medical School as a medical undergraduate in 1964, but after taking an intercalated B.Sc. in Anatomy, she stayed in the Department of Anatomy to complete a Ph.D. and did not return to medicine. She was Head of the Department of Anatomy and Human Sciences at King's College London at the time of her retirement in 2008.

Professor Standring has written prolifically (as Susan Hall) and her research interests have focused on the mechanisms of demyelination and of repair of traumatic injuries in the peripheral nervous system and the neuroinvasive routes for prion protein in acquired transmissible spongiform encephalopathy. In recognition of her work in the field of peripheral nerve regeneration and remyelination, she was appointed Charles Tome Lecturer at the Royal College of Surgeons in 1997, and elected President of the Peripheral Nerve Society, the largest international forum for studying the biology and pathology of the peripheral nervous system, in 2001.

Professor Standring has strong links with the Royal College of Surgeons of England, both as an examiner, a member of the Academic and Research Board and Museums and Archives Committee and as Anatomy Development Tutor, with a remit to ensure that doctors in the early years of their surgical training learn a core of applied surgical anatomy. She is an elected Trustee of the Hunterian Collection at RCSEng, and an Honorary Fellow of the College and she was recently awarded the prestigious Wood Jones Medal of RCSEng for her contribution to anatomy teaching. She delivered the Vicary lecture (a lecture sponsored jointly by RCSEng and the Worshipful Company of Barbers) at the College in October 2013. In 2008, Professor Standring was elected to a Fellowship of King's College London in recognition of her service to the College; she has served on its management board, its Council and on many of its working parties. She has been an External Examiner at almost all of the medical schools in the UK and has advised the General Medical Council on undergraduate medical education. She is a past President of the Anatomical Society.

In 2008, Emmanuel Skordalakes became the first to decode the structure of telomerase, the enzyme that conserves the ends of chromosomes (telomeres) and is a factor in limiting the number of times a normal cell can divide (the Hayflick limit). Skordalakes majored

Figure 368. Susan Standring. http://elsevierauthors. com/susanstandring/.

Figure 369. Robert Tomanek. https://www.icts.uiowa.edu/Loki/research/browseResearch.jsp?id=919272.

in chemistry at the Anglia Ruskin University in Cambridge, and earned a master's degree in chemical research at University College London of the University of London. In 2000, he received his Ph.D. degree from Imperial College, University of London, and obtained a postdoctoral fellowship in the Department of Molecular and Cell Biology at the University of California, Berkeley.

Skordalakes joined The Wistar Institute in 2006. This is where his lab deciphered a key part of beetle telomerase, an enzyme that can enhance or hinder a cell's ability to multiply limitlessly. The catalyst plays an active role in at least 85 percent of all cancers. Skordalakes published the structure of the active region of telomerase, an important milestone in the creation of therapeutics that could inhibit telomerase activity (American Association of Anatomists, 2013).

Robert Tomanek (Fig. 369), in addition to his lifelong achievements in the field of cardiac angiogenesis, published *Coronary Vasculature* in 2012. This book reviews, discusses, and integrates findings from various areas of coronary vasculature research and will be a valuable reference source for cardiovascular scientists and physicians for many years to come. Tomanek earned his B.S. from the University of Omaha and his Ph.D. from the University of Iowa. He stayed at the University of Iowa initially for a fellowship and then as a professor of Anatomy and Cell Biology where he continues his research (American Association of Anatomists, 2013).

Chapter 32

ANATOMICAL SOCIETIES AND ASSOCIATIONS

Anatomists in many countries have established professional organizations and regular meetings for exchange of information and to promote research ideas. It would be difficult to present all existing anatomical associations and in this section only few are presented. More information regarding any anatomical society or anatomy in a particular country can readily be obtained from the Home Page (website e) of the International Federation of Anatomical Associations (IFAA). This umbrella organization (IFFA) was founded in 1903 for,

> ...closer and personal scientific exchange among anatomists, histologists, embryologists, morphologists, anthropologists, veterinarians, dentists, biologists, and zoologists, and professionals of allied health sciences, and their interest in the uniformity of the technological language they all used in teaching

and research, led a group of leaders in the field of Anatomy to found the International Federation of Associations of Anatomists (IFAA).

The IFAA member societies include the American Association of Anatomists, American Association of Clinical Anatomists, Anatomische Gesellschaft, Anatomical Society of Great Britain and Ireland, Italian Association of Anatomy and Histology, Canadian Association for Anatomy, Neurobiology and Cell Biology, Sociedad Anatomica Espanola, Swiss Society for Anatomy, Histology and Embryology, Turkish Society of Anatomy and Clinical Anatomy, Argentina Association of Clinical Anatomy, Australian & New Zealand Association of Clinical Anatomists, Sociedade Brasilieira de Anatomia, Chinese Society for Anatomical Sciences, and the Czech Anatomical Society.

ANATOMISCHE GESELLSCHAFT

The oldest known association of professional anatomists, the *Anatomische Gesellschaft*, was originally a section of the *Gesellshaft Deutscher Natfurforscher und Ärzte* in September 28, 1822. It did not officially separate on its own until September 23, 1986. At that time, it had 40 members who elected Albert von Kölliker as the first president. The first official meeting was held in Leipzig in 1886, and by this point, the organization had grown to 174 members with 74 of them living abroad (website f).

Even though the *Anatomische Gesellschaft* is based in Germany, it has never been a solely German society. By 1902, it had 240 members of which a large portion were from outside of Germany. It had reached 395 official members before the start of World War II. During the war, the organization did not follow the Nazi regimes

plans or wishes. They did not subscribe to the Nazi oriented leadership concept nor did they introduce the exclusion of "non-Aryan members" in their bylaws. Throughout this period, they kept all the members on their listing, even the ones who had been dismissed due to Nazi racial laws (Kühnel, 1989; Schierhorn, 1986). However, the organization did face several controversies after 1933. Over the next few years, some politically active members, including Max Clara, Hans Petersen, Robert Wetzel, August Hirt, and later Eduard Pernkopf, tried to convert the traditionally international society into a purely German entity and asked for the exclusion of "non-Aryan" members. Everyone in the organization did not feel this way and some members even protested the idea. For example, Philipp Stöhr Jr. (1891–1979), Professor at the University of Bonn showed his protest by refusing to visit the society meetings during the later Nazi years (Fleischhauer, 1981). Other members foiled the measures demanded by the National Socialists. For example, Siegfried Mollier of Munich refused to even consider these colleagues' opinions in 1934. That same year, Heinrich von Eggeling, secretary of the society, wrote a personal letter to Hitler explaining why the formation of an exclusively German anatomical society was not desirable. Moreover, when Pernkopf raised the question again in 1943, it was Hermann Stieve who scolded him (Hildebrandt, 2009; Kühnel, 1989). After the war, the *Anatomische Gesellschaft* was founded anew on April 26, 1949.

Since its foundation, the publishing organ of the society has been the *Anatomischer Anzeiger* (Annals of Anatomy). A German-speaking anatomist, Karl von Bardeleben, founded the journal in 1886. Although he was German, he welcomed submissions in English, Italian, Latin, French, and Russian. It has published original papers covering links between anatomy and other areas such as molecular biology, cell biology, reproductive biology, immunobiology, developmental biology, neurobiology, embryology as well as neuroanatomy, neuroimmunology, clinical anatomy, comparative anatomy, modern imaging techniques, evolution, and especially on aging.

ANATOMICAL SOCIETY OF GREAT BRITAIN AND IRELAND

The Anatomical Society of Great Britain and Ireland (ASGBI) was founded in 1887. It was conceived early in 1887 by the surgeon and anatomist Charles Barrett Lockwood. He strategized with George M. Humphry and Alexander Macalister during visits to Cambridge. This resulted in a letter that was sent on April 27, 1887 to various potentially interested colleagues, inviting them to a meeting at the Medical Society of London. This meeting took place on May 6, 1887. At this meeting, Humphry was elected the first president. The two resolutions proposed at the meeting were "*That an Anatomical Society be founded, and that it be called the Anatomical Society of Great Britain and Ireland,*" and "*That the scope and object of the Society be the Anatomy, Embryology and Histology of Man and of Animals*

in so far as they throw light upon the structure of Man" (Garey, 2006). The first ASGBI Annual meeting took place in November 1887 at the University College London. (Garey, 2006).

In 1865, William Turner and George Humphry created the *Journal of Anatomy and Physiology*. In July 1887, the relationship between the journal and ASGBI began. It started with a letter from Turner to Humphry, the acting president of the society, suggesting that the society publish their proceedings at the end of the journal. It continued this way until the Annual Meeting in December 1916, when the society took over the journal. The first action of the society was to change the journal's name to the *Journal of Anatomy*.

AMERICAN ASSOCIATION OF ANATOMISTS

The American Association of Anatomists (AAA) was founded September 17, 1888 at Georgetown University. The first meeting was held in Washington D.C. with the objective of the "*advancement of the anatomical sciences.*" The 14 men who met at this meeting elected Dr. Joseph Leidy of the University of Pennsylvania as their first President, who served until 1891. Leidy was an outstanding American biologist, often called the founder of modern American vertebrate paleontology and parasitology (Warren, 1998).

Although AAA was begun in 1888, the first proceedings of an AAA Annual Meeting were not published until 1901 in the first issue of the *American Journal of Anatomy*, founded by Franklin Paine Mall and Charles S. Minot. In 1906, the *American Journal of Anatomy* established the first volume of the *Anatomical*

Record. It was created in order to "*aid in directing attention to problems affecting the development of the science of Anatomy, the organization for teaching it, and its relations with other branches*" (American Association of Anatomists, 2013).

In 1908, the organization officially changed its name from the Association of American Anatomists to the American Association of Anatomists. By 1917, AAA had reached a membership of over 300 members and in 1924, they adopted a resolution to investigate the status of anatomy. They felt the need to investigate because, "*The status of anatomy, its relations to premedical education and training, and to the preclinical and clinical subjects of the medical college has become a problem of serious importance to us.*" Later presidents for the Association were Ross G. Harrison (1911–13),

Gotthelf C. Huber (1913–15), Henry H. Donaldson (1915–17), Robert R. Bensley (1917–20), C. F. W. McLure (1920–21), Clarence M. Jackson (1921–24), Florence R. Sabin (1924–26), George L. Streeter (1926–28), Charles R. Stockard (1928–30), Herbert M. Evans (1930–32), George E. Coghill (1932–34), Warren Harmon Lewis (1934–36), Frederic T. Lewis (1936–38), S. Walter Ranson (1938–40), Philip E. Smith (1940–42), Edgar Allen (1942–43), J. Parsons Schaeffer (1943–46), George W. Corner (1946–48), George W. Bartelmez (1946–50), Sam L. Clark (1950–52), Leslie B. Arey (1952–54), Samuel R. Detwiller (1954–56), Edward A Boyden (1956–57), Barry J. Anson (1957–58), Davenport Hooker (1958–59), and Normand Louis Hoerr (1959–60), Edward W. Dempsey (1960–61), Harold Cummins (1961–62), Charles P. Leblond (1962–63) Horace W. Magoun (1963–64), Charles Mayo Goss (1964–65), Don Wayne Fawcett (1965–66), Donald Duncan (1966–67), Karl E. Mason (1967–68), Richard J. Blandau (1968–69), Roland H. Alden (1969–70), Raymond Carl Truex (1970–71), David Bodian (1971–72), William U. Gardner (1972–73), John W. Everett (1973–74), Russell T. Woodburne (1974–75), John C. Finerty (1975–76), Carmine D. Clemente (1976–77), Newton B. Everett (1977–78), Berta V. Scharrer (1978–79), Daniel C. Pease (1979–80), Sanford L. Palay (1980–81), Elizabeth D. Hay (1981–82), John E. Pauly (1982–83), Allen C. Enders (1983–84), A. Kent Christensen (1984–85), John V. Basmajian (1985–86), Douglas E. Kelly (1986–87), Henry J. Ralston, III (1987–88), Roger R. Markwald (1988–89), Jerome Sutin (1989–90), Karen Hitchcock (1990–91), William P. Jollie (1991–92), Alan Peters (1992–93), Donald A. Fischman (1993–94), Charles E. Slonecker (1994–95), Michael D. Gershon (1995–96), Gary C. Schoenwolf (1996–97), Bruce M. Carlson (1997–99), Robert D. Yates (1999–2001), John F. Fallon (2001–03), Robert S. McCuskey (2003–05), Kathy Svoboda (2005–07), David Burr (2007–09), Kathryn Jones (2009–11), Jeffrey Laitman (2011–13) (American Association of Anatomists, 2013).

In 1992, *Developmental Dynamics* was first published as a direct descendant of the *American Journal of Anatomy* (American Association of Anatomists, 2013).

AMERICAN ASSOCIATION OF CLINICAL ANATOMISTS

The American Association of Clinical Anatomists (AACA) was officially begun October 17, 1983. It was created with the objectives to advance the science and art of clinical anatomy and to encourage research and publication in the field and to maintain high standards in the teaching of anatomy. These objectives were forged out of frustration with the present state of Anatomy in the United States, with the rapid decline in hours devoted to gross anatomy and subsequently the overall decrease in anatomical knowledge of future physicians (Dawson, 1988).

The founder of the AACA, Dr. Ralph Ger, a professor of Surgery at Stony Brook University, got the idea after attending at meeting of the British Association of Clinical Anatomists (BACA) in 1982. The BACA was formed in England in order to strengthen the ties between clinicians and anatomists. Dr. Ger thought an organization of this nature would have a great impact in the United States and on November 16, 1982 sent a letter to several colleagues inviting them to a founders meeting of a similar society in the United States. The founders meeting took place February 18, 1983 in New York. Eighteen people attended the two-hour meeting. During this meeting, the foundations of AACA were laid down and decisions were made that included the election of the first president, Oliver H. Beahrs. Also, the bylaws, which were modeled after those of the BACA, were drafted (Dawson, 1988).

By the time of the first annual meeting, there were 88 members in AACA. The meeting was held at the Mayo Clinic on May 18–19, 1984. Seventy-seven persons registered. President Beahrs, Local Committee Chairman Dr. Donald Cahill, and Dr. Franklyn G. Knox, Dean of the Mayo Medical School, began the meeting with opening comments. Thirty-seven papers and one panel discussion were presented during the day-and-a-half meeting. Although most were clinically oriented, five papers dealt with anatomy instruction. It was announced that the official journal for the AACA would be *The American Surgeon* (Dawson, 1988).

The second annual meeting was held at the University of Nebraska Medical School on June 7–8, 1985. During the day-and-a-half meeting, 26 papers were presented. Three of these were directly related to anatomy instruction. In addition, a symposium on "*Gross Anatomy in Medical Education*" was held (Beahrs et al., 1986). Additionally, the new president was elected, Ralph Ger, M.D., who would go on to establish a membership committee, an education committee and an honored member committee (Dawson, 1988).

By 1986, the membership of the AACA had increased to 283 members. A year later, in 1987, a meeting was called to discuss the state of the official journal of the AACA. By the end of this meeting, the council had voted unanimously to sever ties with *The American Surgeon* and to create their own independent journal. After this meeting, Dr. Cahill had met with representatives of Alan R. Liss, Inc. to lay the groundwork for the new journal. Meanwhile, Dr. Ger, the selected editor, was in communication with BACA concerning joint sponsorship of the new journal. By mid-August, it was unofficially confirmed that AACA and BACA would join together to create an independent journal, *Clinical Anatomy*.

Chapter 33

MODERNIZATION OF ANATOMY CURRICULA IN MEDICAL SCHOOLS

Anatomy has been the backbone of medical education for hundreds of years, but over the past few decades, it has been subjected to an unexpected controversial debate among anatomists and medical schools. The decrease in hours and focus on anatomy in the basic sciences of medical education over the years has been contentious. With a decline in hours devoted to human anatomy, a debate has arisen over prosection versus dissection. Lastly, the addition of modern technologies such as CT, MRI, ultrasound, and anatomical computer applications in the teaching of anatomy has been a topic of considerable interest.

DECREASED FOCUS/HOURS

A respectable understanding of anatomy provides a platform of knowledge suitable for all medical careers. This concept coupled with the seven-fold increase in medical claims associated with anatomical errors between 1995–2000 has sparked further debate on the decrease in hours committed to anatomy (Ellis, 2002; Turney, 2007). Over the past few decades, the designated teaching time for anatomy has significantly decreased. For example, in 1955, the average time devoted to gross anatomy at 87 schools in the United States and Canada was 330 hours (Hoerr, 1956). Over 40 years later, 123 schools averaged only 182 hours (Collins et al., 1994; Yeager, 1996). It has been said that this decrease has come in order to lessen the factual burden on students and make time for teaching other disciplines such as genetics and molecular biology. Unfortunately, the evidence suggests that the curriculum has decreased too much, to an extent where safety and clinical practice might be compromised (Turney, 2007).

The decrease in time allotted to anatomy curricula is further burdened by the lack of an agreed upon core curriculum. Without guidelines set in place to emphasize what students need to learn in such shortened curricula, one cannot be certain that medical students are receiving the basic anatomical knowledge needed to practice medicine. However, getting anatomists and clinicians to agree upon a core curriculum seems to be a daunting task that only a few have attempted to tackle (Bergman et al., 2011). A recent publication, *The place of anatomy in medical education*, is an example of a crucial contribution to this discussion. This publication describes an ideal course that is "*principle based and problem oriented.*" The authors focus on the definition of and distinction between "general" and "specific" anatomies, and how these can be taught using different learning materials, teaching methods, and assessment programs (Bergman et al., 2011; Louw et al., 2009; Inuwa et al., 2012).

PROSECTION VERSUS DISSECTION

Anatomy has always been the foundation of medical education and until recently was taught solely by dissection and didactic lecture. However, with changes in teaching methodology, the role of dissection in learning, and teaching of practical anatomy has become a contentious issue. Consequently, institutions

356

in North America and Europe have dramatically reduced or eliminated the time students spend in the dissection room and replaced it with prosected or plastinated specimens (plastinates) (McLachlan et al., 2004; McLachlan & Patten, 2006; Reidenberg & Laitman, 2002; Sugand et al., 2010).

In favor of dissection, it is argued that exploration of palpable human anatomy in the dissection laboratory facilitates the learning of the three-dimensional organization of gross anatomical structures. It has been claimed that the understanding of these three-dimensional relationships are crucial in clinical application and practice of medicine (Dinsmore et al., 1999). Additional benefit of cadaveric dissection is the enhancement of knowledge retention. One study showed that students who dissected performed better on a comprehensive final examination, which may indicate that dissection might aid in retention of their knowledge (Yeager, 1996). Also, the act of dissection may facilitate dexterity and the skills needed to use certain instruments necessary for clinical tasks (Ellis, 2001; Moore, 1998; Topp, 2004). Lastly, many argue that the cadaver plays an intricate part in instilling empathy in medical students. The cadaver is often the medical student's first patient and first encounter with pathology and death (Topp, 2004). The argument against dissection is that it is costly, hazardous, and that there are better methods for teaching anatomy such as using prosected specimens and plastinates.

There are multiple arguments in the literature supporting prosection over dissection. For example, it has been suggested that in order to understand anatomical spatial relationships, one should start with visually simplified symmetrical patterns and fundamental lines and build up to the more complex organization (Miller, 2000). This can be accomplished in the anatomy laboratory through careful study of prosected specimens. Moreover, prosections give the students the ability to see multiple variations of anatomy rather than the single variation they are dissecting (Topp, 2004). Recent studies have shown that the use of prosection is an excellent means of maximizing the utilization of scarce materials while increasing educational efficiency (Dinsmore et al., 1999; Nnodim et al., 1996; Yeager, 1996). Collins (2008) goes as far to argue that prosections are sufficient to aid medical students learning anatomy so that dissection could be reserved for only postgraduate surgical trainees.

Plastinates allow for a three-dimensional view of the complexities of the human body that cannot be fully appreciated from a textbook. These specimens allow students to not only see complete organ systems but also their relative positions and relationships to one another. Cadaveric dissection should not be dismissed as obsolete since exposure to dissection develops manual dexterity and important cognitive skills required by all physicians (Granger, 2004; McLachlan, 2004; Moore, 1998; Slotnick & Hilton, 2006; Sugand et al., 2010). Lempp (2005) sums up dissection as an opportunity to reinforce familiarization and respect for the body and integration of theory into clinical practice.

ADDITION OF MODERN TECHNOLOGICAL METHODS

Over the past few decades, anatomical education has been revolutionized. It is now more reliant on the addition of adjuncts to dissection such as radiological imaging and web-based learning. Anatomy courses that incorporate these modalities are arguably the ideal training environment for medical students (Howe et al., 2004; McGhee, 2010). The benefits of using radiology in anatomy instruction range from improved anatomy comprehension to appreciation of medical imaging and radiologists' vital contributions to better healthcare delivery. It has also been shown that using cadavers and imaging together improves the students' ability to identify anatomical structures and provides long-term knowledge retention (Miles, 2005). These radiological images offer in-vivo visualization of anatomy as well as insight into pathological processes that cannot be visualized during dissection (Sugand et al., 2010). Also, Branstetter et al., (2008) found that students had greater interest in the specialty of radiology and fewer negative stereotypes about radiologists than their predecessors after initial exposure to radiology in their anatomy course. Lastly, radiological imaging has become a common diagnostic tool used on a daily basis by the clinician. Therefore, it has become a necessity that medical students develop the basic ability to read and understand radiological images.

A subset of radiology that has an interesting advantage over all the rest is ultrasound. It is the one modality that provides visualization of structures and their movement in a noninvasive manner, coined "*living anatomy*." This coupled with the compact portability of newer ultrasound machines make it an excellent method for enhancing and facilitating the learning of anatomy in a clinically relevant manner. Previous studies have shown that ultrasound has been a helpful educational resource with high student satisfaction for teaching cardiovascular and renal anatomy and for demonstrating organs, forearm muscles and vessels (Miles, 2005; Swamy & Searle, 2012). Additionally, the use of ultrasound in the clinical setting has drastically increased. Its use is already well documented in settings like emergency medicine, cardiology, and anesthesia with many more specialties following suit (Patten et al., 2010). Therefore, there is an increased chance of medical students encountering ultrasound images in clinical practice. It seems only logical then to expose medical students to ultrasound technology, images, and training early in their studies (Swamy & Searle, 2012).

In today's computer age, it is not surprising that students who accessed a web-based computer aided instructional resource for anatomy scored significantly higher on examinations than those who never accessed the online content (McNulty et al., 2009). It is findings like these that have pushed the future of anatomy teaching to rely more on visual aids and technology outside the dissection laboratory. The rationale is that students will naturally forget topics covered in dissection class and resources like web-streamed lectures and instructional videos may prove vital in helping retain such knowledge (Sugand et al., 2010). Tools like these allow reliable self-study that help overcome the decline in hours dedicated to anatomical studies. Additionally, versatile practice tests and self-assessment programs can be integrated into these software to further student learning (Sugand et al., 2010).

BIBLIOGRAPHY

AAA. (1950). American Association of Anatomists. Abstracts of Sixty-Third Annual Session, Louisiana State University School of Medicine, New Orleans, April 5-7 1950. *Anat Rec, 106*:167-318.

AAA. (2000). American Association of Anatomists. In: Proceedings of the American Association of Anatomists One Hundred Thirteenth Meeting, San Diego, CA, April 15-18, 2000. Bethesda, MD: *American Association of Anatomists*, 56 p.

Aciduman, A., Arda, B., Kahya, E., & Belen, D. (2010). The royal book by Haly Abbas from the 10th century: One of the earliest illustrations of the surgical approach to skull fractures. *Neurosurgery, 67*: 1466-74.

Ackerknecht, E. H. (1982). *A short history of medicine.* Baltimore: Johns Hopkins University Press.

Acland, R.D. (2003). Acland's atlas of human anatomy (DVD-ROM). Version 1.0. Baltimore: Lippincott William & Wilkins.

Adams, J. (1817). *Memoirs of the life and doctrines of the late John Hunter, Esq.* London: J Callow.

Adams, F. (1939). *The genuine works of Hippocrates. Translated from the Greek.* Baltimore: Williams & Wilkins Co.

Adelmann, H. (1942). (Trans.): *The embryological treatise of Hieronymous Fabricius of Aquapendente. A facsimile edition, with an introduction, a translation and a commentary.* Ithaca, NY: Cornell University Press.

Agrifoglio, L. (1961). Anatomia e fisiologia del corpo umano in Cicerone. *Pag Stor Med, 5*:32.

Ahmed, R. U. (1982). Status of anatomy and surgery in different civilizations and the contribution of Arabs in this field. In A. R. El-Gindy & H. M. Z. Hassan (Eds.), *Proc 2nd Int. Conf Islamic Med,* Kuwait, pp. 229-234.

Akihito. (1992). Early cultivators of science in Japan. *Science, 258*:578.

Albar, M. A. (1983). Embryology as revealed in the *Quran* and *Hadith* (Parts I & II). *IWMJ, 1*:46.

Allen, E. (1974). *Hunterian museum.* Royal College of Surgeons of England, London.

Allen, E., Turk, J., & Murley, R. (1993). *The case books of John Hunter F.R.S.* London, Royal Society of Medicine Services Ltd.

American Association of Anatomists. (2013). *The many faces of anatomy.* Boston, pp. 1-17.

Amezquita, J. A., Bustamante, M. E., Picazos, A. L., & del Castillo, F. F. (1960). *Historia de la Salubridad y de la Asistencia en Mexico.* Vols I-IV. Mexico, Secretaria de Salubridad y Asistencia, Los Talleres Graficos de la Nacion.

Andrews, B. F. (1981). William Harvey in perinatal perspective. In G. F. Smith & D. Vidyasagar (Eds.), *Historical review and recent advances in neonatal and perinatal medicine.* Volume II, Evanston, IL: Mead Johnson Nutritional Division, pp. 165-178.

Andrews, C. A. R. & Hamilton-Patison, J. (1978). *Mummies.* London: British Museum.

Angetter, D. (2000). Anatomical science at University of Vienna 1938-45. *Lancet, 355*:1454-1457.

Anon. (1962). The man who gave his name. The aqueduct of Sylvius. *The Crest, 5*:7.

Arcieri, J. P. (1970). *Why Alcmaeon of Croton is the father of experimental or scientific medicine.* New York: Alcmaeon Editions.

Arey, L. B. (1962). Barry J. Anson. A biographical sketch. *Q Bull Northwest Univ Med Sch, 36*:185-188.

Aristotle. (1749). *Compleat master piece. Displaying the secrets of nature in the generation of man.* London.

Aristotle. (1831). *The works of Aristotle, the famous philosopher, in four parts.* New England.

Aristotle. (1982). *On the parts of animals: Translated, with introduction and notes by W. Ogle.* London: Kegan Paul, Trench & Co.

Aristotle. (2004). *On the parts of animals.* Kessinger, p. 70.

Asher, L. (1902). *Albrecht von Hallers bedeutung in der biologie der gegenwart,* Bern.

Ashoor, A. A. (1983). The history of Islamic medicine. *IWMJ, 1*:51.

Ashoor, A. A. (1984a). Islamic medicine: Its influence on the Latin West. *IWMJ, 1*:44.

Ashoor, A. A. (1984b). Muslim medical scholars and their works. *IWMJ, 1*:49-50.

Ashrafian, H. (2013). Leonardo da Vinci and the first portrayal of pectus excavatum. *Thorax, 68*:1081.

Aubrey, J. (1898). *Brief lives, chiefly of contemporary set down by John Aubrey between the years 1669 and 1696.* Vol 2. Clark, A. (Ed.). Oxford, pp. 302-4.

Aümuller, G. (1972). Zur Geschichte des Mainzer Anatomischen Instituts. In Jahrbuch d Vereinigung "Freunde der Universitat Mainz," pp. 49–74.

Aümuller, G. & Grundmann, K. (2002). Anatomy during the Third Reich – The Institute of Anatomy at the University of Marburg, as an example. *Ann Anat, 184*:295–303.

Baer, Karl Ernst von. 1828–37. Über Entwickelungsgeschichte der Thiere. Beobachtung und Reflexion. Königsberg, Bei den Gebrüdern Bornträger.

Bailey, J. B. (1896). *The diary of a resurrectionist, 1811–1812.* London: Swan, Sonnenschein & Co., Lim.

Bainton, R. H. (1953). *Hunted heretic. The life and death of Michael Servetus. 1511–1553.* Boston: Beacon Press.

Ball, J. M. (1910). *Andreas Vesalius. The reformer of anatomy.* St. Louis: Medical Science Press.

Ball, J. M. (1928). *The sack-'em-up men. An account of the rise and fall of the modern resurrectionists.* Edinburgh & London: Oliver and Boyd.

Ballestriero, R. (2010). Anatomical models and wax Venuses: Art masterpieces or scientific craft work? *J Anat, 216*:223–34.

Bani, G. (1986). Felice Fontana – His life and works. *Z Mikrosk Anat Forsch, 100*:337–46.

Banister, J. (1578). *The historie of man, sucked from the sappe.* London: John Daye.

Baskett T. F. (1996). *On the shoulders of giants. Eponyms and names in obstetrics and gynaecology.* London: RCOG Press.

Bayon, H. P. (1938a). William Harvey, physician and biologist: His precursors, opponents and successors. Parts I & II. *Ann Sci, 3*:59–118.

Bayon, H. P. (1938b). William Harvey, physician and biologist: His precursors, opponents and successors. Part III. *Ann Sci, 3*:435–456.

Bayon, H. P. (1939a). William Harvey, physician and biologist: His precursors, opponents and successors. Part IV. *Ann Sci, 4*:65–106.

Bayon, H. P. (1939b). William Harvey, physician and biologist: His precursors, opponents and successors. Part V. *Ann Sci, 4*:329–389.

Barlow, D. & Durand, V. M. (2011). *Abnormal psychology: An integrative approach* (6th ed.). Wadsworth, pp. 10–11.

Beahrs, O. H., Chase, R. A., & Ger, R. (1986). Gross anatomy in medical education. *Am Surg, 52*:227–32.

Beasley, A. W. (1982). Ortopaedic aspects of mediaeval medicien. *J R Soc Med, 75*:970–975.

Beckmann, E. C. (2006). CT scanning the early days. *Br J Radiol, 79*:5–8.

Beekman, F. (1935). Bidloo and Cowper, anatomists. *Ann Med Hist, 7*:113–129.

Bellary, S. S., Walters, A., Gielecki, J., Shoja, M. M., Tubbs, R. S., & Loukas, M. (2012). Jacob B. Winslow (1669–1760). *Clin Anat, 25*:545–7.

Belt. E. (1955). *Leonardo the anatomist.* Lawrence, Kansas: University of Kansas Press.

Beltran, A. (1982). *Rock art of the Spanish Levant. Translated by Margaret Brown: The imprint of man.* Cambridge: Cambridge University Press.

Bensley, E. H. (1958). Sculduggery in the dead house. *CACHB (Calgary Associate Clinic Historical Bulletin), 22*:245.

Bentivoglio, M. (2013). Life and discoveries of Camillo Golgi. Nobelprize.org. Nobel Media AB 2013. Web. 28 Sep 2013. http://www.nobelprize.org/nobel_prizes/medicine/laureates/1906/golgi-article.html.

Bergman, E. M., van der Vleuten, C. P., & Scherpbier, A. J. (2011). Why don't they know enough about anatomy? A narrative review. *Med Teach, 33*:403–409.

Bergmann, M. & Wendler, D. (1986). Caspar Bauhin (1560–1624). *Gegenbaurs Morphol Jahrb, 132*:173–81.

Bergmann, M. & Wendler, D. (1987). Zum 300. Todestag von Niels Steno (1638–1686). *Mitt BI Ges exper Med (DDR), 24*:1.

Bergstrasser, G. (1925). Hunain ibn Ishaq uber die syrischen und arabischen Galen-Ubersetzungen. *Abh Kunde Morgenlandes, 17*:21.

Bettany, G. T. (1885–1900). John Bell (1763–1820). *Dictionary of national Biography,* Volume 4.

Beukers, H. (1992). Leiden's medical faculty during its first two centuries. In B. P. Kennedy & D. Coakley (Eds.), *The anatomy lesson: Art and medicine.* Dublin: The National Gallery of Ireland, pp. 121–131.

Biddis, M. D. (1976). The politics of anatomy: Dr. Robert Knox and victorian racism. *Proc Roy Soc Med, 69*:245–50.

Bierbaum, M. & Faller, A. (1979). *Niels Steno. Anatom, Geologe and Bischof, 1638–1686.* Munster: Aschendorff.

Bittar, E. E. (1955). A study of Ibn Nafis. *Bull Hist Med, 29*:353–68, 429.

Blake, C. C. (1870–1871). The life of Dr. Knox. *J Anthropol, 1*:332.

Bonner, T. N. (1963). *American doctors and German universities.* Lincoln, NE & London: University of Nebraska Press.

Bonuzzi, L. & Ruggeri, F. (1988). Appunti preliminari ad un'indagine sulle cere anatomiche. In: *Le Cere Anatomiche Bolognesi Del Settecento.* Bologna: Universita Degli Studi Di Bologna.

Booth, N. B. (1960). Empedocles' account of breathing. *J Hellen Stud, 80*:10–15.

Bordley, III. J. & Harvey, A. M. (1976). *Two centuries of American medicine 1776–1976.* Philadelphia: W. B. Saunders.

Borowitz, H. O. (1986). The scapel and the brush: Anatomy and art from Leonarado da Vinci to Thomas Eakins. *Cleveland Clinic Q, 53*:61–73.

Bower, A. (1817–1830). *The history of the University of Edinburgh* (3 Vols.). Edinburgh: Oliphant, Waugh and Innes.

Bowers, J. Z. (1972). *Western medicine in a Chinese palace.* Philadelphia: Wm. F. Fell.

Boyle, R. (1665–1666). Tryals proposed by Mr. Boyle to Dr. Lower to be made by him, for the improvement of transfusing blood out of live animals into another. *Phil Trans, 22*:385–388.

Bracegirdle, B. (1985). Famous microscopists: Robert Hooke, 1635–1702. *Proc Roy Micr Soc, 20*:305.

Bracegirdle, B. (1986). Famous microscopists: Antoni van Leeuwenhoek, 1632–1723. *Proc Roy Micr Soc, 21*:367.

Brahce, C. I. (1967). Faculty focus: On Russell T. Woodburne. *Univ Mich Med Cent J, 33*:293-5.

Brain, P. (1986). *Galen on bloodletting.* Cambridge: Cambridge University Press.

Brain, P. (2009). *Galen on bloodletting: A study of the origins, development and validity of his opinions, with a translation of the three works.* Cambridge: Cambridge University Press, p. 112.

Brand, R. A. (2009). John Hilton, 1805–1878. *Clin Orthop Relat Res, 467*:2208–9.

Branstetter, B. F., Humphrey, A. L., & Schumann, J. B. (2008). The long-term impact of preclinical education on medical students' opinions about radiology. *Acad Radiol, 15*:1331–1339.

Brass, A. (1984). The art of medicine. Frank H. Netter, MD. *Med J Aust, 141*:880.

Breasted, J. H. (1930). *The Edwin Smith surgical papyrus.* Chicago: University of Chicago Press.

Broadie (2011). *Nature and divinity in Plato's Timaeus.* Cambridge: Cambridge University Press, p. 1.

Brock, C. H. (1983). William Hunter. A reassessment. In C. H. Brock (Ed.), *William Hunter 1718-1783. A memoir by Samuel Foart Simmons and John Hunter.* Glasgow: University of Glasgow Press, pp. 45–49.

Brock, C. H. (1985). The happiness of riches. In W. F. Bynum & R. Porter (Eds.), *William Hunter and the eighteenth-century medical world.* Cambridge: Cambridge University Press, pp. 35–54.

Brock, H. (1994). The many facets of Dr. William Hunter (1718–83). *Hist Sci, 32*:385–408.

Brookes, J. (1828). *A catalogue of the anatomical and zoological museum of Joshua Brookes.* London: R. Taylor.

Browne, E. G. (1921). *Arabian medicine.* Cambridge: Cambridge University Press.

Brunschwig, Hieronymus. (1497). *Buch der Cirurgia.* Strasbourg.

Buckman, Jr., R. F. (1987). The surgical principles of John Hunter. *Surg Gynec & Obstet, 164*:479–484.

Buckman, R. F. & Futrell, J. V. V. (1986). William Cowper. *Surgery, 99*:582–590.

Bujalkova, M., Straka, S., & Jureckova, A. (2001). Hippocrates' humoral pathology in Nowaday's reflections. *Bratisl Lek Listy, 102*:489–492.

Bullock, T. H., Bennett, M. V. L., Johnston, D., Josephson, R., Marder, E., & Fields, R. D. (2005). The neuron doctrine, redux. *Science, 310*:791–793.

Bullough, V. L. (1958). Medieval Bologna and the development of medical education. *Bull Hist Med, 32*:201–15.

Burkhardt, F. (Ed.) (1996). *Charles Darwin's letters. A selection 1825–1859.* Cambridge: Cambridge University Press.

Burn, A. R. (1982). *The pelican history of Greece.* Middlesex: Penguin Books.

Burrell, B. (2003). The strange fate of Whitman's brain. *Walt Whitman Quarterly Review, 20*:107–133.

Burton, W. (1743). *An account of the life and writings of Hermann Boerhaave.* London.

Butterfield, H. (1965). *The origins of modern science, 1300–1800.* London: G. Bell and Sons Ltd., pp. 37–54.

Bylebyl, J. J. (1979). *William Harvey and his age.* Baltimore: Johns Hopkins University Press.

Bynum, W. F. & Porter, R. (1985). *William Hunter and the eighteenth century medical world.* Cambridge: Cambridge University Press.

Cabieses, F. (1979). Diseases and the concept of disease in ancient Peru. In J. Z. Bowers & E. F. Purcell (Eds.), *Aspects of the history of medicine in Latin America.* New York: Independent Publishers Group, pp. 16–53.

Calkins, C. M., Franciosi, J. P., & Kolesari, G. L. (1999). Human anatomical science and illustration: The origin of two inseparable disciplines. *Clin Anat, 12*:120–129.

Cameron, C. (1916). *History of the college of surgeons in Ireland.* Dublin: Fannin.

Cameron, E. (1991). *The european reformation.* Oxford: Oxford University Press.

Campbell, D. (1926). *Arabian medicine and its influence on the middle ages.* London: Kegan Paul, Trench, Trubner and Co., Ltd.

Canfora, L. (1989). *The vanished library.* London: Hutchinson Radius.

Canniff, W. (1894). *The medical profession in upper Canada, 1783–1850.* Toronto: William Briggs.

Capener, N. (1959). John Sheldon, F.R.S., and the Exeter Medical School. *Proc Royal Soc Med, 52*:231–238.

Capo, J. A. & Spinner, R. J. (2007). Compendium of anatomical variants. The levator claviculae muscle. *Clin Anat, 20*:968–969.

Cappelletti, A. J. (1975). Las doctrinas anatomo-fisiologicas de Diogenes de Apollonia. *Riv Storia Med, 75*:11–25.

Cardini, F. & Beonio-Brocchieri, M. T. F. (1991). *Universitaten im Mittelalter.* Munchen: Sudwest Verlag.

Carlino, A. (1999). Paper bodies: A catalogue of anatomical fugitive sheets, 1538–1687. (Trans. Noga Arikha). *Medical History*, Suppl. No 19, London, Wellcome Institute for the History of Medicine.

Carlson, E. C. (1999). Frank N. Low: Gentle giant of electron microscopy (1911–1998). *Anat Rec, 257*:48–49.

Carter, K. C. (1985). Koch's postulates in relation to the work of Jacob Henle and Edwin Klebs. *Med Hist, 29*:353–74.

Cassirer, E. A. (1943). The place of Vesalius in the culture of the Renaissance. *Yale J Biol Med*, 76: 109–20.

Castiglioni, A. (1940). Aulus Cornelius Celsus as a historian of medicine. *Bull Hist Med*, 8:857–73.

Castiglioni, A. (1941). *A history of medicine*. Translated by E. B. Krumbhaar. New York: Alfred A. Knopf.

Celsus. (1935). *De medicina*. Translated by W. G. Spencer. 2 Volumes. London and New York: Loeb Classical Library.

Chadwick, J. & Mann, W. N. (1950). *The medical works of Hippocrates*. Oxford: Blackwell Scientific Publications.

Chauvois, L. (1957). *William Harvey, his life and times: His discoveries: His methods*. London: Hutchinson Medical Publications.

Chiera, E. (1938). *They wrote on clay*. Chicago: University of Chicago Press.

Chisholm, H. (Ed.). (1911). *John Bell (surgeon)*. In Encyclopaedia Britannica (11th ed.), Cambridge: Cambridge University Press.

Chitwood, A. (2004). *Death by philosophy*. Ann Arbor, MI: University of Michigan Press, p. 47.

Choulant, L. (1962). Geschichte und Bibliographie der Anatomischen Abbildung nach ihrer Beziehung auf Anatomische Wissenschaft und Bildene Kunst. Leipzig, Rudolph Weigel.

Choulant, L. (1920). *History and bibliography of anatomic illustration*. Translated and edited with notes and a biography by Mortimer Frank. Chicago: University of Chicago Press.

Cilliers, L. & Retief, F. P. (2006). Medical practice in Graeco-Roman antiquity. *Curationis*, 29:34–40.

Clark, G. (1964). *A history of the Royal College of Physicians of London*. Volume 1. Oxford: Clarendon Press.

Clark, G. N. (1964–1972). *A history of the Royal College of Physicians of London*. Volume 3. Oxford: Clarendon Press.

Clark, K. (1935a). *A catalogue of the drawings of Leonardo da Vinci*. Cambridge: Cambridge University Press.

Clark, K. (1935b). *Catalogue of the drawings by Leonardo da Vinci in the collection of His Majesty the King at Windsor Castle*. London: Macmillan.

Clark, K. (1968). *The drawings of Leonardo da Vinci in the collection of Her Majesty the Queen at Windsor Castle*. Vols. 1–3. London: Phaidon Press Ltd.

Clark, M. E., Nimis, S. A., & Rochefort, G. R. (1978). Andreas Cesalpino, Quaestionum penpateticarum, libri V, liber v, quaestio iv, with translation. *J Hist Med Allied Sci*, 33:185–213.

Clarke, E. (1963). Aristotelian concepts of the form and function of the brain. *Bull Hist Med*, 37:1–14.

Clarke, E. & Dewhurst, K. (1972). *An illustrated history of brain function*. Berkeley, CA: University of California Press, p. 154.

Clarke, E. & Stannard, J. (1963). Aristotle on the anatomy of the brain. *J Hist Med*, 75:130–48.

Clemente, C. D. (1985). William Fredrick Windle 1898–1985. *Exp Neurol*, 90:1–20.

Clendinnen, I. (1993). *Aztecs. An interpretation*. Cambridge: Cambridge University Press.

Coakley, D. (1988). *The Irish school of medicine*. Dublin: Town House.

Coakley, D. (1992). Anatomy and art: Irish dimensions. In B. P. Kennedy & D. Coakley (Eds.), *The anatomy lesson: Art and medicine*. Dublin: The National Gallery of Ireland, pp. 57–78.

Cockburn, A. & Cockburn, E. (1980). *Mummies, disease and ancient cultures*. Cambridge: Cambridge University Press.

Codellas, P. S. (1932). Alcmaeon of Croton: His life, work, and fragments. *Proc Roy Soc Med*, 25:1041–46.

Cohen, D. (1975). *The body snatchers*. Philadelphia: J. B. Lippincott.

Cohen, H. F. (1994). *The scientific revolution*. Chicago: University of Chicago Press.

Cohen, I. B. (1992). What Columbus "saw" in 1492. *Sci Am*, 267:100–106.

Cohen, M. & Trott, N. G. (1973). Wilhelm Conrad Röntgen, 1845–1923. *Br J Radiol*, 46:81–2.

Cohen, J. & Werner, R. M. (2009). On medical research and human dignity. *Clin Anat*, 22:161–162.

Cole, F. J. (1975). *A history of comparative anatomy. From Aristotle to the eighteenth century*. New York: Dover.

Collins, F. S. (2003). An interview with Francis S. Collins, M.D., Ph.D. Director, National Human Genome Research Institute. *Assay Drug Dev Technol*, 1:119–25.

Collins, J. P. (2008). Modern approaches to teaching and learning anatomy. *BMJ*, 337:a1310.

Collins, T. J., Given, R. L., Hulsebosch, C. E., & Miller, B. T. (1994). Status of gross anatomy in the U.S. and Canada: Dilemma for the 21st century. *Clin Anat*, 7:275–296.

Colombo, R. (1559). *De re anatomica*. Venice: N. Bevilacquae.

Comrie, J. D. (1932). *History of Scottish medicine* (2nd ed.) (2 Vols.). London: Baillere, Tindall & Cox.

Comrie, J. D. (1972). *History of Scottish medicine to 1800*. London: Baillere, Tindall & Cox.

Conklin, E. G. (1939). Henry Hubert Donaldson. Biographical memoirs of the National Academy of Sciences, 29:229–243.

Conrad, L., Neve, M., Porter, R., Nutton, V., & Wear, A. (1995). *Western medical tradition. 800 B.C. to 1800*. Cambridge: Cambridge University Press.

Cooke, R. A. (1984). A visit to pathology museums in Italy, France and the U.K. *Bull Roy College Path*, 48:4–5.

Cooper, S. (1930). The medical school of Montpellier in the fourteenth century. *Ann Med Hist*, 2:164–95.

Cooper, B. B. (1843). *The life of Sir Astley Cooper*. 2 Vols. London: J. W. Parker.

Cooper, S. K. (2006). *Aristotle: Philosopher, teacher, and scientist.* Compass Point Books.

Cope, Z. (1959). *The Royal College of Surgeons of England: A history.* London: Anthony Blond.

Cope, Z. (1966). The private medical schools in London (1746–1914). In F. N. L. Poynter (Ed.), *The evolution of medical rducation in Britain.* London: Pitman Medical.

Coppola, E. D. (1957). The discovery of the pulmonary circulation: A new approach. *Bull Hist Med, 31:* 44–77.

Corner, B. C. (1951). *William Shippen, Jr. Pioneer in American medical education.* Philadelphia: American Philosophical Society.

Corner, G. W. (1930). *Clio medica. Anatomy.* New York: Paul B. Hoeber.

Corner, G. W. (1954). *George Linius Streeter 1873–1948. Biographical memoirs.* Vol XXVII. Washington D.C.: National Academy of Sciences.

Corney, B. G. (1914). Some physiological phantasies of third century repute. *Proc Roy Soc Med, 7:*217–27.

Cornford, F. M. (1971). *Plato's cosmology. The Timaeus of Plato translated with a running commentary.* London: Routledge and Kegan Paul.

Cowdry, E. V. (1920). Anatomy in China. *Anat Rec, 20:*31–60.

Cresswell R. (1862). *Aristotle's history of animals: In ten books by Aristotle,* Johann Gottlob Schneider. London: Henry G Bohn, Chapter 8, p. 18.

Creswell, C. H. (1926). *The Royal College of Surgeons of Edinburgh. Historical notes from 1505 to 1905.* Edinburgh: Oliver & Boyd.

Crivellato, E., Mallardi, F., & Ribatti, D. (2006). Diogenes of Apollonia: A pioneer in vascular anatomy. *Anat Rec B New Anat, 289:*116–120.

Crivellato, E., & Ribatti, D. (2006a). A portrait of Aristotle as an anatomist. *Clin Anat, 20:*447–485.

Crivellato, E., & Ribatti, D. (2006b). Aristotle: The first student of angiogenesis. *Leukemia, 20:*1209–1210.

Crivellato, E., & Ribatti, D. (2006c). Mondino de' Liuzzi and his anothomia: A milestone in the development of modern anatomy. *Clin Anat, 19:*581–587.

Crummer, L. (1923). Early anatomical fugitive sheets. *Ann Med Hist, 5:*189–209.

Crummer, L. (1925). Further information on early anatomical fugitive sheets. *Ann Med Hist, 7:*1–5.

Cruz, M. de la. (1940). The Badianus manuscript: An Aztec herbal of 1552. Translated from Nahuatl into Latin by Juan Badiano. Published in facsimile. Introduced, translated into English and annotated by Emily Walcott Emmart. Baltimore: Johns Hopkins University Press.

Cullen, G. M. (1918a). The passing of Vesalius. *Edin Med J, 75:*324.

Cullen, G. M. (1918b). Madrid to Zante. *Edin Med J, 73:*388.

Cunningham, A. (1993). English manuscripts of Francis Glisson. Vol. 1. From anatomia hepatis (The anatomy of the liver). Cambridge Wellcome Texts and Documents, No. 3. Cambridge: Wellcome Unit for the History of Medicine.

Currie, A. S. (1933). Robert Knox, anatomist, scientist, and martyr. *Proc Roy Soc Med, 26:*39–46.

Cushing, H. (1943). *A Bio-bibliography of Andreas Vesalius.* New York: Schuman's.

Damadian, R. (1936). http://crev.info/?scientists= raymond-damadian.

Dawson, D. L. (1988). The American Association of Clinical Anatomists: The beginnings and first five years. *Clin Anat, 1:*237–253.

Dawson, W. R. (1930). *ClioMedica: The beginnings: Egypt and Assyria.* New York: P.B. Hoeber.

Debernardi, A., Sala, E., D'Aliberti, G., Talamonti, G., Franchini, A. F., & Collice, M., (2010). Alcmaeon of Croton. *Neurosurgery, 66:*247–52.

Debus, A. G. (1978). *Man and nature in the Renaissance.* Cambridge: Cambridge University Press.

De Castro Santos Filho, L. (1979). Medicine in colonial Brazil: An overview. In J. Z. Bowers & E. F. Purcell (Eds.), *Aspects of the history of medicine in Latin America.* New York: Independent Publishers Group, 97–111.

De Lint, J. G. (1932). Beitrage zur Kenntnis der Anatomischen Namen im Alten Agypten. *Sudhoff Arch 25:*382–390.

Desmond, A. (1989). *The politics of evolution. Morphology, medicine, and reform in radical London.* Chicago: University of Chicago Press.

Dhorme, E. (1949). *Les Religions de Babylonie et d'Assyrie.* Paris.

Diller, H. (1938). Die Lehre vom Blutkreislauf, eine verschlossene Entdeckung der Hippokratiker. *Sudhoff Arch, 31:*201.

Dinsmore, C. E., Daugherty, S., & Zeitz, H. J. (1999). Teaching and learning gross anatomy: Dissection, prosection, or both of the above? *Clin Anat, 12:*110–114.

Dixon, A. F. (1904). Professor Wilhelm His. *J Anat Physiol, 38:*503–505

Dobell, C. (1932). *Antony van Leeuwenhoek and his "little animals."* London: John Bale, Reprinted: Dover Books, New York, 1960.

Dobson, J. (1952). Eighteenth century anatomists: Joshua Brookes. *Practitioner, 169:*180–4,

Dobson, J. (1954). *William Clift.* London: William Heinemann Medical Books, Ltd.

Dobson, J. (1962). *Anatomical eponyms.* London: E. & S. Livingstone, Ltd.

Dobson, J. (1968). *A guide to the Hunterian Museum.* Edinburgh: E. & S. Livingstone, Ltd.

Dobson, J. (1969). *John Hunter, F.R.S.* Edinburgh: E. & S. Livingstone.

Dobson, J. F. (1925). Herophilus of Alexandria. *Proc Roy Soc Med, 18*:19–32.

Dobson, J. R. (1927). Erasistratus. *Proc Roy Soc Med, 20*:825–32.

Dobson, J. & Walker, R. M. (1980). *Barbers and barber-surgeons of London: A history of the barbers' and barber-surgeons' companies.* Oxford: Oxford University Press, 1980.

Dolby, T. & Alker, G. (1997). *Origins and development of medical imaging.* Carbondale and Edwrdsville, IL: Southern Illinois University Press, p. 6.

Donley J. E. (1946). Harvey, Riolan, and the discovery of the circulation of the blood. *Yale J Biol Med, 18*:319–31.

Doolin, W. (1951). Dublin's surgeon-anatomists. *Ann R Coll Surg Engl, 8*:1–22.

Dorwart, R. A. (1958). Medical education in Prussia under the early Hohenzollern, 1685–1725. *Bull Hist Med, 32*:335–47.

Drake, R. L. & Pawlina, W. (2013). The American Association of Anatomists celebrates 125 years. *Anat Sci Educ, 6*:1–2.

Drimmer, F. (1981). *Body snatchers.* New York: Fawcett Gold Medal Books.

Druml, W. (1999). Fortschritt kann nur entstehen aus einer umfassenden Prasenz des Vergangenen. Editorial. *Wien Klin Wochenschr, 111*:739–740.

Duckworth, W. L. H. (1962). *Galen: On anatomical procedures. The later books.* (Translation.) M. C. Lyons & B. Towers (Eds.). Cambridge: Cambridge University Press.

Dukes, D. E. (1960). London medical societies in the eighteenth century. *Proc R Soc Med, 539*:699–706.

Dungelova, E. & Barinka, L. (1987). Jan Evangelista Purkyne – The scientist and humanist. *Acta Chin Plast, 29*:125–9.

Ebbell. (1937). *The papyrus ebers: The greatest Egyptian medical document.* Copenhagen: Levin & Munksgaard.

Edelstein, L. (1935). The development of Greek anatomy. *Bull Hist Med, 3*:235–249.

Edelstein, L. (1967). The history of anatomy in antiquity. O. Temkin & C. L. Temkin (Eds.), *Ancient medicine.* Baltimore: Johns Hopkins University Press.

Edwards, J. J. & Edwards, M. J. (1959). *Medical museum technology.* Oxford: Oxford University Press.

Edwards, L. F. (1957). The famous Harrison case and its repercussions. *Bull Hist Med, 31*:162.

El-Tatawi, M. E. D. (1924). Der Lungenkreislauf nach El-Koraschi. *Inaug-Diss Freiburg i Br.*

Ellis, H. (2001). Teaching in the dissecting room. *Clin Anat, 14*:149–151.

Ellis, H. (2002) Medico-legal litigation and its links with surgical anatomy. *Surgery, 20*:i–ii.

Engelbach, R. & Derry, D. (1942). Mummification. *Ann du Serv des Antiquités de l'Egypte, 41*:235–265.

Erhard, H. (1941a). Alkmaion, der erste Experimental biologe. *Sudhoff Arch Gesch Med, 34*:77–89.

Erhard, H. (1941b). Diogenes von Apollonia als Biologe. *Sudhoff Arch Gesch Med, 34*:335–6.

Erlam, H. D. (1954). Alexander Monro, primus. *Univ Edinburgh J, 17*:77–105.

Erolin, C., Shoja, M. M., Loukas, M., Shokouhi, G., Rashidi, M.R., Khalili, M., Tubbs, R. S. (2013). What did Avicenna (Ibn Sina, 980–1037 A.D.) look like? *Int J Cardiol, 167*:1660–3.

Estes, J. W. (1989). *The medical skills of ancient Egypt.* Canton, MA: Watson.

Farfan, F. A. (1579). *Tractado Breue de Chirurgia y del Conocimiento y cora de algvnas enfermedades q. en esta tierra mas comunmente suelen auer.* (A brief treatise on medicine) Mexico: Casa de Antonio Ricardo, 1579. (Cited by Schendel, 1968).

Febres-Cordero, F. (1987). *Historia de la medicina en Venezuela y America.* 2 Vols. Caracas: Consejo de Profesores Universitarios Jubilados UCV.

Feely, J. (1992). The Royal College of Physicians and Irish medicine. In B. P. Kennedy & D. Coakley (Eds.), *The anatomy lesson: Art and medicine.* Dublin: The National Gallery of Ireland, pp. 89–105.

Feindel, W. (1964). *Thomas Willis: The anatomy of the brain and nerves.* Montreal: McGill University Press.

Feindel, W. (Ed.). (1978). *The anatomy of the brain and nerves.* Birmingham: The Classics of Medicine Library.

Ferngren, G. B. (1982). A Roman declamation on vivisection. *Trans Coll Physicians Phila, 4*:272–90.

Feyfer, F. M. G. de. (1933). Jan Steven van Calcar (Joannes Stephanus), 1499–1546. *Nederl T Geneesk, 77*:3562.

Field, E. J. & Harrison, R. J. (1968). *Anatomical terms. Their origin and derivation* (3rd ed.). Cambridge: W. Heffer & Sons Ltd.

Finger, S. (2004). *Minds behind the brain: A history of the pioneers and their discoveries.* Oxford: Oxford University Press, pp. 28–33.

Fishman, A. P. & Richards, D. W. (Eds.). (1982). *Circulation of the blood. Men and ideas.* Bethesda, MD: American Physiological Society, pp. 32–37.

Fleetwood, J. (1951). *History of medicine in Ireland.* Dublin: Browne and Nolan, Limited.

Fleetwood, J. (1988). *The Irish body snatchers. A history of body snatching in Ireland.* Dublin: Tomar.

Fleischhauer, K. (1981). In memoriam Philipp Stohr Jr. *Anat Anz, 150*:239–247.

Fleming, D. (1955). Galen on the motions of the blood in the heart and lungs. *Isis, 46*:14–21.

Flemming, P. (1957). The medical aspects of the mediaeval monastery in England. In Z. Cope (Ed.), *Sidelights on the history of medicine.* London: Butterworth pp. 23–36.

Flexner, A. (1910). Medical education in the United States and Canada. *Bulletin of the Carnegie Foundation for the Advancement of Teaching,* No. 4, New York.

Flexner, A. (1937). Introduction. In J. Morgan (Ed.), *A discourse upon the institution of medical schools in America.* (Reprinted from the first edition, Philadelphia, 1765.) Baltimore: Johns Hopkins University Press.

Foot, J. (1794). *The life of John Hunter.* London: T. Becket.

Ford, B. J. (1983). What were the missing Leeuwenhoek's microscopes really like? *Proc Roy Micr Soc, 18*:118–124.

Ford, B. J. (1991). *The Leeuwenhoek legacy.* London: Farrand Press.

Frank, Jr., R. G. (1976). Thomas Willis. In C. G. Gillispie, (Ed.), *Dictionary of scientific biography.* New York: Charles Scribner's Sons.

Frank, Jr., R. G. (1980). *Harvey and the Oxford physiologists.* Berkeley, CA: University of California Press.

Franklin, K. J. (1961). *William Harvey. Englishman, 1578–1657.* London: MacGibbon & Kee.

French, R. K. (1975). *Anatomical education in a Scottish university, 1620. An annotated translation of the lecture notes of John Moir.* Aberdeen: Equipress.

French, R. K. (1979). The history of the heart. In *Thoracic physiology from ancient to modern times.* Aberdeen: Equipress.

French, R. (1993). The anatomical tradition. In W. F. Bynum & R. Porter (Eds.), *Companion encyclopedia of the history of medicine.* Vol. 1. London: Rutledge, pp. 81–125.

French, R. (1994). *William Harvey's natural philosophy.* Cambridge: Cambridge University Press.

Frost, S. B. (1979). *The history of McGill in relation to Montreal and Quebec.* Montreal: McGill University Press.

Fulton, J. F. (1953). *Michael Servetus. Humanist and martyr.* Boston: Beacon Press.

Furley, D. J. & Wilkie, J. S. (Eds.). (1984). *Galen on respiration and the arteries.* Princeton, NJ: Princeton University Press.

Galen, C. (1956). Galen on anatomical procedures. *Proc R Soc Med, 49*:833.

Galen, C. (2003). On the usefulness of the parts of the body. *Clin Orthop Relat Res, 411*:4–12.

Garey, L. (2006). The early days of the Anatomical Society of Great Britain and Ireland. Newsletter Winter 2006 http://www.anatsoc.org.uk/.

Garrison, F. H. (1929). *An introduction to the history of medicine* (4th ed.). Philadelphia: W. B. Saunders.

Gerrits, P. & Veening, J. (2013). Leonardo da Vinci's "A Skull Sectioned": Skull and dental formula revisited. *Clin Anat, 26*:430–435.

Ghalioungui, P. (1982). The West denies Ibn al-Nafis's contribution to the discovery of the circulation. In A. R. El-Gindy & H.M.Z. Hassan (Eds.), *Proc. 2nd Int. Conf. Islamic Med,* Kuwait, pp. 299–304.

Gibbons, A. (2009). Ardipithecus ramidus. A new kind of ancestor: Ardipithecus unveiled. *Science, 326*:36–40.

Gomez, J. M. (1979). Historical synthesis of medical education in Mexico. In J. Z. Bowers & E. F. Purcell (Eds.), *Aspects of the history of medicine in Latin America.* New York: Independent Publishers Group, pp. 88–96.

Goodman, N. M. (1944). The supply of bodies for dissection: A historical review. *Brit Med J, 1*:807.

Gordon, B. L. (1959). *Medieval and Renaissance medicine.* New York: Philosophical Library.

Gordon-Taylor, G. & Walls, E. W. (1958). *Sir Charles Bell. His life and times.* Edinburgh: E. & S. Livingstone Ltd.

Graham, H. (1956). *Surgeons all* (2nd ed.). London: Rich & Cowan.

Granger, N. A. (2004). Dissection laboratory is vital to medical gross anatomy education. *Anat Rec, 281B*:6–8.

Grapow, H. (1935). *Uber die Anatomischen Kenntnisse der Altagyptischen Arzte.* Leipzig: J. C. Hinrichs.

Gray, H. (1999). *Gray's anatomy: The unabridged edition.* Courage Books, p. 15.

Grayson, A. K. (1980). Babylonia. In A. Cotterell (Ed.), *The encyclopedia of ancient civilizations.* London: Rainbird Publishing Group Limited, pp. 89–101.

Green, R. M. (1951). *A translation of Galen's hygiene* (De Sanitate Tuenda). Springfield, IL: Charles C Thomas, 1951.

Green, R. M. (1955). *Asclepiades: His life and writings.* New Haven, CT: E. Licht, pp. 73–76.

Greenblatt, S. H. (1995). Phrenology in the science and culture of the 19th century. *Neurosurgery, 37*:790–804.

Grene, M. (1972). Aristotle and modern biology. *J Hist Ideas, 33*:395–424.

Grensemann, H. (1968). Der Artz Polybus als Verfasser hippokratischer Schriften. Mainz, Akad. Wissensch, Lit.

Gross, C. G. (1997). Leonardo da Vinci on the brain and eye. *Neuroscientist, 3*:347–354.

Gross, C. G. (1999). *Brain, vision, memory: Tales in the history of neuroscience.* Bradford.

Gruner, O. C. (1930). *A treatise on the canon of medicine of Avicenna. Incorporating a translation of the first book.* London: Luzac & Co.

Gudger, E. W. (1923). Pliny's "Historia naturalis": The most popular natural history ever published. *ISIS, 6*:269–281.

Guerra, F. (1979). Pre-Columbian medicine: Its influence on medicine in Latin America today. In J. Z. Bowers & E. F. Purcell (Eds.), *Aspects of the history of medicine in Latin America.* New York: Independent Publishers Group, pp. 1–15.

Guild, S. R. (1935). G. Carl Huber 1865–1934 In memoriam. *Anat Rec, 62*:1–6.

Gurunluoglu, R., Shafighi, M., Gurunluoglu, A., & Cavdar, S. (2011). Giulio Cesare Aranzio (Arantius) (1530–89) in the pageant of anatomy and surgery. *J Med Biogr, 19*:63–9.

Guthrie, D. (1959). The contribution of Holland and Scotland to the evolution of medical education in America. In *History of American medicine, A symposium.* New York: MD Publications, p. 165.

Guthrie, D. (1956). The three Alexander Monros and the foundation of the Edinburgh Medical School. *J R Coll Surg Edinb, 2*:24–33.

Guthrie, R. D. (2005). *The nature of paleolithic art.* Chicago: University of Chicago Press

Guthrie, W. K. (2012). *The Greek philosophers: From Thales to Aristotle.* Routledge, p. 47.

Gütz, W. (1986). Der Anatom der Gynaäkologen Von einhundert Jahren wurde Hermann Steive geboren. *Notabene medici, 9*:598–600.

Haberling, W. (1924). Johannes Muller. Das Leben des rheinischen Naturforschers. Leipzig: Akademische Verlagsgesellschaft.

Haeger, K. (1989). *The illustrated history of surgery.* London: Harold Starke.

Hale, J. (1993). *The civilization of Europe in the Renaissance.* London: HarperCollins, pp. 542–559.

Hall, A. P. (1960). Studies on the history of the cardiovascular system. *Bull Hist Med, 34*:391.

Hall, R. A. (1974). Medicine and the Royal Society. In A. G. Debus (Ed.), *Medicine in seventeenth century England.* Los Angeles: University of California Press, pp. 421–452.

Hamilton, D. (1987). *The healers. A history of medicine in Scotland.* Edinburgh: Canongate.

Hancock, G. (1995). *Fingerprints of the gods.* New York: Crown.

Hansen, J. T. (2006). Frank H. Netter, M.D. (1906–1991): The artist and his legacy. *Clin Anat, 19*: 481–6.

Hare, H. A. (1923). Leidy and his influence on medical science. *J Acad Nat Sci Phila, 75*:73–87.

Harrington, J. (1920). *The School of Salernum.* New York: Paul B. Hoeber.

Harris, C. R. S. (1973). *The heart and the vascular system in ancient Greek medicine from Alcmaeon to Galen.* Oxford: Clarendon.

Harris, J. E. & Weeks, K. R. (1973). *X-raying the Pharaohs.* New York: Charles Scribner's Sons.

Hartman, C. G. (1962). *Science and the safe period.* Baltimore: Williams and Wilkins.

Hartwell, E. M. (1881a). The hindrances of anatomical study in the United States. *Ann Anat Surgery, 3*:209.

Hartwell, E. M. (1881b). The present legal status of the study of human anatomy in the United States. *Ann Anat Surg, 4*:8–14.

Harvey, E. N. (1953). Edwin Grant Conklin: 1863–1952. *Science, 117*: 703–705.

Harvey, W. (1957). *Exercitatio anatomica de motu cordis et sanguinis in animalibus.* Sumptibus Guilielmi Fitzeri, Francofurti, 1628. (Movement of the heart and blood in animals. An anatomical essay.) Translated by Kenneth J. Franklin. Oxford: Blackwell.

Hayek, S. (1984). How Al-Zahrawi reached the occident. *IWMJ 1*:49.

Heagerty, J. J. (1928). *Four centuries of medical history in Canada* (2 vols.). Toronto: MacMillan Company of Canada Limited.

Heckscher, W. S. (1958). *Rembrandt's anatomy of Dr. Nicholas Tulp. An iconological study.* Washington Square, NY: New York University Press.

Heilemann, H. A. (2011). Influence of the Casserius Tables on fetal anatomy illustration and how we envision the unborn. *J Med Libr Assoc, 99*:23–30.

Hein, J. (1976). Zur Geschichte der Anatomie und Chirurgie. Christian Heinrich Stinger 1782–1842. Anatom und Chirurg in Marburg. Mannheim, Großdruckerei und Verlag GmbH.

Herrlinger, R. (1953). Didactic originality of the anatomical paintings by Leonardo Da Vinci. *Anat Anz, 99*:366–95.

Herrlinger, R. (1970). *History of medical illustration.* London: Pitman.

Herzog, B. F. (2007). Chester B. McVay: Small-town surgeon, world-famous herniologist. *Surgery, 141*:119–20.

Hildebrand, R. (1988). Un beau monument iconographique de la science de l'homme. *Medizinhist J, 23*:291–318.

Hildebrandt, S. (2008). Capital punishment and anatomy: History and ethics of an ongoing association. *Clin Anat, 21*:5–14.

Hildebrandt, S. (2009a). Anatomy in the Third Reich: An outline, Part 1. National Socialist politics, anatomical institutions, and anatomists. *Clin Anat, 22*:883–93.

Hildebrandt, S. (2009b). Anatomy in the Third Reich: An outline, Part 2. Bodies for anatomy and related medical disciplines. *Clin Anat, 22*:894–905.

Hildebrandt, S. (2009c). Anatomy in the Third Reich: An outline, Part 3. The science and ethics of anatomy in National Socialist Germany and postwar consequences. *Clin Anat, 22*:906–15.

Hildebrandt, S. & Redies, C. (2012). Anatomy in the Third Reich. *Ann Anat, 194*:225–226.

Hilloowala, R. (1984). Illustrations from the Wellcome Institute Library. The origins of the Wellcome anatomical waxes. *Med Hist, 28*:432–7.

Hilloowala, R. & Goldstein, C. H. (1982) Tulane Medicine, (April) 9.

Hilton, J. (1950). *Rest and pain* (Edited by E. W. Walls, E. E. Philipp & H. J. B. Atkins). London: G. Bell and Sons.

Hinsey, J. C. (1943). Stephen Walter Ranson: 1880–1942. *Science, 97*:245–56.

Hintzsche, E. (1948). Albrecht von Haller als Anatom und seine Schule. *Z Anat Entw Gesch, 10*:4068.

Hirsch, A. (Ed.). (1884–1888). *Biographisches Lexikon der hervorragenden Aerzte aller Zeiten und Volker* (6 vols.). Vienna and Leipzig: Urban and Schwarzenberg.

Hoernle, A. F. R. (1907). Studies in the medicine of ancient India. Part I. *Osteology or the bones of the human body.* Oxford: Clarendon Press.

Hodges, D. L. (1985), *Renaissance fictions of anatomy.* Amherst, MA: University of Massachusetts Press, pp. 1–19, 89–106.

Hoerr, N. I. (1956). The role of the anatomical disciplines in medical education. *J Med Educ, 31*:7–24.

Hoizey, D. & Hoizey, M. J. (1993). *A history of Chinese medicine.* (Translated by Paul Bailey.) Edinburgh: University Press.

Holloway, S. W. F. (1964). Medical education in England, 1830–1858: A sociological analysis. *History, 49*:299–324.

Holt, S. B. (1976). Harold Cummins 1894–1976. *J Med Genet, 13*:540.

Hooper, A. (1980). Further information on the prehistoric representations of human hands in the cave of Gargas. *Med Hist, 24*:214–6.

Hoss, R., Broad, P., & Kremer, J. P. (1984). *KL Auschwitz seen by the SS. Selection, elaboration and notes by Bezwinska, Jadwiga and Czech, Danuta.* New York: Howard Fertig, pp. 1–331.

Hoyt, Jr., W. D. (1942). A young Virginian prepares to practice medicine, 1796–1800. *Bull Hist Med, 11*:582–6.

Howe, A., Campion, P., Searle J., & Smith, H. (2004). New perspectives - approaches to medical education at four new UK medical schools. *BMJ, 329*:327–331.

Hsieh, E. T. (1920). A review of ancient Chinese anatomy. *Anat Rec, 20*:97.

Huard, P. & Wong, M. (1968). *Chinese medicine.* (Translated from the French by B. Fielding.) London: Weidenfeld and Nicolson.

Hubbard, C. (2001). Eduard Pernkopf's atlas of topographical and applied human anatomy: The continuing ethical controversy. *Anat Rec, 265*:207–11.

Hubotter, F. (1929). *Die chinesische medizin.* Leipzig: Verlag der Asia Major.

Hughes, J. T. (1992). *Thomas Willis, 1621–1675: His life and work.* London: Royal Society of Medicine Services.

Hun, H. (1883). *A guide to American medical students in Europe.* New York: William Wood.

Hunter, J. (1786a). *A treatise on the venereal disease.* London: G. Nichol.

Hunter, J. (1786b). *Observations on certain parts of the animal oeconomy.* London: Longmans.

Hunter, J. (1794). *A treatise on the blood, inflammation, and gun-shot wounds, by the Late John Hunter. To which is prefixed a short account of the author's life, by his brother-in-law Everard Home.* London: for Georg Nicol, Bookseller to His Majesty.

Hunter, J. (1835). *The works of John Hunter, F.R.S., with notes.* Edited by James F. Palmer (4 vols.). London: Longmans, With a Life, by Drewry Ottley.

Hunter, J. (1861). *Essays and observations on natural history, anatomy, physiology, psychology, and geology...by Richard Owen* (2 vols.). London: John Van Voorst.

Hunter, J. (1976). *Letters from the past. From John Hunter to Edward Jenner.* London: Royal College of Surgeons of England.

Hunter, T. B., Peltier, L. F., & Lund, P. J. (2000). Radiologic history exhibits. Musculoskeletal eponyms: Who are these guys? *Radiographics, 20*:819–316.

Hunter, W. (1774). *The anatomy of the human gravid uterus exhibited in figures.* Birmingham: Joannes Baskerville.

Huntley, F. L. (1951). Sir Thomas Browne, M.D., William Harvey, and the metaphor of the circle. *Bull Hist Med, 25*:236–47.

Hyrtl, J. (1879). Das Arabische und Hebraische in der Anatomie. Wien.

Ilberg, J. (1889). Uber die Schriftstellerei des Klaudios Galenos (I). *RheinMus Phil, 44*:207–39.

Ilberg, J. (1892). Uber die Schriftstellerei des Klaudios Galenos (II). *Rhein Mus Phil, 47*:489–514.

Ilberg, J. (1896). Uber die Schriftstellerei des Klaudios Galenos (III). *Rhein Mus Phil, 57*:165–96.

Ilberg, J. (1897). Uber die Schriftstellerei des Klaudios Galenos (IV). *Rhein Mus Phil, 52*:591–623.

Ilberg, J. (1902). Uber die Schriftstellerei des Klaudios Galenos. *Rhein Mus Phil, 47*:489.

Ilberg, J. (1930). Wann ist Galenos geboren? *Sudhoff Arch, 23*:289–292.

Imai, M. (2011). Herophilus of Chalcedon and the Hippocratic tradition in early Alexandrian medicine. *Hist Sci* (Tokyo), *21*:103–122.

Inuwa, I.M., Taranikanti, V., Al-Rawahy, M., Roychoudhry, S., & Habbal, O. (2012). "Between a rock and a hard place": The discordant views among medical teachers about anatomy content in the undergraduate medical curriculum. *Sultan Qaboos Univ Med J, 12*:19–24.

Isler, H. (1968). *Thomas Willis 1621–1675: Doctor and scientist.* London: Hafner.

Iturbe, U., Pereto, J., & Lazcano, A. (2008). The young Ramon y Cajal as a cell-theory dissenter. *Int Microbiol, 11*:143–145.

Izquierdo, J. J. (1937). A new and more correct version of the views of Servetus on the circulation of the blood. *Bull Hist Med 5*:914–932.

Jack, D. (1981) *Rogues, rebels, and geniuses.* Toronto: Doubleday.

Jaeger, W. (1948). *Aristotle* (2nd ed.). London: Oxford University Press.

Janick, J., Paris, H. S., & Parrish, D. C. (2007). The cucurbits of Mediterranean antiquity: Identification of taxa from ancient images and descriptions. *Ann Bot, 100*:1441–1457.

Janson, H. W. (1977). *History of art.* New York: Harry N. Abrams.

Janssens, P. A. (1957). Medical views on prehistoric representations of human hands. *Med Hist, 7*:318–322.

Jastrow Jr., M. (1914). The medicine of the Babylonians and Assyrians. *Proc Roy Soc Med, 7*:109–176.

Jee, B. S. (1978). *A short history of Aryan medical science.* Delhi: New Asian Publishers.

Johanson, D. C. & White, T. D. (1980). On the status of Australopithecus Afarensis. *Science, 207*:1104–5.

Johanson, D. C. & Edey, M. A. (1982). *Lucy: The beginnings of human kind.* New York: Warner Books.

Johansen, T. K. (1997). *Aristotle on the sense-organs*. Cambridge, pp. 54–5.

Jolin, L. (2013). A physician's library. *CAM* (Cambridge Alumni Magazine), *68*:32–37.

Jones, W. H. S. (1945). Hippocrates and the corpus Hippocraticum. *Proc Brit Acad, 31*:103–25.

Jouanna, J. (1969). Le medecin Polybe est-il l'auteur de plusieurs ouvrages de la collection hippocratique. *Rev Etud Grec, 82*:552–62.

Kambara, H. (1974). The wooden models of the human skeleton made during the Edo period in Japan. Proceedings of the XXIII International Congress of the History of Medicine. London: Wellcome Institute of the History of Medicine, Vol. 2, pp. 981–983.

Kapitza, B., Göbbel, L., & Schultka, R. (2002). Dr. Gustav Wilhelm Münter (1804–1870) – the "unknown" assistant of the Meckel collection. *Ann Anat, 184*:547–50.

Kaufman, M. H. (2005). John Bell (1763–1820), the father of surgical anatomy. *J Med Biogr, 13*:73–81.

Kapferer, R. (1951). *Dieanatomischen Schriften. Die Anatomie des Herz. Die Adern inder Hippokratischen Sammlung*. Stuttgart: Hippokrates Verlag.

Kardel T. (1994). Steno on muscles. Dedication to elementorum myologiae specimen (1667), p. 85.

Karnovsky, M. J. (1979). Dedication to Albert H Coons 1912–1978. *J Histochem Cytochem, 27*:1117–8.

Keele, K. D. (1951). Leonardo da Vinci, and the movement of the heart. *Proc R Soc Med, 44*:209–213.

Keele, K. D. (1952). *Leonardo da Vinci on movement of the heart and blood*. London: Harvey and Blythe Ltd.

Keele, K. D. (1961). Three masters of experimental medicine Erasistratus, Galen and Leonardo da Vinci. *Proc Roy Soc Med, 54*:577–88.

Keele K. D. (1973). Leonardo da Vinci's views on arteriosclerosis. *Med Hist, 7*:304–308.

Keele, K. D. (1979). Leonardo da Vinci's "Anatomia Naturale." *Yale J Biol Med, 52*:369–409.

Keele, K. D. (1983). *Leonardo da Vinci's elements of the science of man*. New York: Academic Press.

Keevil, J. J. (1957). The seventeenth century English medical background. *Bull Hist Med, 31*:408–24.

Kell, P. E. (2004–2013). Joshua Brookes (1761–1833) anatomist. *Oxford dictionary of national biography*. Oxford: Oxford University Press.

Kemp, M. (2006). Leonardo Da Vinci: *The marvellous works of nature and man*. Oxford: Oxford University Press.

Kemp, M. (2010). Style and non-style in anatomical illustration: From renaissance humanism to Henry Gray. *J Anat, 216*:192–208.

Kennedy, M. (2003). Benjamin Franklin's house: The naked truth. In "*The Guardian*," 11 August.

Kennedy, B. P. & Coakley, D. (Eds.). (1992). *The anatomy lesson: Art and medicine*. Dublin: The National Gallery of Ireland National Gallery of Ireland.

Keswani, N. H. (1970). Medical education in India since ancient times. In C. D. O'Malley (Ed.), *The history of medical education*. Berkeley, CA: University of California Press, pp. 329–366.

Keswani, N. H. (1973). Susruta, the pioneer anatomist and the father of surgery. In G. D. Singhal & L. V. Guru (Eds.), *Anatomical and obstetric considerations in ancient Indian surgery*. Banaras: Banaras Hindu University Press.

Keynes, G. (Ed.). (1928). The anatomical exercises of Dr. William Harvey De Motu Gordis 1628: De Circulatione Sanguinis 1649: The first English text of 1653 now newly edited. London: The Nonesuch Press.

Keynes, G. (1966). *The life of William Harvey*. Oxford: Clarendon Press.

Ketham, J. (1925). The Fasciculo di medicina, Venice, 1493. With an introduction, etc. by Charles Singer (Monumenta Medica, II), 2 vols, facsims. fol. Florence: R. Lier.

Ketham, J. (1941). *Fasciculus medicine*. (Copy in the British Library, London).

Kevorkian, J. (1959). *The story of dissection*. New York: Philosophical Library.

Kühn, C. G. (1821–1833) Claudii Galeni Opera omnia (20 vols.). Medicorum Graecorum Opera quae extant. Leipzig.

King, L. S. (1954). Plato's concepts of medicine. *J Hist Med, 9*:38.

King, R. (1994). John Hunter and the natural history of the human teeth: Dentistry, digestion, and the living principle. *J Hist Med Allied Sci, 49*:504–520.

Klee, E. (2003). Das Personenlexikon zum Dritten Reich We war was vor und nach 1945. Frankfurt/M, p. 310.

Klein-Franke, F. (1982). Vorlesungen uber dieMedizin imIslam. Beiheft 23. Sudhoffs Archiv. Zeitschrift fur Wissenschaftsgeschichte. Wiesbaden: Franz Steiner Verlag GMBH.

Knight, B. (1980). *Discovering the human body*. New York: Lippincott & Crowell.

Knox, R. (1852). *Great artists and great anatomists; A biographical and philosophical study*. London: John Van Voorst.

Kobler, J. (1960). *The reluctant surgeon. The life of John Hunter*. London: Heinemann.

Koelbing, H. (1968). Zur Sehtheorie im Altertum: Alkmeon und Aristoteles. *Gesnerus* (Aarau), *25*:5–9.

Kopsch, F. (1913). Zweihundert Jahre Berliner Anatomie. *Dtsch Med Wochenschr, 39*:948–1003.

Kornell, M. (1989). Fiorentino and the anatomical text. *The Burlington Magazine, 131*:843–847.

Köckerling, D., Köckerling, D., & Lomas, C. (2013). Cornelius Celcus – ancient encyclopedist, surgeon-scientist, or master of surgery? *Langenbecks Arch Surg, 398*:609–16.

Kramer, S. N. (1961). *Sumerian mythology*. New York: Harper.

Kramer, S. N. (1963) *The Sumerians: Their history, culture and character.* Chicago, University of Chicago Press.

Krumbhaar, E. B. (1922). The early history of anatomy in the United States. *Ann Med Hist, 4*:271.

Kühnel, W. (1989). 100 Jahre anatomische Gesellschaft. *Verh Anat Ges, 82*:31–75.

Ladell, M. (1983). Burke and Hare almost lost Dr. Knox his life. *Medical Post, 19*:60.

Ladman, A. J. (1981). R. C. Truex memorial. *Anat Rec, 201*:3–11.

Lachman, E. (1977). Anatomist of infamy: August Hirt. *Bull Hist Med, 51*:594–602.

Lagunoff, D. (2002). A Polish, Jewish scientist in 19th century Prussia. *Science, 298*:2331.

Laitko, H. (Ed.). (1987). Wissenschaft in Berlin. Von den Anfangen bis zum Neubeginn nach 1945. Berlin: Dietz Verlag.

Lambert, S. W. (1936a). A reading from Andreae Vesalii, de Corporis Humani Fabrica Liber VII De Vivorum sectione nonnulla caput XIX. *Bull NY Acad Med, 12*:346–386.

Lambert, S. W. (1936b). The physiology of Vesalius. *Bull NY Acad Med, 12*:387–415.

Langdon-Brown, W. (1946). *Some chapters in Cambridge medical history.* Cambridge: Cambridge University Press, pp. 1–19.

Lanza, B., Puccetti, M. L. A., Poggesi, M., & Martelli, A. (1979). *Le Cere Anatomione Della Specola.* Florence: Arnaud Editore Firenze.

Lassek, A. M. (1958). *Human dissection: Its drama and struggle.* Springfield, IL: Charles C Thomas.

Lawrence, D. G. (1958). "Resurrection" and legislation on body-snatching in relation to the anatomy act in the Province of Quebec. *Bull Hist Med, 32*: 408–24.

Le Fanu, W. R. (1946). *John Hunter: A list of his books.* London: Royal College of Surgeons of England.

Leake, C. D. (1952). *The old Egyptian medical papyri.* Lawrence, KS: University of Kansas Press.

Leakey, R. E. (1981). Die Suchen nach dem Menschen. Wie wir wurden, was wirsind. Umschau Verlag, Sigma Press und PR, Frankfurt (M).

Leblond, S. (1966). Anatomistes et resurrectionnistes au Canada et plus particulierement dans la Province de Quebec. *Can Med Assoc J, 95*:1193–7 & 1247–51.

Leca, A. P. (1981). *The Egyptian way of death: Mummies and the cult of the immortal.* New York: Doubleday.

Leiser, G. (1983). Medical education in Islamic lands from the seventh to the fourteenth century. *Hist Med, 38*:48–75.

Lempp, H. K. (2005). Perceptions of dissection by students in one medical school: Beyond learning about anatomy. A qualitative study. *Med Educ, 39*:318–325.

Leroi-Gourhan, A. (1982). *The dawn of European art.* Translated by Sara Champion. Cambridge: Cambridge University Press.

Lett, H. (1943). Anatomy at the barber-surgeon's hall. *Brit J Surg, 31*:101–111.

Lewin, R. (1983). Fossil Lucy grows younger, again. *Science, 219*:43–44.

Lind, L. R. (1975). *Studies in pre-Vesalian anatomy. Biography, translations, documents.* Philadelphia: The American Philosophical Society.

Lindeboom, G. A. (1959). *Bibliographia Boerhaaviana.* Leiden: E. J. Brill.

Lindeboom, G. A. (1968). *Herman Boerhaave. The man and his work.* London: Methuen.

Lindeboom, G. A. (1974). Boerhaave's influence in British medicine. Proceedings of the XXIII International Congress of the History of Medicine. London: Wellcome Institute of the History of Medicine. London, Vol. 1, pp. 734–738.

Lindner, H. H. (1989). *Clinical anatomy.* Norwalk, CT: Appleton & Lange.

Linss, W. (1981). Karl von Barderleben-der Begründer des Anatomischen Anzeigers. *Anat Anz, 150*:5.

Lint, J. G. de. (1924). Fugitive anatomical sheets. *Janus, 28*:78–91.

Lipsett, W. G. (1961). Celsus – First medical historian. *Hamdard Med Dig, 5*:13.

Lloyd, G. E. R. (1968). *Aristotle: The growth and structure of his thought.* Cambridge: Cambridge University Press.

Lloyd, G.E.R. (1975a). Alcmaeon and the early history of dissection. Sudhoff Arch 59:113-147.

Lloyd, G. E. R. (1975b). The Hippocratic question. *Class Quart, 25–26*:177–92.

Lloyd, G. E. R. (1975c). A note on Erasistratus of Ceos. *J Hellen Stud, 95*:172–175.

Locy, W. A. (1911). Anatomical illustrations before Vesalius. *J Morph, 22*:945–988.

Logan, C. A. (1999). The altered rationale for the choice of a standard animal in experimental psychology: Henry H. Donaldson, Adolf Meyer, and "the" albino rat. *Hist Psychol, 2*:3–24.

Longrigs, J. (1976). The "roots of all things." *ISIS, 67*:420.

Lonie, I. M. (1964). Erasistratus, the Erasistrateans and Aristotle. *Bull Hist Med, 55*:426–443.

Lonsdale, H. (1870). *The sketch of the life and writings of Robert Knox the anatomist.* London: MacMillan.

Loukas, M., Tubbs, R. S., Louis, R. G., Pinyard, J., Vaid, S., & Curry, B. (2007a). The cardiovascular system in the pre-Hippocratic era. *Int J Cardiol, 120*:145–149.

Loukas, M., Clarke P., Tubbs, R. S., & Kapos, T. (2007b). Raymond de Vieussens. *Anat Sci Int, 82*:233.

Loukas, M., Clarke, P., Tubbs, R. S., & Kolbinger, W. (2008a). Adam Christian Thebesius, a historical perspective. *Int J Cardiol, 129*:138–40.

Loukas, M., Lam, R., Tubbs, R. S., Shoja, M. M., & Apaydin, N. (2008b). Ibn al-Nafis (1210–1288): The first description of the pulmonary circulation. *Am Surg, 74*:440–2.

Loukas, M., Lanteri, A., Ferrauiola, J., Tubbs, R. S., Maharaja, G., Shoja, M. M., Yadav, A., & Rao, V. C. (2010a). Anatomy in ancient India: A focus on the Susruta Samhita. *J Anat, 217*:646–50.

Loukas, M., Saad, Y., Tubbs, R. S., & Shoja, M. M. (2010b). The heart and cardiovascular system in the Qur'an and Hadeeth. *Int J Cardiol, 140*:19–23.

Loukas, M., Colborn, G. L., & Tubbs, R. S. (2010c). John Elias Skandalakis, MD, PhD, FACS (1920–2009). *Clin Anat, 23*:332–4.

Loukas, M., Hanna, M., Alsaiegh, N., Shoja, M. M., & Tubbs, R. S. (2011a). Clinical anatomy as practiced by ancient Egyptians. *Clin Anat, 24*:409–415.

Loukas, M., Tubbs, R. S., Mirzayan, N., Shirak, M., Steinberg, A., & Shoja. M. M. (2011b). The history of mastectomy. *Am Surg, 77*:566–71.

Loukas, M., Bellary, S., Kuklinsk, M., Ferrauiola, J., Yadav, A., Shoja, M., Shaffer, K., & Tubbs, R. S. (2011c). The lymphatic system: A historical perspective. *Clin Anat, 24*:807–816.

Loukas, M. & Tubbs, S. (2012). Gene L. Colborn (1935–2011). *Clin Anat, 25*:793–4.

Louw, G., Eizenberg, N., & Carmichael, S. W. (2009). The place of anatomy in medical education: AMEE guide no 41. *Med Teach, 31*:373–386.

Lovejoy, C. O. (1981). The origin of man. *Science, 277*:341–350.

Low, F. N. (1953). The pulmonary alveolar epithelium of laboratory animals and man. *Anat Rec, 117*:241–264

Lower, R. (1932). Tractus de corde (London, 1669). Translated by Franklin, K. J. in Gunther, R. T. (Ed.), *Early science in Oxford*. Oxford: Oxford University Press.

Luyendijk-Elshout, A. M. (1992). The image of the human body. In B. P. Kennedy & D. Coakley (Eds.), *The anatomy lesson: Art and medicine*. Dublin: The National Gallery of Ireland, pp. 79–87.

Lydiatt, D. & Bucher, G. (2011). Historical vignettes of the thyroid gland. *Clin Anat, 24*:1–9.

Lyons, A. S. & Petrucelli, II, R. J. (1978). *Medicine. An illustrated history*. New York: Harry N. Abrams.

Macalister, A. (1884). A sketch of the history of anatomy. *Dublin J Med Sc, 4*:1–19.

Macalister, A. (1891). *The history of the study of anatomy in Cambridge*. London: Clay.

Macalister, A. (1898) The oldest anatomical memoranda extant. *J Anat, 32*:775–78.

Macalister, A. (1900). *James Macartney*. London: Hodder & Stoughton.

Macdonald, T. (2003). *The social significance of health promotion*. London: Routledge, p. 13.

MacGillivray, R. (1988). Body-snatching in Ontario. *CBMH* (Canadian bulleting of Medical History), *5*:51.

MacGregor, G. (1884). *The history of Burke and Hare and of resurrectionist times. A fragment from the criminal annals of Scotland*. Glasgow: Thomas D. Morrison.

MacKinney, L. C. (1962). The beginnings of western scientific anatomy: New evidence and a revision in interpretation of Mondeville's role. *Med Hist, 6*:233.

MacKinney, L. C. (1965). *Medical illustrations in medieval manuscripts*. London: Wellcome Historical Medical Library.

Magner, L. N. (1979). *A history of the life sciences*. New York: Marcel Dekker.

Magner, L. N. (1992). *A history of medicine*. New York: Marcel Dekker.

Magoun, H. W. & Marshall, L. (2003). *American neuroscience in the twentieth century*. The Netherlands: Swets and Zeitlinger B.V. Lisse, pp: 96–98.

Majno, G. (1975). *The healing hand*. Cambridge, MA: Harvard University Press.

Major, R. H. (1954). *A history of medicine* (2 vols.). Springfield, IL: Charles C Thomas.

Malina, P. & Spann, G. (1999). Das Senatsprojekt der Universitat Wien "Untersuchungen zur Anatomischen Wissenschaft in Wien 1938–1945." *Wien Klin Wochenschr, 111*:743–753.

Mallowan, M. E. L. (1965). *Early Mesopotamia and Iran*. New York: McGraw-Hill.

Marandola, P., Musitelli, S., Jallous, H., Speroni, A., & de Bastiani, T. (1994). The Aristotelian kidney. *Am J Nephrol, 14*:302–306.

Margolis, I. B., Carnazzo, A. J, & Finn, M. P. (1976). Intramural hematoma of the duodenum, *Am J Surg, 132*:779–783.

Marks, G. & Beatty, W. K. (1973). *The story of medicine in America*. New York: Charles Scribner's Sons.

Marketos, S. G. & Skiadas, P. (1999). Hippocrates: The father of spine surgery. *Spine, 24*:1381–7.

Mariette, A., Délié, H., & Béchard, É. (1872). "Album du Musée de Boulaq." Mourès & Cie, Imprimeurs-Editeurs: Le Caire, p. 61.

Marti-Ibanez, F. (Ed.). (1958). *History of American medicine*. New York: MD Publications.

Matsen, H. S. (1969). Allessandro Achillini and his doctrine of "Universals"and "Transcendentals." Ph.D. Diss., Columbia University (Cited in Lind, 1975).

Matthiae, P. (1981). *Ebla: An empire rediscovered*. New York: Doubleday.

May, M. T. (1968) *Galen: On the usefulness of the parts of the body. De Usu Partium*. Vols. 1 & 2. Translated from the Greek with an introduction and commentary. New York: Cornell University Press.

McCurdy, E. & Cape, J. (1932). *The mind of Leonardo da Vinci*. London, p. 147.

McGhee, J. (2004). 3-D visualization and animation technologies in anatomical imaging. *J Anat, 216*:264–270.

McGirr, E. M. & Stoddart, W. (1991). Changing theories in 18th-century medicine. The inheritance and legacy of William Cullen. *Scot Med J, 36*:23–6.

McKenna, M. (1987a). William Harvey, 1. That incomparable invention of Dr. Harvey's. *Canadian J Surg, 30*:139–41.

McKenna, M. (1987b). William Harvey, 2. Harvey and the Royal College versus the "Empirics." *Canadian J Surg, 30*:215–7.

McLachlan, J. C. (2004). New path for teaching anatomy: Living anatomy and medical imaging vs. dissection. *Anat Rec, 281B*:4–5.

McLachlan, J. C., Bligh, J., Bradley, P., & Searle, J. (2004). Teaching anatomy without cadavers. *Med Educ, 38*:418–424.

McLachlan J. C. & Patten, D. (2006). Anatomy teaching: Ghosts of the past, present and future. *Med Educ, 40*:243–253.

McMinn, R. M. H. (1990). *Last's anatomy. Regional and applied.* Edinburgh: Churchill Livingstone.

McMurrich, J. P. (1930). *Leonardo da Vinci, the anatomist (1452–1514).* Baltimore: Williams & Wilkins.

McNulty, J. A., Sonntag, B., & Sinacore, J. M. (2009). Evaluation of computer-aided instruction in a gross anatomy course: A six-year study. *Anat Sci Educ, 2*:2–8.

McPhedran, N. T. (1993). *Canadian medical schools. Two centuries of medical history 1822 to 1992.* Montreal: Harvest House Ltd.

Meissner, B. (1920–1925). *Babylonien und Assyrien.* 2 Vols. Heidelberg.

Messbarger, R. (2010). *The lady anatomist: The life and work of Anna Morandi Manzolini.* Chicago: University of Chicago Press.

Metzner, M. T. (1954). Beitrage zu Samuel Sommerrings Leben und Wirken in Mainz, nach Briefen und unveroffentlichten Urkunden. Med Diss, Mainz.

Meyer, A. W. (1936). *An analysis of the De Generatione Animalium of William Harvey.* Stanford, CA: Stanford University Press.

Meyer, A. W. (1939). *The rise of embryology.* Stanford, CA: Stanford University Press.

Meyer, A. W. (1971). *Historical aspects of cerebral anatomy.* London: Oxford University Press.

Meyerhoff, M. (1926). New light on Hunain ibn Ishaq and his period. *Abh Kunde Morgenlandes, 8*:695.

Meyerhof, M. (1928). The book of the ten treatises on the eye ascribed to Hunain ibn is-haq (809–877 A.D.). Cairo, Government Press.

Meyerhof, M. (1935). Ibn an-Nafis (XIIIth cent) and his theory of the lesser circulation. *Isis, 23*:100–20.

Michels, N. A. (1955). *The American Association of Anatomists. A tribute and brief history commemorating the sixty-eighth annual session 1955.* New York: Blakiston Division, McGraw-Hill.

Michels, N. A. (1987). For the advancement of anatomical science. In J. E. Pauly (Ed.), *The American Association of anatomists, 1888–1987.* Baltimore: Williams & Wilkins, pp. 14–29.

Miinster, L. (1933). Allessandro Achillini, anatomico e filosofo, professore dello studio di Bologna 1463–1512. *Rivista distoria dellescienze mediche enaturali, 24*:7–22; 54–77.

Miles, K. A. (2005). Diagnostic imaging in undergraduate medical education: An expanding role. *Clin Radiol 60*:742–745.

Miller, H. W. (1962). The aetiology of disease in "Plato's Timaeus." *Trans Amer Philol Ass, 93*:175–187.

Miller, R. (2000). Approaches to learning spatial relationships in gross anatomy: Perspective from wider principles of learning. *Clin Anat, 13*:439–443.

Mitscherlich, A. & Mielke, F. (1960). Medizin ohne Menschlichkeit. Dokumente des Nurnberger Arzteprozesses. Frankfurt/Main: Fischer.

Modanlou, H. D. (2008). Tribute to Zakariya Razi (865–925 AD), an Iranian pioneer scholar. *Arch Iran Med, 11*:673–7.

Moll, A. A. (1944). *Aesculapius in Latin America.* Philadelphia: W. B. Saunders.

Moir, D. M. (1831). *Outlines of the ancient history of medicine.* Edinburgh: William Blackwood.

Mondino, da Luzzi. (1493). Anathomia, emendata per Melerstat. Leipzig, Landsberg, (Copy in the Library of the Royal College of Physicians, London).

Monro, P. A. G. (1996). The professor's daughter. An essay on female conduct by Alexander Monro (primus). *Proceedings of the Royal College of Physicians of Edinburgh, 26*:237.

Moon, R. O. (1909). *The relation of medicine to philosophy.* London: Longmans, Green.

Moose, C. J. (1998). *Dictionary of world biography*, Volume 1. Salem Press, p. 48.

Moore, K. L. (1982). Highlights of human embryology in the Koran and Hadith. *Proc 7th Saudi Medical Meeting, 51*–58.

Moore, K. L. (1992). *Clinically oriented anatomy* (3rd ed.). Baltimore: Williams & Wilkins.

Moore, K. L. (2012). My 60 years as a clinical anatomist. *Int J Anat Variat, 5*:1–4.

Moore, N. (1918). *The history of St. Bartholomew's Hospital.* London: C. Arthur Pearson Limited.

Moore, N. (1998). To dissect or not to dissect? *Anat Rec, 253*:8–9.

Moore, W. (2005a). John Hunter – surgeon and resuscitator. *Resuscitation, 66*:3–6.

Moore, W. (2005b). *The knife man.* London: Bantam.

Morgan, J. (1765). *A discourse upon the institution of medical schools in America.* Baltimore: William Bradford, 1765 (Reprinted with an Introduction by A. Flexner. Baltimore: Johns Hopkins Press, 1937).

Morley, J. (1971). *Death, heaven and the victorians.* London: Studio Vista.

Morton, L. T. (1991). London's last private medical school. *J Roy Soc Med, 84*:682.

Mosconi, T. & Kamath, S. (2003). Bilateral asymmetric deficiency of the pectoralis major muscle. *Clin Anat, 16*:346–349.

Morsink, J. (1982). *Aristotle on the generation of animals: A philosophical study.* Lanham: University Press of America.

Mundino. (1500). *Anathomia* (Copy in the British Library, London).

Myers, J. A. (1955). Edward Allen Boyden, author, renowned anatomist, and beloved teacher. *J Lancet, 75*:150–152.

Nasr, S. H. (1968). *Science and civilization in Islam.* Cambridge: Harvard University Press.

Nasr, S. H. (1976). *Islamic science. An illustrated study.* World of Islam Festival Publishing (Cited in Uddin, 1982).

Nauck, E. T. (1959). Zur Chronologie and Topographie der Lehranatomien in Mitteleuropa bis zum Jahre 1700. *Anat Anz, 106*:409–29.

Needham, J. (1959). *A history of embryology* (2nd ed.). Cambridge: Cambridge University Press.

Needham, J. (1970). *Clerks and craftsmen in china and the west. Lectures and addresses on the history of science and technology.* Cambridge: At the University Press, pp. 78, 8c, 407.

Nemir, P. Jr. (1980). Memoir of Oscar V. Batson 1894–1979. *Trans Stud Coll Physicians Phila, 2*:67–70.

Nemri, A. (2010). Santiago Ramon y Cajal. *Scholarpedia, 5*:8577.

Netter, F. (1956). Medical illustration – Its history and present day practice. *J Int Col Surg, 26*:505–513.

Netter, F. M. (2013). *Medicine's Michelangelo.* Quinnipiac University Press.

Nicholas, J. S. (1961). *Ross Granville Harrison 1870–1959. Biographical memoirs.* National Academy of Sciences, Washington D.C.

Nichols, R. (1995). *The diaries of Robert Hooke, The Leonardo of London 1635–1703.* Lewes, East Sussex: The Book Guild Limited.

Nielsen, A. E. (1942). A translation of Olof Rudbeck's Nova Excercitatio Anatomica announcing the discovery of the lymphatics (1653). *Bull Hist Med, 11*:304.

No authors listed. (1884). *Canada Med Surg J, 12*:504–506.

No authors listed. (1943). Dr. Lewellys Franklin Barker. *Bull Med Libr Assoc, 31*:372–373.

No authors listed. (1951). Oliver S. Strong, 1864–1951. *J Comp Neurol, 94*:179–80.

No authors listed. (1954). Edward Allen Boyden. *Anat Rec, 118*:3–18.

No authors listed. (1964a). In memoriam: Olof Larsell, 1886–1964. *J Comp Neurol, 123*:1–4.

No authors listed. (1964b). *Nobel lectures, physiology or medicine 1942–1962.* Amsterdam: Elsevier.

No authors listed. (1971). In memoriam: Bradley M. Patten. *Dev Biol, 26*:659.

No authors listed. (2003). Leonard Hayflick Bio. University of California, San Francisco http://www.ageless animals.org/hayflickbio.htm

No authors listed. (2005). Book review: Gray's anatomy, 39th Edition: The anatomical basis of clinical practice. *Am J Neuroradiol, 26*:2703–2704.

No authors listed. (2010). Pioneer of cloning Ian Wilmut Date of birth: July 7, 1944. Academy of Achievement, http://www.achievement.org/autodoc/page/wil0bio-1.

Northcote, W. (1772). *A concise history of anatomy from the earliest ages of antiquity.* London: T. Evans.

Norman, J. M. (Ed.). (1986). *The anatomical plates of Pietro da Cortona.* New York: Dover,

Norman, J. M. (Ed.) (1991). *Morton's medical bibliography* (5th ed.). Aldershot, Hampshire: Scolar Press.

Nnodim, J. O., Ohanaka, E. C., & Osuji, C. U. (1996). A follow-up comparative study of two modes of learning human anatomy. *Clin Anat, 9*:258–262.

Nuland, S. (1992). *Medicine. The art of healing.* New York: Hugh Lauter Levin Associates.

Nussbaum, W. (1968). Kleine Medizinalgeschichte Berns. *Berner Zt fur Geschichte u Heimatkunde, 30*:55.

Nutton, V. (1985). John Caius and the Eton Galen: Medical philology in the Renaissance. *Medizinhist J, 20*:227–52.

Nutton, V. (1988). *From Democedes to Harvey.* London: Variorum Reprints.

Nutton, V. (1993a). Greek science in sixteenth century Renaissance. In J. V. Field & F. A. J. L. James (Eds.), *Renaissance and revolution.* Cambridge: Cambridge University Press, pp. 15–28.

Nutton, V. (1993b). Wittenberg anatomy. In O. P. Grell & A. Cunningham (Eds.), *Medicine and the reformation.* London: Routledge, pp. 11–32.

Obermaier, H. & Kuhn, H. (1930). Buschmannkunst: Felsmalereien aus Sudwestafrika. Leipzig: G. Schmidt & C. Gtinther.

O'Brien, D. (1970). The effect of a simile: Empedocles' theories of seeing and breathing. *J Hellen Stud, 90*:140–179.

O'Brien, E. (1983). *The Royal College of Surgeons in Ireland.* The Irish Heritage Series: 40. Dublin: Eason & Son Ltd.

O'Dowd, M. J. & Philipp, E. E. (1994). *The history of obstetrics and gynaecology.* New York: Parthenon.

Oelhafen, K. I., Shayota, B., Muhleman, M., Klaassen, Z., & Loukas, M. (2013). Benjamin Alcock (1801–?) and his canal. *Clin Anat, 26*:662–6.

Ogilvie, M. & Harvey, J. (2000). *The biographical dictionary of women in science: Pioneering lives from ancient times to the mid-20th century,* Volume 1 A–K. New York: Routledge pp, 305, 1140–1141.

Ogle, W. (1882). *Aristotle on the Parts of Animals.* Translated with introduction and notes. London: Kegan Paul, Trench & Co.

Olry, R. (1995). *Dictionary of anatomical eponyms.* Stuttgart: Gustav Fischer Verlag.

Olry, R. (1999). Antonio Pacchioni and Giovanni Fantoni on the anatomy and functions of the human cerebral dura mater. *J Int Soc Plastination, 14*:9–11.

Olsen. (2009). *Technology and science in ancient civilizations.* Praeger, p.163.

O'Malley, C. D. (1953). *Michael Servetus. A translation of his geographical, medical and astrological writings with introduction and notes.* Philadelphia: American Philosophical Society.

O'Malley, C. D. (1964). *Andreas Vesalius of Brussels 1514–1564.* Berkeley and Los Angeles: University of California Press.

O'Malley, C. D. (1970). *The history of medical education.* Berkeley, CA: University of California Press.

O'Malley, C. D., Poynter, F. N. L., & Russell, K. F. (1961). *William Harvey lectures on the whole of anatomy.* Berkeley and Los Angeles, CA: University of California Press.

O'Malley, C. D., & Saunders, J. B. de C. M. (1982). *Leonardo da Vinci on the human body: The anatomical, physiological, and embryological drawings of Leonardo da Vinci.* New York: Greenwich House.

O'Neill, Y. V. (1969). The Fünfbilderserie reconsidered. *Bull Hist Med, 43*:236–245.

O'Neill, Y. V. (1977). The Fünfbilderserie – a bridge to the unknown. *Bull Hist Med, 51*:538–549.

O'Neill, Y. V. (1982). Tracing Islamic influences in an illustrated anatomical manual. In A. R. El-Gindy & H. M. Z. Hassan (Eds.), *Proc 2nd Int Conf Islamic Med,* Kuwait, pp. 154–162.

Oppenheimer, J. M. (1946). *New Aspects of John and William Hunter.* London: W. M. Heinemann.

Oppenheimer, J. M. (1975). Reflections on fifty years of publications on the history of general biology and special embryology. *Q Rev Biol, 50*:373–87.

Oransky, I. (2004). Sir Godfrey N. Hounsfield. *Lancet, 364*:1032.

Orlandini, G. E. & Paternostro, F. (2010). Anatomy and anatomists in Tuscany in the 17th century. *Ital J Anat Embryol, 115*:167–74.

Ottley, D. (1839). *The life of John Hunter.* Philadelphia: Haswell, Barrington, and Haswell.

O'Rahilly R. (1947). Benjamin Alcock, anatomist. *Ir J Med Sci, 262*:622.

O'Rahilly, R (1993). Anatomy, The writing of a textbook. *Clin Anat, 6*:366–369.

Orth, E. (1925). Cicero und die Medizin. Kaisersesch, Martintal.

Packard, F. R. (1920). History of the School of Salernum. In J. Harrington (Ed.), *The School of Salernum.* New York: Paul B. Hoeber, pp. 7–52.

Packard, F. R. (1931). *History of medicine in the United States.* New York: Paul B. Hoeber.

Packard, F. R. (1973). *History of medicine in the United States* (2 vols.). New York: Hafner Press.

Pagel, W. (1970). Review: Galen and the usefulness of the parts of the body. *Med Hist, 14*:406–408.

Pagel, J. (1901). Biographisches Lexikon hervorragender Ärzte des neunzehnten Jahrhunderts. Berlin: Wien.

Paget, S. (1897). *John Hunter. Man of science and surgeon (1728-1793).* London: T. Fisher Unwin.

Palay, S. L. (1995). A memoir in appreciation of Alan Peters on his sixty-fifth birthday. *J Comp Neurol, 355*:2–5.

Parent, A. (2007). Félix Vicq d'Azyr: Anatomy, medicine and revolution. *Can J Neurol Sci, 34*:30.

Parker, R. G. (1983). Academy of Fine Arts. Print from a copper engraving, 1578, by Cornelis Cort after a drawing by Jan Van Der Straet, 1573. *Hist Med, 38*:16.

Parkinson, D. (1995). Lateral sellar compartment: History and anatomy. *J Craniofac Surg, 6*:55–68.

Perloff, J. K. (2013). Human dissection and the science and art of Leonardo da Vinci. *Am J Cardiol, 111*:775–7.

Parsons, E. A. (1952). *The Alexandrian Library: Glory of the Hellenic world.* Amsterdam: Elsevier.

Patten, D., Donnelly, L., & Richards, S. (2010). Studying living anatomy: The use of portable ultrasound in the undergraduate medical curriculum. *Inter J Clin Skills, 4*:72.

Paul, F. (1937). Die Leichenkonservierung in der Neuzeit. *Ciba Zeitschrift, 4*:1488–96.

Peachey, G. C. (1924). *A memoir of William and John Hunter.* Plymouth: William Brendon and Son.

Perry, M. O. (1993) John Hunter – triumph and tragedy. *J Vasc Surg, 17*:7–14.

Persaud, T. V. N. (1983) A brief history of anatomy at the University of Manitoba. *Anat Anz, 153*:3–31.

Persaud, T. V. N. (1984). *Early history of human anatomy.* Springfield, IL: Charles C Thomas.

Persaud, T. V. N. (1989). Historical development of the concept of a pulmonary circulation. *Can J Cardiol, 5*:12–16.

Persaud, T. V. N. (1996). John Hunter. In L. N. Magner, (Ed.), *Doctors, nurses, and practitioners: A bio-bibliographic sourcebook.* Westport, CT: Greenwood Press.

Peterson, D. W. (1977). Observations on the chronology of the Galenic corpus. *Bull Hist Med, 57*:484–95.

Pettinato, G. (1981). *The archives of Ebla.* New York: Doubleday.

Pfannenstiel, M. (1949). Die Entdeckung des menschlichen Zwischenkiefers durch Goethe und Oken. *Naturwissenshaften, 36*:193–8.

Philip, M. (2012). Professor R. M. H. McMinn, MBCHB, PhD, MD, FRCS (eng) (20 September 1923–11 July 2012). *Clin Anat, 25*:1097–8.

Piatt A. (1912). *Aristotle. Des generations animalium.* Oxford: Oxford University Press.

Polson, C. J. & Marshall, T. K. (1975). *The disposal of the dead* (3rd ed.). London: The English Universities Press Limited.

Porzionato, A., Macchi, V., Stecco, C., Parenti A., & De Caro, R. (2012). The anatomical school of Padua. *Anat Rec* (Hoboken), *295*:902–16.

Potter, P. (1976). Herophilus of Chalcedon: An assessment of his place in the history of anatomy. *Bull Hist Med, 50*:45–60.

Potter, P. & Soltan, H. (1997). Murray Llewellyn Barr, O. C. *Biogr Mem Fellows R Soc, 43*:32–46.

Poulson, J. E. & Snorrason, E. (1986). *Nicolaus Steno 1638–1686, A re-consideration by Danish scientists.* Denmark: Gentofte.

Power, D. A. (1857). The rise and fall of the private medical schools in London. *Brit Med J, 1*:1388, 1451–53.

Power, D. (1897). *William Harvey.* London: T. Fisher Unwin.

Power, D. (1927). *The birth of mankind or the woman's book. A bibliographic study.* London: The Bibliographic Society.

Poynter, F. N. L. (Ed.) (1966). *The evolution of medical education in Britain.* London: Pitman Medical.

Pratt, C. W. M. (1981). *The history of anatomy in Cambridge.* Cambridge: Cambridge University Press.

Preda, A. (1990). Medicine in ancient China and Corea. *Family Practice.*

Premuda, L. (1974). Padua's 750 years of medicine. Image. *Medical Ilustrate Roche, 54*:33.

Premuda, L. (Ed.). (1986). I Secoli Doro Della Medizina. 700 Anni Di Scienza Medica A Padova. Modena: Edizioni Panini.

Prendergast, J. (1928). Galen's view of the vascular system in relation to that of Harvey. *Proc R Soc Med, 21*:1839–1848.

Prendergast, J. S. (1930). The background of Galen's life and activities and its influence on his achievements. *Proc Roy Soc Med, 23*:1131–48.

Pressac, J-C. (1985). The Struthof Album: Study of the gassing at Natzweiler-Struthof of 86 Jews whose bodies were to constitute a collection of skeletons. A photographic document. New York: The Beate Klarsfeld Foundation. p 1–88.

Preuss, A. (1970). Science and philosophy in Aristotle's "Generation of Animals." *Hist Biol, 3*:1.

Pringle, H. (2010). Anatomy. Confronting anatomy's Nazi past. *Science, 329*:274–5.

Puccetti, M .L., Perugi, L., & Scarani, P. (1995). Gaetano Giulio Zumbo. The founder of anatomic wax modeling. *Pathol Annu, 30*:269–81.

Puccetti, M. L. (1997). Human anatomy in wax during Florentine enlightenment. *Ital J Anat Embryol, 102*:77–89.

Punt, H. (1983). *Bernhard Siegfried Albinus (1697–1770) on "Human Nature."* Amsterdam: B. M. Israel.

Putscher, M. (1972). Geschichte der medizinischen Abbildung. Von 1600 bis zur Gegenwart (2nd ed.). Munchen: Moos.

Qatagya, S. (1982). Ibnul-Nafis had dissected the human body. In A. R. El-Gindy & H. M. Z. Hassan (Eds.), *Proc 2nd Int Conf Islamic Med,* Kuwait, pp. 306–312.

Qvist, G. (1981). *John Hunter 1728–1793.* London: William Heinemann.

Rae, I. (1964). *Knox the anatomist.* Edinburgh: Oliver and Boyd.

Randall, J. H. (1940). The development of scientific method in the school of Padua. *J Hist Ideas, 1*:177.

Ranke, H. (1933). Medicine and surgery in ancient Egypt. *Bull Hist Med, 7*:237.

Rather, L. J. (1974). Pathology at mid-century: A reassessment of Thomas Willis and Thomas Sydenham. In A. G. Debus (Ed.), *Medicine in seventeenth century England.* Los Angeles, CA: University of California Press, pp. 71–112.

Regenspurger, K. & Heinstein, P. (2003). Justus Christian Loders Tabulae anatomicae (1794–1803): Anatomische Illustrationen zwischen wissenschaftlichem, künstlerischem und merkantilem Anspruch. *Med Hist J, 38*:245–284.

Reidenberg, J. S. & Laitman, J. T. (2002). The new face of gross anatomy. *Anat Rec, 269*:81–88.

Retief, F., Stulting, A., & Cilliers, L. (2008). The eye in antiquity. *S Afr Med J, 98*:697–700.

Richardson, B. W. (1885). Vesalius, and the birth of anatomy. *Asclepiad, 2*:132.

Richardson, R. (1987). *Death, dissection and the destitute.* London: Routledge & Kegan Paul Ltd.

Richardson, R. (1991). "Trading assassins" and the licensing of anatomy. In R. French & A. Wear (Eds.), *British medicine in an age of reform.* London: Routledge, pp. 74–91.

Richer, P. (1903). Du role de l'anatomie dans l'histoire de l'art. *France Med, 50*:318.

Ricketts, H. T. (1955). Robert R. Bensley. *Diabetes, 4*:334–5.

Richter, G. (1937). Die bauliche Entwicklung des anatomischen Theaters. *Z Anal Entwgesch, 106*:138.

Riesman, D. (1935). *The story of medicine in the middle ages.* New York: Paul B. Hoeber.

Rifkin B. A., Ackerman, M. J., & Folkenberg, J. (2006). *Human anatomy. A visual history from the Renaissance to the Digital Age.* New York: Abrams.

Riva, A., Conti, G., Solinas, P., & Loy, F. (2010). The evolution of anatomical illustrations and wax modeling in Italy from the 16th to early 19th centuries. *J Anat, 216*:209–22.

Robb-Smith, A. H. T. (1971). Anatomy and physiology. In W. C. Gibson (Ed.): *British contributions to medical science.* London: Wellcome Institute of the History of Medicine, pp. 55–72.

Robb-Smith, A. H. T. (1974). Cambridge medicine. In A. G. Debus (Ed.): *Medicine in seventeenth century England.* Los Angeles, CA: University of Los Angeles Press, pp. 327–369.

Roberts, K. B. & Tomlinson, J. D. W. (1992). *The fabric of the body.* Oxford: Clarendon Press.

Rock, A. (1969). Medicine in Cambridge, 1660–1760. *Med Hist, 13*:107–122.

Rolleston, H. D. (1932). John Caius (1510–1573), M.D., F.R.C.P. In *The Cambridge medical school.* Cambridge: Cambridge University Press, pp. 190–198.

Rolleston, H. (1939). The early history of the teaching of: I Human anatomy in London; II. Morbid anatomy and pathology in Great Britain. *Ann Med Hist*, *1*:203–238.

Rollins, C. P. (1943). Oporinus and the publication of the Fabrica. *Yale J Biol Med*, *16*:129–134.

Rose, F. C. (2009). Cerebral localization in antiquity. *J Hist Neurosci*, *18*:239–47.

Ross, W. D. (1952). *Aristotle works.* 12 Vols. Oxford: Clarendon.

Rosito, P., Mancini, A. F., & Paolucci, G. (2004). Anna Morandi Manzolini (1716–1774) master sculptress of anatomic wax models. *Pediatr Blood Cancer*, *42*:388–9.

Rossetti, L. (1983). *The University of Padua.* Trieste: Edizioni Lint.

Roth, M. (1892). *Andreas Vesalius Bruxellensis.* Berlin: G. Reimer.

Roth, M. (1895). Vesaliana. *Arch Path Anat*, *141*:462–478.

Rothstein, W. G. (1972). *American physicians in the nineteenth century. From sects to science.* Baltimore: Johns Hopkins University Press.

Rothstein, W. G. (1987). *American medical schools and the practice of medicine.* New York: Oxford University Press.

Russel, G. A. (1982). The anatomy of the eye: Ibn Al-Haytham and the Galenic tradition. In A. R. El-Gindy & H. M. Z. Hassan (Eds.), *Proc 2nd Int Conf Islamic Med*, Kuwait, p. 176.

Russell, K. F. (1940). John Browne and his treatise on the muscles. *Aust N Z J Surg*, *10*:113–116.

Russell, K. F. (1949). A bibliography of anatomical books published in English before 1800. *Bull Hist Med*, *23*:268–306.

Russell, K. F. (1959a). The anatomical plagiarist. *Med J Aust*, *46*:249–52.

Russell, K. F. (1959b). John Browne, 1642–1702. A seventeenth century surgeon, anatomist and plagiarist. *Bull Hist Med*, *33*:503–525.

Russell, K. F. (1987). *British anatomy 1525–1800* (2nd ed.). Winchester: St. Paul's Bibliographies.

Rutkow, I. M. (1993). *Surgery: An illustrated history.* St. Louis, MO: Mosby-Year Book.

Saadat S. (2009). Human embryology and the Holy Quran: An overview. *Int J Health Sci*, *3*:103–109.

Sahagun, Fray Bernardino de. (1905). Historia General de las Cosas de Nueva Espana. Edicion partial en facsimile de los Codices Matitenses en lengua Mexicana que se custodian en las Bibliotecas del Palacio Real y de la Real Academia de la Historia. Madrid: Fototipia de Hauser y Menet.

Sabin, F. R. (1918). Franklin Paine Mall: A review of his scientific achievement. *Science*, *47*:254–261.

Said, H. M. (1965). *Medicine in China.* Karachi: Hamdard Academy.

Sandelowsky, B. H. (1983). Archaeology in Namibia. *Amer Sci*, *77*:606–615.

Sarton, G. (1954). *Galen of Pergamon.* Lawrence, KS: University of Kansas Press.

Saunders, J. B. de C.M. & O'Malley, C. (1982). *The anatomical drawings of Andreas Vesalius.* New York: Crown.

Savage-Smith E. (1995). Attitudes toward dissection in medieval Islam. *J Hist Med Allied Sci*, *50*:67–110.

Savitt, T. L. (1990). *Fevers, agues, and cures. Medical life in Old Virginia.* Richmond, VA: Virginia Historical Society.

Scarborough, J. (1976). Celsus on human vivisection at Ptolemaic Alexandria. *Clio Med*, *77*:25–38.

Schacht, J. (1957). Ibn al-Nafis, Servetus and Colombo. *Al-Andalus*, *22*:317.

Sheldon, J. (1784). *The history of the absorbent system.* London.

Schendel, G. (1968). *Medicine in Mexico. From Aztec herbs to betatrons.* Austin, TX: University of Texas Press.

Scherer, H. (1990). Rembrandt's Dr. Tulp's anatomy lesson. *Minn Med*, *73*:24–8.

Scherz, G. (1964). Niels Steno. Denker und Forscher im Barock, 1638–1686. Stuttgart: Wissenschaftliche Verlagsgesellschaft M.B.H.

Schierhorn, H. (1978). Zur Frühgeschichte anatomischer Zeitschriften und die bedeutung des Anatomischen Anzeigers. *Anat Anz*, *143*:10–20.

Schierhorn, H. (1986). Mitglieder der Anatomischen Gesellschaft im antifaschistischen Exil. *Verh Anat Ges*, *80*:957–963.

Schmiedebach, H. P. (1990). Robert Remak (1815–1865). A Jewish physician and researcher between recognition and rejection. *Z Arztl Fortbild* (Jena): *84*:889–94.

Schmitt, S. (2009). From physiology to classification: A comparative anatomy and Vicq d'Azyr's plan of reform for life sciences and medicine (1774–1794). *Sci Context*, *22*:145–93.

Schnalke, T. (1995). *Diseases in wax. The history of the medical moulage.* Illinois: Quintessence.

Schneider, D. (1997). Profile: Raymond V. Damadian. Scanning the horizon. *Sci Am*, *276*:32–34.

Schumacher, G. H. (2007). Theatrum anatomicum in history and today. *Int J Morphol*, *25*:15–32.

Schumacher, G. H. & Wischhusen, H. (1970). *Anatomia Rostochiensis.* Berlin: Akademie Verlag.

Schupbach, W. (1982). The paradox of Rembrandt's 'Anatomy of Dr. Tulp.' *Med Hist*, *2*:1–110.

Shanks, N. J. & Al-Kalai, D. (1984). Arabian medicine in the Middle Ages. *Roy Soc Med*, *77*:60–5.

Sethi, A. G., Sharma, P., Mohta, M., & Tyagi, A. (2003). Shock – a short review. *Indian J Anaesth*, *47*:345–359.

Shah, M. (1992). Premier Chirurgien du Roi: The life of Ambroise Pare (1510–1590). *J Roy Soc Med*, *85*:292–294.

Sharer, R. J. (1994). *The ancient Maya* (5th ed.). Cambridge: Cambridge University Press.

Shenker, N., & Ellis, H. (2007). John Hilton (1805–78): Anatomists and surgeon. *J Med Biogr, 15*:219–26.

Shepherd, F. J. (1919). *Reminiscences of student days and dissecting room.* Montreal: Private Printing.

Shoja, M. M., Tubbs, R. S, Loukas, M., & Ardalan, M. R. (2008). The Aristotelian account of hearts and veins. *Int J Cardiol, 125*:304–310.

Shoja, M. M., Tubbs, R. S., Shokouhi, G., & Loukas, M. (2010). Wang Qingren and the 19th century Chinese doctrine of the bloodless heart. *Int J Cardiol, 145*:305–6.

Shoja, M. M., Rashidi, M. R., Tubbs, R. S., Etemadi, J., Abbasnejad, F., & Agutter, P. S. (2011). Legacy of Avicenna and evidence-based medicine. *Int J Cardiol, 150*:243–6.

Shoja, M. M., Agutter, P. S., Loukas, M., Benninger, B., Shokouhi, G., Namdar, H., Khalili, M., & Tubbs, R. S. (2013). Leonardo da Vinci's studies of the heart. *Int J Cardiol, 167*:1126–33.

Shryock, R. H. (1966). *Medicine in America. Historical essays.* Baltimore: Johns Hopkins University Press.

Shultz, S. M. (1992). *Body snatching.* Jefferson, NC: McFarland.

Siegel, R. E. (1959). Theories of vision and color perception of Empedocles and De mocritus; some similarities to the modern approach. *Bull Hist Med, 33*:145–59.

Siegel, R. E. (1968). *Galen's system of physiology and medicine. An analysis of his doctrines and observations on blood flow, respiration, humors and internal diseases.* Basel: Karger.

Sigerist, H. E. (1933). *Amerika und die medizin.* Leipzig: G. Thieme.

Sigerist, H. E. (1952). Alkmaion von Kroton und die Anfange der europaischen. *Physiologic Schweiz Med Woch, 52*:964–968.

Simili, A. (1951). Un referto medico-legale inedito e autografo di Bartolomeo da Varignana. *Policlinico* (Prat), *58*:150–1.

Simmons, S. F. (1783). *An Account of the life and writings of the late William Hunter, M.D., F.R.S. and S.A.* London: W. Richardson.

Simon, M. (1906). *Sieben Biicher Anatomie des Galen.* 2 Vols. Leipzig.

Simpson, D. (2005). Phrenology and the neurosciences: contributions of F. J. Gall and J. G. Spurzheim. *Anz J Surg, 75*:475–82.

Sinclair, H. M. (1974). Oxford medicine. In A. G. Debus (Ed.), *Medicine in seventeenth century England.* Los Angeles, CA: University of Los Angeles Press, pp. 371–391.

Singer, C. (1952). Galen's elementary course on bones. *Proc Roy Soc Med, 45*:767–76.

Singer, C. (1956). *Galen on anatomical procedures: Translation of the surviving books with introduction and notes.* Oxford: Oxford University Press.

Singer, C. (1957). *A short history of anatomy and physiology from the Greeks to Harvey.* New York: Dover.

Singer, C. & Holloway, S. W. F. (1960). Early medical education in England in relation to the pre-history of London University. *Med Hist, 4*:1–17.

Singer, C. & Underwood, E. A. (1962). *A short history of medicine* (2nd ed.). Oxford: Clarendon Press.

Singhal, G. D. & Guru, L. V. (1973). *Anatomical & obstetric considerations in ancient Indian surgery.* Banaras: Banaras Hindu University Press.

Siraisi, N. G. (1981). *Taddeo Alderotti and his pupils.* Princeton, NJ: Princeton University Press.

Siraisi, N. G. (1990). *Medieval & early Renaissance medicine: An introduction to knowledge and practice.* Chicago: University of Chicago Press.

Slonecker, C. E. (1998). A personal remembrance: Richard J. Blandau, PhD, MD (1911–1998). *Anat Rec, 253*:130–131.

Slotnick, H. B.,& Hilton, S. R. (2006). Proto-professionalism and the dissecting laboratory. *Clin Anat, 19*:429–436.

Smellie, W. (1968). A set of anatomical tables, with explanations, and an abridgment of the practice of midwifery. Facsimile reprint. Philadelphia: W. B. Saunders.

Smith, C. V. M. (1992). Richard Owen: An eminent Victorian. *Biologist, 39*:212–216.

Smith, S. B. (2006). From ars to scientia: The revolution of anatomic illustration. *Clin Anat, 19*:382–388.

Smith, C. U. (2010). The triune brain in antiquity: Plato, Aristotle, Erasistratus. *J Hist Neurosci, 19*:1–14.

Smith, C. U., Frixione, E., Finger, S., & Clower, W. (2012). *The animal spirit doctrine and the origins of neurophysiology.* Oxford: Oxford University Press, p. 22.

Snellen, H.A. (1984). *History of cardiology.* Rotterdam: Donker Academic.

Sorabji, R. (1970). Aristotle on demarcating the five senses. *Philos Rev, 80*:55–79.

Souques, A. (1934). Que doivent a Herophile et a Erasistrate l'anatomie et la physiologie due systeme nerveux. *Bull Soc Franc Hist Med, 28*:357–65.

Sournia, J-C. (1992). *The illustrated history of medicine.* London: Harold Starke.

Spector, B. (1955). The Massachusetts anatomy act of 1831: A beneficent, humanizing instrument. *Bull New Eng Medical Center, 1*:72–80.

Speransky, L. S., Bocharov, V. J., & Goncharov, N. I. (1983). The personage of Jan Stephan Van Calcar's frontispiece to Andreas Vesalius' book "On the structure of the human body." *Anat Anz, 153*:465–79.

Spinner, R. J., Vincent J. F., & Wolanskyj, A. P. (2011). Discovering the elusive Beauchene: The originator of the disarticulated anatomic technique. *Clin Anat, 24*:797–801.

Sprigge, S. S. (1899). *The life and times of Thomas Wakley.* London: Longman.

Stakhov, A. P. (1989). The golden section in the measurement theory. *Computers Math Applic, 17*:613–638.

Steiger, G., & Flaschendrager, W. (Eds.). (1981). Magister und Scholaren. Geschichte Deutscher Universitaten und Hochschulen im Uberblick. Leipzig: Urania Verlag.

Stephen, L. (1888). *James Drake (DNB00). Dictionary of national biography, 1885–1900*, Volume 15. London: Elder Smith.

Stern C. D. (1986). A historical perspective on the discovery of the accessory duct of the pancreas, the ampulla 'of Vater' and pancreas divisum. *Gut, 27*:203–12.

Stierlin, H. (1978). *The world of India.* New York: Mayflower.

Stone, T. (1929). *Observations on the phrenological development of Burke, Hare, and other atrocious murderers; Measurements of the heads of the most notorious thieves....* Glasgow: Robertson and Atkinson.

Stroppiana, L. (1963). Eanatomia nel "corpus Hippocraticum." *Riv Stor Med, 158*:9–17.

Struthers, J. (1867). *Historical sketch of the Edinburgh anatomical school.* Edinburgh: Maclachlan and Stewart.

Stuart, G. S. (1987). *The mighty Aztecs.* Washington, DC: National Geographic Society.

Stürzbecher, M. (1958). Beitrag zur Geschichte der Berliner Anatomie im 18.Jahrhundert. *Dtsch Med Journ, 9*:439–42.

Stürzbecher, M. (1959). Zur Geschichte des Anatomischen Theaters in Berlin im 18. Jahrhundert. *Med Mitt Schering, 20*:102.

Stürzbecher, M. (1963). Aus der Frühgeschichte der Berliner Anatomie. *Dtsch Med Journ, 14*:121–6.

Sudhoff, K. (1907). *Studien zur Geschichte der Medizin.* Leipzig: J. A. Barth.

Sudhoff, K. (1908). Ein Beitrag zur Geschichte der Anatomie im Mittelalter speziell der anatom ischen Graphik nach Handschriften des 9. bis 15. Jahrhunderts. Leipzig: T. A. Barth.

Sudhoff, K. (1916). Ein unbekannter Druck von Johann Peyligks aus Zeite "Compendiosa capitis physici declaratio" auch "Anatomia totius corporis humani" genannt. *Arch Gesch Med, 9*:309–14.

Sudhoff, K. (1930). Konstantin der Afrikaner und die Medizinschule von Salerno. *Sudhoff Arch, 23*:293–8.

Sudhoff, K. (1964). Ein Beitrag zur Geschichte der Anatomie im Mittelalter, speziell der anatomischen Graphik nach Handschriften des 9. bis 15. Jahrhunderts. Hildesheim, Georg Olms Verlagsbuchhandlung, Hildesheim.

Sugand, K., Abrahams, P., & Khurana, A. (2010). The anatomy of anatomy: A review for its modernization. *Anat Sci Educ, 3*:83–93.

Swamy, M., & Searle, R. F. (2012). Anatomy teaching with portable ultrasound to medical students. *BMC Med Educ, 12*:99.

Symonds, C. (1955). The Circle of Willis. *Brit Med J, 1*:119–24.

Talbott, J. H. (1961). Friedrich Gustave Jacob Henle. *JAMA, 177*:779–81.

Tarshis, J. (1969). *Father of modern anatomy: Andreas Vesalius.* New York: Dial, p. 103

Tascioglu, A. O. & Tascioglu, A. B. (2005). Ventricular anatomy: Illustrations and concepts from antiquity to renaissance. *Neuroanatomy, 4*:57–63.

Temkin, O. (1935). Celsus on medicine and the ancient medical sects. *Bull Hist Med, 3*:249–64.

Temkin, O. (1940). Was Servetus influenced by Ibn al-Nafis? *Bull Hist Med, 8*:731–734.

Temkin, O. (1973). *Galenism: Rise and decline of a medical philosophy.* Ithaca, NY: Cornell University Press.

Thacher, T. (1677/8) A brief rule to guide the common-people of New-England. How to order themselves and theirs in the small pocks, or measles. [First published in 1677/8, reprinted in 1702 and 1721–22.] Boston: John Foster, 1677.

Thacher, J. (1828). *American medical biography.* Boston: Richardson, vol. i, p. 52.

Thomas, K. B. (1971). General medicine. In W. C. Gibson (Ed.), *British contributions to medical science.* London: Wellcome Institute of the History of Medicine, pp. 1–10.

Thomas, K. B. (1974). The great anatomical atlases. *Proc Roy Soc Med, 67*:223–32.

Thompson, I. M. (1954). The golden age of anatomy in Edinburgh. *Manitoba Med Rev, 34*:156.

Thomson, S. C. (1942). The Great Windmill Street School. *Bull Hist Med, 12*:377–91.

Thomson, S. C. (1943). The surgeon-anatomists of Great Windmill Street School. *Bull Soc Med Hist, 5*:55–75.

Tichacek, B. (1987). Jan Evangelista Purkyne. 200th anniversary of birthday. *J Hyg Epidemiol Microbiol Immunol, 31*:233–5.

Tipton, J. A. (2006). Aristotle's fish: The case of the kobios and phucis. *Perspect Biol Med, 49*:369–83.

Tobias, P. V. (1993). The contributions of J. C. Boileau Grant to the teaching of anatomy. *S Afr Med J, 83*:352–53.

Toledo-Pereyra, L. H. (2011). *Surgical revolutions: A historical and philosophical view.* World Scientific Publishing, p. 145.

Topp, K. S. (2004). Prosection vs. dissection, the debate continues: Rebuttal to Granger. *Anat Rec B New Anat, 281*:12–4.

Tubbs, R. S., Shoja, M. M., Loukas, M., & Oakes, W. J. (2007). Abubakr Muhammad Ibn Zakaria Razi, Rhazes (865–925 AD). *Childs Nerv Syst, 23*:1225–6.

Tubbs, R. S., Loukas, M., Shoja, M. M., Ardalan, M. R., & Oakes,W. J. (2008). Richard Lower (1631–1691) and his early contributions to cardiology. *Int J Cardiol, 128*:17.

Tubbs, R.S., Riech, S., Verma, K., Chern, J., Mortazavi, M., & Cohen-Gadol, A. A. (2011a). China's first surgeon: Hua Tuo (c. 108–208 AD). *Childs Nerv Syst, 27*:1357–60.

Tubbs, R. S., Mortazavi, M. M., Shoja, M., Loukas, M., & Cohen-Gadol, A. A. (2011b). The bishop and anatomist Niels Steno (1638–1686) and his contributions to our early understanding of the brain. *Childs Nerv Syst, 27*:1–6.

Tubbs, R. S., Loukas, M., Shoja, M. M., Mortazavi, M. M., & Cohen-Gadol, A. A. (2011c). Félix Vicq d'Azyr (1746–1794): Early founder of neuroanatomy and royal French physician. *Childs Nerv Syst, 27*:1031–4.

Tubbs, R.S., Rompola, O., Verma, K., Malakpour, M., Shoja, M. M. Mortzavi, M. M., & Loukas, M. (2012). James Drake (1667–1707): Anatomist and political activist. *Clin Anat, 25*:295.

Turney, B. W. (2007). Anatomy in a modern medical curriculum. *Ann R Coll Surg Engl, 89*:104–107.

Uddin, J. (1982). Ibn Sina's viewpoint of human anatomy. In A. R. El-Gindy & H. M. Z. Hassan. (Eds.), *Proc 2nd Int Conf Islamic Med*, Kuwait, pp. 163–175.

Underwood, E. A. (1963). The early teaching of anatomy at Padua. *Ann Sci, 19*:1–26.

Unschuld, P. U. (1985). *Medicine in China. A history of ideas.* Berkeley, CA: University of California Press.

Uragoda, C. G. (1987). *A history of medicine in Sri Lanka – from the earliest times to 1948.* Colombo: Sri Lanka Medical Association.

Uschmann, G. (1955). *Caspar Friedrich Wolff. Ein Pionier der modernen Embryologie.* Berlin: Urania-Verlag.

Vacek, Z. (1987). Jan Evangelista Purkyne and his contribution to the development of histology and embryology. *Folia Morphol* (Praha), *35*:338–60.

Valadez, F. (1974). Anatomical studies at Oxford and Cambridge. In A. G. Debus (Ed.), *Medicine in seventeenth century England.* Los Angeles, CA: University of California Press, pp. 393–420.

Valverde de Hamusco, J. (1556). Historia de la composicion del cuerpo humano. Roma: Antonio Salamanca y Antonio Lafrery.

Van der Ben, N. (1975). *Empedocles: The poem of Empedocles "Peri physios." Towards a new edition of all the fragments.* Amsterdam: B. R. Gruner.

Van Praagh, R. & Van Praagh, S. (1983). Aristotle's "triventricular" heart and the relevant early history of the cardiovascular system. *Chest, 84*:462–468.

Van Wyhe, J. (2002). The authority of human nature: The Schädellehre of Franz Joseph Gall. *Br J Hist Sci, 35*:17–42.

Van Zuylen, J. (1981). The microscopes of Antoni van Leeuwenhoek. *J Microsc, 121*:309–28.

Vesalius, A. (1543). *De Humani corporis fabrica.* Basel: Johannes Oporinus, pp. 15–39.

Vesling J. (2013). Complete dictionary of scientific biography, 2008. Encyclopedia.com. 28 May. http://www.encyclopedia.com.

Viets, H. R. (1937). The life of Thomas Thacher; Smallpox and other epidemic diseases in England and New England. In T. Thacher (Ed.), *A brief rule to guide the common people of New England.* Baltimore: John Hopkins Press.

Von Grunebaum, G. E. (1963). Der Einfluss des Islam auf die Entwicklung der Medizin. *Bustan, 3*:19–22.

Von Hagens, G. (2002). Anatomy and plastination. In G. Von Hagens, & A. Whalley (Eds.), *Body worlds: Catalogue of the exhibition.* Heidelberg: Institute for Plastination.

von Lüdinghausen, M. (2004). In remembrance Johannes Lang, MD (1923–2003). *Clin Anat, 17*:526.

Von Staden, H. (1992). The discovery of the body: Human dissection and its cultural contexts in ancient Greece. *Yale J Biol Med, 65*:223–41.

Von Toply, R. (1903). Aus der Renaissancezeit (neue Streiflichler uber die Florentiner Akademie und die anatomischen Zeichnungen des Vesal) *Janus, 8*:130–140.

Voneida, T. J. (1997). Roger Wolcott Sperry, 20 August 1913–17 April 1994. *Biogr Mem Fellows R Soc, 43*:461–70.

Vrettos, T. (2001). *Alexandria: City of the western mind.* New York: The Free Press.

Wackwitz, A. (1985). Abraham Vater (1684–1751). *Anat Anz, 160*:77–79.

Wake, W. C. (1952). The corpus hippocraticum. Ph.D. Thesis, London University.

Walsh, J. J. (1908). Pierre-Joseph Desault. In *The Catholic Encyclopedia.* Volume 4, New York: Robert Appleton.

Walsh, J. J. (1911). *Old-time makers of medicine: The story of the students of teachers and the medieval medicine.* New York: Fordham University Press.

Walsh, J. (1934a). Galen's writings and influences inspiring them. *Ann Med Hist, 6*:1–30.

Walsh, J. (1934b). Galen's writings and influences inspiring them. *Ann Med Hist, 6*:143–49.

Waltenbacher, T. (2008). Zentrale Hinrichtungsstatten. Der Vollzug der Todesstrafe in Deutschland von 1937–1945. Scharfrichter im Dritten Reich. Berlin: Zwilling, pp. 1–263.

Warren, L. (1998). *Joseph Leidy: The last man who knew everything.* New Haven, CT: Yale University Press, p. 320.

Website a. http://en.wikipedia.org/wiki/Zuo_Zhuan.

Website b. http://baillement.com/lettres/mueller_bio.html.

Website c. onlinelibrary.wiley.com/doi/10.3322/canjclin.23.5.305/pdf.

Website d. Munk William, http://munksroll.rcplondon.ac.uk/Biography/Details/950.

Website e. http://www.ifaa.net.

Website f. http://anatomische-gesellschaft.de/.

Website g. http://www.anatsoc.org.uk/Home.aspx.

Webesite h. http://www.clinical-anatomy.org/honored-members.html.

Wegner, R. N. (1939). Das Anatomenbildnis. Seine Entwicklung im Zusammenhang mit der ana tomischen Abbildung. Basel: B. Schwalbe.

Wegner, R. N. (1956). Die Geschichte des anatomischen Instituts und Museums der Universitat Greifswald. Band II, pp. 282–295.

Wellman, M. (1913). A Cornelius Celsus: Eine Quellenuntersuchung. Berlin: Weidmann.

Wellman, M. (1913). A Cornelius Celsus: Eine Quellenuntersuchung. Berlin: Weidmann. In German.

Wellman, M. (1924). A. Cornelius Celsus. *Sudhoff Arch,* *16*:209.

Wendler, D. (1984). Johannes Willer – Leben und werk. In J. Grosser (Ed.), Charite Annalen. N.F. 3, Berlin: Akademie-Verlag, pp. 265-274.

Wendler, D. (1986). Nathaniel Highmore (1613–1685) und die Oberkieferhohle. *Anat Anz, 162*:375–80.

Wendt, W. E. (1976). Art mobilier from the Apollo II cave, South West Africa: Africa's oldest dated works of art. *S Afr Archaeol Bull, 31*:5–11.

West, J. B. (2011). History of respiratory gas exchange. *Compr Physiol, 1*:1509–1523.

Weyers, W. (2009). Jacob Henle – a pioneer of dermatopathology. *Am J Dermatopathol, 31*:6–12.

Wharton, E. (1902). Vesalius in Zante (1564). *N Amer Rev, 175*:625–631.

Whitlock, D. G. (2007). The Cajal Club: Its origin, originator and benefactor, Wendell J. S. Krieg. *Brain Res Rev, 55*:450–62.

Whitteridge, G. (1964). *The anatomical lectures of William Harvey. Prelectiones Anatomie Universalis De Musculis.* Edinburgh and London: E. & S. Livingstone Ltd.

Whitteridge, G. (1990). From Da Vinci to Harvey. *J Roy Soc Med, 83*:674–5.

Wilford, F. A. (1968). Embryological analogies in Empedocles' cosmogony. *Phronesis, 73*:108–18.

Williams, D. (1988). This history of Eduard Pernkopf's Topographische Anatomie des Menschen. *Biocommunication, 15*:2–12.

Willis, R. (Ed.). (1847). *The works of William Harvey.* London: Sydenham Society.

Wilson, A. (1995). *The making of man-midwifery. Childbirth in England, 1660–1770.* Cambridge: Harvard University Press.

Wilson, L. (1989). The performance of the body in the renaissance theater of anatomy. *Representations, 17*:62–95.

Wilson, L. G. (1959). Erasistratus, Galen and the pneuma. *Bull Hist Med, 33*:293–314.

Winau, R. (1987). *Medizin in Berlin.* Berlin: Walter de Gruyter.

Winkelmann, A. & Schagen, U. (2009). Hermann Stieve's clinical-anatomical research on executed women during the "Third Reich." *Clin Anat, 22*:163–71.

Winkelmann, A. & Noack, T. (2010). The Clara cell: a "Third Reich eponym"? *Eur Respir J, 36*:722–727.

Wischhusen, H. & Schumacher, G.-H. (1968). Von den ersten anatomischen Lehrsektionen bis zu systematischen Praparieriibungen in Rostock. *WZ Rostock, 77*:45.

Withington, E. T. (1922). Galen's anatomy. 3rd Int Congr Hist Med, London.

Wolf, S. L. (2008). John V. Basmajian, MD June 21, 1921– March 18. http://www.aapb.org/files/publications/ BasmajianTribute.pdf

Wolf-Heidegger, G. & Cetto, A. M. (1967). Die anatomische Sektion in bildlicher Darstellung. Basel: S. Karger.

Wolpert, L. (2004). Much more from the chicken's egg than breakfast – A wonderful model system. *Mech Dev, 121*:1015–1017.

Wotton, W. (1694). *Reflections upon ancient and modern learning.* London: Peter Buck.

Woywodt, A., Lefrak, S. & Matteson, E. (2010). Tainted eponyms in medicine: The "Clara" cell joins the list. *Eur Respir J, 36*:706–8.

Wright, J. (1925). A medical essay on the Timaeus. *Ann Med Hist, 7*:117–127.

Wright-St. Clair, R. E. (1964). *Doctors Monro.* London: The Wellcome Historical Medical Library.

Wyatt, R. B. H. (1924). *William Harvey (1578–1657).* London: Leonard Parsons.

Yeager, V. L. (1996). Learning gross anatomy: Dissection and prosection. *Clin Anat, 9*:57–9.

Zimmer, H. R. (1948). *Hindu medicine.* Baltimore: Johns Hopkins University Press.

Zimmerman, L. M. & Veith, I. (1961). *Great ideas in the history of surgery.* Baltimore: Williams & Wilkins.

Zimmerman, L. M. (1974). Surgery. In A. G. Debus (Ed.), *Medicine in seventeenth century England.* Los Angeles, CA: University of California Press, pp. 49–69.

Zwiener, S. (2004). Johann Samuel Eduard d'Alton (1803–1854): Leben und Wirken Dissertation zur Erlangung des akademischen Grades Doktor der Zahnmedizin (Dr. med. dent.) vorgelegt an der Medizinischen Fakultät der Martin-Luther-Universität Halle-Wittenberg.

Zwiener, S., Göbbel, L., & Schultka, R. (2002). Der Anatom Johann Samuel d'Alton (1803–1845) – Leben und Wirken in Hallee (Saale). *Ann Anat, 184*:555–9.

INDEX